Organic–Inorganic Composite Polymer Electrolyte Membranes

Inamuddin · Ali Mohammad
Abdullah M. Asiri
Editors

Organic–Inorganic Composite Polymer Electrolyte Membranes

Preparation, Properties, and Fuel Cell Applications

Springer

Editors
Inamuddin
Faculty of Engineering & Technology
Aligarh Muslim University
Aligarh
India

Abdullah M. Asiri
Department of Chemsitry
King Abdulaziz University
Jeddah
Saudi Arabia

Ali Mohammad
Faculty of Engineering & Technology
Aligarh Muslim University
Aligarh
India

ISBN 978-3-319-52738-3 ISBN 978-3-319-52739-0 (eBook)
DOI 10.1007/978-3-319-52739-0

Library of Congress Control Number: 2017933866

Printed on acid-free paper

This Springer imprint is published by Springer Nature
The registered company is Springer International Publishing AG
The registered company address is: Gewerbestrasse 11, 6330 Cham, Switzerland

Preface

Organic–Inorganic Composite Polymer Electrolyte Membrane: Preparation, Properties and Fuel Cell Applications is intended to explore the recent developments in the area of polymer electrolyte membranes for high-temperature polymer electrolyte membrane fuel cells. This book presents a unified viewpoint of the operating principles of fuel cells, fuel cells thermodynamics and efficiency, various methodologies used for the fabrication of polymer electrolyte membranes, issues related to the chemical and mechanical stabilities of the membranes. Special attention is paid to the fabrication of electrospun nanocomposite membranes.

This book discusses the developments in the fast-growing and promising area of PEM materials obtained by using hygroscopic inorganic fillers, solid proton conductors, titanium oxide, zirconia and sulphated zirconia, silica, zeolite, montmorillonite, heterocyclic solvents, ionic liquids, anhydrous H_3PO_4 blends, heteropolyacids, etc.

This book is of great importance not only to the learners but also to the more experienced fuel cell researchers, enabling a deep understanding of the organic–inorganic membranes for fuel cells, methods for the fabrication of membranes as well as properties and applications. Based on thematic topics, the book edition contains the following 17 chapters:

Chapter 1 deals with the use of ionic liquids (ILs) as fuel cell electrolytes. ILs with good mechanical properties and high conductivity has great potential to fabricate polymer membranes. This chapter focuses on the recent advances in the application of ionic liquid-based polymer membranes for the development of fuel cell technology. The key factors affecting membrane performance have been nicely discussed.

In Chap. 2, a review on the recent developments about organic/TiO_2 nanocomposite membranes is presented. The unique properties, such as electrical property and processability together with thermal and chemical stabilities as well as high proton conductivity at high temperatures of organic–inorganic composite systems are highlighted.

In Chap. 3, the importance of nanoscience and nanotechnology for producing new materials with enhanced properties and potential applications are discussed. Among the variety of nanoparticles (NPs), silica NP is considered of a particular interest due to its ease of synthesis, functionalization and precise controlling of size and distribution of particles. Superior features of polymeric membranes are summarized.

Chapter 4 describes the basic requirements of polymer electrolyte membrane (PEM) fuel cells. This chapter is divided into seven sections. The first section presents some statistical data of publications concerning fuel cell, PEM fuel cells (PEMFC) and PEMFC with zeolites. The second section exhibits some concepts about zeolites types, structure, properties, and industrials applications. The third section presents the role of the zeolite properties on the performance of the PEMFC. The fourth section describes the main technique used for producing zeolite/polymer nanocomposite membrane for PEMFCs. The two following sections outline the state of the art of using the zeolite for PEMFC applications, being the fifth and sixth sections dedicated to the synthetic and natural polymers, respectively.

Chapter 5 covers the use of heteropolyacids (HPAs) in the preparation of proton exchange membranes (PEM) for polymer electrolyte membrane fuel cells (PEMFCs). The fundamental aspects of HPAs and their corresponding salts in addition to various structural configurations such as Keggin, Wells–Dawson, and Lacunar are discussed. The use of HPAs for preparation of membranes for high-temperature PEMFC and direct methanol fuel cell (DMFC) based on the immobilization on various substrates including perfluorinated sulfonic acids (PFSAs), aromatic hydrocarbons, poly(vinyl alcohol) (PVA), and polybenzimidazole (PBI) are reviewed.

In Chap. 6, organic/montmorillonite nanocomposite membrane and membrane fabrication techniques are discussed. The fabrication technique, properties of the fabricated membranes, and performance are explained in detail and compared.

In Chap. 7, the application of electrospun nanofibers from organic, inorganic, and composite organic–inorganic is extensively reviewed. The interesting features of electrospun nanofibers to improve fuel cell performance in terms of power density, ionic conductivity, interfacial resistance, and chemical stability, as well as mechanical strength are discussed. The development and evolution of fuel cells as one of the advanced energy conversion systems is suggested.

Chapter 8 describes the basic overview of fuel cell technology in order to understand the construction and the working principle of this eco-friendly technology. The fuel cells thermodynamics and efficiency are discussed in detail.

Chapter 9 presents an overview of the commonly used polymer hosts and inorganic additives. The available literature on sulfated zirconia (S–ZrO_2) nanohybrid membrane technology is discussed in view of catalyzing the future research to develop more suitable polymer electrolyte membranes for fuel cell.

In Chap. 10, literature data on the promotional role of under-rib convection for polymer electrolyte membrane fuel cells (PEMFCs) fueled by hydrogen and methanol have been structured and analyzed, with the aim of providing a guide to improve fuel cell performance through the optimization of flow field interaction.

Chapter 11 presents a brief overview of various techniques used for the peparation of organic–inorganic nanocomposite polymer electrolyte membranes for fuel cells and discusses the encountered challenges, the problems to be overcome, the major findings and guidance for future developments. With the advances in nanomaterials and polymer chemistry, the innovative nanocomposite membranes with superior properties can be designed by blending of nanoparticles in a polymer matrix, doping or infiltration and precipitation of nanoparticles and precursors, self-assembly of nanoparticles, layer-by-layer fabrication method, and nonequilibrium impregnation-reduction.

In Chap. 12, important attributes of PEM, such as chemical and mechanical stabilities, are described and reviewed. Efforts have been made to highlight the responses of chemical and mechanical stabilities of membrane at different temperatures and relative humidities of the fuel cell operation that lead to cell failure. A literature review regarding the chemical and mechanical degradation of membrane as well as the mitigation for the membrane degradation is also presented.

The focus of Chap. 13 is on the membrane of polymeric fuel cells and the methods of using electrospinning process to prepare these membranes. Three important objectives of this chapter with respect to using electrospinning in membranes of polymeric fuel cells include reducing methanol crossover, improving proton conductivity, and suppressing water swelling.

Chapter 14 summarizes the various fabrication techniques for the development of polymer electrolyte membrane fuel cells (PEMFC). A brief overview of different fabrication techniques, such as plasma method, phase inversion method, sol-gel method, direct copolymerization, ultrasonic coating technique, ultra-violet polymerization, in situ reduction, decal transfer method and catalyst coated membrane method, is presented in this chapter.

Chapter 15 reviews structure and property of chitosan with respect to its applications in fuel cells. In addition, different synthetic strategies to prepare chitosan-based polymer electrolyte membranes and their properties for the use in fuel cells are critically examined. These strategies include chemical modifications of chitosan, blending with other polymers and their composites for polymer electrolyte membranes among fuel cell applications. Recent achievements and prospect of its applications have also been discussed in this chapter.

Chapter 16 emphasizes about the basics information of fuel cells. It also summarizes the available information details about the types of fuel cells, the design and constructions of fuel cells, catalysts in fuel cells, materials and the methods used for preparation of fuel cells.

Chapter 17 presents details about fuel cell technology and production of proton exchange membranes developed through electrospinning technique along with recent progress made in research of new materials, such as Nafion, poly(vinylidene fluoride) (PVDF), poly(ethylene oxide) (PEO), and poly(vinyl alcohol) (PVA).

Composite membranes composed of highly conductive and selective layer-by-layer (LbL) films and electrospun fiber mats have been examined for mechanical strength and electrochemical selectivity. The present status and future prospectus of electrospun nanofibres for fuel cell applications are highlighted.

Aligarh, India Inamuddin
Aligarh, India Ali Mohammad
Jeddah, Saudi Arabia Abdullah M. Asiri

Acknowledgements

We are most indebted to the grace of the Almighty "One Universal Being" who inspires the entire Humanity to knowledge and who has given us the required favor to complete this work.

This book is the outcome of the remarkable contribution of experts from various interdisciplinary fields of science and cover the most comprehensive, in-depth, and up-to-date research and reviews. We are thankful to all the contributing authors and their coauthors for their esteemed work. We would also like to thank all publishers, authors, and others who granted me permission to use their figures, tables, and schemes.

We would like to express our deep gratitude to Prof. T. Urushadze (Georgia State Agriculture University, Georgia), Prof. K. Aoki (Toyohashi University of Technology, Japan), Prof. Chang Liu (Houston Methodist Research Institute, USA), Prof. Rajeev Jain (Jiwaji University, India), Prof. S. Shtykov (Saratov State University, Russia), Prof. M.M. Srivastava (Dayal Bagh University, India), Prof. M.C. Chattopadhyaya (Allahabad University, India), Prof. A.P. Gupta and Prof. B.D. Malhotra (Delhi Technological University, India), Prof. J.K. Rozylo (Maria Curie-Skloclowska, Poland), Prof. P.K. Sharma (JNV University, Jodhpur), Dr. Ajay Taneja (Dr. B.R. Ambedkar University, India), Prof. M.S. Chauhan (Himachal Pradesh University, India), Prof. Anees Ahmad, Prof. Syed Ashfaq Nabi, Prof. Hamid Ali, Engr. Yusuaf Ansari, Dr. S.J.A. Rizvi, Dr. Fazal-ur-Rehman, Dr. P. Tripathi, Dr. Wasi Khan, Dr. Syed Asad Ali, Dr. Riyaz Ahmad, Dr. Anwar Shahzad, Dr. Abdul Qayyum Khan and Dr. Sadaf Zaidi (Aligarh Muslim University, India), Prof. M. Mascini (University of Firenze, Italy), Prof. Ishtiaq Ahmad and Prof. Rakesh Kumar Mahajan (Guru Nanak Dev University, Amritsar, India), Dr. Raju Khan (NEIST, Assam, India), Prof. Seon Jeon Kim (Hanyang University, South Korea), Prof. Kenneth I. Ozoemena (University of Pretoria, South Africa), Prof. Gaber Eldesoky and Prof. Zeid-AL-Othman (King Saud University, Saudi Arabia), Prof. Sheikh Raisuddin (Jamia Hamdard University, India), Prof. Byong-Hun Jeon (Yonsei University, South Korea), Prof. A.I. Yahya and Prof. M. Luqman (A'Sharqiyah University, Sultanate of Oman), Dr. Gaurav Sharma

(Shoolini University, India), Prof. Rajaram Sakharam Mane and Prof. Omprakash Yemul and Prof. Dr. P.K. Zubaidha (Swami Ramanand Teerth Marathwada University, India), Prof. Altaf Hussain Pandith (University of Kashmir, India), Prof. S.D. Sharma and Prof. Vikas Gupta (IFTM University, India), Dr. R.J. Tayade (Central Salt & Marine Chemicals Research Institute, India), Prof. Richard Akinyeye (University of Ado Ekiti, Nigeria), and Prof. Toribio Fernández Otero (Technical University of Cartagena, Spain) for their valuable suggestions, guidance, and constant inspiration.

It is with immense gratitude that Dr. Inamuddin thank to his departmental colleagues Prof. M. Mobin, Prof. Asif Ali Khan, Prof. R.A.K. Rao, Prof. Faiz Mohammad, Dr. M.Z.A. Rafiqui, Dr. Abu Nasar, Dr. Rais Ahmad, Dr. Yasser Azim, Dr. Aiman Ahmad, Dr. Musheer Ahmad, and Dr. Farman Ali, without whose continuous encouragement this book would have not been completed. Dr. Inamuddin cannot thank enough to his friends and colleagues Dr. M.M. Alam (USA), Dr. Amir-Al-Ahmad (KFUPM, Saudi Arabia), Dr. Zafar Alam, Dr. Mu. Naushad, Dr. Salabh Jain, Dr. Hemendra Kumar Tiwari, Dr. Adesh Bhadana, Dr. Shakeel Ahmad Khan, Satish Singh, and others, for their timely help, good wishes, encouragement, and affection. The help received from my research group (Dr. Aiman Jahan Khanam, Dr. Arshi Amin, Tauseef Ahmad Rangreez, Ajhar Khan, Late Sardar Hussain, Beenish, and Sufia-ul-Haque) is appreciatively acknowledged.

Finally, we feel short of words and are full of emotions in thanking our family members for their constant inspiration and gracious support.

Inamuddin
Ali Mohammad
Abdullah M. Asiri

Contents

Editors and Contributors

About the Editors

Dr. Inamuddin is currently working as Assistant Professor in the Department of Applied Chemistry, Aligarh Muslim University (AMU), Aligarh, India. He obtained Master of Science degree in Organic Chemistry from Chaudhary Charan Singh (CCS) University, Meerut, India, in 2002. He received his Master of Philosophy and Doctor of Philosophy degrees in Applied Chemistry from AMU in 2004 and 2007, respectively. He has extensive research experience in multidisciplinary fields of Analytical Chemistry, Materials Chemistry, Electrochemistry and, more specifically, Renewable Energy and Environment. He has worked on different research projects as project fellow and senior research fellow funded by University Grants Commission (UGC), Government of India, and Council of Scientific and Industrial Research (CSIR), Government of India. He has received Fast Track Young Scientist Award from the Department of Science and Technology, India, to work in the area of bending actuators and artificial muscles. He has completed four major research projects sanctioned by University Grant Commission, Department of Science and Technology, Council of Scientific and Industrial Research, and Council of Science and Technology, India. He has published 74 research articles in international journals of repute and 10 book chapters in knowledge-based book editions published by renowned international publishers. He has published six edited books with Springer, United Kingdom, three by Nova Science Publishers, Inc. U.S.A., one by CRC Press Taylor & Francis Asia Pacific, two by Trans Tech Publications Ltd., Switzerland and one by Materials Science Forum, U.S.A. He is the member of various editorial boards of the journals. He has attended as well as chaired sessions in various international and nation conferences. He has worked as a Postdoctoral Fellow, leading a research team at the Creative Research Initiative Center for Bio-Artificial Muscle, Hanyang University, South Korea, in the field of renewable energy, especially biofuel cells. He has also worked as a Postdoctoral Fellow at the Center of Research Excellence in Renewable Energy, King Fahd University of Petroleum and Minerals, Saudi Arabia, in the field of polymer electrolyte membrane fuel cells and computational fluid dynamics of polymer electrolyte membrane fuel cells. He is a life member of the Journal of the Indian Chemical Society. His research interest includes ion exchange materials, sensor for heavy metal ions, biofuel cells, supercapacitors, and bending actuators.

Prof. Ali Mohammad is presently working as UGC-Emeritus Fellow in the Department of Applied Chemistry, Faculty of Engineering and Technology, Aligarh Muslim University, Aligarh, India after serving as chairman of the department of Applied Chemistry for 6 years. His scientific interests include physico-analytical aspects of solid-state reactions, micellar thin layer chromatography, surfactants analysis, and green chromatography. He is the author or coauthor of 250 scientific publications including research articles, reviews, and book chapters. He has supervised 53 students for Ph.D./M.Phil. and M.Tech. degrees. He has also served as Editor of Scientific

Journal, "Chemical and Environmental Research" published from India since 1992–2012 and as the Associate Editor for Analytical Chemistry section of the Journal of Indian Chemical Society. He has published three edited books with Springer, United Kingdom, He has been the member of editorial boards of Acta Chromatographica, Acta Universitatis Cibiniensis Seria F. Chemia, Air Pollution, and Annals of Agrarian Science. He has attended as well as chaired sessions in various international and nation conferences. He is the life member of several Indian scientific and chemical societies. He has also served as Visiting Professor in King Saud University, Riyadh, Saudi Arabia. Dr. Mohammad obtained his M.Sc. (1972), M.Phil. (1975), Ph.D. (1978), and D.Sc. (1996) degrees from Aligarh Muslim University, Aligarh, India.

Prof. Abdullah M. Asiri is the Head of the Chemistry Department at King Abdulaziz University since October 2009 and he is the founder and the Director of the Center of Excellence for Advanced Materials Research (CEAMR) since 2010 till date. He is Professor of Organic Photochemistry.

He graduated from King Abdulaziz University (KAU) with B.Sc. in Chemistry in 1990 and a Ph.D. from University of Wales, College of Cardiff, U.K. in 1995. His research interest covers color chemistry, synthesis of novel photochromic and thermochromic systems, synthesis of novel coloring matters, and dyeing of textiles, materials chemistry, nanochemistry and nanotechnology, polymers and plastics.

Professor Asiri is the principal supervisor of more than 20 M.Sc. and six Ph.D. theses; he is the main author of ten books of different chemistry disciplines. Professor Asiri is the Editor-in-Chief of King Abdulaziz University Journal of Science. A major achievement of Prof. Asiri is the discovery of tribochromic compounds, a class of compounds which change from slightly or colorless to deep colored when subjected to small pressure or when grind. This discovery was introduced to the scientific community as a new terminology published by IUPAC in 2000. This discovery was awarded a patent from European Patent office and from UK patent.

Professor Asiri involved in many committees at the KAU level and also on the national level, he took a major role in the advanced materials committee working for KACST to identify the National plan for science and technology in 2007.

Professor Asiri played a major role in advancing the chemistry education and research in KAU, he has been awarded the best researchers from KAU for the past five years. He also awarded the Young Scientist Award from the Saudi Chemical Society in 2009, and also the first prize for the distinction in science from the Saudi Chemical Society in 2012. He also received a recognition certificate from the American Chemical Society (Gulf region Chapter) for the advancement of chemical science in the Saudi Arabia. Also he received a Scopus certificate for the most Publishing Scientist in Saudi Arabia in chemistry in 2008.

He is also a member of the Editorial Board of various journals of international repute. He is the Vice-President of Saudi Chemical Society (Western Province Branch). He holds four USA patents, more than 800 publications in international journals, seven book chapters, and ten books.

Contributors

Ebrahim Abouzari-lotf Advanced Materials Research Group, Center for Hydrogen Energy, Institute of Future Energy, Universiti Teknologi Malaysia, Kuala Lumpur, Malaysia

Himanshu Agarwal Department of Chemistry, LRPG Collge, Sahibabad, Ghaziabad, India

Arshad Ahmad Advanced Materials Research Group, Center for Hydrogen Energy, Institute of Future Energy, Universiti Teknologi Malaysia, Kuala Lumpur, Malaysia

Shakeel Ahmed Department of Chemistry, Faculty of Natural Science, Jamia Millia Islamia, New Delhi, India

Izazi Azzahidah Amin Nanotechnology & Catalysis Research Centre (NANOCAT), Institute of Graduate Studies, University of Malaya, Kuala Lumpur, Malaysia

Nuha Awang Advanced Membrane Technology Research Centre (AMTEC), Universiti Teknologi Malaysia, Johor Bahru, Johor, Malaysia; Faculty of Chemical and Energy Engineering, Universiti Teknologi Malaysia, Johor Bahru, Johor, Malaysia

Adriana Balasoiu CAICAM, Faculty of Industrial Chemistry and Environmental Engineering, Politehnica University of Timisoara, Timisoara, Romania

Aniruddha Chatterjee Department of Plastic and Polymer Engineering, Marathwada Institute of Technology, Aurangabad, Maharashtra, India

Pritha Chatterjee Department of Civil Engineering, Indian Institute of Technology Kharagpur, Kharagpur, India

Saif Ali Chaudhry Department of Chemistry, Faculty of Natural Science, Jamia Millia Islamia, New Delhi, India

Moganapriya Chinnasamy Department of Mechanical Engineering, Kongu Engineering College, Erode, Tamil Nadu, India

José Raúl Corpas-Martínez Chemical Engineering Department, University of Granada, Granada, Spain

Classius Ferreira da Silva Biotechnology Laboratory of Natural Products, Institute of Environmental Sciences, Chemical and Pharmaceutical, Federal University of São Paulo, Diadema, SP, Brazil; Department of Chemical Engineering, Institute of Environmental Sciences, Chemical and Pharmaceutical, Federal University of São Paulo, Diadema, Brazil

Antonia Pérez de los Ríos Department of Chemical Engineering, Faculty of Chemistry, University of Murcia, Murcia, Spain

Narcis Duteanu CAICAM, Faculty of Industrial Chemistry and Environmental Engineering, Politehnica University of Timisoara, Timisoara, Romania

Romilda Fernandez Felisbino Biotechnology Laboratory of Natural Products, Institute of Environmental Sciences, Chemical and Pharmaceutical, Federal University of São Paulo, Diadema, SP, Brazil

Francisco José Hernández Fernández Department of Chemical and Environmental Engineering, Higher Technical School of Industrial Engineering, Polytechnic University of Cartagena, Cartagena, Spain

María José Salar García Department of Chemical and Environmental Engineering, Higher Technical School of Industrial Engineering, Polytechnic University of Cartagena, Cartagena, Spain

Makarand M. Ghangrekar Department of Civil Engineering, Indian Institute of Technology Kharagpur, Kharagpur, India

Eliezer Ladeia Gomes Department of Chemical Engineering, Institute of Environmental Sciences, Chemical and Pharmaceutical, Federal University of São Paulo, Diadema, Brazil

D.P. Hansora Department of Plastic Technology, University Institute of Chemical Technology, North Maharashtra University, Jalgaon, Maharashtra, India

Saiqa Ikram Department of Chemistry, Faculty of Natural Science, Jamia Millia Islamia, New Delhi, India

Ahmad Fauzi Ismail Advanced Membrane Technology Research Centre (AMTEC), Universiti Teknologi Malaysia, Johor Bahru, Johor, Malaysia; Faculty of Chemical and Energy Engineering, Universiti Teknologi Malaysia, Johor Bahru, Johor, Malaysia

Juhana Jaafar Advanced Membrane Technology Research Centre (AMTEC), Universiti Teknologi Malaysia, Johor Bahru, Johor, Malaysia; Faculty of Chemical and Energy Engineering, Universiti Teknologi Malaysia, Johor Bahru, Johor, Malaysia

Joon Ching Juan Nanotechnology & Catalysis Research Centre (NANOCAT), Institute of Graduate Studies, University of Malaya, Kuala Lumpur, Malaysia; School of Science, Monash University, Bandar Sunway, Malaysia

Mahadevappa Y. Kariduraganavar Department of Studies in Chemistry, Karnatak University, Dharwad, Karnataka, India

Hyung-Man Kim Department of Mechanical Engineering and High Safety Vehicle Core Technology Research Center, INJE University, Gimhae-Si, Gyongsangnam-Do, Republic of Korea

Chin Wei Lai Nanotechnology & Catalysis Research Centre (NANOCAT), Institute of Graduate Studies, University of Malaya, Kuala Lumpur, Malaysia

Víctor Manuel Ortiz Martínez Department of Chemical and Environmental Engineering, Higher Technical School of Industrial Engineering, Polytechnic University of Cartagena, Cartagena, Spain

Antonio Martínez-Férez Chemical Engineering Department, University of Granada, Granada, Spain

Takeshi Matsuura Faculty of Chemical and Energy Engineering, Universiti Teknologi Malaysia, Johor Bahru, Johor, Malaysia; Department of Chemical Engineering, University of Ottawa, Ottawa, ON, Canada; Advanced Membrane Technology Research Centre (AMTEC), Universiti Teknologi Malaysia, Skudai, Johor, Malaysia

Shahram Mehdipour-Ataei Iran Polymer and Petrochemical Institute, Tehran, Iran

Toraj Mohammadi Faculty of Chemical Engineering, Research and Technology Centre for Membrane Separation Processes, Iran University of Science and Technology (IUST), Tehran, Iran

Balappa B. Munavalli Department of Studies in Chemistry, Karnatak University, Dharwad, Karnataka, India

Mohamed Mahmoud Nasef Advanced Materials Research Group, Center for Hydrogen Energy, Institute of Future Energy, Universiti Teknologi Malaysia, Kuala Lumpur, Malaysia; Malaysia–Japan International Institute of Technology, Universiti Teknologi Malaysia, Kuala Lumpur, Malaysia

Vinh Duy Nguyen Department of Mechanical Engineering and High Safety Vehicle Core Technology Research Center, INJE University, Gimhae-Si, Gyongsangnam-Do, Republic of Korea

Javier Miguel Ochando-Pulido Chemical Engineering Department, University of Granada, Granada, Spain

Maryam Oroujzadeh Iran Polymer and Petrochemical Institute, Tehran, Iran

Mohd Hafiz Dzarfan Othman Advanced Membrane Technology Research Centre (AMTEC), Universiti Teknologi Malaysia, Johor Bahru, Johor, Malaysia; Faculty of Chemical and Energy Engineering, Universiti Teknologi Malaysia, Johor Bahru, Johor, Malaysia

Samir Kumar Pal Department of Mining Engineering, Indian Institute of Technology Kharagpur, Kharagpur, West Bengal, India

Mukhlis A. Rahman Advanced Membrane Technology Research Centre (AMTEC), Universiti Teknologi Malaysia, Johor Bahru, Johor, Malaysia; Faculty of Chemical and Energy Engineering, Universiti Teknologi Malaysia, Johor Bahru, Johor, Malaysia

Rathanasamy Rajasekar Department of Mechanical Engineering, Kongu Engineering College, Erode, Tamil Nadu, India

Mashallah Rezakazemi Department of Chemical Engineering, Shahrood University of Technology, Shahrood, Iran

Adnan Ripin Advanced Materials Research Group, Center for Hydrogen Energy, Institute of Future Energy, Universiti Teknologi Malaysia, Kuala Lumpur, Malaysia

Mohtada Sadrzadeh Department of Mechanical Engineering, University of Alberta, Edmonton, AB, Canada

Palaniappan Sathish Kumar Department of Mining Engineering, Indian Institute of Technology Kharagpur, Kharagpur, West Bengal, India

Sathyamangalam Munusamy Senthil Department of Mechanical Engineering, Kongu Engineering College, Erode, Tamil Nadu, India

Patrícia Severino Laboratory of Nanotechnology and Nanomedicine, Institute of Technology and Research, University of Tiradentes, Aracaju, SE, Brazil

Sharf Ilahi Siddiqui Department of Chemistry, Faculty of Natural Science, Jamia Millia Islamia, New Delhi, India

Eliana B. Souto Department of Pharmaceutical Technology, Faculty of Pharmacy, University of Coimbra, Coimbra, Portugal; REQUIMTE/LAQV, Group of Pharmaceutical Technology, Faculty of Pharmacy, University of Coimbra, Coimbra, Portugal

Marco Stoller Department of Chemical Materials Environmental Engineering, University of Rome "La Sapienza", Rome, Italy

Anand I. Torvi Department of Studies in Chemistry, Karnatak University, Dharwad, Karnataka, India

S. Wazed Ali Department of Textile Technology, Indian Institute of Technology Delhi, New Delhi, India

Patrícia Hissae Yassue-Cordeiro Department of Chemical Engineering, Laboratório de Catálise, Cinética e Reatores Químicos (LCCRQ), State University of Maringá, Maringá, Brazil

Masoumeh Zakeri Advanced Materials Research Group, Center for Hydrogen Energy, Institute of Future Energy, Universiti Teknologi Malaysia, Kuala Lumpur, Malaysia

Chapter 1
Organic–Inorganic Membranes Impregnated with Ionic Liquid

Víctor Manuel Ortiz Martínez, María José Salar García, Francisco José Hernández Fernández and Antonia Pérez de los Ríos

Abstract The use of ionic liquids (ILs) as fuel cell electrolytes is gaining growing interest. Their application as fuel cell membranes is fairly recent but much progress has been made in this field. Solid or semisolid conductive electrolytes are generally preferred over fluidic materials in these devices, and ILs have shown great potential to fabricate polymer membranes with good mechanical properties and high conductivity. Thus, they could replace conventional perfluorosulfonic acid membranes for operation at low and high temperatures. This chapter focuses on the recent advances in the application of ionic liquid-based polymer membranes for the development of fuel cell technology. Although special attention is paid to proton exchange membranes fuel cells (PEMFCs), other fuel cell types such as alkaline (AFCs) and microbial fuel cells (MFCs) are also covered. The membrane configurations comprised in this work include several ionic liquid–polymer membrane options such as supported and inclusion membranes, polymerizable ionic liquids and ionic liquid–polymer composites. The key factors affecting membrane performance in each case are discussed.

Keywords Fuel cell · Proton exchange · Ion exchange · Electrolyte · Ionic liquid · Polymer membrane

---

V.M.O. Martínez (✉) · M.J.S. García (✉) · F.J.H. Fernández (✉)
Department of Chemical and Environmental Engineering, Higher Technical School of Industrial Engineering, Polytechnic University of Cartagena,
Campus Muralla del Mar, C/Doctor Fleming S/N, 30202 Cartagena, Spain
e-mail: victor.ortiz@upct.es

M.J.S. García
e-mail: mariajose.salar@upct.es

F.J.H. Fernández
e-mail: fj.herfer@upct.es

A. Pérez de los Ríos
Department of Chemical Engineering, Faculty of Chemistry, University of Murcia,
Campus de Espinardo, 30100 Murcia, Spain
e-mail: aprios@um.es

Inamuddin et al. (eds.), *Organic–Inorganic Composite Polymer Electrolyte Membranes*, DOI 10.1007/978-3-319-52739-0_1

1 Introduction

Fuel cells are a promising technology for use as source of bioenergy from different types of low-cost materials such as air, methanol or even wastes. In recent years, many advances have been made in material science in order to develop novel and efficient membranes for their application in these devices. Among the most important factors in selecting a membrane for fuel cells are: (i) conductivity, (ii) internal resistance to ion transport, (iii) thermal, chemical and mechanical stability, (iv) biofouling trend, (v) fuel permeability and (vi) cost [1–3]. The efficiency of the separator is crucial for the widespread implementation of fuel cell technology. Perfluorosulfonic acid membranes such as Nafion, which are commonly used in fuel technology, present several limitations at high temperature and under anhydrous conditions in addition to being expensive. Since the operation of fuel cells at elevated temperature is receiving much attention because it can greatly enhance reaction kinetics at electrodes, other alternatives need to be found. The development of alternative membranes for low and room temperature fuel cells also poses a challenge in terms of durability and cost.

The current trend towards improving the performance of fuel cell technology includes the development of membranes based on ionic liquids (ILs). Their unique properties make possible to adapt this type of membranes to a specific device, since the chemical and physical properties of ionic liquids can be tailored by varying the nature of the cation and the anion in their structure [4]. Ionic liquids have attracted growing interest as electrolyte membrane due to their high conductivity and physical properties such as thermal stability.

The idea of including ionic liquids in fuel cell membranes is relatively recent and much progress is being made in this field. This chapter focuses on the recent advances in the field of polymer membranes based on ILs for their application in different types of fuel cells, mainly proton exchange membranes fuel cells (PEMFCs), but also other types such as alkaline (AFCs) and microbial fuel cells (MFCs). ILs can be used in their original state or in polymerized form as liquid phase or mixed with a polymer and other composite materials to form fuel cell membranes. All these options will be described in this chapter.

2 Ionic Liquids: General Properties and Applications

Ionic liquids have become one of the most promising materials investigated in the last decades for a broad range of chemical applications due their unique properties. These compounds are mixtures of anions and cations that remain liquid below 100 °C. They generally exhibit moderate polarities, high viscosities, negligible vapour pressure and virtually non-flammability. However, the main advantage of ILs is their task-specific functionality, since their properties can be easily adapted to a specific purpose by altering the substitutive group on the cation or the combined anion in their structure. These properties make them especially attractive for many technological applications, including electrochemical devices [5]. The synthesis of ionic liquids often involves the formation of the desired cation by several techniques such as amine protonation, with a subsequent anion exchange usually via the generation of a Lewis acid ionic liquid or by metathesis method [6].

Ethylammonium nitrate was the first ionic liquid reported in the literature by Walden [7] in 1914, however, it took nearly 70 years after that ILs started to gain much more importance. In 1982, dialkylimidazolium-based ionic liquids began to be deeply studied, especially ILs containing metal halide anions, such as chloroaluminate anions, which easily react with water [8]. Almost ten years later, these types of anions were replaced by halides such as tetrafluoroborate (BF_4^-) or hexafluorophosphate (PF_6^-). Dialkylimidazolium-based ILs often show moderate polarity and hydrophobic behaviour, and can be used as inert solvents [9–11]. Finally, a third generation of ionic liquids comprises those including biodegradable and low toxic cations and anions. They offer similar properties to those displayed by the ILs belonging to the second group but are generally more hydrophilic. This third group is usually characterized by the presence of cations such as choline and quaternary ammonium cations, and anions such as sugars, organic acids or alkylphosphates. This group also includes eutectic solvents, which consist of mixtures of salts (Table 1). The main advantage of third generation ionic liquids is the lower cost of the synthesis and purification processes, especially in the case of eutectic solvents [5, 12].

One of the most promising applications of ionic liquids is their use as green solvents, replacing polar organic solvents, which are volatile and therefore pose environmental risks. Instead, the vapour pressure of ionic liquids is near zero, enabling easy recovery and reusability. ILs have been used as replacement of volatile organic solvents in a wide variety of chemical processes, including separation and purification technology and reaction media in biochemical and chemical catalysis. Other major advantage is that there is a certain class of these compounds that does not denature enzymes and, therefore, are suitable for enzyme-based applications, such as enzymatic reactions or enzymatic fuel cells (EFCs). Ionic liquids have been employed as co-solvents with water, as a secondary phase in different types of mixtures and as non-aqueous solvents [13].

In recent years, ILs have been widely studied in immobilized form in several types of membranes for applications in separation technology and as ion exchange separators in fuel cells [14–17]. The application potential of ILs lies in their simple operation, possibilities for upscaled processes and environmental-friendly properties. IL membranes have been used as selective transport for several compounds such as organic acids, aromatic hydrocarbons, mixed gases and metal ions. They have also gained much attention for CO_2 capture [18]. Since ILs have high selectivity and ionic conductivity, including proton transport properties, separators based on these compounds have proven to offer promising prospects for fuel cell technology under several conditions, including anhydrous, non-anhydrous and alkaline conditions. For example, imidazolium-based ionic liquids have been used to prepare organic–inorganic hybrid alkaline anion exchange membranes (AEMs)

Table 1 General classification of ionic liquids [5]

Ionic liquid	Cations	Anions
First generation	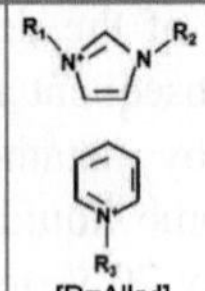 [R=Alkyl]	$AlCl_4^-$ $Al_2Cl_7^-$
Second generation	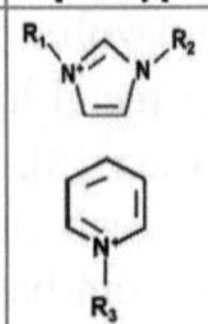 NR_4^+ [R=Alkyl]	PF_6^- BF_4^- $N(SO_2CF_3)_2^-$ $Me_2SO_3^-$
Third generation	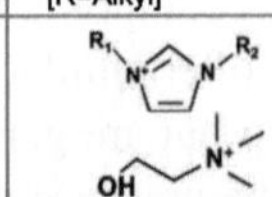 Amino acids [R= Oxygenated alkyl]	Amino acids $R_2PO_4^-$ $ROSO_3^-$
	Eutectic solvents 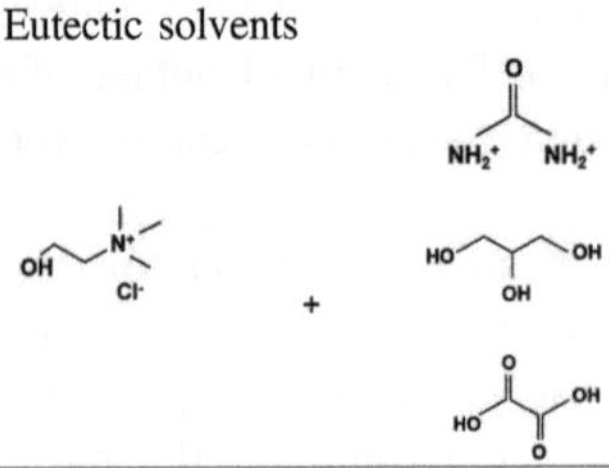	

that, in addition to good conductivity, have exhibited good mechanical properties and excellent stability [19]. Table 2 lists the most common cations and anions in ionic liquids.

3 Ionic Liquids as Electrolytes in Fuel Cells

In recent years, the application of ionic liquids as non-aqueous electrolytes in electrochemical systems has attracted much interest due to their unique properties. These compounds offer wide stability windows with high resistance for electrochemical reduction and oxidation processes besides negligible vapour pressure and non-flammability. Although the specific conductivities of ionic liquids are lower compared to those of conventional aqueous electrolytes solutions yet still provide fairly high conductivities depending on the anions and cations present in their structure, with similar values to those offered by inorganic electrolytes in aprotic organic solvents [20]. The conductivities of ionic liquids at room temperature range

Table 2 Common IL cations and anions

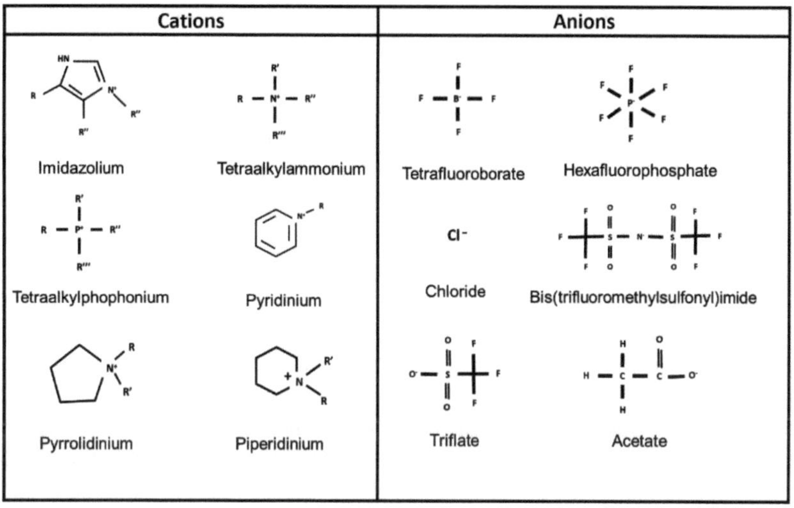

from 0.1 to 18 mS cm^{-1}. Ionic liquids based on the group imidazolium have been reported to display higher overall ionic conductivities compared to other cationic groups. For instance, ILs containing the cation 1-ethyl-3-methylimidazolium, [emim], present an average conductivity of about 10 mS cm^{-1}, while ILs containing cations based on the ammonium, pyrrolidinium, piperidinium and pyridinium groups typically show overall lower conductivities up to 5 mS cm^{-1}. Also, ILs with lower viscosities tend to have higher conductivities [21]. These compounds have a broad stability window typically from 2 to 6 V, the most common value being around 4.5 V. The electrochemical window is defined by the reduction of the organic cation and the oxidation of the anion [22]. Regarding electrochemical stability, ionic liquids including halide anions (e.g. Cl^-, Br^-) offer lower stability values between 2 and 3 V, while ionic liquids based on fluorine-containing anions such as bis(trifluoromethylsulfonyl)imide, [Tf_2N^-], show wider electrochemical window up to 6 V. ILs based on tetraalkylammonium cations also have enhanced stability above 4 V [21–23]. All these properties make ionic liquids ideal as electrolytes for many electrochemical applications including advanced batteries, dye sensitized solar cells, double layer capacitors, actuators, thermocells, water splitting technology and fuel cells [24].

Protic ionic liquids (PILs), which have a proton available for hydrogen bonding, offer great potential as fuel cell electrolytes operating at ambient or high temperatures. ILs as proton carriers allow acid or basic aqueous electrolytes to be avoided when the use of non-aqueous electrolytes is required. PILs can be easily synthesized through the combination of a Brønsted acid and a Brønsted base, being characterized by the proton transfer from the acid to the base and acting as proton acceptor and donor. This provides them with unique properties compared to the rest of ionic liquids. Brønsted acidic ionic liquids (BAILs), which are very close to PILs, are a special case since they can act either as protic or aprotic ILs depending on whether the available proton is on the cation or the anion [25]. The ILs diethylmethylammonium trifluoromethanesulfonate, [dema][TfO], and 1-methylimidazolium tetrafluoroborate, [hmim][BF_4], are examples of protic ionic liquids with cations based on ammonium and imidazolium groups, respectively. The advantage of using PILs as electrolytes for PEMFCs is that they can be operated at temperatures above 100 °C under anhydrous conditions, which is especially important to ensure good performance in high-temperature fuel cells, including operational fuel cell voltage and current output. From a practical point of view, ionic liquids have shown great potential to fabricate polymer membranes. In this sense, solid or semisolid conductive electrolytes with good mechanical properties are preferred over fluidic materials. The proton transport mechanism in PILs can be explained by the combination of the Grotthuss and vehicle mechanisms. While the first mechanism refers to the proton hopping between hydrogen bondings, vehicle mechanism involves diffusion and migration phenomena. The predominant conduction mechanism strongly depends on the composition of each ionic liquid [26].

Although protic ionic liquids are particularly relevant, aprotic ionic liquids have also been used in fuel cells since they can offer high overall ionic conductivity in comparison to protic ILs. Furthermore, aprotic ILs have been of interest for fuel cells whose operation may not depend exclusively on proton exchange but also on the transport of other ion species, as in the case of several types of microbial fuel cells. In fact, several types of separators apart from proton exchange membranes have been investigated for this technology, including cation exchange membranes (CEM) and anion exchange membranes (AEM) [3].

4 Ionic Liquid Polymer Membranes for Fuel Cells

Due to the unique properties of ionic liquids, mainly their high ionic conductivity and excellent thermal stability, these materials have been studied as possible substitute for Nafion membranes. Nafion is a solid polymer electrolyte widely used as proton-conducting material in fuel cells. It consists of a copolymer of tetrafluoroethylene (TFE) and sulfonyl fluoride vinyl ether. The TFE material provides the backbone system for the membrane that contains sulfonic acid pending groups (–SO_3H). As is widely known, Nafion displays featured properties for its application in fuel cell technology such as chemical stability and mechanical durability. Under favourable conditions, Nafion membranes can offer maximum proton conductivities (σ) in the order of 10 mS cm^{-1} [27]. However, several disadvantages related to operational limitations and high cost are associated with Nafion. For instance, it requires complete hydration to ensure a good performance, with a significant loss of conductivity at relative humidity (RH) levels below 100% and at temperatures above 80 °C due to dehydration. This makes Nafion especially suitable for low temperature fuel cells (LTFCs) but poses a problem for the practical implementation of several other types of fuel cells that need to be operated at temperatures higher than 120 °C [25]. Moreover, it is generally accepted that the practical implementation of PEMFCs would require the use of proton exchange membranes with high proton conductivity under anhydrous conditions and above 100 °C [4]. In the specific case of direct methanol fuel cells (DMFCs), Nafion presents very high methanol permeability. Finally, its price remains high enough so that the use of Nafion membranes in fuel cells can still be considered inefficient from an economic perspective [28]. In comparison, ILs display high stability and conductivity at high temperatures. Since many ILs remain liquid at room temperature, they need to be immobilized as liquid phase in membrane materials to form an ideally non-leaching and stable solid film that can be used as separator, ensuring continuous operation with a minimal amount of ionic liquid as active phase. The use of ionic liquid-based polymer membranes is among the approaches adopted to incorporate ILs into the separator in fuel cells. Polymer membranes based on ILs mainly comprise supported ionic liquids membranes (SILMs), polymer inclusion membranes (PILIMs), polymerized ionic liquid membranes (PyILMs) and other ionic liquid/polymer composite materials. Polymer SILMs, which consist of a liquid phase (ionic liquid) supported on a porous material by capillary forces, are favoured by high surface areas, low amount of solvent and high efficiency in comparison with other liquid membranes. However, their main limitation derives from the lack of long-term stability since the IL phase tends to leave the pores. Alternatively, PILIMs are formed by immobilizing the ionic liquid into a polymer matrix, showing improved mechanical and chemical properties compared to SILMs. Polymerized ionic liquid (PyILs) membranes are formed by polymerizing IL monomers, combining the unique properties of ionic liquids and the general features of polymers resulting in solid film structures. PyILMs can be mainly prepared by homopolymerization and copolymerization techniques. Other IL membrane types are IL-based polymer gel membranes and polymer/ionic liquid composite membranes including functionalized materials. Figure 1 summarizes the most important polymer membrane types based on ILs [18].

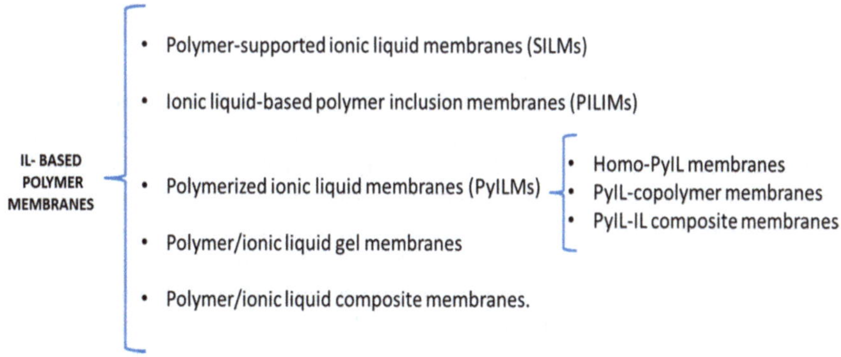

Fig. 1 Classification of IL-based polymer membranes

Parameters such as thermal and chemical stability, compatibility of the polymer with the ionic liquid and polymer/ionic liquid ratio, are important factors that determine the final conductivity and the ionic liquid retention into the polymer matrix, and thus the overall performance of fuel cell devices.

The first attempts by Kreuer et al. [29] to create proton-conductive polymer membranes based on the imidazole molecule were made in 1998. Imidazole was employed as proton carrier in sulfonated poly(ether ether ketone) (SPEEK) membranes, showing high proton conductivity at high temperatures under anhydrous conditions. In 1999, Abu-Lebdeh and Owen [30] used room temperature eutectic mixtures of imidazole along with imidazolium salts in poly(vinylidene difluoride) (PVDF) membranes. The aim of such strategy was to create a network that served as proton transfer pathway by bonding a proton solvent (imidazole) to a polymer backbone. Later, Sun et al. [31] investigated polymer membrane systems based on Nafion and polytetrafluoroethylene (PTFE) impregnated with imidazole compounds. Since then, much progress has been made in order to create alternative polymer membranes based on ionic liquids. The following subsections will focus specifically on the application of ionic liquid-based polymer membranes in fuel cell technology.

4.1 *Ionic Liquid/Polymer Membranes*

Polymer membranes allow ionic liquids to be prepared as solid electrolyte films for their application as separators in fuel cell devices. When ionic liquids are mixed with common polymers, which act as a plasticizer, their transport properties can be enhanced due to the increase of the ion dissociation and local dynamics in their structure [4]. Polymer membranes based on ILs can be relatively easily prepared by casting method, which consists of mixing the inert polymeric support with ionic liquid and an appropriate solvent such as tetrahydrofuran (THF). They can be also prepared by impregnating the polymer of interest with the ionic liquid in pure or solution form and by supporting the ILs onto porous polymer materials by other physical techniques such as ultrafiltration.

ILs including imidazolium cations have been the most commonly used to prepare IL polymer membranes for their application in proton exchange membrane fuel cells. In line with the work by Kreuer et al. [29], other authors have used sulfonated poly(ether ether ketone) as polymer matrix to prepare proton-conductive membranes due to its stability properties and low cost. 1-Ethyl-3-methylimidazolium diethyl phosphate has been combined with this type of polymers by casting method for their application in high-temperature PEMFCs under anhydrous conditions, with a resulting conductivity ranging between 1.25 and 3.16 mS cm^{-1} (145 °C) and a maximum power density of 203 mW cm^{-2} [32]. Another option is the impregnation of SPEEK membranes in aqueous solutions of ionic liquids. For example, SPEEK membranes impregnated with the ionic liquid 1-butyl-3-methylimidazolium tetrafluoroborate, [bmim][BF_4], have proven to reach high conductivity values of up to 9.3 mS cm^{-1} at 140 °C [33] with a maximum power density reported of 130 mW cm^{-2} in PEMFCs at 80 °C [34]. In SPEEK membranes, the IL phase interacts with the sulfonic acid sites on the polymer through an ion exchange process. The conductivity of these membranes impregnated with the ionic liquid greatly improves with temperature when compared with the pure SPEEK membrane. However, membranes with a large amount of embedded ionic liquid can led to a decrease in conductivity due to an increase in the membrane resistance.

Several imidazolium-based IL membranes combined with poly(vinylidenefluoride-co-hexafluoropropylene) (PVdFHFP) have also been investigated. These include, among others, 2,3-dimethyl-1-octylimidazolium trifluoromethylsulfonate ([dmoim][TfO]), 1-butyl-3-methylimidazolium trifluoromethylsulfonate ([bmim][TfO]), *N*-ethylimidazolium trifluoromethyl-sulfonate ([eim][TfO]), 2,3-dimethyl-1-octylimidazolium bis(trifluoromethyl-sulfonyl)imide ([dmoim][TFSI]) and *N*-ethylimidazolium bis(trifluoromethanesulfonyl)imide ([eim][TFSI]) [35–38]. The resulting conductivity in such cases is within the interval 2–20 mS cm^{-1}. The power performance reported for these membranes in non-humidified fuel cells at inmerdiate and high temperatures are in the range 0.2–1.0 mW cm^{-2}. In these types of membranes, the thermal stability improves with higher amounts of polymer but there is a consequent decrease in the conductivity, and the polymer concentration needs to remain below 40 wt% so that the conductivity remains high (10^{-2} S cm^{-1} at room temperature).

Most of the imidazolium-based IL membranes reported in the literature have been prepared using monocationic ionic liquids. Dicationic ILs consisting of a double-charge cation formed by two individual cations bonded to each other through an alkyl chain and linked to respective anions of the same type are a new class of ionic liquids. Diatonic ILs can display superior properties compared to monocationic ILs in terms of thermal stability with conductivities of the order of those offered by monocationic ILs [39]. Cationic ionic liquids have been scarcely studied in membrane form in fuel cells. The performance of the dicationic IL 3-di (3-methylimidazolium) propane bis(trifluoromethylsulfonyl)imide, pr $[mim]_2[Tf_2N]$, has been compared with that of the monocationic 1-hexyl-3-methylimidazolium bis(trifluoromethanesulfonyl)imide, $[hmim][Tf_2N]$, in polybenzimidazole (PBI) membranes in high-temperature PEMFCs (180 °C). PBI is a non-shulphonated polymer with exceptional chemical and thermal stability. The conductivity of the membranes based on the dicationic IL is higher compared with the monocationic IL-based membrane, with a maximum power performance of 44 mW cm^{-2} in the first case. Both membranes offer improved conductivity properties compared to standard phosphoric acid-loaded PBI membranes due to the inclusion of ionic liquid. The proton exchange in these composite membranes is mainly due to the Grotthus transport mechanism [40].

Several disadvantages have been reported for imidazolium-based polymer membranes. On one hand, imidazolium cations offer smaller electrochemical active surface areas compared with, for example, trimethyl amide and pyridinium groups. In addition, the performance of fuel cells based on ILs containing imidazolium cations can also be limited by the presence of impurities such as halide anions [41, 42].

IL liquids containing ammonium protic cations have recently gained more attention for the preparation of proton exchange membranes. The ionic liquid diethylmethylammonium trifluoromethanesulfonate, [dema][TfO], has been among the most studied PILs in fuel cells. Membranes based on this ionic liquid and sulfonated polyimide (SPI) prepared by casting method have been investigated in non-humidified H_2/O_2 fuel cells due to the high compatibility of both components. These membranes display thermal stability above 300 °C and an ionic conductivity higher than 10 mS cm^{-1} (160 °C). The ratio of [dema][TfO] employed in SPI membranes can greatly affect the temperature dependence, being possible to achieve significant ionic conductivity at low temperature with high ratios of ionic liquid (>0.1 mS cm^{-1} at 40 °C for a minimum ratio of 67 wt%). The high conductivity of such membranes has been linked to the strong interaction between the IL [dema][TfO] and the sulfonate groups of SPI. The maximum power densities reported at intermediate temperatures under non-humidified conditions range from 100 to 190 mW cm^{-2} (at 80 and 120 °C, respectively) [43–45]. The same ionic liquid has been combined with PBI by casting method for their application in anhydrous H_2/Cl_2 fuel cells. These membranes showed an ionic conductivity of 20.7 mS cm^{-1} at 160 °C, ensuring good chloride and hydrogen electrode kinetics. The maximum power density achieved so far in this type of fuel cell is 30 mW cm^{-2} at 140 °C [46]. Other ILs with very similar structure to [dema][TfO] such as triethylammonium trifluoromethanesulphonate, [N_{133}][TfO], have been supported on high-porosity polyimide (Matrimid®), exhibiting conductivities in the same order (20 mS cm^{-1} at 130 °C) [47]. The proton-conducting mechanism in this type of ionic liquids has been explained by both proton hopping (Grotthuss mechanism) and diffusion (vehicular mechanisms) [46, 48].

It is also possible to dope polymer membranes with ionic liquid cations or anions in order to improve some of their properties. Both imidazolium and ammonium ILs have been employed to dope Nafion materials to enhance their permeability, thermal stability and proton conductivity at high temperature, which are some of the limitations displayed by Nafion membranes as commented in previous sections. This technique has proven to offer good results when it comes to reducing the permeability of Nafion-based membranes in direct methanol fuel cells (DMFCs), which is a critical factor in this type of fuel cells. For example, the incorporation of the cation *n*-dodecyltrimethylammonium, [DTA], into the structure of a protonated Nafion-112 membrane can reduce the permeability of the membrane towards methanol by 80%. However, this strategy requires the determination of the optimal incorporation degree, since this modification in the polymer structure can lead to a significant increase in the membrane resistance. Optimal values of 20% could give the optimum membrane performance in such cases [49]. Polymer electrolyte membranes doped with PILs are also a promising material option for anhydrous proton conduction. Protic ionic liquids such as triethylammonium methanesulfonate and triethylammonium perfluorobutanesulfonate have been used for anion-doping in Nafion-117 membranes, increasing their high thermal stability up to 140 °C [50]. Other more complex approaches include sequential doping steps such as neutralizing Nafion 117 with e.g. triethylammonium, [TEA], and then doping it with the IL of interest, e.g. the protic ionic liquid triethylammonium trifluoromethanesulfonate, [dema][TfO]. By this technique, the resulting IL polymer membrane can increase its thermal stability up to 155 °C and its conductivity up to 7.3 versus 6.1 mS cm^{-1} for Nafion-117 membrane at high temperature (145 °C) [51]. In such cases, hydrogen bonding interactions appear between the acidic sites (N–H) of the ammonium cations and the SO_3^- groups in Nafion or in the trifluoromethanesulfonate anion. The distribution of these SO_3^- groups around the ammonium cations increases with the concentration of ionic liquid. An increase in temperature and doping level leads to a faster diffusion capacity in the membrane structure and thus to a higher ionic conductivity [52]. Therefore, the IL uptake, the interaction between the ionic liquid and the polymer used and the structuration of the ionic liquid in the membrane matrix are of key importance in such strategies [53].

Other IL groups have shown great potential for their application in PEMFCs. This is the case of ILs containing fluorohydrogenate anions such as *N*-ethyl-*N*-methylpyrrolidinium fluoro-hydrogenate [EMPyr$(FH)_{1.7}$F]. PVdFHFP membranes based on this IL have displayed conductivities above 40 mS cm^{-1} at 120 °C and a maximum power density of 103 mW cm^{-2} at 50 °C under non-humidified conditions in PEMFCs [54].

The application of ionic liquids as exchange membranes has not restricted to intermediate- and high-temperature fuel cells. Very recently, IL-based polymer membranes have been investigated as separators in fuel cells operating at room temperature as is the case of microbial fuel cells (MFCs). Ionic liquids including imidazolium and ammonium cations supported on hydrophilic polyamide (Nylon) by ultrafiltration have been investigated as alternative separators in these devices. This includes ionic liquids based on the cations 1-*n*-alkyl-3-methylimidazolium (*n*-butyl, *n*-octyl) and methyl trioctylammonium, in combination with hexafluorophosphate, tetrafluoroborate, chloride and bis{(trifluoromethyl)sulfonyl}imide anions. Among these SILMs, the membrane based on methyltrioctylammonium chloride, [mtoa][Cl], showed the best performance in terms of power density and removal of organic matter from wastewater used as fuel in MFCs. However, the stability of this type of IL-based membranes is limited in contact with aqueous media as happens in MFCs [55]. Instead, ionic liquid polymer inclusion membranes prepared by casting method have received more attention in microbial fuel cells (Fig. 2). Membranes of polyvinyl chloride (PVC) and ionic liquids with imidazolium, ammonium and phosphonium cations have shown potential application in such devices [56, 57]. Although imidazolium and ammonium-based ILs have been among the most studied types, several ILs containing phosphonium cations offer good conductivity and their combination with an aromatic anion improves the movement of ionic charges. For example, the IL triisobutyl(methyl)phosphonium tosylate, $[P_{I4,I4,I4,1}]$ [TOS], contains a sulfonil group, also present in Nafion, that may act as a proton-conductive group. Phosphonium-based IL polymer membranes have been compared with Nafion in single-chamber MFCs, offering equivalent power densities (794 vs. 755 mW m^{-3}, respectively). The stability of PILIMs in aqueous media is one of the key factors for the long-term operation in microbial fuel cells [58]. In this respect,

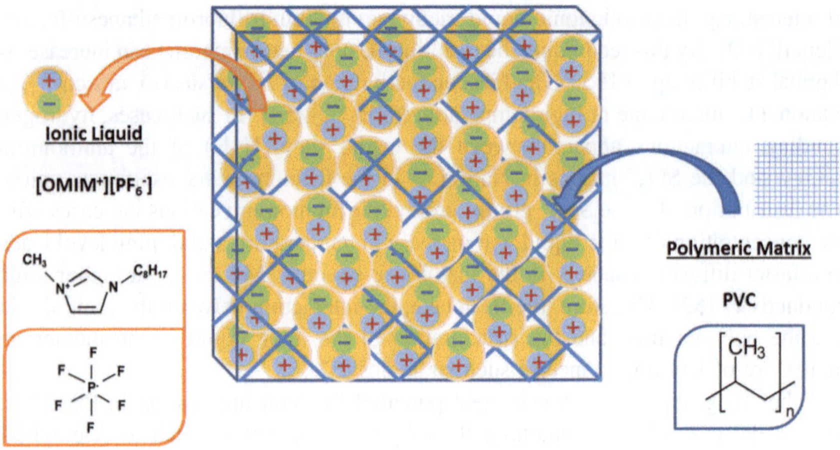

Fig. 2 IL polymer inclusion membranes used in MFCs

ionic liquids containing anions susceptible to hydrolysis can severely affect fuel performance, as occurs with the anion hexafluorophosphate [PF_6^-] [57].

Another advantage of PILIMs is that they can be embedded in several types of conducting materials for fuel cell construction and upscaling. PILIMs can be embedded into cathode materials such as carbon cloth to create cathode–membrane assemblies, reducing the electrode spacing. This option can significantly improve fuel cell performance compared with setting up a PILIM and a carbon-based cathode as separate components (e.g. methyltrioctylammonium chloride) [59]. This configuration based on the ionic liquid triisobutyl(methyl)phosphonium tosylate, [$P_{I4,I4,I4,1}$][TOS], has been employed in upscaled continuous mode MFCs, achieving notable level of power density (12.3 W m^{-3}) and high percentage of chemical oxygen demand (COD) removal when using wastewater as fuel [16]. As MFCs use several types of wastes of different nature as fuel, the possible interaction between the ionic liquid used as electrolyte and the composition of the medium need to be carefully studied in order to ensure an optimal performance.

4.2 *Polymerized Ionic Liquid Membranes*

The innovative idea of applying polymerized ionic liquids in fuel cell technology is quite young and has opened up a promising path for fuel cell membrane development. Poly(ionic liquid)s (PyILs) or polymerized ionic liquids are polyelectrolytes whose structure is made of repeating units based on ionic liquids. Although ILs are often liquid at room temperature, the complex macromolecular structure of

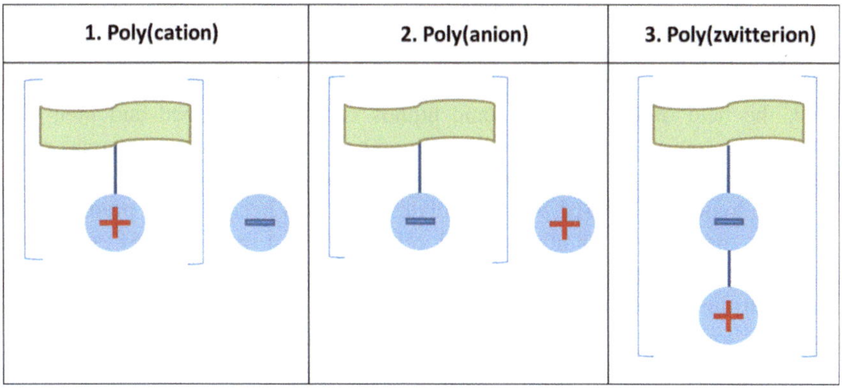

Fig. 3 Most common monomer units of polymerized ionic liquids [4]

PyILs makes most of them solid [60, 61]. PyILs are usually synthetized by direct radical polymerization of ILs. They can be grouped in polycations bearing a cation, polyanions bearing an anion and polyzwitterions (bearing a cation and an anion), although other structures are possible [62] (see Fig. 3).

Polymerized ionic liquids can be prepared by two different techniques: (i) direct polymerization of ionic liquid monomers, and (ii) chemical modification of polymers. While ILs are characterized by high conductivity (up to 10 mS cm^{-1}), Polymerized ionic liquids by these techniques exhibit ionic conductivities below 10^{-3} mS cm^{-1} since the glass transition temperature increases after the polymerization process, with a lower number of mobile ions available. However, several methods have been developed to enhance the ionic conductivity of PyILs such as: (i) using a higher concentration of the carrier ions and (ii) modifying the structure of the electrolyte [4]. Their solid state reduces the possibility of leakage in fuel cells compared with non-polymerized ionic liquids. PyILs have also been used as dispersant to stabilize different types of materials. Finally, their porous structure makes them useful for many commercial applications such as ion exchange materials, metal recovery or liquid chromatography [60].

The ionic conductivity of PyILs makes them suitable for preparing different types of exchange membranes. A wide variety of Polymerized ionic liquid-based membranes can be synthesized depending on their specific applications. Díaz et al. [63] performed the polymerization of 1-(4-sulfobutyl)-3-vinylimidazolium trifluoromethanesulfonate ([HSO_3-bvim][TfO]) to obtain a solid electrolyte membrane. The imidazolium cation was selected due to its high electrochemical stability while the triflate anion was chosen because of its high conductivity. This material was used as separator in proton exchange membrane fuel cells (PEMFCs), with a maximum power output of 33 mW cm^{-2} at room temperature. They also observed that the thickness of the membranes plays an important role in terms of proton transfer resistance. Later, Díaz et al. [64] photopolymerized 1-(4-sulphobutyl)-3-vinylimidazolium trifluoromethanesulphonate, [HSO_3-bvim][TfO], and the ionic liquid 1-(4-sulphobutyl)-3-methylimidazolium trifluoromethanesulphonate, [HSO_3-bmim][TfO], for achieving values of conductivity of 290 S cm^{-1} and a maximum power density of 49 mW cm^{-2} operating at 40 °C. All these works show how quick the field of polymerized ionic liquids is being developed and their high potential as separators in fuel cells. Thus, the replacement of commercial membranes such as Nafion could take place in the near future.

These ionic liquid-based polyelectrolytes have also been used to prepare anion exchange membranes for alkaline fuel cells (Fig. 4). Lin et al. [65] crosslinked 1-vinyl-3-methylimidazolium iodide ([vmim][I]) with styrene and acrylonitrile to obtain an anion exchange membrane with hydroxide ions in its structure. The hydroxide ion conductivity was higher than 10 mS cm^{-1} and good mechanical and chemical properties were obtained in alkaline conditions at 60 °C. Qiu et al. [66] also crosslinked styrene and acrylonitrile with 1-allyl-3-(6-(1-butyl-2-methylimidazol-3-ium-3-yl)hexyl)-2-methylimidazol-3-ium bromide ([abmhm][$Br]_2$), to fabricate AEMs. This material showed a hydroxide conductivity of approximately 20 mS cm^{-1} and long-term stability in alkaline solutions even at temperatures above 50 °C. Qiu et al. [67] also compared the performance of two types of copolymer membranes based on ionic liquids as AEMs. Membranes based on 1-methyl-3-(4-vinylbenzyl)imidazolium chloride ([mvbim][Cl]), and *N*,*N*,*N*-trimethyl-1-(4-vinylphenyl) methanaminium chloride ([tmvpma][Cl]), exhibited

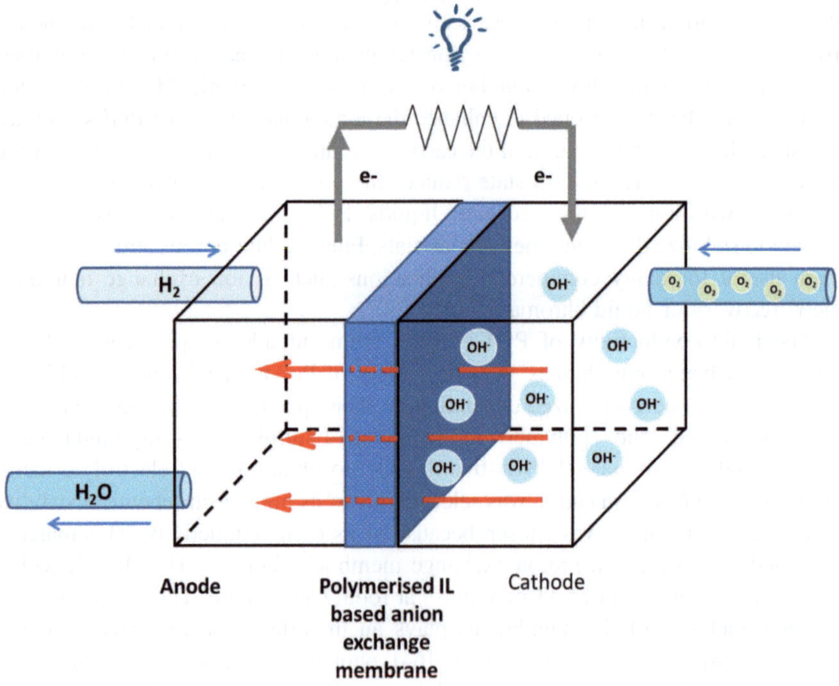

Fig. 4 Scheme of a double chamber alkaline fuel cell using a polyionic liquid separator

hydroxide conductivities up to 10 mS cm^{-2}. In general, imidazolium-based ILs have shown better chemical stability under alkaline conditions versus ammonium ILs, since membranes based on quaternary ammonium salts can be subjected to degradation processes in alkaline solutions.

Finally, there are several research works reporting promising results on the use of PyILs in quasi-solid state dye-sensitized solar cells (DSSCs). The application of ILs in DSSCs has increased their efficiency. For instance, the use of imidazolinium cations in DSSCs leads to the decrease of recombination at the working electrode due to the multilayer adsorption, thus improving the DSSC performance [68]. The long-term stability of DSSCs has also been enhanced by replacing liquid electrolytes by solid state PyILs electrolytes [69–72].

4.3 IL Gel and Composite Polymer Membranes

This subsection includes other strategies to prepare polymer membranes based on ionic liquids such as gel polymer membranes, polymerization techniques and other IL-based polymer membrane composites that also contain non-polymer functionalized materials.

The sol–gel transition of polymers in ILs is another option to simply fabricate polymer electrolyte membranes. Imidazolium-based ILs such as 1-ethyl-3-methylimidazolium trifluoromethanesulfonate ([C_2mim][TfO]) can be used with poly [(vinylidene fluoride)-co-hexafluoropropylene] P(VdF-co-HFP), to form polymer gels with still high ionic conductivity. The resulting gel membrane has a porous structure which has been associated with the presence of spherulites formed during the sol–gel transition of the polymer in the IL [26]. Also, vinyl monomers such as methyl methacrylate can be easily polymerized by free radical polymerization when solubilized in ILs. The ionic liquid 1-ethyl-3-methylimidazolium bis(trifluromethanesulphone) imide ([C_2mim][Tf_2N]) has been employed to fabricate polymers gel of high ionic conductivity in the order of 10 mS cm^{-1} using this technique [4]. These gel membranes showed high potential for their applications in fuel cells.

The polymerization of ionic liquid microemulsions, which are stable dispersions containing two immiscible liquid phases stabilized by surfactants, has proven to be a suitable technique for the preparation of IL gel polymer electrolyte membranes. IL-based microemulsions are gaining growing interest due to the outstanding properties of both the microemulsion system and the ionic liquid. Several examples of protic ionic liquid microemulsions can be found in the literature. Yan et al. [73] fabricated proton-conducting membranes via polymerization of microemulsions of PILs dispersed in polymerizable oil (mixture of styrene and acrylonitrile). The resulting membranes based on the PILs *N*-methylimidazolium trifluoromethanesulfonate ([mim][TfO]), *N*-ethylimidazolium trifluoromethanesulfonate ([eim][TfO]) and 1,2-dimethylimidazole trifluoromethanesulfonate ([dmim][TfO]), showed similar proton conductivities up to 100 mS cm^{-1} at 160 °C.

Polymerized IL microemulsions pose, however, several challenges such as their long-term stability, since the ionic liquid contained in the membrane structure can be released to the media. In line with this work, Chu et al. [74] fabricated proton-conducting membranes based on polyamidoamine dendrimers based on PILs. Polyamidoamine dendrimers are monodisperse polymers characterized by high concentration of primary and tertiary amine groups. This type of dendrimers can be easily protonated to form macromolecular cationic centres of PILs via neutralization with Brønsted acids. The membrane composites can be fabricated by cross-linking polymerizable oils containing this polyamidoamine dendrimer type, showing very high thermal stability (up to 350 °C) and proton conductivity of the order of 12 mS cm^{-1} at 160 °C. Membranes based on this type of macromolecular PILs display higher conductivity compared with the same type of membranes containing small molecule PILs. Other alternative materials for proton-conducting membranes include silica hybrid separators based on PILs. Lin et al. [75] prepared PIL-based hybrid membranes by polymerizing a mixture of 1-ethylimidazolium trifluoromethane-sulfonate ([eimm][TfO]), and a polymer precursor, poly (styrene-co-acrylonitrile), in which silica fillers were dispersed (silica nanoparticles and mesoporous silica nanospheres). Under anhydrous conditions, these hybrid membranes exhibit proton conductivities up to 10 mS cm^{-1} at 160 °C. However, the power performance of these membranes in fuel cells has not been thoroughly investigated yet.

As commented above, polybenzimidazole has been extensively investigated doped with several inorganic acids for high temperature applications up to 200 °C. Recently, PBI membranes based on different inorganic fillers and ionic liquids have been fabricated in order to improve crossover properties of PBI membranes at high temperatures. PBI gel electrolytes can be synthesized containing acid along with ionic liquid and polymer material as potential proton exchange membranes for high-temperature PEMFCs. For example, composites consisting of the acid H_3PO_4, the ionic liquid 1-methyl-3-propyl-methylimidazolium dihydrogen phosphate and PBI display ionic conductivities in the order of 2 mS cm^{-1} (150 °C) under anhydrous conditions. These membrane composites show high thermal stability (up to 200 °C). A three-dimensional hydrogen bonding network has been proposed for this type of composite, with predominance of the Grotthuss hopping mechanism [76].

Another recent approach consists of the development of polymer composites based on ion exchange materials functionalized with ionic liquid in order to improve their proton conductivity. Xu et al. [77] developed a composite membrane formed by poly(benzimidazole) loaded with H_3PO_4 and graphite oxide doped with the ionic liquid 1-(3-aminopropyl)-3-methylimidazolium bromide, [apmim][Br]. The inclusion of graphite oxide functionalized with the cation group 1-(3-aminopropyl)-3-methylimidazolium allows the proton conductivity to be improved from 25 to 35 mS cm^{-1}. The performance of this type of membranes in PEMFCs can reach 320 mW at 175 °C. The improvement of proton conductivity in comparison with PBI membranes solely loaded with H_3PO_4 can be explained by the presence of hydrogen bonds in the IL functionalized graphite, forming acidic functional groups, such as carboxylic acid, epoxy oxygen and especially the aminopropyl-methyl imidazolium group, which facilities proton conduction.

Finally, hybrid IL polymer membranes have been prepared by adding large pore zeolites functionalized with protic ionic liquids in H_2PO_4-doped PBI membranes via casting solution (Fig. 5). NH_4-BETA and NaY zeolite types have a porous structure that allows accommodating a wide variety of cations. They have been functionalized with imidazolium and ammonium-based ionic liquids such as trimethylammoniun dimethyl phosphate, *N*,*N*-dimethyl-*N*-(2-hydroxyethyl)ammonium bis(trifluoromethanesulfonyl)imide and 1-H-3-methylimidazolium bis (trifluoromethanesulfonyl)imide to fabricate proton exchange composite membranes. The IL/zeolite/PBI composites offer satisfactory performance for long-term operation with sufficient conductivity (32 mS cm^{-1} at 150 °C). The inclusion of this type of filler in the PBI membrane leads to a decrease in the fuel crossover due to molecular sieving properties of the inorganic loading and the improved organic–inorganic interface owing to the presence of new functional groups. Moreover, the promotion of the ionic conductivity is the result of a higher phosphoric acid uptake and the presence of new conduction pathways [78].

All these advances show the great possibilities of ionic liquids in the field of fuel cell polymer membranes. Given that the use of these compounds in fuel cell

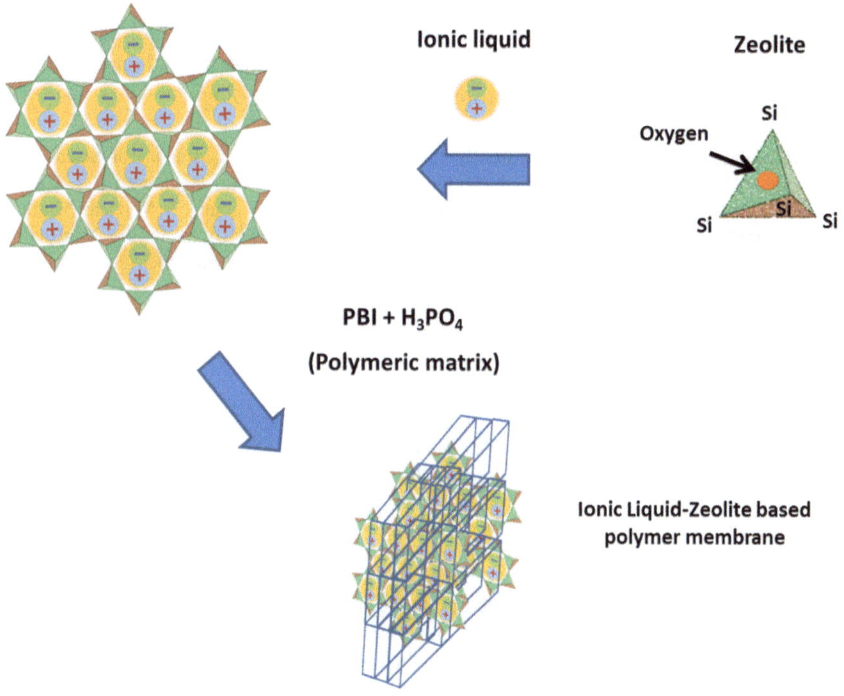

Fig. 5 Ionic liquid–zeolites polymer membranes

technology is fairly recent, much progress on the development of IL polymer membranes for fuel cell applications is expected to be made in next years.

5 Conclusions

The main advantages of using ionic liquids as fuel cell electrolytes are their liquid near-zero vapour pressure, thermal and chemical stability, task-specific properties and high ion and proton conductivities. Polymer membranes based on ILs are a promising alternative to perfluorosulphonic acid membranes such as Nafion, which is one of the most used electrolytes in fuel cells but pose operational and cost limitations. In comparison, IL-based polymer membranes can display high conductivity and stability at high temperature under non-anhydrous conditions. Several techniques have been developed to fabricate IL-based polymer membranes, including supported membranes, impregnated membranes, polymer inclusion membranes by casting method and, more recently, the polymerization of ionic liquids as one of the most innovative techniques. Other types are IL gel and hybrid IL polymer composite materials. The compatibility between the polymer and the ionic liquid used as well as the ionic liquid/polymer ratio are key factors for ensuring a good conductivity and mechanical stability. IL polymer membranes have been successfully investigated mainly in proton exchange membrane fuel cells, but also in other types of fuel cells such as alkaline and microbial fuel cells, which operate at room temperature. All these facts make ionic liquids ideal candidates for fuel cell separators and thus for promoting fuel cell technology.

Acknowledgements This work is partially supported by the Spanish Ministry of Science and Innovation (MICINN), by the FEDER (Fondo Europeo de Desarrollo Regional), Ref. CICYT ENE2011-25188 and by the SENECA Foundation 18975/JLI/2013 grants. V.M. Ortiz-Martínez (Ref. FPU12/05444) and M.J. Salar-García (Ref. BES-2012-055350) thank the Ministry of Education and the Ministry of Economy and Competitiveness for supporting their doctoral theses.

References

1. Daud SM, Kim BH, Ghasemi M, Daud WRW (2015) Separators used in microbial electrochemical technologies: current status and future prospects. Bioresour Technol 195:170–179. doi:10.1016/j.biortech.2015.06.105
2. Pintauro PN (2015) Perspectives on membranes and separators for electrochemical energy conversion and storage devices. Polym Rev 55:201–207. doi:10.1080/15583724.2015.1031378
3. Li WW, Sheng GP, Liu XW, Yu HQ (2011) Recent advances in the separators for microbial fuel cells. Bioresour Technol 102:244–252. doi:10.1016/j.biortech.2010.03.090
4. Díaz M, Ortiz A, Ortiz I (2014) Progress in the use of ionic liquids as electrolyte membranes in fuel cells. J Membr Sci 469:379–396. doi:10.1016/j.memsci.2014.06.033

5. Gorke J, Srienc F, Kazlauskas R (2010) Toward advanced ionic liquids. Polar; enzyme-friendly solvents for biocatalysis. Biotechnol Bioprocess Eng 15:40–53. doi:10.1007/s12257-009-3079-z
6. Pérez de los Ríos A, Hernandez-Fernandez FJ, Martinez FJ, Rubio M, Villora G (2007) The effect of ionic liquid media on activity, selectivity and stability of Candida antarctica lipase B in transesterification reactions. Biocatal Biotransfor 25:151–156. doi:10.1080/10242420701379213
7. Walden P (1914) Molecular weights and electrical conductivity of several fused salts. Bull Acad Imp Sci Saint Petersburg 1800:405–422
8. Wilkes JS, Levisky JA, Wilson RA, Hussey CL (1982) Dialkylimidazolium chloroaluminate melts: a new class of room-temperature ionic liquids for electrochemistry, spectroscopy and synthesis. Inorg Chem 21:1263–1264. doi:10.1021/ic00133a078
9. Cull SG, Holbrey JD, Vargas-Mora V, Seddon KR, Lye GJ (2000) Room-temperature ionic liquids as replacements for organic solvents in multiphase bioprocess operations. Biotechnol Bioeng 69:227–233. doi:10.1002/(SICI)1097-0290(20000720)69:2<227:AID-BIT12>3.0.CO;2-0
10. Sheldon RA, Lau RM, Sorgedrager MJ, van Rantwijk F, Seddon KR (2001) Biocatalysis in ionic liquids. Green Chem 4:147–151. doi:10.1039/B110008B
11. Wasserscheid P, Welton T (eds) (2008) Ionic liquids in synthesis, 2nd edn. Wiley-VCH, Weinheim
12. Abbott AP, Boothby D, Capper G, Davies DL, Rasheed RK (2004) Deep eutectic solvents formed between choline chloride and carboxylic acids: versatile alternatives to ionic liquids. J Am Chem Soc 126:9142–9147. doi:10.1021/ja048266j
13. Lozano LJ, Godínez C, de los Rios AP, Hernández-Fernández FJ, Sanchez-Segado S, Alguacil FJ (2011) Recent advances in supported ionic liquid membrane technology. J Memb Sci 376:1–14. doi:10.1016/j.memsci.2011.03.036
14. Bara JE, Camper DE, Gin DL, Noble (2010) DR Room-temperature ionic liquids and composite materials: platform technologies for CO_2 capture. Acc Chem Res 41:152–159. doi:10.1021/ar9001747
15. Ortiz-Martínez Gajda I, Salar-García MJ, Greenman J, Hernández-Fernández FJ, Ieropoulos I (2016) Study of the effects of ionic liquid-modified cathodes and ceramic separators on MFC performance. Chem Eng J 291:317–324. doi:10.1016/j.cej.2016.01.084
16. Salar-García MJ, Ortiz-Martínez VM, Baicha Z, de los Ríos AP, Hernández-Fernández FJ (2016) Scaled-up continuous up-flow microbial fuel cell based on novel embedded ionic liquid-type membrane-cathode assembly. Energy 101:113–120. doi:10.1016/j.energy.2016.01.078
17. Scovazzo P, Havard D, McShea M, Mixon S, Morgan D (2009) Long-term, continuous mixed gas dry fed CO_2/CH_4 and CO_2/N_2 separation performance and selectivities for room temperature ionic liquid membranes. J Membr Sci 327:41–48. doi:10.1016/j.memsci.2008.10.056
18. Dai Z, Noble RD, Gin DL, Zhang X, Deng L (2016) Combination of ionic liquids with membrane technology: a new approach for CO_2 separation. J Membr Sci 497:1–20. doi:10.1016/j.memsci.2015.08.060
19. Feng T, Lin B, Zhang S, Yuan N, Chu F, Hickner MA, Wang C, Zhu L, Ding J (2016) Imidazolium-based organic–inorganic hybrid anion exchange membranes for fuel cell applications. J Membr Sci 508:7–14. doi:10.1016/j.memsci.2016.02.019
20. Freemantle M (2010) An introduction to ionic liquids, 1st edn. The Royal Society of Chemistry, Cambridge
21. Galiński M, Lewandowski A, Stępnia I (2006) Ionic liquids as electrolytes. Electrochim Acta 51:5567–5580. doi:10.1016/j.electacta.2006.03.016
22. Zhang S, Sun N, He X, Lu X, Zhang X (2006) Physical properties of ionic liquids: database and evaluation. J Phys Chem Ref Data 35:1475–1517. doi:10.1063/1.2204959
23. Greaves TL, Drummond CJ (2008) Protic ionic liquids: properties and applications. Chem Rev 108:206–237. doi:10.1021/cr068040u

24. MacFarlane DR, Tachikawa N, Forsyth M, Pringle JM, Howlett PC, Elliott GD, Davis JH, Watanabe M, Simonf P, Angellg CA (2014) Energy applications of ionic liquids. Energy Environ Sci 7:232–250. doi:10.1039/c3ee42099j
25. Martinelli A (2013) Ionic liquids for green energy applications. In: Kadokawa J (ed) Ionic liquids—new aspects for the future, 1st edn. Intech, Rijeka. doi:10.5772/52863
26. Ueki T, Watanabe M (2008) Macromolecules in ionic liquids: progress, challenges, and opportunities. Macromolecules 413739. doi:10.1021/ma800171k
27. Bautista-Rodríguez CM, Rosas-Paleta A, Rivera-Márque Z (2009) Study of electrical resistance on the surface of nafion 115® membrane used as electrolyte in PEMFC Technology. Part I: statistical inference. Int J Electrochem Sci 4:43–59
28. Zaidi JR (2009) Trends in polymer electrolyte membranes for PEMFC. In: Zaidi J, Matsuura T (eds) Polymer membranes for fuel cells, 1st edn. Springer, Heidelberg, pp 7–25
29. Kreuer KD, Fuchs A, Ise M, Spaeth M, Maier J (1998) Imidazole and pyrazole-based proton conducting polymers and liquids. Electrochim Acta 43:1281–1288. doi:10.1016/S0013-4686(97)10031-7
30. Abu-Lebdeh Y, Owen J (1999) Extended abstracts of the 12th international conference on solid state ionics. In: 12th international conference on solid state ionics (SSI-12), Halkidiki, June 1999. International Society for Solid State Ionics, p 419
31. Sun JZ, Jordan LR, Forsyth M, MacFarlane DR (2001) Acid-organic base swollen polymer membranes. Electrochim Acta 46:1703–1708. doi:10.1016/S0013-4686(00)00774-X
32. Jothi PR, Dharmalingam S (2014) An efficient proton conducting electrolyte membrane for high temperature uel cell in aqueous-free medium. J Membr Sci 450(2014):389–396. doi:10.1016/j.memsci.2013.09.034
33. Zhang H, Wu W, Wang J, Zhang T, Shi B, Liu J, Cao S (2015) Enhanced anhydrous proton conductivity of polymer electrolyte membrane enabled by facile ionic liquid-based hoping pathways. J Membr Sci 476:136–147. doi:10.1016/j.memsci.2014.11.033
34. Guerreiro da Trindade L, Becker MR, Celso F, Petzhold CL, Martini EMA, de Souza RF (2016) Modification of sulfonated poly (ether ether ketone) membranes by impregnation with the ionic liquid 1-butyl-3-methylimidazolium tetrafluoroborate for proton exchange membrane fuel cell applications. Polym Eng Sci 56:1037–1044. doi:10.1002/pen.24334
35. Malis J, Mazur P, Schauer J, Paidar M, Bouzek K (2013) Polymer-supported 1-butyl-3-methylimidazolium trifluoromethanesulfonate and 1-ethylimidazolium trifluoromethanesulfonate as electrolytes for the high temperature PEM-type fuel cell. Int J Hydrog Energ 38:4697–4704. doi:10.1016/j.ijhydene.2013.01.126
36. Schauer J, Sikora A, Plıskova M, Malis J, Mazur P, Paidar M (2011) Ion-conductive polymer membranes containing 1-butyl-3-methylimidazolium trifluoromethanesulfonate and 1-ethylimidazolium trifluoromethanesulfonate J Membr Sci 367:332–339. doi:10.1016/j.memsci.2010.11.018
37. Sekhon SS, Krishnan P, Singh B, Yamada K, Kim CS (2006) Proton conducting membrane containing room temperature ionic liquid. Electrochim Acta 52:1639–1644. doi:10.1016/j.electacta.2006.03.095
38. Wang JT, Hsu SL (2011) Enhanced high-temperature polymer electrolyte membrane for fuel cells based on polybenzimidazole and ionic liquids. Electrochim Acta 56:2842–2846. doi:10.1016/j.electacta.2010.12.069
39. Pitawala J, Matic A, Martinelli A, Jacobsson P, Koch V, Croce F (2009) Thermal properties and ionic conductivity of imidazolium Bis(trifluoromethanesulfonyl)imide dicationic ionic liquids. J Phys Chem B 113:10607–10610. doi:10.1021/jp904989s
40. Hooshyari K, Javanbakht M, Adibi M (2016) Novel composite membranes based on PBI and dicationic ionic liquids for high temperature polymer electrolyte membrane fuel cells. Electrochimacta 205:142–152. doi:10.1016/j.electacta.2016.04.115
41. Gao J, Guo Y, Wu B, Qi L, Li B, Liu J, Wang Z, Liu W, Gu J, Zou Z (2014) Impact of cation selection on proton exchange membrane fuel cell performance with trimethylethyl amide, ethyl pyridinium and ethylmethyl imidazolium ionic liquid carried by poly(vinylidene

fluoride) membrane as electrolyte. J Power Sour 251:432. doi:10.1016/j.jpowsour.2013.11.038
42. Gao J, Liu J, Liu W, Li B, Xin Y, Yin Y, Gu J, Zou Z (2012) An efficient and green approach to prepare hydrophilic imidazolium ionic liquids free of halide and its effect on oxygen reduction reaction of Pt/C catalyst. Int J Hydrog Energ 37:13167–13177. doi:10.1016/j.ijhydene.2012.03.135
43. Lee S-Y, Ogawa A, Kanno M, Nakamoto H, Yasuda T, Watanabe M (2010) Nonhumidified intermediate temperature fuel cells using protic ionic liquids. J Am Chem Soc 132:9764–9773. doi:10.1021/ja102367x
44. Lee S-Y, Yasuda T, Watanabe M (2010) Fabrication of protic ionic liquid/sulfonated polyimide composite membranes for non-humidified fuel cells. J Power Sour 195:5909–5914. doi:10.1016/j.jpowsour.2009.11.045
45. Yasuda T, Nakamura S, Honda Y, Kinugawa K, Lee S-Y, Watanabe M (2012) Effects of polymer structure on properties of sulfonated polyimide/protic ionic liquid composite membranes for nonhumidified fuel cell applications. ACS Appl Mater Interfaces 4:1783–1790. doi:10.1021/am300031k
46. Liu S, Zhou L, Wang P, Zhang F, Yu S, Shao Z, Yi B (2014) Ionic-liquid-based proton conducting membranes for anhydrous H_2/Cl_2 fuel-cell applications. ACS Appl Mater Interfaces 6:3195–3200. doi:10.1021/am404645c
47. Langevin D, Nguyen QT, Marais S, Karademir S, Sanchez J-Y, Iojoiu C, Martinez M, Mercier R, Judeinstein P, Chappey C (2013) High-temperature ionic-conducting material: advanced structure and improved performance. J Phys Chem C 117:15552–15561. doi:10.1021/jp312575m
48. Yasuda T, Watanabe M (2013) Protic ionic liquids: fuel cell applications. MRS Bull 38:560–566. doi:10.1557/mrs.2013.153
49. de Yuso MdVM, Cuberes MT, Romero R, Neves L, Coelhoso I, Crespo JG, Rodríguez-Castellón E, Benavente J (2014) Modification of a Nafion membrane by n-dodecyltrimethylammonium cation inclusion for potential application in DMFC. Int J Hydrog Energ 39:4023–4029. doi:10.1016/j.ijhydene.2013.05.046
50. Di Noto V, Piga M, Giffin GA, Lavina S, Smotkin ES, Sanchez J, Iojoiu C (2012) Influence of anions on proton-conducting membranes based on neutralized Nafion 117, triethylammoniummethane sulfonate, and triethylammonium perfluorobutanesulfonate. 1. Synthesis and properties. J Phys Chem C116:1361–1369. doi:10.1021/jp204241y
51. Di Noto V, Negro E, Sanchez J, Iojoiu C (2010) Structure-relaxation interplay of a new nanostructured membrane based on tetraethylammonium trifluoromethanesulfonate ionic liquid and neutralized nafion 117 for high-temperature fuel cells. J Am Chem Soc 132:2183–2195. doi:10.1021/ja906975z
52. Sunda AP (2015) Ammonium-based protic ionic liquid doped Nafion membranes as anhydrous fuel cell electrolytes. J Mater Chem 3:12905–12912. doi:10.1039/C5TA02315G
53. Schmidt C, Glück T, Schmidt-Naake G (2008) Modification of Nafion membranes by impregnation with ionic liquids. Chem Eng Technol 31:13–22. doi:10.1002/ceat.200700054
54. Kiatkittikul P, Nohira T, Hagiwara R (2015) Nonhumidified fuel cells using n-ethyl-n-methyl-pyrrolidinium fluorohydrogenate ionic liquid-poly(vinylidene fluoride-hexafluoropropylene) composite membranes. Energies 8:6202–6214. doi:10.3390/en8066202
55. Hernández-Fernández FJ, de los Ríos AP, Mateo-Ramírez F, Godínez C, Lozano-Blanco LJ, Moreno JI, Tomás-Alonso F (2015) New application of supported ionic liquids membranes as proton exchange membranes in microbial fuel cell for waste water treatment. Chem Eng J 279:115–119. doi:10.1016/j.cej.2015.04.036
56. Hernández-Fernández FJ, de los Rios AP, Salar-García MJ, Ortiz-Martínez VM, Lozano-Blanco LJ, Godinez C, Alonso-Tomas F, Quesada-Medina J (2015) Recent progress and perspectives in microbial fuel cells for bioenergy generation and wastewater treatment. Fuel Process Technol 138:284–297. doi:10.1016/j.fuproc.2015.05.022

57. Salar-García MJ, Ortiz-Martínez VM, de los Ríos AP, Hernández-Fernández FJ (2015) A method based on impedance spectroscopy for predicting the behavior of novel ionic liquid-polymer inclusion membranes in microbial fuel cells. Energy 89:648–654. doi:10.1016/j.energy.2015.05.149
58. Hernández-Fernández FJ, de los Ríos AP, Mateo-Ramírez F, Juarez MD, Lozano-Blanco LJ, Godínez C (2016) New application of polymer inclusion membrane based on ionic liquids as proton exchange membrane in microbial fuel cell. Sep Purif Technol 160:51–58. doi:10.1016/j.seppur.2015.12.047
59. Ortiz-Martínez VM, Salar-García MJ, Hernández-Fernández FJ, de los Ríos AP (2015) Development and characterization of a new embedded ionic liquid based membrane-cathode assembly for its application in single chamber microbial fuel cells. Energy 93:1748–57. doi:10.1016/j.energy.2015.10.027
60. Yuan J, Antonietti M (2011) Poly(ionic liquid)s: polymers expanding classical property profiles. Polymer 52:1469–1482. doi:10.1016/j.polymer.2011.01.043
61. Yuan J, Mecerreyes D, Antonietti M (2013) Poly(ionic liquids)s: an update. Prog Polym Sci 38:1009–1036. doi:10.1016/j.progpolymsci.2013.04.002
62. Mecerreyes D (2011) Polymeric ionic liquids: broadening the properties and applications of polyelectrolytes. Prog Polym Sci 36:1629–1648. doi:10.1016/j.progpolymsci.2011.05.007
63. Díaz M, Ortiz A, Vilas M, Tojo E, Ortiz I (2014) Performance of PEMFC with new polyvinyl-ionic liquids based membranes as electrolytes. Int J Hydrog Energ 39:3970–3977. doi:10.1016/j.ijhydene.2013.04.155
64. Díaz M, Ortiz A, Isik M, Mecerreyes D, Ortiz I (2015) Highly conductive electrolytes based on poly([HSO_3-BVIm][TfO])/[HSO_3-BMIm][TfO] mixtures for fuel cell applications. Int J Hydrog Energ 40:11294–11302. doi:10.1016/j.ijhydene.2015.03.109
65. Lin B, Qiu L, Lu J, Yan F (2010) Cross-linked alkaline ionic liquid-based polymer electrolytes for alkaline fuel cell applications. Chem Mater 22:6718–6725. doi:10.1021/cm102957g
66. Qiu B, Lin B, Qiu L, Yan F (2012) Alkaline imidazolium—and quaternary ammonium—functionalized anion exchange membranes for alkaline fuel cell applications. J Mater Chem 22:1040–1045. doi:10.1039/C1JM14331J
67. Qiu B, Lin B, Si Z, Qiu L, Chu F, Zhao J, Yan F (2012) Bis-imidazolium-based anion-exchange membranes for alkaline fuel cells. J Power Sour 217:329–335. doi:10.1016/j.jpowsour.2012.06.041
68. Lee C-P, Chu T-C, Chang L-Y, Lin J-J, Ho K-C (2013) Solid-state ionic liquid based electrolytes for dye-sensitized solar cells. In: Kadokawa J (ed) Ionic liquids—new aspects for the future, 1st edn. Intech, Rijeka. doi:10.5772/53647
69. Priya ARS, Subramania A, Jung YS, Kim KJ (2008) High-performance of quasi-solid-state dye-sensitized solar cell based on an electrospum PVdF-HFP membrane electrolyte. Langmuir 24:9816–9819. doi:10.1021/la801375s
70. Yang CC, Wey JY, Liou TH, Li YJ, Shih JY (2012) A quasi solid state dye-sensitized solar cell based on poly(vinylidenefluoride-co-hexafluoropropylene)/SBA-15 nanocomposite membrane. Mater Chem Phys 132:431–437. doi:10.1016/j.matchemphys.2011.11.049
71. Zhao J, Shen X, Yan F, Qiu L, Lee S, Sun B (2001) Solvent-free ionic liquid/poly(ionic liquid) electrolytes for quasi-solid-state dye-sensitized solar cells. J Mater Chem 21:7326–7330. doi:10.1039/C1JM10346F
72. Cheng X, Zhao J, Zhang J, Qiu L, Xu D, Zhang H, Han X, Sun B, Fu G, Zhang Y, Yan F (2012) Bis-imidazolium based poly(ionic liquid) electrolytes for quasi-solid-state dye-sensitized solar cells. J Mater Chem 22:18018–18024. doi:10.1039/C2JM33273F
73. Yan F, Yu S, Zhang X, Qiu L, Chu F, You J, Lu J (2009) Enhanced proton conduction in polymer electrolyte membranes as synthesized by polymerization of protic ionic liquid-based microemulsions. Chem Mater 21:1480–1484. doi:10.1021/cm900098r
74. Chu F, Lin B, Yan F, Qiu L, Lu J (2011) Macromolecular protic ionic liquid-based proton-conducting membranes for an hydrous proton exchange membrane application. J Power Sour 196(7979):7–984. doi:10.1016/j.jpowsour.2011.05.064

75. Lin B, Cheng S, Qiu L, Yan F, Shang S, Lu J (2010) Protic ionic liquid-based hybrid proton-conducting membranes for anhydrous proton exchange membrane application. Chem Mater 22:1807–1813. doi:10.1021/cm9033758
76. Ye H, Huang J, Xu J, Kodiweera N, Jayakody J, Greenbaum S (2008) New membranes based on ionic liquids for PEM fuel cells at elevated temperatures. J Power Sour 178:651–660. doi:10.1016/j.jpowsour.2007.07.074
77. Xu C, Liu X, Cheng J, Scott K (2015) A polybenzimidazole/ionic-liquid-graphite-oxide composite membrane for high temperature polymer electrolyte membrane fuel cells. J Power Sour 274:922–927. doi:10.1016/j.jpowsour.2014.10.134
78. Eguizabal A, Lemus J, Pina MP (2013) On the incorporation of protic ionic liquids imbibed in large pore zeolites to polybenzimidazole membranes for high temperature proton exchange membrane fuel cells. J Power Sour 222:483–492. doi:10.1016/j.jpowsour.2012.07.094

Chapter 2
Organic/TiO_2 Nanocomposite Membranes: Recent Developments

Javier Miguel Ochando-Pulido, José Raúl Corpas-Martínez, Marco Stoller and Antonio Martínez-Férez

Abstract Fuel cells may become a key energy management, but technical and economic feasibility still need to be sensibly improved. Many studies in order to overcome the limits of the technology are nowadays in progress. A promising and interesting development solution appears to be the improvement of the membrane properties used in fuel cells by nanotechnologies. In this book chapter, a review on the recent developments about organic/TiO_2 nanocomposite membranes will be presented, and the results obtained in the recent years will be discussed. As a main issue, polymer composites containing a small amount of inorganic materials lead to a significant increment in the interfacial area of the organic–inorganic phases, enhancing a considerable volume fraction of the interfacial polymer. Moreover, these composite systems may be capable to provide unique combination of organic properties, such as electrical property and processability, together with inorganic, comprising thermal and chemical stability and minor fuel permeability. To sum up, the organic–inorganic composite systems might also provide improved chemical and mechanical stability, as well as high proton conductivity at high temperatures.

Keywords Fuel cells · Nanocomposite membranes · Titanium dioxide · Novel energy technology · Polymer electrolyte membranes

---

J.M. Ochando-Pulido (✉) · J.R. Corpas-Martínez · A. Martínez-Férez
Chemical Engineering Department, University of Granada,
Avenida Fuentenueva s/n, 18071 Granada, Spain
e-mail: jmochandop@ugr.es

M. Stoller
Department of Chemical Materials Environmental Engineering,
University of Rome "La Sapienza", Via Eudossiana 18, 00184 Rome, Italy

Inamuddin et al. (eds.), *Organic–Inorganic Composite Polymer Electrolyte Membranes*, DOI 10.1007/978-3-319-52739-0_2

1 Introduction

Fuel cells are nowadays being intensely investigated as novel energy source. Briefly explained, fuel cells are electrochemical devices in which an electrocatalytic process leads to a direct transformation of chemical energy of the fuel to electricity.

The high efficiency and the fact that fuel cells do not emit pollutants such as SO_x, NO_x and CO_2, compared to coal combustion engines, makes fuel cell a very promising technology. On the other hand, the high costs involved with the commercialization of fuel cells make imperative to carry on further investigation to cope with them, mainly focusing on the development of new materials and optimization of prior or novel fabrication processes.

In particular, among the various fuel cell system types available nowadays, polymer electrolyte membrane fuel cells (PEMFC) appears to be promising, since they satisfy technical and economical requirements and results to be very versatile, and can be used for a wide range of small-scale applications such as automotive, stationary equipment and portable power generation, among others Kim et al. [1].

In PEMFCs, the power-generating system consists of an electrochemical reaction which involves gases such as hydrogen, methanol and ethanol. The main reactions are briefly summarized below:

On the anode: $2H_2 \rightarrow 4H^+ + 4e^-$
On the cathode: $O_2 + 4e^- + 4H^+ \rightarrow 2H_2O$
Resulting in the overall reaction: $2H_2 + O_2 \rightarrow 2H_2O$

A scheme of the operating mechanism of polymer electrolyte membrane fuel cell presented is in Fig. 1. The key component of PEMFCs is the polymer electrolyte membrane (PEM). This PEM behaves as a semipermeable barrier and thus provides a barrier for the passage of the electrons and fuel whereas functions as an electrolyte to transfer the protons from the anode to the cathode. Hydrogen ions (H^+) at the anode move to the cathode through the PEM, which produces as a consequence the electrical current and water as a by-product Jang and Hang [2].

In the 1970s, DuPont developed the first cation-exchange membranes based on polymer fluorosulfonic acid ionomers, named Nafion® series. Nafion® PEM was chosen as a standard for polymeric electrolyte fuel cells owing to thermal and chemical stability and high ionic conductivity (0.1 S cm^{-1} at fully hydrated condition). However, Nafion® membranes associated with several drawbacks, mainly a decrease in the ionic conductivity and low humidity at high temperatures, which poses a barrier for their definite commercialization according to expected cost-efficiency.

For these reasons, intensive research on novel proton-conducting membranes has been performed recently to improve the performance of the existing fuel cells. During the past few years, a plethora of researchers have focused on the development of reliable, high performance PEMs which could be able to provide high proton conductivity and, at the same time, ensure a series of key features comprising mainly:

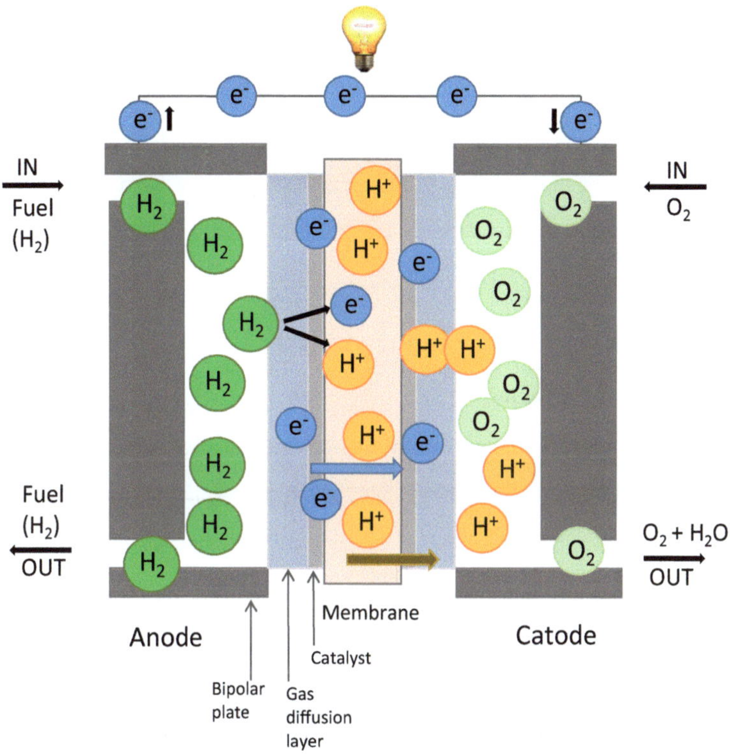

Fig. 1 Polymer electrolyte membrane fuel cell operating mechanism scheme

- low fuel permeability,
- high oxidative stability,
- good mechanical stability,
- low membrane electrolyte and assembly fabrication costs.

An interesting review on new polymer options for their implementation in these electrolyte membranes for fuel cells was published by Lee et al. [3]. For example, modified Nafion® and several hydrocarbon-based polymers such as sulfonated aromatic polymers based on poly(ether ketone) (PEEK), poly(ether sulfone) (PES), and polybenzimidazole (PBI) have attracted interest for use in PEMFCs in the recent years.

An illustrative summary of the principal polymers used in the fabrication of polymer electrolyte membrane fuel cells is given in Fig. 2.

In addition to this, it is especially relevant to highlight the role that inorganic–organic composite membranes have recently gained for these purposes, which is the main focus of the present chapter. Inorganic–organic composite membranes have drawn much attention for the development of new proton-conducting electrolytes in fuel cell applications in order to overcome such existing drawbacks.

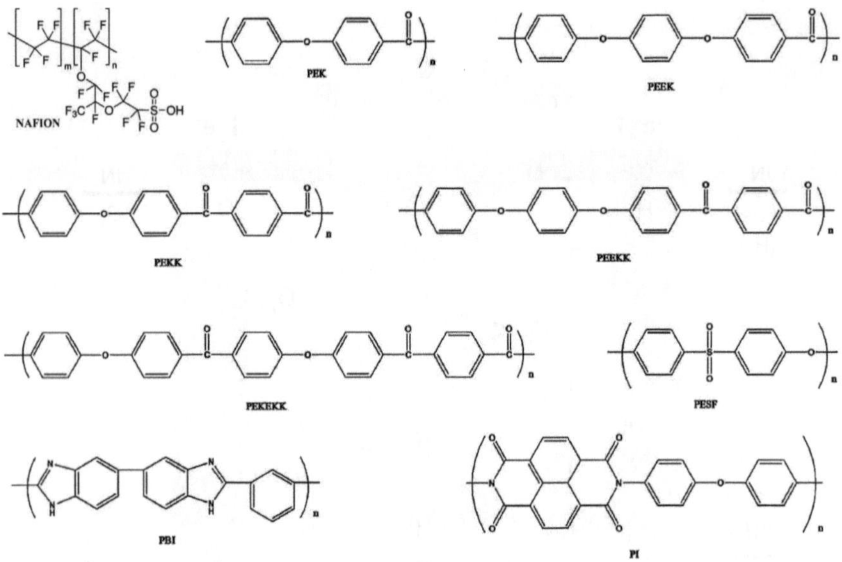

Fig. 2 Main polymers used in the fabrication of polymer electrolyte membrane fuel cells

Despite the advances accomplished on composite membranes research during the past few years, technical and economic challenges in the commercialization of fuel cells still hold on. Moreover, composite membranes can be prepared in various forms such as precursors and particles. There is a wide range of different materials and modification methods available to develop composite membranes. The most typical materials used extensively in recent research for composite fabrication include solid acids, silica and clays [4, 5], metals [6, 7] and blends using a second polymer because of unique properties and surface chemistry as well as excellent mechanical strength for long term use. The most widely used techniques to modify the inorganic particles by adding functional ionic sites, to improve the electronic properties, are chemical synthesis, sol–gel method and the doping or infiltration of inorganic nanoparticles and precursors [8].

In this book chapter, a review on the recent developments about organic/TiO_2 nanocomposite membranes will be presented, and the results obtained in the recent years will be discussed. As a main issue, polymer composites containing a small amount of inorganic materials lead to a significant increment in the interfacial area of the organic–inorganic phases, enhancing a considerable volume fraction of the interfacial polymer. Moreover, these composite systems may be able to provide a unique combination of organic properties, such as electrical property and processability, together with inorganic, comprising thermal and chemical stability and minor fuel permeability. To sum up, the organic–inorganic composite systems might also provide improved chemical and mechanical stabilities, as well as high proton conductivity at high temperatures.

2 TiO_2-Polymer Electrolyte Membranes (PEMs)

In polymer electrolyte membrane (PEM), methanol is converted into CO_2 by moving to the cathode, such that reduction of functional methanol results in the formation of a multiple potential at the cathode.

Also, composite materials formed by polymers as alternative to Nafion® such as polyethersulfone (PES), poly(etheretherketone) (SPEEK), and chitosan (CS) and inorganic fillers such as Fe_3O_4, TiO_2, montmorillonite (MMT), carbon nanofiber (CNF), and activated carbon nanofiber (ACNF), as well as carbon nanotubes (CNT) have been investigated for use as microbial fuel cells (MFC) membranes with promising results. The cost of these composite membranes is lower than that of Nafion®. Furthermore, the addition of inorganic fillers to the polymer electrolyte membrane in MFCs helps to reduce the substrate oxygen crossover and improves the performance of the membrane during the microbial fuel cell (MFC) operation. Incorporation of a variety of nanosized materials may produce synergistic effects.

A brief summary of the main organic–inorganic nanocomposite TiO_2 polymer electrolyte membrane (PEMs) systems proposed for PEMFCs up to the date is given in Table 1. Liccocia et al. [9] carried out the synthesis of nanocrystalline ceramic oxides added to Nafion® in order to achieve the goal of operating at temperatures above 120 °C. Commercial Nafion® 115 and Nafion®/TiO_2 composite membranes were tested in polymer electrolyte fuel cells, both for direct hydrogen (DH-PEFC) between 80 and 130 °C. The best performance was with Nafion®/TiO_2 composite membranes in whole range, obtaining a power density of up to 0.514 mW cm^{-2} for the composite membrane versus 0.354 mW cm^{-2} obtained with Nafion® 115 at 0.56 V and 110 °C. The stability of the composite membrane above 130 °C was the fascinating part of this report, reaching a power density of 0.254 mW cm^{-2} at 0.5 V.

In a similar way Slade et al. [10] tested a series of cation-exchange membranes made with Nafion® modified with inert filler particles (SiO_2, ZrO_2 or TiO_2; 5–20 wt%). Results showed improvement in ion-exchange capacity/equivalent weight, water take-up, thickness change on hydration and ionic and electrical conductivity.

2.1 Perfluorinated Organic–Inorganic Nanocomposite Polymer Electrolyte Membranes (PEMs)

Although perfluorinated polymers present some advantages, they also exhibit a series of drawbacks, mainly due to the loss of fuel and proton conductivity at high temperatures given by their relative low water content. As a consequence, extensive research has been performed in the recent past towards the development and fabrication of high performance alternative membranes. A comprehensive summary of the perfluorinated organic–inorganic nanocomposite PEMs is presented in Table 2.

Table 1 Main organic–inorganic nanocomposite TiO_2 polymer electrolyte membranes (PEMs) used for PEMFC

Membrane type	Membranes synthetized
Perfluorinated organic–inorganic nanocomposite PEMs	TiO_2/PVDF-HFP, TiO_2/Nafion®, TiO_2-RSO_3H/Nafion® and TiO_2 aerogels/Nafion® doped with Nb, Ta and V
Acid–base polymer complex-based organic–inorganic PEMs	TiO_2/SO_3H/Nafion®, F–TiO_2–NT/Nafion®, PVA/TiO_2/Nafion®, PVA/nt-TiO_2/PSSA and P_2O_5–SiO_2–TiO_2/PVA
Poly(ether ether ketone)-based nanocomposite PEMs	$(RSiO_{1.5})n$/SPEEK, TiO_2/SPEEK, TiO_2/PANI/SPEEK, TiO_2/SO_3H/SPEEK, functionalized TiO_2 submicrospheres/SPEEK and TiO_2 (Rutile)/SPEEK
TiO_2-modified polytetrafluoroethylene, PANI, polyethersulfone (PES) and polysolfone (PS) nanocomposite PEMs	TiO_2/PAA/PTFE, PANI/mesoporous TiO_2, TiO_2/SPES, (S–TiO_2)/SPSEBS SPSU and sulfated TiO_2/NMPA
TiO_2 solar cells	TiO_2/Nafion®/Pt and DSSC by photocured polymer electrolyte membrane
Carbon materials and metal–carbon nanotube (CNTs)–TiO_2 composites	$Ru_{85}Se_{15}/TiO_2/C$, $Pt/TiO_2/C$, $Pd/TiO_2/C$, Pt/$(Ti_xSn_{1-x}O_2)$/C, TNTs/Pt/C, TiO_2 (Rutile)/C (semi-graphitic), TiO_2/GN and TiO_2 nanowire/GN

PVDF-HFP poly(vinylidene fluoride-co-hexafluoropropylene); *PVA* poly(vinyl alcohol); *PTFE* polytetrafluoro-ethylene; *PAA* polyacrylamide; *SPEEK* sulfonated poly(ether ether ketone); *PANI* polyaniline; *SPSEBS* polystyrene ethylene butylene polystyrene; *NMPA* nitrilotri(methyl triphosphonic acid); *SPSU* sulfonated polysulfone; *DSSC* flexible dye-sensitized solar cell; *TNTs* TiO_2 nanotubes; *GN* Graphene

Researchers focused on the use of novel polymers, and in parallel to this on the introduction of inorganic materials as fillers in the polymer matrix. As an example, modified perfluorinated polymer-based membranes containing hydrophilic metal oxides including titanium dioxide (TiO_2), zirconium dioxide (ZrO_2) or silicon dioxide (SiO_2) were produced. Other studies were reported on the use of heteropoly acids (phosphotungstic acid and silicatungstic acid). Mohammadi et al. [11] investigated Nafion® composite membranes containing ZrO_2 and TiO_2 nanoparticles. The TiO_2 additive (average size diameter around 25 nm) enhanced the water retention properties of these membranes. Furthermore, these nanocomposite membranes were found to exhibit better electrochemical performance of the cell even at a high temperature (110 °C) and humidity (30%). The metal–oxide-based inorganic fillers improved the mechanical properties, and also contributed to the blocking of the fuel such as methanol by increasing the transport pathway tortuousness and improving the proton conductivity, which resulted in better cell performance.

Deposition of TiO_2 nanoparticles (NPs) from a sol solution was carried out by Liu et al. [12] to prepare modified Nafion® 112 membrane films via spin coating, for which they observed a maximum cell voltage by introducing 0.009 mg cm^{-2} TiO_2 NPs. Similar results were obtained by Santiago et al. [13] with Nafion®-TiO_2

Table 2 Perfluorinated organic–inorganic nanocomposite PEMs

Author/s	Membrane type	Achievement
Li et al. [14]	TiO_2/PVDF-HFP	Ionic conductivity and ions transport activation energy enhanced (0.94×10^{-3} S cm^{-1} and 18.71 kJ mol^{-1}) at 9.0 wt% mass fraction
Liccocia et al. [9]	TiO_2/Nafion® 115	Power density of up to 0.514 mW cm^{-2} for TiO_2/ Nafion® versus 0.354 mW cm^{-2} obtained with Nafion® 115 at 0.56 V and 110 °C
Liu et al. [12]	TiO_2/Nafion® 112	Maximum cell voltage by introducing 0.009 mg cm^{-2} TiO_2 NPs
Santiago et al. [13]	TiO_2/Nafion® hybrid membranes	Increase in the ohmic drop when TiO_2 content was increased
Amjadi et al. [15]	TiO_2/Nafion®	Water uptake of Nafion®/TiO_2 membrane with 3 wt% doping found more than double with respect to pure Nafion® membrane
Zhengbang et al. [16]	TiO_2/Nafion®	Conductivity of fabricated composite membranes was higher than that of original Nafion®; TiO_2 nanowire (5 wt%) also improved the mechanical properties of the composite membrane at different temperatures and humidity conditions
Cozzi et al. [17]	TiO_2–RSO_3H/Nafion®	Proton conductivity was 0.08 S cm^{-1} at 140 °C for the composite membrane containing 10 wt% TiO_2–RSO_3H; also higher ion-exchange capacity (IEC) was registered
Beauger et al. [18]	TiO_2 aerogels/Nafion® doped with Nb, Ta and V	Improved electronic conductivity from 10^{-6} S cm^{-1} for pure TiO_2 aerogels to 8.3×10^{-2} S cm^{-1} after doping TiO_2 aerogels at 10 at.% for Nb

PVDF-HFP poly(vinylidene fluoride-co-hexafluoropropylene)

hybrid membranes, which showed an increasing ohmic drop when the TiO_2 content was increased. Li et al. [14] developed a composite TiO_2/PVDF-HFP membrane, and observed that the ionic conductivity of the membrane at room temperature and the ions transport activation energy were enhanced (0.94×10^{-3} S cm^{-1} and 18.71 kJ mol^{-1}, respectively) upon 9.0 wt% mass fraction.

Nafion®/TiO_2 nanocomposite membranes were prepared by in situ sol–gel and casting methods, in order to compare both synthesis methods. The former was performed by soaking the Nafion® membranes in tetrabutylortotitanate (TBT) and methanol solution, whereas a Nafion®/TiO_2 composite membrane was fabricated with 3 wt% of TiO_2 particles by the solution casting method. The sol–gel method was found to lead to more uniform distribution of Ti particles in the Nafion®/TiO_2 composite membrane than the ones produced by casting. Water uptake of Nafion®/ TiO_2 membrane with 3 wt% doping level was found to be more than double with respect to that of the pure Nafion® membrane. The membrane electrode assembly (MEA) prepared from Nafion®/Titania nanocomposite membrane showed the highest PEMFC performance in terms of voltage *versus* current density at high temperature (110 °C).

Furthermore, Amjadi et al. [15] investigated the synthesis of Nafion®/TiO_2 membranes for proton-exchange membrane fuel cell (PEMFC) operating at high temperatures.

Zhengbang et al. [16] developed and studied the performance of TiO_2 nanowire-reinforced Nafion® composite membranes, by self-assembly of positively charged TiO_2 nanowires and negatively charged Nafion® molecules at low pH conditions. They found that the conductivity of the fabricated composite membranes was higher than that of the original Nafion® at low temperatures. The TiO_2 nanowires associated with the $-SO_3$ hydrophilic sites through self-assembly influenced the proton transport, which induced a stable cell performance. TiO_2 nanowire (5 wt%) also improved the mechanical properties of the composite membrane at different temperatures and humidity conditions.

Cozzi et al. [17] performed extensive research on organically functionalized titanium oxide/Nafion® composite proton-exchange membranes. They synthesized an organically modified ceramic material (TiO_2–RSO_3H) by covalently grafting propylsulfonic acid groups on the surface of TiO_2 nanoparticles, and subsequently used it as filler in Nafion-based composite membranes. The synthetic pathway for the surface functionalization of TiO_2 nanoparticles is shown in Fig. 3.

The authors reported superior performance of Nafion®/TiO_2–RSO_3H composite membranes compared to Nafion® (improvement up to 40%). In this regard, proton conductivity of the hybrid material was confirmed to be up to 0.08 S cm^{-1} at 140 °C for the composite membrane containing 10 wt% TiO_2–RSO_3H, and also higher ion-exchange capacity (IEC) was registered. Furthermore, the filler led to cell response enhancement, both in terms of higher delivered power density and lower methanol crossover.

Beauger et al. [18] synthetized doped TiO_2 aerogels as alternative catalyst supports for PEMs. Nb-, Ta- and V-doped TiO_2 aerogels were tested to increase the electronic conductivity of TiO_2. All dopants improved the electronic conductivity and whatever the dopant, the electronic conductivity was increased with the dopant level. The pure TiO_2 aerogel had about 10^{-6} S cm^{-1} electronic conductivity whereas the maximum value was obtained after doping at 10 at.% for Nb

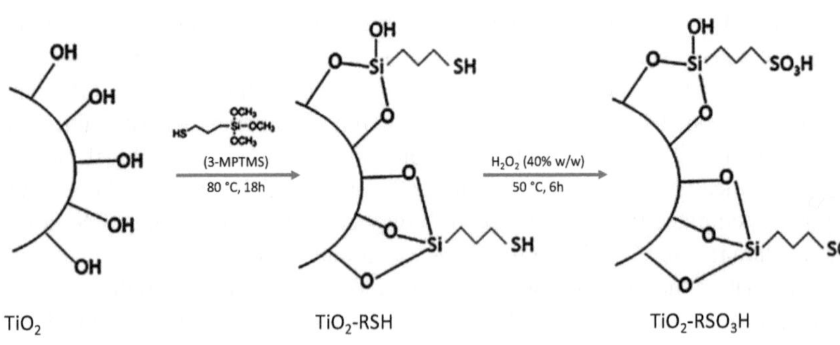

Fig. 3 Synthetic pathway for surface functionalization of TiO_2 nanoparticles [14]

(8.3×10^{-2} S cm^{-1}). The results showed, at all levels of performance, that the best dopants were Nb, Ta and finally V. Within this context Gojković et al. [19] used Pt/Nb–TiO_2 and Pt–Ru/Nb–TiO_2 as anode catalysts supporters for PEMFCs. Results indicated good conductivity of the supporting material compared with Pt/TiO_2 membranes.

2.2 *Acid–Base Polymer Complex-Based Organic–Inorganic Nanocomposite PEMs*

Acid–base polymer complex-based organic–inorganic nanocomposites have been recently introduced for PEMs Table 3. The main acid–base polymer based organic–inorganic composites lists.

Houn-Rhee et al. [20] grafted thiol and sulfone groups onto the surface of titanate nanosheets to perform organic acid functionality (HSO_3^-) to their surfaces. Results showed higher ionic conductivity values for nanocomposite membranes containing 3–7 wt% HSO_3^- titanate than composite membranes containing 3 wt% TiO_2/P25 but lower than Nafion® 115. However, these authors found out that methanol and water permeability rates decreased with increasing the amount of HSO_3^- titanate added into the Nafion® matrix, as can be seen in Table 4.

Jun et al. [21] studied the functionalization of titania nanotube composite membranes (TiO_2–NT) using 3-mercaptopropyl-tri-methoxysilane (MPTMS) as a sulfonic acid functionalization agent. Results showed a considerable increase in the

Table 3 Acid–base polymer complex-based organic–inorganic PEMs

Author/s	Membrane type	Achievement
Houng-Rhee et al. [20]	TiO_2/ SO_3H/Nafion®	Higher ionic conductivity values for nanocomposite membranes containing 3–7 wt% HSO_3^- titanate compared with composite membranes containing 3 wt% TiO_2/P25
Jun et al. [21]	F–TiO_2–NT/Nafion®	Improvement in water permeability: water uptake of F–TiO_2–NT/Nafion® composite was about 7 and 4% higher than Nafion® 112 and recast Nafion® respectively
Yang et al. [22]	PVA/TiO_2/ Nafion®	Ionic conductivity of PVA/TiO_2 composite polymer membrane proportional to amount of TiO_2, achieving 0.0901 S cm^2 at 70 °C with 15 wt% of TiO_2
Yang et al. [23]	PVA/nt–TiO_2/ PSSA	Good electrochemical performance, of air-breathing direct methanol fuel cell
Mroczkowska-Szerszen et al. [24]	P_2O_5–SiO_2–TiO_2/PVA	Stable values of activation energy over examined temperature range

PVA poly(vinyl alcohol); *PSSA* polystyrene sulfonic acid; *NT* nanotubes

Table 4 Methanol and water permeability rate of nanocomposite membranes containing 3–7% HSO_3-titanate and Nafion® 115

Membrane	H_2O permeability (μmol/min)	MeOH permeability (μmol/min)
Nafion® 115	49	11
3% HSO_3-titanate	43	9.1
5% HSO_3-titanate	38	8.6
7% HSO_3-titanate	35	8.4
10% HSO_3-titanate	33	8.3

proton conductivity of F–TiO_2–NT in respect of TiO_2–NT. At the temperature of 80 °C, the conductivity of F–TiO_2–NT was as high as 0.08 S cm^{-1} versus 0.03 S cm^{-1} for the TiO_2–NT membranes. Moreover, these authors presented F–TiO_2–NT/Nafion® composite membranes as potential replacements for commercial Nafion® membranes in high temperature PEMFCs because their improvement in water permeability. The water uptake of F–TiO_2–NT/Nafion® composite is about 7 and 4% higher than Nafion® 112 and recast Nafion® respectively.

Yang et al. [22] studied poly(vinyl alcohol)/TiO_2 composite polymer membranes on alkaline direct alcohol fuel cell (DAFC). The DAFC was formed of a cathode based on MnO_2/C catalysts, while the anode was made of Pt/Ru and PVA/TiO_2 composite polymer membrane. Results showed that the ionic conductivity of PVA/TiO_2 composite polymer membrane was proportional to the amount of TiO_2, achieving 0.0901 S cm^2 at 70 °C with 15 wt% of TiO_2. Finally, the authors reported, in terms of the maximum power density, that the electrochemical performance of alkaline DAFC with methanol was better than that of DAFC with ethanol.

Yang et al. [23] developed a PVA/nt-TiO_2/PSSA composite polymer membrane with good electrochemical performance, for air-breathing direct methanol fuel cell based.

Mroczkowska-Szerszen et al. [24] studied the physicochemical properties of alkoxy compound and nanopowder P_2O_5–SiO_2–TiO_2/PVA composite membrane. They fabricated a composite glassy membrane by means of an optimized sol–gel process followed by post-thermal treatment of the resultant hydrogel. The electr tests of the samples in the fuel cells performed using hydrogen showed stable values of the activation energy over the entire temperature range under study.

2.3 *TiO_2-Modified Polytetrafluoroethylene Membranes*

Polytetrafluoroethylene (PTFE) exhibits superior properties to many other polymers in terms of chemical stability, thermal resistance, low water absorption, potential biocompatibility, good mechanical strength and erosion resistance, making it an attractive material for anti-corrosion coatings and fuel cell composite membranes. PTFE can be processed to porous fibres or thin films through solvent-free melt spinning or extrusion and post-stretching strategies. Qian et al. [25] fixed TiO_2 functional thin layer on super hydrophobic PTFE UF membrane via plasma-enhanced graft of poly acryl acid (PAA) prior to coating TiO_2. TiO_2 attached on the PTFE-based UF membranes through the chelating bidentate coordination between surface-grafted carboxyl group and Ti^{4+}. The novel TiO_2/PAA/PTFE membranes also exhibited excellent antifouling and self-cleaning performance due to the intrinsic hydrophilicity and photocatalytic properties of TiO_2.

2.4 Poly(ether ether ketone)-Based Nanocomposite PEMs

The good mechanical and thermal stabilities as well as low cost of sulfonated PEEK (SPEEK) qualify it as a very promising PEM. Moreover, it also allows direct electrophilic sulfonation and solution casting in organic solvents and shows high proton conductivity, governed by the degree of hydration and functionalization. A brief summary of SPEEK-based nanocomposite PEMs is provided in Table 5.

Various authors have reported studies in respect SPEEK preparation [26, 27]. Too highly sulfonated SPEEK membranes (above 60% degree sulfonation, where 100% sulfonation corresponds to one sulfonic acid group per repeat unit) appears to suffer from excessive swelling under the humidified fuel cell conditions and lose their dimensional stability with high methanol crossover. Membrane dehydration is also a serious problem for shrinkage and fuel crossover.

TiO_2 has also been used as a filler to modify the properties of SPEEK membranes, specifically to enhance the transport phenomena. Karthikeyan et al. [28] used 3-aminopropyltriethoxysilane (APTMS) and imidazole glycidoxypropyl

Table 5 Poly(ether ether ketone)-based nanocomposite PEMs

Author/s	Membrane type	Achievement
Karthikeyan et al. [28]	$(RSiO_{1.5})_n$/SPEEK	Methanol and water permeabilities lower than pristine SPEEK
Dou et al. [30]	TiO_2/SPEEK	Improvement in water uptake and reduction in methanol permeability
Tripathi et al. [31]	TiO_2/PANI/SPEEK	Low methanol permeability and slow water dehydration
Ayyaru et al. [32]	TiO_2/SO_3H/SPEEK	Higher peak power density compared to that of cells with SPEEK-TiO_2 and SPEEK membranes
Wu et al. [33]	Functionalized TiO_2 submicrospheres/SPEEK	Higher selectivity and proton conductivity (5.98×10^{-3} S cm^{-1} at 20 °C, 100% RH) than TiO_2/SPEEK membranes
Narayanaswamy et al. [34]	TiO_2 (Rutile)/SPEEK	Power density 98.1 mW m^{-2} and current density 300 mA m^{-2} yielded by SPEEK +7.5% TiO_2 composite membrane

SPEEK sulfonated poly(ether ether ketone); *PANI* polyaniline trimethoxysilane as precursors to generate $(RSiO_{1.5})_n$ in a SPEEK matrix through a sol–gel process. Methanol and water permeabilities of these membranes were noted to be lower than pristine SPEEK. The APTMS precursor derived membrane showed lower permeability, but the direct methanol fuel cell (DMFC) performance for the membrane with imidazole glycidoxypropyl trimethoxysilane precursor was found to be better, as a result of good proton conductivity.

Vona et al. [29] used either a basic catalyst (pyridine) or a chelating agent (2, 4-pentandione) to control the inorganic network features in the SPEEK matrix. Sol–gel procedure with varied nanosized TiO_2 content was also used to obtain SPEEK-TiO_2 hybrid membranes, which showed both enhanced water uptake was well as retention along with reduction in methanol permeability [30].

On the other hand, Triphati et al. [31] carried out the modification of polyaniline (PANI) surface SPEEK composite membranes with nanosized Si, Zr and Ti oxides by means of sol–gel procedure. The developed nanocomposite membrane exhibited low methanol permeability and slow water dehydration, whereas no membrane stability loss, explained due to a synergetic effect.

Titanium dioxide (TiO_2) is a semiconductor with photocatalytic properties, which improves membrane fouling performance and hydrophilic characteristics. Ayyaru et al. [32] fabricated a nanocomposite SPEEK membrane with incorporated sulfonated TiO_2 particles (TiO_2–SO_3H) in a MFC. It was found that the oxygen mass transfer coefficient and the internal resistance of the SPEEK membrane decreased with increasing the TiO_2–SO_3H percentage. Moreover, the MFC with the prepared SPEEK-TiO_2–SO_3H membrane showed sensibly higher peak power density if compared to that of cells with SPEEK-TiO_2 and SPEEK membranes.

Wu et al. [33] prepared SPEEK organic–inorganic hybrid proton-exchange membranes with aminoacid functionalized titania submicrospheres. For the synthesis and functionalization of pristine TiO_2 (uniform particle size around 200 nm), the authors used four kinds of aminoacids including oxidized L-cysteine (TiO_2–Scys), o-phospho-L-serine (TiO_2–Pser), aspartic acid (TiO_2–Asp) and histidine (TiO_2-His). The methanol crossover of the hybrid membranes was reported to be reduced by two folds, along with the enhancement of the anti-swelling property and thermal stability of the hybrid membranes. In particular, the TiO_2–Pser membrane exhibited the highest selectivity. Whereas, the TiO_2–Scys embedded membrane exhibited the highest proton conductivity (about 5.98×10^{-3} S cm^{-1} (20 °C, 100% RH) at 15 wt% of the filler. In contrast, the lowest conductivity found for the TiO_2–His embedded membrane, may be due to the strong acid–base interaction between the basic imidazole groups of histidine and the sulfuric acid groups of SPEEK.

Narayanaswamy et al. [34] prepared a polymer composite electrolyte membrane of TiO_2 (Rutile) + SPEEK by a sol–gel method. The composite membranes prepared by solvent casting method have antifouling properties as well as enhanced oxygen mass transfer coefficients and are transport capable to cations other than protons. These membranes also showed improved proton conductivity and electrochemical properties, along with ion exchange and water absorption capacity. In addition, the prepared composite membranes showed higher maximum power density and current density (highest power density 98.1 mW/m^2 and current density 300 mA/m^2 yielded by SPEEK +7.5% TiO_2 composite membrane) than commercial Nafion®117s.

2.5 PANI Based Membranes

PANI-doped mesoporous metal oxides have been tested as MFC anode materials [35, 36]. A unique nanostructured PANI/mesoporous TiO_2 composite was investigated as an anode in *E. coli* MFC [35]. According to these authors the best performance was observed for the composite with 30 wt% PANI, providing two-fold higher power density (1495 mW/m^2).

Table 6 presents the summary of the leading TiO_2-modified polytetrafluoroethylene, PANI, polyethersulfone (PES) and polysolfone (PS) nanocomposite PEMs.

2.6 PES Based Membranes

PES or sulfonated PES (SPES)-based membranes are widely used for electrochemical processes, and in particular for PEMFCs, mainly because of their excellent workability and mechanical strength. Devrim et al. [37] synthetized SPES-TiO_2

Table 6 TiO_2-modified polytetrafluoroethylene, PANI, polyethersulfone (PES) and polysolfone (PS) nanocomposite PEMs

Author/s	Membrane type	Achievement
Qian et al. [25]	TiO_2/PAA/PTFE	Excellent antifouling and self-cleaning performance
Qiao et al., Wang et al. [35, 36]	PANI/mesoporousTiO_2	Power density of 1495 mW/m^2 for the composite with 30 wt% PANI
Devrim et al. [37]	TiO_2/SPES	Conductivity in the range 10^{-3}–10^{-2} S cm^{-1} and 300 mA cm^{-2} current density at 0.6 V for H_2–O_2/PEMFC at 1 atm and 85 °C
Ayyaru et al. [39]	(S–TiO_2)/SPSEBS	Improved proton conductivity respect of SPSEBS membrane and exhibited the highest peak power density (1345 ± 17 m Wm^{-2})
Aslan et al. [40]	SPSU and sulfated, TiO_2/NMPA	Maximum proton conductivity of 0.002 S cm^{-1} at 150 °C, and methanol permeability of the composite membranes lower than commercial Nafion® 112

PTFE polytetrafluoroethylene; *PAA* poly acryl acid; *SPEEK* sulfonated poly(ether ether ketone); *PANI* polyaniline; *SPSEBS* polystyrene ethylene butylene polystyrene; *NMPA* nitrilotri(methyl triphosphonic acid); *SPSU* sulfonated polysulfone composite membranes by blending of TiO_2 with SPES. The prepared composite membranes showed a conductivity in the range 10^{-3}–10^{-2} S cm^{-1} and 300 mA cm^{-2} current density at 0.6 V for H_2–O_2/PEMFC at 1 atm and 85 °C. Also, brittle nature of these membranes was explained to be due to low miscibility of both components.

Park et al. [38] fabricated proton-conducting membranes for high temperature fuel cells by introducing TiO_2 NPs in a poly (styrene sulfonic acid) polymer matrix. Ayyaru et al. [39] developed sulfonated TiO_2 (S–TiO_2)/polystyrene ethylene butylene polystyrene (SPSEBS) nanocomposite membranes by solution casting. The obtained results demonstrated that the incorporation of sulfonated TiO_2 improved the proton conductivity of the SPSEBS membrane and exhibited the highest peak power density (1345 ± 17 m Wm^{-2}). Moreover, the composite membrane was found to provide more than fourfold higher power density than Nafion® in MFCs. The SPSEBS-S-TiO_2 membrane (7.5%) exhibited the highest ion-exchange capacity (IEC), water uptake and proton conductivity capacity.

2.7 *Polysulfone-Based Membranes*

Aslan et al. [40] conducted investigation on nanocomposite membranes based on sulfonated polysulfone and sulfated nano-titania/nitrilotri(methyl triphosphonic acid) (NMPA) for proton-exchange membrane fuel cells. They prepared proton-conducting nanocomposite membranes via ternary mixtures comprising sulfated nanotitania (TS), sulfonated polysulfone (SPSU) and nitrilotri(methyl triphosphonic acid) (NMPA). The produced membranes were thermally stable up to 270 °C.

The maximum proton conductivity of SPSU-TS-NMPA was 0.002 S cm^{-1} at 150 °C, and methanol permeability of the composite membranes was lower than that of commercial Nafion® 112. These results were explained on the basis of complexation between SPSU/TS and NMPA, which inhibited the exclusion of NMPA on swelling in excess water, as supported by spectroscopic measurements and water uptake studies. The proton conductivity results of Aslan et al. [40] have been compared with those obtained by other authors (Fig. 4).

2.8 *TiO_2 Solar Cells*

Seger et al. [41] constructed a proton-exchange membrane under UV excitation. The PEM was assembled using TiO_2 photocatalyst at the anode, Pt electrocatalyst at the cathode and both pressed on a Nafion® membrane. Hydrogen was continuously generated by the irradiation of UV light when MEA was inserted under a methanol fuel cell.

The results showed an increase in the photocurrent on increasing the amount of TiO_2. However, there was an optimal point around 3.0 mg/cm^2 of TiO_2 upon which 37 μA/cm^2 photocurrent could be achieved. The photocurrent approached to maximum value at 0.2 M methanol concentration. Nevertheless, concentrations above 0.1 M of methanol may cause some problems.

The previous work reported TiO_2 solar cells are summarized in Table 7.

Bella et al. [42] demonstrated that electrodes and electrolytes in the form of membranes can be successfully coupled, ensuring better handling and processability in contrast with traditional gel electrolytes systems and ex situ fabricated photoanodes. They fabricated a flexible dye-sensitized solar cell (DSSC) by means of an innovative combination of vertically aligned TiO_2 nanotubes, grown on to bendable titanium mesh, and a photocured polymer electrolyte membrane

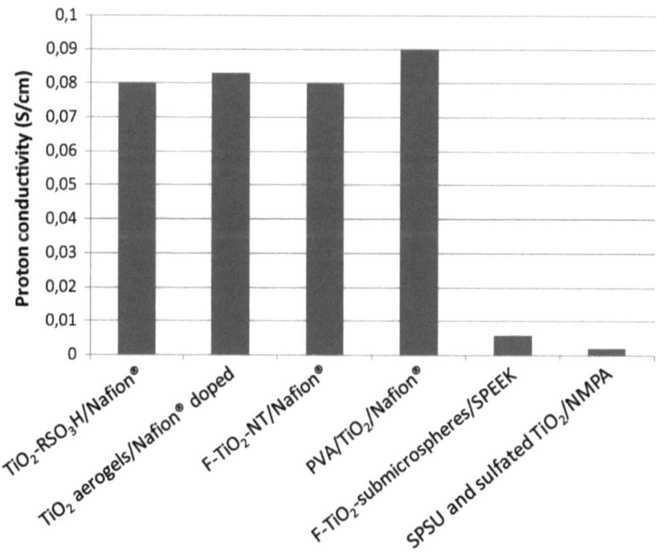

Fig. 4 Reported proton conductivity provided by different organic-TiO_2 nanocomposite membranes for PEMFC: TiO_2–RSO_3H/Nafion® [14], TiO_2 aerogels/Nafion® doped [16], F–TiO_2-NT/Nafion® [11], PVA/TiO_2/Nafion® [22], F–TiO_2 submicrospheres/SPEEK [33], SPSU and sulfated TiO_2/NMPA [40]

Table 7 TiO_2 solar cells

Author/s	Membrane type	Achievement
Seger et al. [41]	TiO_2/Nafion®/Pt	Hydrogen continuously generated by UV light irradiation; 37 μA cm^{-2} photocurrent with 3.0 mg cm^{-2} of TiO_2
Bella et al. [42]	DSSC by photocured polymer electrolyte membrane	Conversion efficiency equal to 3.4% under sun irradiation

DSSC flexible dye-sensitized solar cell

(synthetized of difunctional BEMA and monofunctional PEGMA). The authors reported sunlight to electricity conversion efficiency equal to 3.4% under sun irradiation.

2.9 Carbon Materials and Metal–Carbon Nanotube (CNTs)–TiO_2 Composites

The role of metals in the metal–carbon nanotube (CNT) composite depends on the nature of metal. The main role of the metal oxides such as MnO_2, SnO_2 and TiO_2 is to facilitate the electron transfer between the microbes and the anode. The MFCs with metal-CNT composite modified anodes show significantly improved performance than those with pure CNTs and the bare anode support. A terse summary of the investigations related with carbon materials and metal–carbon nanotube composites is presented in Table 8.

Table 8 Carbon materials and metal–carbon nanotube (CNTs)-TiO_2 composites

Author/s	Membrane type	Achievement
Xu et al. [43]	$Ru_{85}Se_{15}/TiO_2/C$	Positive influence on the electrochemical stability of the membrane, reaching power densities of 171 and 124 mW cm^{-2} at 500 mA cm^{-2} for $Ru_{85}Se_{15}/TiO_2/C$ and $Ru_{85}Se_{15}/C$ MEA respectively
Chao et al. [44]	$Pt/TiO_2/C$	MEA with 5% Pt/TiO_2 found to provide the highest current density
Matos et al. [45]	Pd/ TiO_2/C	Higher conductivity than membranes based on TiO_2/C
Li et al. [46]	$Pt/(Ti_xSn_{1-x}O_2)/C$	The $Pt/Ti_{0.9}Sn_{0.1}O_2$–C catalyst exhibited the highest activity in contrast with commercial Pt/C catalysts and Pt/TiO_2–C
Zhang et al. [47]	TNTs/Pt/C	Maximum power density of 206–305 mW cm^{-2} upon ultralow Pt loading
Garcia-Gomez et al. [48]	TiO_2 (Rutile)/C (semi-graphitic)	Maximum current density obtained in a half microbial fuel cell equal to 0.8 mA/cm^2
Zhao et al. [50]	TiO_2/GN	Improvement of the performance of GN as MFC anode material
Zhou et al. [51]	TiO_2 nanowire/GN	Internal resistance reduction in the sulphur cathode and physically immobilize the dissolved lithium polysulfides for Li/dissolved polysulfide batteries; 1327 mA hg^{-1} specific capacity at 0.2 °C rate, high coulombic efficiency (~100%) and long life cycle (over 200 cycles)

TNTs TiO_2 nanotubes; *GN* graphene

2.9.1 Carbon-TiO_2 Composites

Xu et al. [43] synthesized $Ru_{85}Se_{15}/TiO_2/C$ (5 wt% TiO_2 and 20 wt% $Ru_{85}Se_{15}$) to use as electrocatalysts in membrane fuel cells. The cathode of the membrane electrode assembly (MEA) was formed by a mix of Nafion®, $Ru_{85}Se_{15}/TiO_2/C$ and ethanol to form homogeneous slurry, whereas the anode was made of commercial 28.4 wt% Pt/C catalyst. Finally, the MEA (5 cm^2 active area) was made by hot-pressing the anode and the cathode to the Nafion® 212 membrane. This fuel cell was evaluated with hydrogen on the anode side and nitrogen on the cathode, under a potential cycling accelerated aging test (AAT) between 0.6 and 1.0 V for 1000 cycles. Results revealed that TiO_2 provided a positive effect on the electrochemical stability to the membrane, reaching power densities of 171 and 124 mW cm^{-2} at 500 mA/cm^{-2} for $Ru_{85}Se_{15}/TiO_2/C$ and $Ru_{85}Se_{15}/C$ MEA respectively.

On the other hand, Chao et al. [44] investigated the addition of platinum (Pt)/TiO_2 particles by the impregnation method on C/TiO_2 for the anode of MEA. The anode was made of 20 wt% Pt/C catalyst, 5 wt% Nafion®, 20 wt% Pt/TiO_2 and alcohol whereas the cathode was prepared with the same reagents without Pt/TiO_2. Finally, the Nafion® 112 membrane was pressed between the anode and the cathode. Results revealed that the electrochemically active surface areas decreased as the Pt/TiO_2 addition increased. Nevertheless, the water contact angle become smaller as the amount of Pt/TiO_2 particles was increased. MEA with 5% Pt/TiO_2 was found to provide the highest current density at different anode humidifier temperatures ranging from 25 to 75 °C.

Matos et al. [45] reported promising results on hybrid inorganic/organic materials for the direct formic acid fuel cells with palladium (Pd)-based catalysts. They developed Pd-based catalysts supported on hybrid TiO_2–C materials from different carbon origins by solvothermal and slurry synthesis. The authors tested them by electrooxidation of formic acid. The authors noted up to three times higher activity per Pd mass for the carbonized hybrid TiO_2–C supports with mesopore texture and high anatase/rutile ratio in the TiO_2 framework than the catalyst prepared on a commercial Vulcan XC-72 carbon black. Also, during Pd deposition, hybrid TiO_2–C supports acquire high conductivity. These behaviours were explained by their higher hydrophilicity, which ensures good access of formic acid to the Pd crystallites in the catalyst layer, and high enough conductivity to enhance electron flow from the active sites of Pd to the current collector. This fact would permit reducing the amount of expensive noble metal at the anode.

A new type of SnO_2–TiO_2 solid solution ($Ti_xSn_{1-x}O_2$) support was prepared by Li et al. [46] via a solvothermal method with substitution of Ti^{4+} by Sn^{4+} in the TiO_2 lattice, for methanol oxidation. In addition, $Ti_xSn_{1-x}O_2$ was combined with conventional carbon black (Vulcan XC-72) to synthetize a hybrid support ($Ti_xSn_{1-x}O_2$–C) for depositing Pt nanoparticles. The well-dispersed Pt nanoparticles (2–3 nm) were observed to get mostly deposited on the boundaries of $Ti_{0.9}Sn_{0.1}O_2$ and carbon blacks. The Pt/$Ti_{0.9}Sn_{0.1}O_2$–C catalyst exhibited the highest activity in contrast to commercial Pt/C and Pt/TiO_2–C catalysts, in terms of methanol oxidation and CO stripping. This was due to its triple junction structure, which can help improve Pt utilization with increased electrochemical active surface areas. Moreover, the high OH content on $Ti_{0.9}Sn_{0.1}O_2$ and the strengthened metal support interactions, both promoting the oxidation of poisoning CO absorbed on Pt active sites.

Zhang et al. [47] fabricated an oriented ultrathin catalyst layer (UTCL) for developing low cost fuel cells. The UTCL comprising high electrical conductivity TiO_2 nanotubes (TNTs) via a plasma-enhanced chemical vapour deposition, which helped in improving the electrical conductivity of the TNTs. Subsequently, Pt catalysts were deposited on the TNTs by the radio frequency (RF) sputtering, leading to oriented C-TNTs-Pt electrode. The maximum power density of 206–305 mW cm^{-2} upon ultralow Pt loading was observed. Figure 5 demonstrates the comparative magnitudes of power density realized from different organic TiO_2 nanocomposite membranes.

Garcia-Gomez et al. [48] synthetized a new dual nanofiber material based on TiO_2 (Rutile)-C(semi-graphitic)/C(semi-graphitic) mats with a side by side configuration for its use in microbial fuel cells. The prepared nanostructured material exhibited high surface area and biocompatibility, and it was found to provide excellent electrical performance with a maximum current density obtained in a half microbial fuel cell equal to 0.8 mA/cm^2.

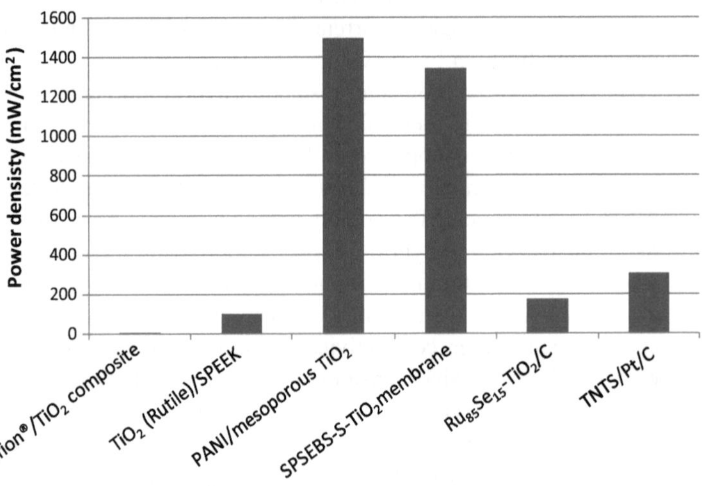

Fig. 5 Power density provided by different organic-TiO_2 nanocomposite membranes for PEMFC: Nafion®/TiO_2 [9], TiO_2 (Rutile)/SPEEK [34], PANI/mesoporous TiO_2 [35], SPSEBS-S-TiO_2 membrane [38], $Ru_{85}Se_{15}/TiO_2/C$ [43], TNTs/Pt/C [47]

2.9.2 Graphene (GN)-TiO_2 Composites

It has been reported in the literature that GN as MFC anode is effective in achieving improved electron transfer between the microorganisms and the anode. Zhang et al. [49] reported graphene modification on the anode significantly increased electricity generation in MFCs. On the other hand, to improve the performance of GN as MFC anode material Zhao et al. [50] synthetized hybrid metal–GN (M = TiO_2) by a microwave-assisted hydro (solvo) thermal process, in which TiO_2 nanoparticles were assembled on graphene.

Zhou et al. [51] developed free-standing TiO_2 nanowire/graphene hybrid membrane for Li/dissolved polysulfide batteries. In these kinds of batteries, graphene membrane with high electrical conductivity serves as a current collector in order to effectively reduce the internal resistance in the sulphur cathode and physically immobilize the dissolved lithium polysulfides. In this case, the TiO_2 nanowires inserted into the graphene membrane conferred it a hierarchical composite structure, in which the TiO_2 nanowires offer strong chemical binding with the lithium polysulfides along with a strong catalytic effect for polysulfide reduction and oxidation. This latter fact promotes fast redox reaction kinetics with high capacity and low voltage polarization. The developed membrane hybrid electrode is known for its high specific capacity (1327 mA hg^{-1} at 0.2 °C rate), high Coulombic efficiency (approaching 100%) and long life cycle (over 200 cycles).

3 Conclusions

In the future, fuel cells appear to become a key technology within the energy management framework of many devices, machines and energy networks, if technical and economic feasibility are improved sensibly.

Many studies in order to overcome the limits of the technology are nowadays in progress. A promising and interesting development path appears to be the improvement of the membrane properties used in fuel cells by nanotechnologies.

All membranes already used in the past appear to sensibly benefit from the application of nanotechnologies, with an increase of the fuel cell efficiency and durability. The advantage of this approach is to improve existing and well-known products, already used in the past in similar devices. On the other hand, novel membranes produced with the help of nanotechnologies for fuel cell applications appear to be more promising, and are now performing their test in time.

A definitive solution to provide a perfect membrane to a fuel cell does not exist, but the intensive efforts of many research groups worldwide and the achieved research progresses may surely provide in the very next years an answer to this problem.

Acknowledgements The Spanish Ministry of Economy and Competitiveness is gratefully acknowledged for project CTM2014-61105-JIN fund.

References

1. Kim MS, Lee SJ, Kang JU, Bae KJ (2005) Preparations of polypropylene membrane with high porosity in supercritical CO_2 and its application for PEMFC. J Ind Eng Chem 11:187–193
2. Jang SY, Han SH (2012) Preparation of high styrenic sulfonated polySEPS/clay composite film for proton exchange membranes (PEMs). J Ind Eng Chem 18:1280–1285
3. Lee SJ, Quan ND, Hwang JM, Lee SD, Kim HG, Lee HJ, Kim HS (2006) Polymer electrolyte membranes for fuel cells. J Ind Eng Chem 12:175–183
4. Ayad MM, El-Nasr AA, Stejskal J (2012) Kinetics and isotherm studies of methylene blue adsorption onto polyaniline nanotubes base/silica composite. J Ind Eng Chem 18:1964–1969
5. Liou TH, Lin HS (2012) Synthesis and surface characterization of silica nanoparticles from industrial resin waste controlled by optimal gelation conditions. J Ind Eng Chem 18:1428–1437
6. Hosseini SM, Madaeni SS, Zendehnam A, Moghadassi AR, Khodabakhshi AR, Sanaeepur H (2013) Preparation and characterization of PVC based heterogeneous ion exchange membrane coated with Ag nanoparticles by (thermal-plasma) treatment assisted surface modification. J Ind Eng Chem 19:854–862
7. Khajenoori M, Rezaei M, Nematollahi B (2013) Preparation of noble metal nanocatalysts and their applications in catalytic partial oxidation of methane. J Ind Eng Chem 19:981–986
8. Peighambardoust SJ, Rowshanzamir S, Amjadi M (2010) Review of the proton exchange membranes for fuel cell applications. Int J Hydrogen Energy 35:9349–9384
9. Licoccia S, Traversa E (2006) Increasing the operation temperature of polymer electrolyte membranes for fuel cells: from nanocomposites to hybrids. Power Sour 159:12–20
10. Slade SM, Smith JR, Campbell SA, Ralph TR, Ponce De León C, Walsh FC (2010) Characterisation of a re-cast composite Nafion® 1100 series of proton exchange membranes incorporating inert inorganic oxide particles. Electrochim Acta 55:6818–6829
11. Mohammadi G, Jahanshahi M, Rahimpour A (2013) Fabrication and evaluation of Nafion nanocomposite membrane based on ZrO_2–TiO_2 binary nanoparticles as fuel cell MEA. Int J Hydrogen Energy 38:9387–9394
12. Liu Z, Guo B, Huang J, Hong L, Han M, Gan LM (2006) Nano-TiO_2-coated polymer electrolyte membranes for direct methanol fuel cells. J Power Sour 157:207–211
13. Santiago EI, Isidoro RA, Dresch MA, Matos BR, Linardi M, Fonseca FC (2009) Nafion–TiO_2 hybrid electrolytes for stable operation of PEM fuel cells at high temperature. Electrochim Acta 54:4111–4117
14. Li ZH, Zhang HP, Zhang P, Li GC, Wu YP, Zhou XD (2008) Effects of the porous structure on conductivity of nanocomposite polymer electrolyte for lithium ion batteries. J Membr Sci 322:416–422
15. Amjadi M, Rowshanzamir Peighambardoust SJ, Hosseini MG, Eikani MH (2010) Investigation of physical properties and cell performance of Nafion/TiO_2 nanocomposite membranes for high temperature PEM fuel cells. Int J Hydrogen Energy 5:9252–9260
16. Zhengbang W, Tang H, Mu P (2011) Self-assembly of durable Nafion/TiO_2 nanowire electrolyte membrane temperature PEM fuel cells. J Membr Sci 369:250–257
17. Cozzi D, Bonis C, Epifanio A, Mecheri B, Tavares AC, Licoccia S (2014) Organically functionalized titanium oxide/Nafion composite proton exchange membranes for fuel cells applications. J Power Sour 248:1127–1132
18. Beauger C, Testut L, Berthom-Fabry S, Georgi F, Guetaz L (2016) Doped TiO_2 aerogels as alternative catalyst supports for proton exchange membrane fuel cells: a comparative study of Nb, V and Ta dopants. Microporous Mesoporous Mater 232:109–118
19. Gojković SL, Babić BM, Radmilović VR, Krstajić NV (2010) Nb-doped TiO_2 as a support of Pt and Pt–Ru anode catalyst for PEMFCs. J Electroanal Chem 639:161–166

20. Houn-Rhee C, Kim Y, Sung J, Kyung H, Chang H (2006) Nanocomposite membranes of surface-sulfonated titanate and Nafion® for direct methanol fuel cells. Power sour 159: 1015–1024
21. Jun Y, Zarrin H, Fowler M, Chen Z (2011) Functionalized titania nanotube composite membranes for high temperature proton exchange membrane fuel cells. Hydrogen energy 36:6073–6081
22. Yang C, Chiu S, Lee K, Chien W, Lin C, Huang C (2008) Study of poly(vinyl alcohol)/ titanium oxide composite polymer membranes and their application on alkaline direct alcohol fuel cell. Power sour 184:44–51
23. Yang CC, Chien WC, Li YJ (2010) Direct methanol fuel cell based on poly(vinyl alcohol)/ titanium oxide nanotubes/poly(styrene sulfonic acid) (PVA/nt–TiO_2/PSSA) composite polymer membrane. J Power Sour 195(11):3407–3415
24. Mroczkowska-Szerszen M, Siekierski M, Letmanowski R, Zabost D, Piszcz M, Zukowska G, Sasim E, Wieczorek W, Dudek M, Struzik M (2013) Synthetic preparation of proton conducting polyvinyl alcohol and TiO_2-doped inorganic glasses for hydrogen fuel cell applications. Electrochim Acta 104:487–495
25. Qian Y, Chi L, Zhou W, Yu Zhang Z, Zhang Z, Jian Z (2016) Fabrication of TiO_2-modified polytetrafluoroethylene ultrafiltration membranes via plasma-enhanced surface graft pretreatment. Surf Sci 360:749–757
26. Hogarth WHJ, Diniz da Costa JC, Lu GQ (2005) Solid acid membranes for high temperature proton exchange membrane fuel cells. J Power Sour 142:223–237
27. Tripathi BP, Shahi VK (2007) SPEEK–zirconium hydrogen phosphate composite membranes with low methanol permeability prepared by electro-migration and in situ precipitation. J Colloid Interface Sci 316(2):612–621
28. Karthikeyan CS, Nunes SP, Schulte K (2006) Permeability and conductivity studies on ionomer-polysilsesquioxane hybrid materials. Macromol Chem Phys 207:336–341
29. Vona MLD, Ahmed Z, Bellitto S, Lenci A, Traversa E, Licoccia S (2007) SPEEK–TiO_2 nanocomposite hybrid proton conductive membranes via in situ mixed sol-gel process. J Membr Sci 96(2):156–161
30. Dou Z, Zhong S, Zhao C, Li X, Fu T, Na H (2008) Synthesis and characterization of a series of SPEEK/TiO_2 hybrid membranes for direct methanol fuel cell. J Appl Polym Sci 109: 1057–1062
31. Tripathi BP, Shahi VK (2009) Surface redox polymerized SPEEK–MO_2–PANI (M = Si, Zr and Ti) composite polyelectrolyte membranes impervious to methanol. Colloids Surf A 340:10–19
32. Ayyaru S, Dharmalingam S (2013) Improved performance of microbial fuel cells using sulfonated polyether ether ketone (SPEEK) TiO_2–SO_3H nanocomposite membrane. RSC Adv 3:25243–25251
33. Wu H, Cao Y, Shen X, Li Z, Xu T, Jiang Z (2014) Preparation and performance of different amino acids functionalized titania-embedded sulfonated poly(ether ketone) hybrid membranes for direct methanol fuel cells. J Membr Sci 463:134–144
34. Narayanaswamy V, Dharmalingam S (2015) Development of cation exchange resin-polymer electrolyte membranes for microbial fuel cell application. J Membr Sci 50:6302–6312
35. Qiao Y, Bao SJ, Li CM, Cui XQ, Lu ZS, Guo J (2008) Nanostructured polyaniline/titanium dioxide composite anode for microbial fuel cells. ACS Nano 2:113–119
36. Wang Z, Tang H, Pan M (2011) Self-assembly of durable Nafion/TiO_2 nanowire electrolyte membranes for elevated-temperature PEM fuel cells. J Membr Sci 369:250–257
37. Devrim Y, Erkan S, Bac N, Eroglu I (2009) Preparation and characterization of sulfonated polysulfone/titanium dioxide composite membranes for proton exchange membrane fuel cells. Int J Hydrogen Energy 34:3467–3475
38. Park JT, Koh JH, Roh DK, Shul YG, Kim JH (2011) Proton-conducting nanocomposite membranes based on P(VDF-co-CTFE)-g-PSSA graft copolymer and TiO_2–PSSA nanoparticles. Int J Hydrogen Energy 36:1820–1827

39. Ayyaru S, Dharmalingam S (2015) A study of influence on nanocomposite membrane of sulfonated TiO_2 and sulfonated polystyrene-ethylene-butylene-polystyrene for microbial fuel cell application. Energy 88:202–208
40. Aslan A, Ayhan B (2014) Nanocomposite membranes based on sulfonated polysulfone and sulfated nano-titania/NMPA for proton exchange membrane fuel cells. Solid State Ionics 255:89–95
41. Seger B, Kamat PV (2009) Fuel cell geared in reverse: photocatalytic hydrogen production using a TiO_2/Nafion/Pt membrane assembly with no applied bias. J Phys Chem C 113: 18946–18952
42. Bella F, Lamberti A, Sacco A, Bianco S, Chiodoni A, Bongiovanni R (2014) Novel electrode and electrolyte membranes: towards flexible dye-sensitized solar cell combining vertically aligned TiO_2 nanotube array and light-cured polymer network. J Membr Sci 470:125–131
43. Xu T, Zhang H, Zhong H, Ma Y, Jin H, Zhang Y (2010) Improved stability of TiO_2 modified $Ru_{85}Se_{15}/C$ electrocatalyst for proton exchange membrane fuel cells. Power sour 195(24): 8075–8079
44. Chao WK, Huang RH, Huang CJ, Hsueh JL, Shieu FS (2010) Effect of hygroscopic platinum/titanium dioxide particles in the anode catalyst layer on the PEMFC performance. J Electrochem Soc B1012–B1018
45. Matos J, Borodzinski A, Zychora AM, Kedzierzawski P, Mierzwa B, Juchniewicz K, Mazurkiewicz M, Hernández-Garrido JC (2015) Direct formic acid fuel cells on Pd catalysts supported on hybrid TiO_2–C materials. Appl Catal B 163:167–178
46. Li Y, Liu C, Liu Y, Feng B, Li L, Pan H, Kellogg W, Higgins D, Wu G (2015) Sn-doped TiO_2 modified carbon to support Pt anode catalysts for direct methanol fuel cells. J Power Sour 286:354–361
47. Zhang C, Yu H, Fu L, Xiao Y, Gao Y, Li Y, Zeng Y, Jia J, Yi B, Shao Z (2015) An oriented ultrathin catalyst layer derived from high conductive TiO_2 nanotube for polymer electrolyte membrane fuel cell. Electrochim Acta 153:361–369
48. Garcia-Gomez NA, Balderas-Renteria I, Garcia-Gutierrez DI, Mosqueda HA, Sánchez EM (2015) Development of mats composed by TiO_2 and carbon dual electrospun nanofibers: a possible anode material in microbial fuel cells. Mater Sci Eng B 193:130–136
49. Zhang Y, Mo G, Li X, Zhang W, Zhang J, Ye J, Huang X, Yu C (2011) Graphene-modified electrodes for enhancing the performance of microbial fuel cells. J Power Sour 196:5402–5407
50. Zhao CE, Wang WJ, Sun D, Wang X, Zhang JR, Zhu JJ (2014) Nanostructured graphene/TiO_2 hybrids as high-performance anodes for microbial fuel cells. Chem Eur J 20:7091–7097
51. Zhou G, Zhao Y, Zu C, Manthiramn A (2015) Free-standing TiO_2 nanowire-embedded graphene hybrid membrane for advanced Li/dissolved polysulfide batteries. Nano Energy 12:240–249

Chapter 3
Organic/Silica Nanocomposite Membranes

Palaniappan Sathish Kumar, Samir Kumar Pal, Moganapriya Chinnasamy and Rathanasamy Rajasekar

Abstract Nanoscience and nanotechnology has become a versatile and promising subject for producing new materials with enhanced properties and potential applications. In this regard, nanoparticles (NPs) have received growing attention in every sector of science and technology. The size, shape, structure, and chemical properties of engineered NPs open a vast range of technical applications and novel approaches in basic research science. Among the variety of NPs, silica NP is of a particular interest due to its ease of synthesis, functionalization, and precise controlling of size and distribution of particles. Superior features of polymeric membranes especially flexibility and processability have made them one of the best candidates for commercial applications. Meanwhile, despite of the outstanding characteristics of these membranes, their application is still limited due to the trade-off trend between gas permeability and selectivity.

Keywords Nanoparticles · Phase inversion · Sol-gel · Nafion membranes

List of Abbreviations

AEM	Anion-exchange membrane
AgNPs	Silver nanoparticles
AFM	Atomic force microscopy
CO_2	Carbon dioxide

P. Sathish Kumar (✉) · S.K. Pal
Department of Mining Engineering, Indian Institute of Technology Kharagpur, Kharagpur 721302, West Bengal, India
e-mail: sathishcad91@gmail.com

S.K. Pal
e-mail: pal.samir09@gmail.com

M. Chinnasamy · R. Rajasekar
Department of Mechanical Engineering, Kongu Engineering College, Erode 638052, Tamil Nadu, India
e-mail: mogana98@gmail.com

R. Rajasekar
e-mail: rajasekar.cr@gmail.com

Inamuddin et al. (eds.), *Organic–Inorganic Composite Polymer Electrolyte Membranes*, DOI 10.1007/978-3-319-52739-0_3

CTAB	Cetyl trimethyl ammonium bromide
CEM	Cation-exchange membrane
CMC	Carboxy methyl cellulose
DMA	Dynamic mechanical analysis
DSC	Differential scanning calorimetry
EG	Ethylene glycol
EDX	Energy dispersive X-ray spectroscopy
Fe_3O_4	Iron oxide
FS	Fumed silica
FTIR	Fourier transform infrared
FFV	Fractional free volume
FE	Field emission
FO	Forward osmosis
H_2	Hydrogen
HfO_2	Hafnium dioxide
IPDI	Isophorone diisocyanate
IP	Interfacial polymerization
ILMS	Ionic liquid-modified silica
MMMs	Mixed matrix membranes
MPTES	Mercapto propyl triethoxy silane
NPs	Nanoparticles
NR	Natural rubber
N_2	Nitrogen
NMR	Nuclear magnetic resonance
O_2	Oxygen
OPBI	Poly (oxyphenylene benzimidazole)
PEBA	Poly (ether block amide)
PVDF-HFP	Poly (vinylidene fluoride-hexafluoropropylene)
PBI	Polybenzimidazole
PEEK	Polyether ether ketone
PES	Polyethersulfone
PMMA	Poly (methyl methacrylate)
PDMC	Poly (2-methacryloyloxy ethyl trimethyl ammonium chloride)
PDMOS	Polydimethoxysiloxane
PEBAX	Poly (amide-6-b-ethylene oxide)
PI	Phase inversion
PP	Polypropylene
PE	Polyethylene
PEG	Poly (ethylene glycol)
PPG	Poly (propylene glycol)
PEMFC	Proton-exchange membrane fuel cell
PVA	Poly (vinyl alcohol)
PEC	Polyelectrolyte complex
PDMS	Polydimethylsiloxane

PWA	Phosphotungstic acid
PHEMA	Poly (2-hydroxy ethyl methacrylate)
PEI	Poly (ethylenimine)
PPO	Poly (phenylene oxide)
RO	Reverse osmosis
SiO_2	Silicon dioxide
SBR	Styrene-butadiene rubber
SBA	Santa Barbara Amorphous
SEM	Scanning electron microscopy
SAXS	Small-angle X-ray scattering
SA-SNP	Sulfonated silica nanoparticles
SDS	Sodium dodecyl sulfate
TiO_2	Titanium dioxide
$TiSiO_4$	Titanium (IV) orthosilicate
Tg	Glass transition temperature
TEM	Transmission electron microscopy
TGA	Thermogravimetric analysis
TEOS	Tetraethyl orthosilicate
UV	Ultraviolet
UF	Ultrafiltration
VRFBS	Vanadium redox flow batteries
WAXD	Wide-angle X-ray diffraction
XRD	X-ray diffraction
ZnO	Zinc oxide
ZrO_2	Zirconium dioxide

1 Introduction

The fabrication of nano-based systems is an active area of research, where the properties of the parent/host materials are tailored by incorporating a guest component. Organic–inorganic nanocomposite materials are recognized as a new class of advanced materials which can be synthesized or processed using versatile approaches and own tunable properties.

Polymer matrix nanocomposite membranes are advanced membranes with nanomaterials dispersed in their polymer matrices. Such membranes could be used for gas–gas, liquid–liquid, and liquid–solid separations [1]. Polymer nanocomposites have drawn a great deal of interest in recent years since these materials possess high potential to achieve great property improvement by the addition of small amount of NPs in the polymer matrices. The combination of nanoscale inorganic species with organic polymers has a high potential for future applications. These materials are widely used in automotive, aerospace, construction, and electronic industries because they provide improved mechanical, electrical, thermal, and physical properties over native polymers. However, nanoscale inorganic species (filler) are hydrophilic and for that reason, it is difficult to homogeneously disperse them in organic polymers, especially polymers of low polarity. The advantage of nanofillers can only be achieved if the particles are finely dispersed in the polymer matrix. A critical challenge in the design of these hybrid inorganic–organic species is the control of proper mixing between the two dissimilar phases. Nevertheless, physical mixtures of organic polymers and inorganic particles may lead to separation in discrete phases or result in poor mechanical, optical, and electrical properties. Inorganic materials, such as silica, are usually incorporated into various polymer matrixes by different approaches to enhance the thermal and mechanical stability of neat polymers. Silica is a typical example of inorganic filler that according to their particle size, shape, distribution, amount, and orientation within nano- or micro-composite membranes can meet various technological requirements for electrolyte fuel cells, such as enhanced proton conductivity, reduced methanol permeability, and improved mechanical stability.

Among the variety of NPs, silica NP is of a particular interest due to its ease of synthesis, functionalization, and precise controlling of size and distribution of particles. The successful sol–gel method of Stober reported in 1968 for preparing mono-dispersed silica NPs based on the hydrolysis of silicon alkoxides in alcohol media had been used for the preparation of inorganic materials. The ability to effectively modify the properties of silica NPs by controlling their structure at nanoscale level makes them extremely attractive candidates for many applications, from fundamental scientific studies to commercially realizable technologies. On the other hand, polymers as common organic materials have many applications, but they often require certain properties in order to satisfy their numerous application requirements. NPs are a type of nanofiller that offer the opportunity of developing polymers with new or advanced properties without any impact on their process abilities. Generally, nanoscaled particles exhibit an enormous surface area being several orders of magnitude larger than the surface of conventional fillers; therefore, low loadings of NPs require efficient dispersing in polymer matrix. An efficient dispersion of NPs in polymer matrix, however, is vital to fabricate desirable polymer nanocomposite.

It had been reported that addition of silica NPs into polymer matrix could offer many advantages to the resulting composites; however, it is difficult to homogenously disperse these particles into hydrocarbon polymers matrix. In general, the polar fillers have inherently low compatibility with nonpolar polymer matrix, especially hydrocarbons such as polypropylene (PP) and polyethylene (PE). As reported by several authors, hydrogen bonds formed between the hydrophilic particles may cause the particles to agglomerate into bundles and unevenly distribute throughout the nonpolar polymer matrix during compounding processing. However, by changing species of grafting monomer and conditions of grafting, interfacial interaction between NPs and polymer matrix in nanocomposites can be purposely adjusted. Therefore, covalent linking is needed to improve the effective dispersion and interaction between silica particles and organic functional groups. It means that there is a need to introduce active inorganic groups such as amines, thiols, isocyanates, and vinyl groups on the silica particles. Many efforts had been carried out for organically modification of NPs using various organo-alkoxysilanes. In order to improve the dispersion stability of zinc oxide (ZnO) NPs and better dispersibility of iron oxide (Fe_3O_4) NPs, treatment with a silane-coupling agent is required. In continuation, to avoid silica agglomerations during mixing, the silica particles were treated with bis (triethoxysilyl-propyl) tetrasulfane and four different silane-coupling agents. Also, some authors have used vinyl triethoxysilane as a coupling agent to allow the incorporation of reactive groups before the grafting of poly (ethylene glycol) (PEG) and poly (propylene glycol) (PPG) onto silica NPs via ultraviolet (UV)-photopolymerization.

Recently, the nanoscale silica-based mesoporous materials have become popular for their potentially useful catalytic, magnetic, optical, and semi-conducting properties, which are exploited in technological and biomedical applications. Furthermore, the nanoscale silica-based mesoporous materials are good candidates for the inorganic component of polymer mesocomposite because of their excellent thermal and mechanical stability and inherent air pore inside the nanochannels. Certain equations have already proposed to well adapt the membranes containing exclusively exfoliated structures and become less accurate as the dispersion of the silicate sheets is more complex, i.e., when different types of structures from exfoliated ones to intercalated ones co-exist. High delamination properties allow the formation of intercalated and exfoliated structures of silicate layers into wide range of organic species affecting the overall performance of the nanocomposites. In reality, the silicate layers tend to form larger and complicated structure when they are poorly dispersed in the nanocomposite. Exfoliated individual dispersed silicate layers exist when there is no observable contact with each silicate layer within the nanocomposite.

The concept of making nanocomposite membranes was originally developed to overcome the Robeson upper boundary in the field of gas separation in 1990s, where highly selective zeolites were incorporated into polymers to improve both permeability and selectivity. Besides gas separation, many other applications had also been examined using nanocomposite membranes. Over the last decade, efforts to overcome the drawbacks of Nafion membranes have proposed different studies. A useful approach to develop this type of membranes is to modify Nafion membranes with micron or submicron inorganic/organic additives such as silica, titanium dioxide (TiO_2), titanium (IV) orthosilicate ($TiSiO_4$), and zirconium dioxide (ZrO_2). These additives exhibit a high water retention capability, which allows the membrane to retain proton conductivity at elevated temperatures. Such composite membranes have exhibited improvements in water retention and proton-exchange membrane fuel cell (PEMFC) performance at elevated temperatures. The addition of an inorganic material into polymer membrane can alter and improve physical and chemical polymer properties of interest [such as elastic modulus, solvent permeation rate, tensile strength, hydrophilicity, and glass transition temperature (Tg)] while retaining its important polymer properties to enable operation in the fuel cell.

The membrane is at the heart of every membrane process and can be considered as a permselective barrier or interface between two phases. Membrane separation processes are currently applied in various fields such as water and waste water treatment, medicine, pharmacy, food, and beverage industries. Recently, membrane-based processes gained importance in biotechnology due to their ability for size and/or charge-based protein separations. Pressure-driven processes such as microfiltration, ultrafiltration, and nanofiltration are usually applied for protein separation and purification. Collagen-based proteins have recently attracted great attention due to their extensive applications in biotechnology, tissue engineering, and biomedical science. However, there are few reports about collagen proteins purification using membranes. Hence, in the last two decades, extensive efforts have been made to improve the separation performance of these membranes for various applications. Incorporation of an inorganic phase within the polymer matrix is one of the simplest methods that have been widely employed in order to enhance separation performance of polymeric membranes. The first step to prepare a proper membrane for a special application is selection of the membrane material. Researchers have formulated material selection guidelines for preparation of membranes which is applicable for carbon dioxide (CO_2) removal. Based on this guideline, selection of polymers with polar groups increases the affinity of CO_2 to the polymer matrix and consequently improves separation performance of CO_2 from nonpolar gases.

Nowadays, many research activities have been associated with organic–inorganic nanocomposites for membrane preparation and separation processes. The incorporation of silica particles into poly (vinyl alcohol) (PVA) caused desirable changes in the morphology and crystalline structure of the PVA-silicon dioxide (SiO_2) nanocomposite membrane which significantly enhanced the thermal stability and stability of the membranes in ethylene glycol (EG) aqueous solutions. For pervaporation dehydration of ethanol–water mixtures, silica NPs in chitosan-silica complex membranes served as spacers between the polymer chains to provide extra space for water permeation, so as to bring about high permeation rates within the complex membranes. Polyelectrolyte complex (PEC)/SiO_2 nanohybrid membranes showed very high performance in isopropanol dehydration as compared with other polymeric nanohybrid membranes. The selectivity of these membranes is slightly better than that of pristine PEC membranes due to the fine dispersion of SiO_2. Moreover, the incorporation of SiO_2 also improves the processability and mechanical properties of PEC. For the pervaporation of ethanol–water mixtures using organophilic nano-silica filled polydimethylsiloxane (PDMS) composite membranes, the solubility selectivity and the diffusion selectivity was enhanced with increasing organophilic nano-silica concentration. In addition, the composite membranes exhibited striking advantages in total flux and separation factor as compared with unfilled PDMS membrane. In some research work, natural rubber (NR) as a green polymer matrix was used for the NR/styrene-butadiene rubber copolymer (SBR)-SiO_2 nanocomposite membranes due to their low Tg values. Furthermore, NR can suppress the swelling of the polymeric structure and maintain the integrity of the membrane.

Polymer composite membranes and barrier films have attracted much interest in recent years; some of the reasons can be easily understood in terms of composite theory particularly for cases where the filler has a high aspect ratio. Composite theories assume that the properties of each phase are the same as if the other phases were not there; however, smaller surface area of the filler particles, the polymer chain packing, and dynamics may be altered. This makes very little impact on overall performance when the particles are large. As the particles become much smaller, on the other hand, the amount of surface area increases and the distance between particles decreases. In this case, effects of the interface become very important and the matrix phase may not effectively behave as it supposed to do in the bulk. The particles of interest may be in the form of spheres, rods, and platelets of which fumed silica, carbon nanotubes, and clays are examples. Another interesting case is where the particles are permeable but highly selective, like zeolites, and may improve the selectivity of the composite, i.e., so-called "mixed matrix membranes (MMMs)". However, performance of such composite membranes often seems to be compromised by the formation of voids or defects caused by dewetting of the polymer at the surface of the zeolite particles. These problems have hindered the development of such membrane materials. Polymers filled with nanosized particles often show filler agglomeration and network formation in the polymer matrix which are undesirable to resolve NP agglomeration or defects. To improve the filler particle dispersion in the matrix, a great deal of attention has been devoted to modify the interface between the filler and the polymer matrix.

The effect of silica particles on the gas separation properties of high performance polyimides had been investigated and it had been observed that incorporation of silica particles into the polymer matrix mainly improved the gas separation properties of polymer-silica membranes, especially for oxygen (O_2)/nitrogen (N_2) separation. The main improvements were the attainment of high O_2/N_2 selectivity and O_2 permeability of some silica-polymer hybrids, simultaneously. Therefore, O_2/N_2 separation properties of some silica-polymer membranes were above the Robeson's upper bound line, while that of the pure polymer was below the Robeson's upper bound line. This upper bound was expressed as a line on a log–log plot of separation factor versus permeability of the more permeable gas above which virtually no data exist. The effect of silica particles on the CO_2 permeation of silica–polyimide composite membranes was studied. The obtained results showed that, in the presence of silica in the polymer matrix, the CO_2 permeability of composite membranes increases by tenfold compared to polymer. The experimental results also showed 50–100% increase in gas solubility of membranes composed of silica in comparison with the pure polyimide. Several authors investigated the effect of the distribution of silica nanoparticles (SNP) on the gas separation performance of polyether-imide membranes and polyimide-silica membranes. Amino silicate was used as a compatible agent between organic and inorganic phases. The permeability of hybrid membranes was also investigated for methane, butane, and pentane gases at 80 bar. The obtained results showed that the gas permeability reduction in the presence of silica particles of the polymer matrix was only 10%, while, under the same condition, permeability of pure polymer dropped 60% during the experiments.

Polymers with high degree of hydrophilicity are practically utilized as the water-selective membranes. Using hybrid polymer/inorganic (nano) composites is another approach to improve properties of the membranes. Addition of inorganic (nano) particles such as alumina, titanium oxide, silica, zirconia, or clay to a polymer matrix and preparing membranes such as of PVA/SiO_2-phosphotungstic acid (PWA) composite, PVA/sulfonated santa Barbara amorphous (SBA-15) mesoporous silica, PVA/montmorillonite, Nafion/[(ZrO_2)-hafnium dioxide (HfO_2) 0.25], Nafion/[(SiO_2)–(HfO_2)0.28], and Nafion/fluoroalkylated silica has improved mechanical properties (via acting as a reinforcing agent) of the membrane and decreased methanol crossover. Proton conduction has been observed to be dependent on the size of the particles (nano or micro), their surface properties (acidic or basic), and their functionality. Moreover, proton conduction in the presence of NPs seems to be complex and additional effects such as interactions between the polymer and NP can also act at the same time.

Due to its preference to water, the PVA membranes suffer from the excessive swelling which ultimately leads to a loss of membrane integrity. A common practice to limit swelling is to cross-link PVA via chemical bonding with the cross-linker (e.g., glutaraldehyde). Alternatively, NPs which interact with PVA molecules through H-bonding, e.g., fumed silica (FS), can serve as the physical cross-linker for PVA membranes. An advantage of adding FS particles in the PVA membrane is to enhance water permeability and selectivity of the PVA nanocomposite membrane. Incorporation of FS particles was found to increase the fractional free volume (FFV) of PVA nanocomposite membranes which results in the promotion of water diffusion. Moreover, owing to the hydrophilicity of FS particles, water uptake of the FS filled membrane was increased compared to unfilled membrane because, additional water absorption can occur by trapping of water molecules on the surface of FS particles. In fact, FS particles in the PVA nanocomposite membrane have a two-fold benefit in providing the membrane stability and increasing water permeability, subject to condition that NPs are well distributed within the polymer matrix. To improve the dispersion of FS particles in the PVA matrix, the compatibility between them must be enhanced. As mentioned earlier, to improve membrane performance, several methods such as cross-linking, blending, and grafting of a chemical agent to the original polymer matrix had been employed for membrane modification. Among these, cross-linking has been widely employed for improving membrane physicochemical properties, such as hydrophobicity, crystallinity, and mechanical strength. Inorganic fillers have been integrated into a membrane matrix to improve permeability by increasing the selective sorption of favorable components and/or reducing undesirable diffusion. The effects of different inorganic fillers such as alumina, silica, zeolites, carbon black, carbon nanotubes, and activated carbon had been investigated on the performance of polymeric membranes. Usually, membranes are made from organic polymers, which are cost effective and have high process flexibility but such polymer membranes are mechanically weak. If, however, inorganic materials, which often have high mechanical strength, high thermal stability, and special gas separation properties, are added as functional fillers, the resulting MMMs can be outstanding membranes for gas separation processes. Therefore, in recent years, MMMs that combine favorable properties of both organic polymers and inorganic fillers have received significant attention. However, there are still many problems restricting the wide application of MMMs. These problems can be grouped into two categories: (1) poor dispersion of the inorganic fillers in the organic matrix. Generally, inorganic fillers are synthesized from their precursor solutions by sol–gel process and are mixed into the polymer-casting solution. However, fillers easily aggregate via the van der Waals force between particles. Poor dispersion of the fillers contributes to the ineffectiveness of the membrane and also decreases the mechanical strength of the membrane. (2) Interfacial defects. Weak physical or chemical bonds between the inorganic fillers and the organic polymers result in phase separation and the formation of interfacial voids between the two components. These defects limit the separation performance of the MMMs due to the short-circuiting of gas flow through interfacial voids. Several technologies have recently been adopted to overcome the problems mentioned above. These technologies include the silane-coupling method to reduce interfacial voids, the priming method to ensure that the polymers effectively coat the zeolite fillers, and the in situ sol–gel method. The in situ sol–gel method is usually used to prepare silica-polymer MMMs. In this method, the inorganic fillers are formed in situ in the polymer matrix by adding the precursor solution of silica directly to the polymer-casting solution. However, the incorporation of silica appeared to reduce the crystallinity of the poly (phenylene oxide) (PPO)-silica MMMs and limited their improvement in gas selectivity. In general, MMMs are fabricated using semi-crystalline polymers, such as PPO, polyimide, and polyurethane as a matrix. These types of polymers exhibit good gas selectivity due to their crystalline molecular chain arrangement. However, the inorganic filler in the mixed matrix can disrupt the structural arrangement of the polymer chains, thus decreasing the crystallinity of the matrix. With low crystallinity, the MMMs are more prone to plasticization during high-pressure permeation by highly soluble gases, such as CO_2. Consequently, the selectivity of the MMMs suffers because of the decrease in the crystalline structure in the polymer matrix. In all the cases, NPs offer a significantly higher surface area-to-volume ratio which could be led to much greater extraction capacity and efficiency. The fabrication of conventional nanocomposite membranes using phase inversion (PI) process with the effect of nanofillers on final products and thin film nanocomposite (TFN) membranes through interfacial polymerization (IP) process are displayed and explained in Figs. 1 and 2.

In most of the applications, the NPs should necessarily be chemically stable, uniform in size, and well dispersed in the media. Silica has been reported to be a suitable supporting matrix to immobilize noble metal NPs such as silver NPs (AgNPs) and to avoid their aggregation. Synthesis of a nanocomposite of SiO_2

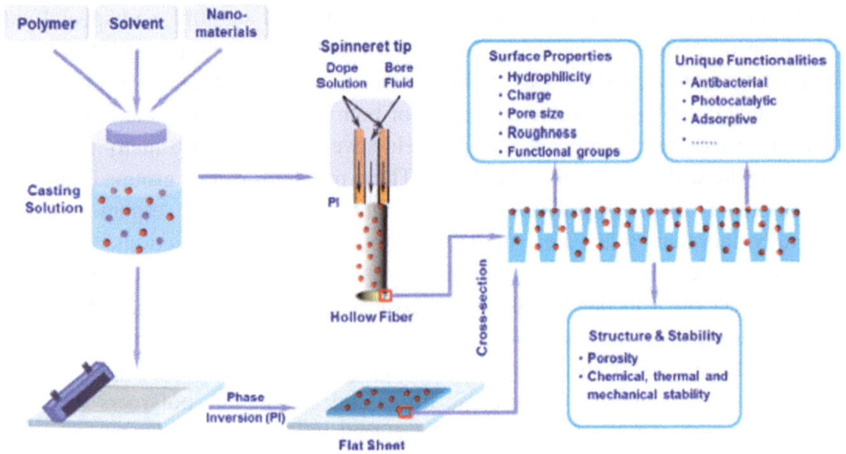

Fig. 1 Fabrication of conventional nanocomposite membranes through the PI process and the main effects of nanofillers on final products (reproduced from [1])

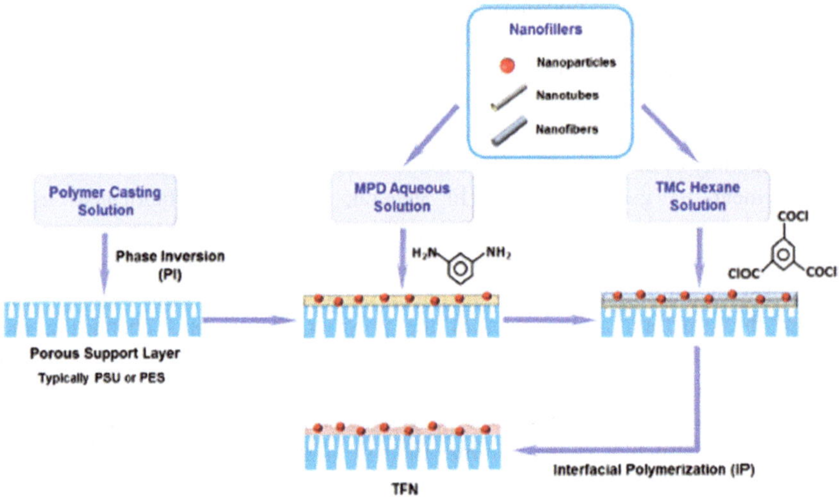

Fig. 2 Fabrication of TFN membranes through the IP process (reproduced from [1])

colloidal particles containing homogeneously dispersed AgNPs appears to be a useful approach. Also, SiO_2-conductive polymer could be used as a substrate to stabilize the AgNPs. Silica NPs were chosen as the fillers for the nanofibrous substrate due to their high thermal and chemical stabilities, non-toxicity, commercial viability, etc. The effects of different silica NP loadings on the membrane characteristics and their performances investigated by some researchers are summarized below.

2 Silica Nanoparticle-Based Membranes

Knoonsap et al. [2] prepared a nanocomposite membranes by incorporating poly (2-hydroxyethyl methacrylate) (PHEMA)-grafted FS particle onto the PVA membrane. Transmission electron microscopy (TEM) analysis confirmed the dispersion of FS particle by means of grafted PHEMA within the PVA matrix. Mechanical property of the prepared nanocomposite membrane in both dry and swollen states is shown in Fig. 3. Strong adhesion between the particles shifts Tg of the

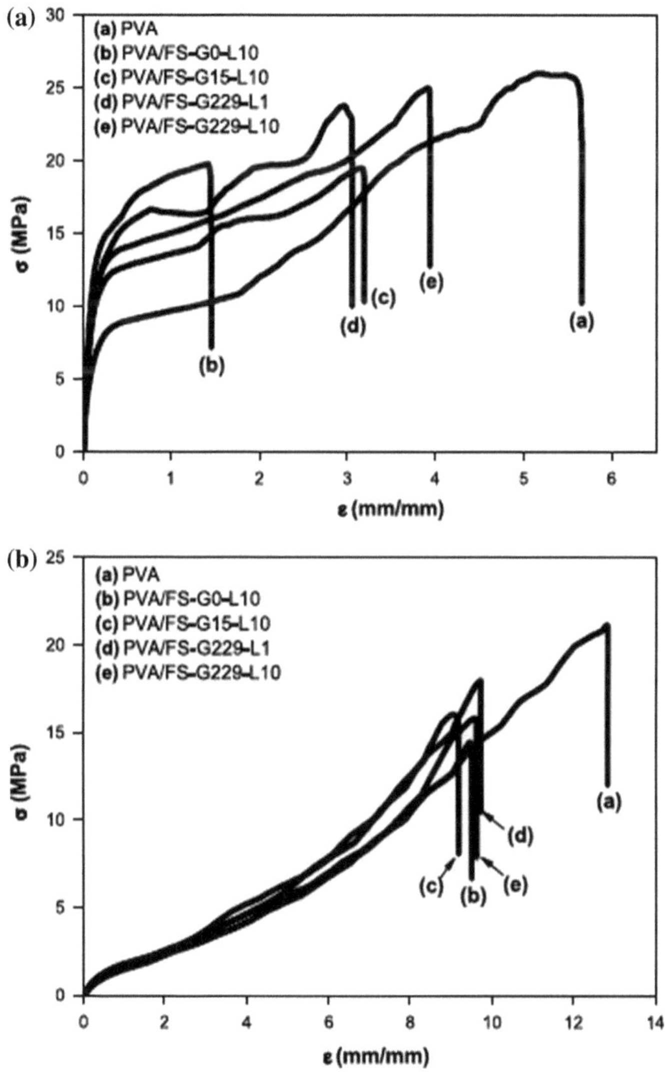

Fig. 3 Stress–strain curves of the nanocomposite membranes in dry state (**a**) and water-swollen state at 30 °C (**b**) (reproduced from [2])

nanocomposite membranes to higher temperature compared to neat. On complete swelling, its Tg decreased and dynamic swelling analysis shows enhanced water diffusion coefficient with increase in addition of FS content. Enhanced water uptake was observed compared to unfilled PVA membrane. Dehydration of water–acetone mixture was also studied by adopting pervaporation technique to enhance the water permeability in the presence of FS NPs.

The gas transport properties of poly (amide-6-b-ethylene oxide) (PEBAX) in the presence of silica were studied by a group of researchers. It revealed that the presence of silica in the polymer matrix has an important influence on the membrane morphology and this effect leads to favorable changes in the permeation properties of the membranes. Many other research teams also reported that in the presence of silica, the degree of crystallinity of polyamide section markedly decreases. Therefore, the percentage of the amorphous phase of PEBAX increases and the permeability of gases also enhanced with respect to silica content, as it was observed experimentally. The results of gas absorption in the composite membranes showed an enhanced uptake values with increase in the amount of silica. Two factors were introduced for such an increase in the absorption of gases in composite membranes containing silica: (i) the obtained changes in the morphology of interface between silica, polyamide, polyethylene oxide segments and ultimate increase in the percentage of the amorphous region in the composite membrane; and (ii) interaction of additional hydroxy groups of silica with polar molecules like CO_2.

Chen et al. used in situ polymerization of silica NPs during the preparation of polyether-based polyurethane/silica nanocomposites. First, silica NPs were added to the polyether polyol by in situ polymerization and then polymerization reaction was completed by isophorone diisocyanate (IPDI). Based on the obtained Fourier transform infrared (FTIR) spectra and thermogravimetric analyzer (TGA) thermograms, it was shown that in the case of in situ polymerization method, silica and polyether segments interact more strongly than that of simple blending method. This phenomenon further increases the Tg of hybrid polymers and the homogeneous distribution of silica NPs in the polymer. The obtained results also showed that the mechanical and physical properties of polyurethane-silica hybrids in the presence of silica particles (28 nm size) were the optimum. In addition, polyurethane-silica hybrids produced by the method of in situ polymerization had better mechanical properties than those made by simple blending method.

Chen et al. also investigated the effect of silica NPs on the mechanical and thermal properties of polyether-based polyurethane prepared by in situ polymerization. The obtained results showed that adding silica NPs into the polymer improved the hardness, Tg, and adhesion between different layers of polyurethane, but the resin viscosity increased.

Tian et al. [3] embedded silica NPs in poly (ethylenimine) (PEI) nanofibrous support for making thin film nanocomposite polyamide forward osmosis (FO) membrane. The results depicted that addition of silica NPs enhanced the pore size and porosity of the prepared FO membrane. Figure 4 displays the field emission (FE)-scanning electron microscopy (SEM) and cross-sectional image of the

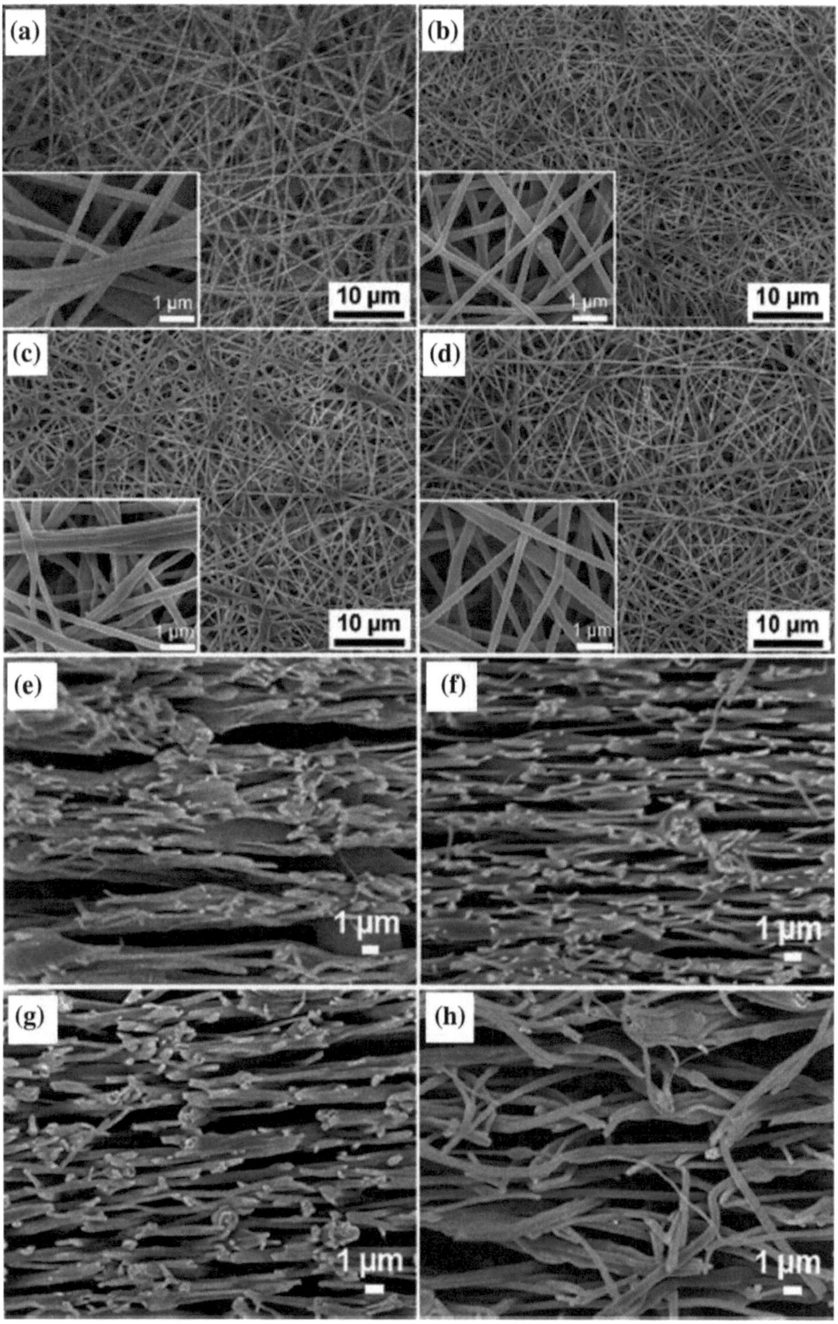

Fig. 4 FESEM micrographs showing the surface morphologies of the membrane substrates **a** control, **b** 1S, **c** 2S, **d** 3S, and the cross section of the substrate **e** control, **f** 1S, **g** 2S, and **h** 3S. The *insets* demonstrate a higher magnification (reproduced from [3])

developed nanocomposite membrane. The prepared membrane containing those substrates normally has higher osmotic water flux, which arise from the reduced structural parameter of the support layer. The silica loading of 1.6 wt% had the maximum substrate porosity, small structural parameter, and most promising water flux in the active layer. The active layer faced the draw solution orientation by applying draw solution and deionized feed water onto it.

Leung et al. [4] reported the preparation of sol–gel derived silica nanocomposite anion-exchange membrane for vanadium redox flow batteries (VRFBs). Energy-dispersive X-ray (EDX) spectroscopy and FTIR studies confirm the presence and uniformity of silica particles formed in the membrane via in situ sol–gel process. Ion-exchange capacity, area resistance, and water uptake were evaluated and compared with the pristine anion-exchange membrane (AEM) as well as with Nafion cation-exchange membrane. AEM shows inferior properties compared to pristine AEM and Nafion cation-exchange membrane (CEM). In addition, rates of self-discharge and capacity fading also showed lower values. The Coulombic and energy efficiencies were comparatively high as 92 and 73%, respectively. The comparison of permeability values of the four vanadium ions is shown in Fig. 5.

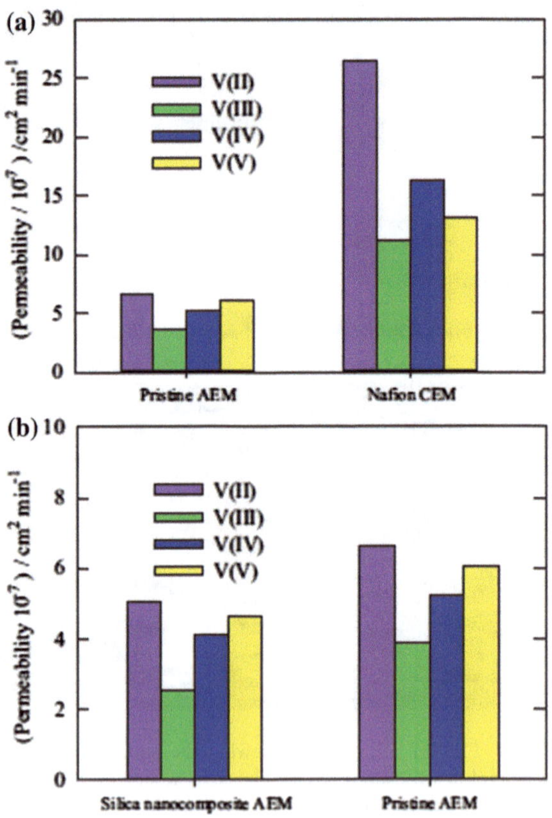

Fig. 5 Comparison of the permeability of the four vanadium ions through the membrane between (**a**) the Nafion CEM and the pristine AEM (**b**) and between the silica nanocomposite AEM and the pristine AEM (reproduced from [4])

Akbari et al. [5] synthesized the high dispersible hydrophobic PEG/vinyl silane-grafted silica NPs to fabricate protein-repellent PE nanocomposites. Fabrication of the same is a challenge for researchers in manufacturing biocompatible composites and low fouling membranes. The samples were prepared using one-pot/one-step sol–gel method. High density PE was used as a polymer to fabricate and analyze the protein-repellent property.

Nadargi et al. [6] reported a facile synthesis of super hydrophobic silica–iron oxide nanocomposites via a co-precursor sol–gel process. The use of silica precursor altered the pore structure.

Takahashi and Paul [7, 8] attempted to understand the effect of nanosized particles on the local properties and specifically permselectivity characteristics of glassy polymer matrix composites. As a result, matrix properties cannot be changed by the incorporation of filler. Nanocomposite membrane based on an amorphous, glassy PEI was developed using treated FS through solution casting and melt processing techniques. The presence of voids or defects increases the gas permeability and decreases the selectivity of the prepared membrane. The relationship between magnitude of voids and the permeation properties was examined in detail. As a whole, the technique for dispersing nanosized particles, FS in the polymer matrix, characterization of morphology of this mixture using TEM and SEM, and their local properties based on gas permeation were the most important while preparing nanocomposite membranes without chemical coupling.

The same group of researchers used coupling agent to improve the gas permeation properties of prepared nanosized SiO_2 incorporated polymer matrix membrane. The solubility coefficients computed by dividing the experimental permeability by the diffusivity obtained from observed time lag enhanced with the addition of silica content. TEM and SEM studies were used to examine the morphology of these prepared nanocomposite membranes.

Maity et al. [9] studied the effect of surface modifier molecules on morphological and structural changes of poly (oxyphenylene benzimidazole) (OPBI) nanocomposite membrane which is prepared by solution blending method. This would enhance the proton-exchange membrane properties like acid retention capability. The chemical interactions between them were also analyzed using FTIR, nuclear magnetic resonance (NMR) spectroscopy, and wide-angle X-ray diffraction (WAXD) studies. The developed composites displayed high thermal, mechanical, and oxidative stabilities, whereas TEM disclosed the formation of self-assembled clusters in the OPBI matrix. The incorporation of 15% ionic liquid-modified silica (ILMS) showed two-fold increment in proton conductivity compared to neat OPBI and this will attribute to the self-assembled clusters of ILMS NP in the matrix.

Devrim and Devrim [10] developed an air-cooled PEMFC containing Nafion/silica nanocomposite membrane to improve the stack performance at relative humidity conditions. SEM image, EDX spectra, and silica mapping of the Nafion-silica composite membrane are shown in Fig. 6. The samples were prepared using solution casting method and gas diffusion electrodes were produced by ultrasonic spray coating method. The short stack was tested in dead-end mode with controlling temperature by air on–off control system.

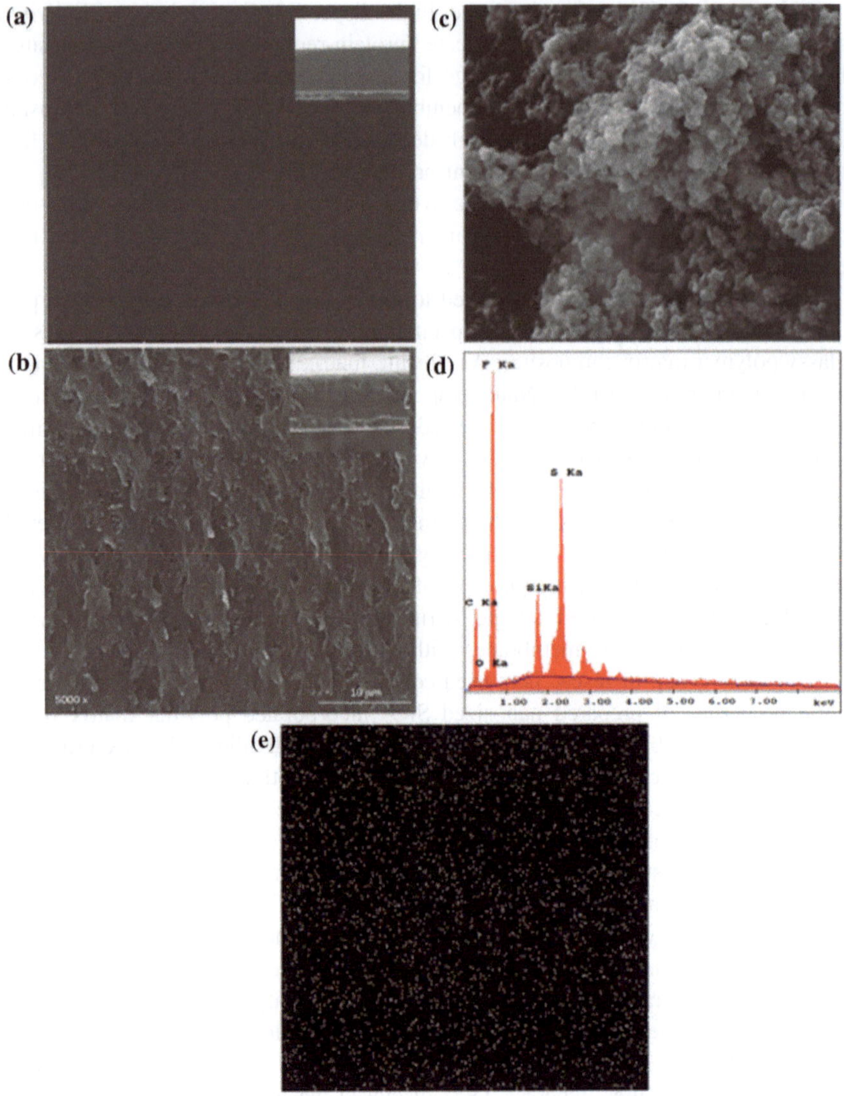

Fig. 6 SEM image of the **a** Nafion membrane, **b** silica particle, **c** Nafion/silica composite membrane, **d** EDX spectra of the Nafion/silica composite membrane, and **e** silica mapping of Nafion/silica composite membrane (reproduced from [10])

Ghadimi et al. [11] investigated the separation performance of poly (ether block amide) (PEBA) membrane by incorporating silica NPs into it. The chemical surface modification was performed on silica NPs via esterification reaction. Permeation, sorption, and diffusion coefficients through the prepared neat and nanocomposite membranes were studied. Increase in loading of NPs content enhanced the ideal permeation selectivity to 17, 45, and 124, respectively.

Roelofs et al. in [12] developed interconnected silica network using Aerosil 380, tetraethyl orthosilicate (TEOS) or a combination of both to prepare and characterize sulfonated polyether ether ketone (PEEK)-based mixed matrix membranes. The thermal characteristics of the membranes with A-system were analyzed using TGA and presented in Fig. 7. The behavior was studied using ethanol–water systems and directly applied in an ethanol fuel cell. Proton conductivity of the samples was strongly correlated with the water content in the membrane. With respect to the polymeric reference, enhanced fuel cell performance was observed based on ratio of proton diffusion coefficient to ethanol permeability coefficient, whereas the presence of inorganic phase lead to decrease in their value for the same.

Zeng et al. [13] analyzed the intrinsic relationship among proton conductivity, thermal stability, and structural symmetries for PWA-meso-silica nanocomposite membranes and their performance in direct methanol fuel cells. The 3D mesostructures displayed higher proton conductivity and higher stability as a function of relative humidity compared to 2D mesostructures.

Naghsh et al. [14] analyzed the performance of cellulose acetate silica nanocomposites membranes in the separation of ethylene/ethane and propylene/propane. Silica NPs were prepared via hydrolysis of TEOS and membranes were fabricated using solution casting method. The prepared membranes were characterized using FTIR, SEM, and TGA techniques. Pure gas permeation experiments were done which showed enhanced permeability with the addition of 30 wt% silica particles onto the composite membranes. Time lag method was used to determine the diffusion coefficient, whereas solubility coefficient was calculated indirectly from permeability and diffusion coefficients. Addition of more amount of silica mass fraction increases the solution coefficient and decreases the diffusion coefficient. No plasticization effects were found throughout the experiments.

Huang et al. [15] synthesized mesoporous silica particles as inorganic fillers and fabricated polyethersulfone (PES) nanocomposite membranes having antifouling properties by phase inversion method. The chemical group's presents in PES and PES-mesoporous silica membrane were investigated using FTIR spectroscopy

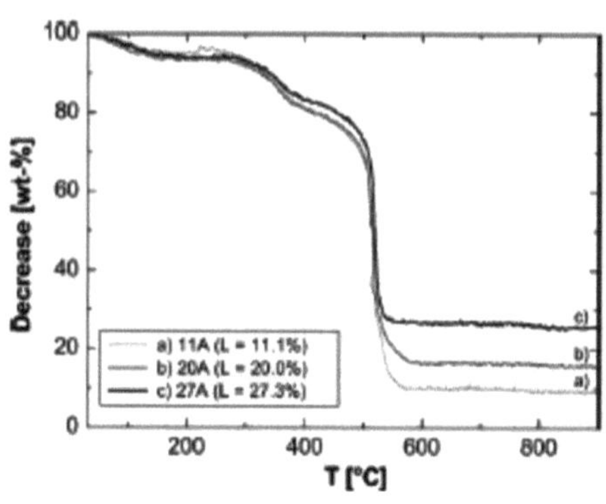

Fig. 7 TGA curves for membranes of the A-system (reproduced from [12])

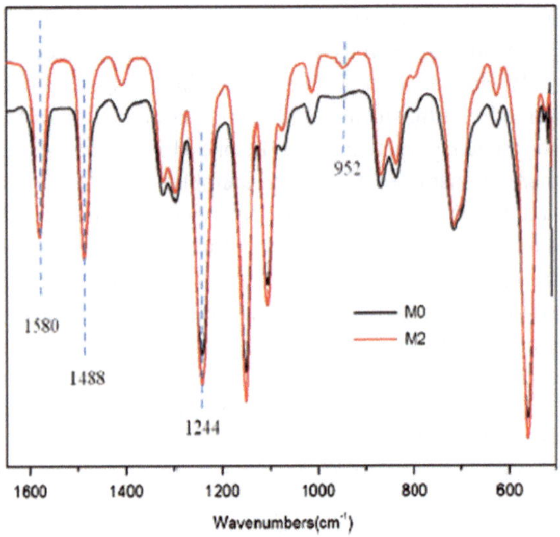

Fig. 8 FTIR spectra of pure PES (M0) membrane and PES-mesoporous silica nanocomposite membrane (M2) (reproduced from [15])

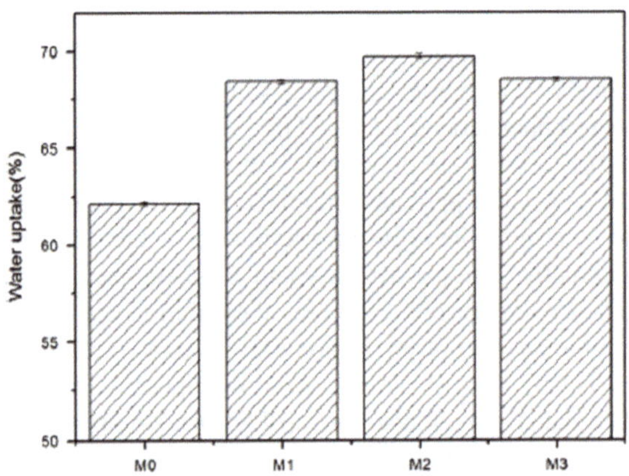

Fig. 9 Water uptake of PES membrane with different mesoporous silica loadings (reproduced from [15])

(Fig. 8). The effects of silica particles on the microstructure and properties of hybrid membranes were investigated by SEM, TGA, and ultrafiltration (UF) experiments. The membranes with 2% silica depicted excellent hydrophilicity, water uptake (Fig. 9), water permeability, thermal stability, and good antifouling performance. The protein adsorption on the membrane surface decreased when the silica content increased to 2%. No benefit was gained from higher silica content (4%) which resulted in particle agglomeration.

Treekamol et al. [16] adopted melt extrusion and solvent casting processes to prepare functionalized silica-incorporated Nafion nanocomposite membranes for medium-temperature application in PEMFC. Ion-exchange capacity, conductivity, water uptake, dimensional stability, thermal stability, and morphology were examined. Addition of silica beyond 5 wt% causes filler agglomeration. The prepared composite membranes displayed excellent thermal stability but decreased fuel cell performances and higher hydrogen crossover which may be possibly due to porosity resulting from suboptimal particle-matrix compatibility. Addition of 10 wt% silica and experiments at 34% relative humidity at 90 °C resulted good fuel cell performance for the prepared nanocomposite membranes.

Tancharernrat et al. [17] prepared styrene-butadiene copolymer-silica nanocomposites via differential microemulsion polymerization and NR/SBR–SiO_2 membranes for pervaporation of water–ethanol mixtures. Modified SiO_2 was completely encapsulated with SiO_2 as core and SBR as shell. High conversion and small particle size were obtained at low sodium dodecyl sulfate (SDS) concentration. NR/SBR–SiO_2 composites could be used as a novel membrane for pervaporation. NR/SBR–SiO_2 membranes had high mechanical properties and water selectivity.

Torabi and Ameri [18] attempted to produce methyl acetate by coupled esterification reaction process using synthesized cross-linked PVA/silica nanocomposite membranes. Solution casting technique was used to prepare the samples and coupled pervaporation reaction using fumaric acid as cross-linking agent was used to dehydrate the reaction mixture. The membranes were characterized using FTIR, SEM, TGA, and X-ray diffraction (XRD). SEM results revealed nanoscale distribution of silica particles in the polymer matrix. Increase in silica content showed reduction in crystallinity of the samples. The effects of different experimental variables revealed enhancement in acid conversion and water permeance throughout the membranes with the increasing silica content of the membranes.

Huang et al. [19] prepared poly (methyl methacrylate) (PMMA)-silica mesocomposite and nanocomposite membranes. The comparative studies on the preparation and physical properties of resultant membranes were investigated in detail. Vinyl-modified mesopore silica and silica particles were used in this study. PMMA silica mesocomposite was found to show enhanced mechanical strength, thermal stability, thermal insulation, gas permeability, optical clarity, surface hydrophobic, and surface roughness properties compared to PMMA silica nanocomposite based on the studies of dynamic mechanical analysis (DMA), differential scanning calorimetry (DSC), TGA, Hot Disk, UV-vis transmission microscopy, contact angle measurements, and atomic force microscopy (AFM).

Liu et al. [20] attempted in situ incorporation of modified silica during ionic complexation between carboxymethyl cellulose (CMC) and poly (2-methacryloyloxy ethyl trimethyl ammonium chloride) (PDMC) to prepare CMC-modified silica nanocomposites. PEC nanocomposites and their membranes were characterized with FTIR, TGA, SEM, and tensile testing. The mechanical properties were superior to pristine PEC membranes.

Li et al. [21] fabricated semi-aromatic polyamide/spherical mesoporous silica nanocomposite reverse osmosis membrane (Fig. 10) with superior permeability. The experimental results revealed that the microstructures and surface features are significantly different, including the surface morphology (Fig. 11), polymer framework structure, surface charge density, hydrophilicity, and the thickness of barrier layer. The excellent separation performance of the nanocomposite reverse

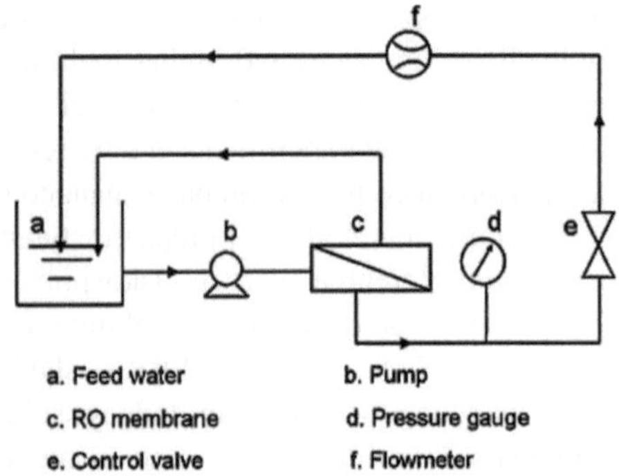

Fig. 10 Schematic of RO membrane performance testing apparatus (reproduced from [21])

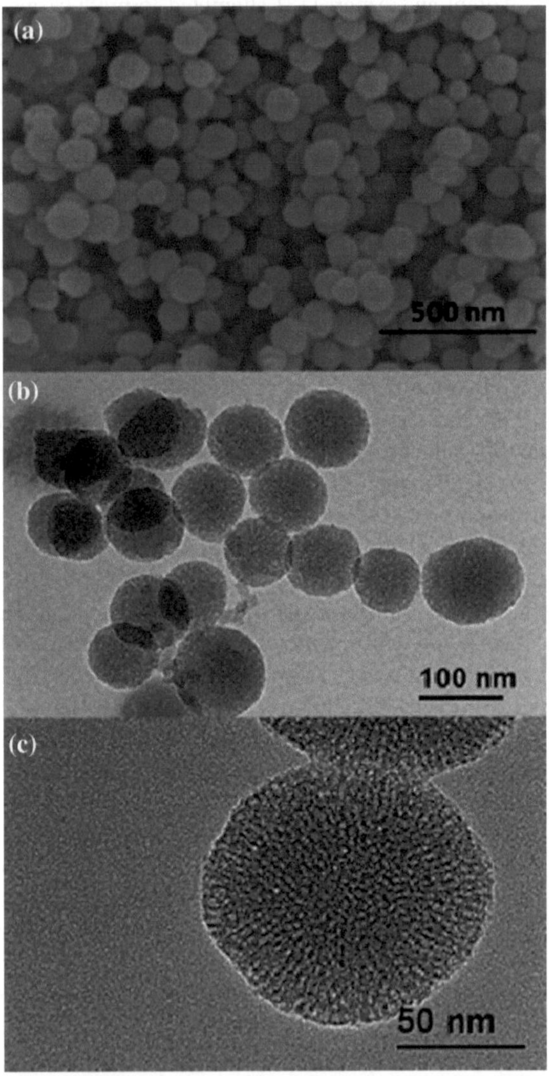

Fig. 11 **a** SEM and **b** and **c** TEM images of the synthesized mesoporous silica spheres (reproduced from [21])

osmosis (RO) membrane was attributed to the higher freedom-degree semi-aromatic framework, the incorporation of organic solution, the improved surface hydrophilicity, the thinner barrier layer, and the enhanced surface negative charge density.

Deng and coworkers synthesized the aromatic polyamide/porous-silica-NP nanocomposite membrane by IP method and the results demonstrated the increase in permeate flux from 28.5 to 46.6 L/m^2h with the incorporation of the silica NPs and the salt rejection can be stably maintained at a relatively high level. It has been reported that the incorporation of nanomaterials (silica, zeolite, carbon nanotube, etc.) can increase the surface hydrophilicity of polyamide RO membrane, optimize the polymer cross-linking degree, change the surface charge distribution, and even introduce certain drawbacks on interface between the incorporated nanomaterials and polymer membrane. These factors are all favorable for the increase of permeability of aromatic polyamide RO membrane. In comparison with other traditional nanomaterial additives employed in RO membrane fabrication, mesoporous silica spheres synthesized using cetyl trimethyl ammonium bromide (CTAB) as soft template not only have fine hydrophilicity, but also controllable spherical morphology (90–750 nm) and abundant mesopore channels (1.5–3.6 nm). As a result, the mesoporous silica-incorporated RO membranes usually possess higher permeate flux and comparable selectivity compared with other nanocomposite RO membranes. However, there are only few reports on the fabrication of polyamide RO membranes containing mesoporous silica.

Seck et al. [22] synthesized poly (vinylidene fluoride-hexafluoropropylene) (PVDF-HFP)/thiol functionalized silica nanocomposite membranes through reactive extrusion and in situ sol–gel reactions in the presence of inorganic precursor polydimethoxysiloxane (PDMOS) and mercapto propyl triethoxy silane (MPTES). Morphology of the prepared membrane was examined by SEM. By controlling the oxidation reaction of the thiol groups into sulfonic acid functions, promising ion-exchange capacity and conductivity values were obtained.

Ahsani and Yegani [23] adopted thermally induced phase separation method to prepare microporous PP membrane. Hydrophilic SiO_2 NPs were synthesized by in situ and ex situ approaches using Stober method. Addition of silica NPs onto the membrane enhanced the surface hydrophilicity, pure water flux, and mechanical properties, whereas reversible fouling ratio increased due to higher flux recovery of the membrane. The combined fouling models revealed that cake filtration standard blockage fouling mechanism was operative for all membranes.

Zhuang et al. [24] developed PPO-silica mixed matrix membranes via in situ sol–gel method for hydrogen (H_2)/CO_2 separation. The density and crystallinity properties of the same on the effect of thermal annealing treatment were characterized and investigated in detail through TGA, XRD, FE-SEM, and FTIR. Annealing method improves the transport properties of developed PPO based membranes and induce polymer chains to recrystallize. Addition of 5 wt% silica enhanced the gas separation, thermal stability, and plasticization resistance after the annealing process.

Nambi Krishnan et al. [25] attempted to develop sulfonated PES-based silica nanocomposite membranes for high-temperature polymer electrolyte fuel cell applications. The preparation method is illustrated in Fig. 12. The SEM analyses depict reduced anisotropy and TEM proved well-embedded NPs in the polymer matrix. Small-angle X-ray scattering (SAXS) analysis determined the separation length between the ion-rich domains. Increase in addition of silica enhanced the tensile strength and Young's modulus of the prepared membrane. Enhanced ex situ in plane proton conductivity was displayed 3.4 times higher current density than pure PES 70.

Salarizadeh et al. [26] investigated the preparation, characterization, and properties of proton-exchange nanocomposites membranes based on PVA and polysulfonic acid-grafted silica NPs. The samples were prepared using solution casting process and characterized using FTIR, impedance spectroscopy, TGA, water uptake, tensile strength, and SEM analysis. Addition of 5 wt% NPs enhanced the higher proton conductivity of the polymer electrolyte membranes.

Shahabadi et al. [27] developed polymer electrolyte nanocomposites membranes by adding silica NPs modified through surface-initiated atom transfer radical polymerization. FTIR, TGA, SEM, and X-ray mapping were used to characterize the prepared membranes. Water uptake, proton conductivity, and ion-exchange

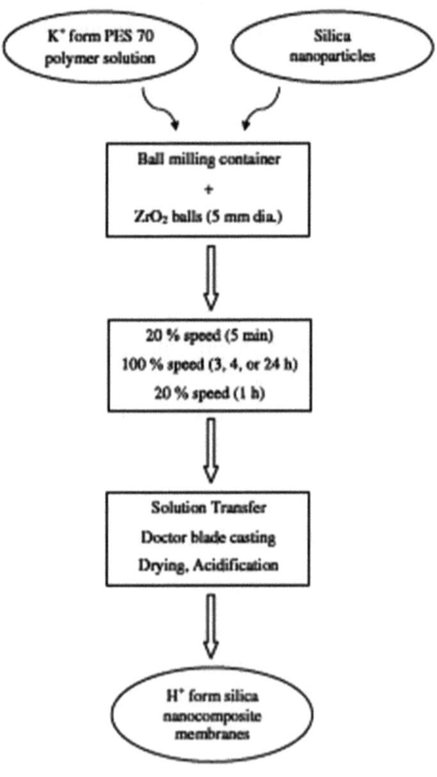

Fig. 12 Preparation method of sulfonated poly (ethersulfone)-based silica nanocomposite membranes (reproduced from [25])

capacity of the prepared PVA-based membranes enhanced by adding the silica NPs. Hydrophilic nature and grafting efficiency of the sulfonated hydrophilic monomers played an important role in the proton conductivity of the PVA-based membranes. Adding silica NPs onto the PVA matrix enhanced the mechanical properties and decreased the methanol uptake and permeability for PVA-based polymer electrolyte membranes.

Suryani [28] prepared and analyzed the properties of polybenzimidazole (PBI)/sulfonated silica nanocomposites membranes for proton exchange. The chemical structure of sulfonated silica NPs (SA-SNP) and PBI are shown in Fig. 13.

The compatibility between SA-SNP and PBI was increased to enhance the mechanical strength of the prepared nanocomposite membranes. Figure 14 explains the water uptake property of PBI-silica and PBI-sulfonated silica nanocomposite membranes at 25–80 °C. The selectivity is about 1.3-folds that of Nafion 117 membrane as evident from Fig. 15. The obtained results indicated that the prepared nanocomposite membranes could be utilized as the proton-exchange membrane for direct methanol fuel cells.

Liu et al. [29] developed novel nanocomposite membranes based on sulfonated mesoporous silica NPs as inorganic nanofillers, modified with sulfonated polyimides for direct methanol fuels. The microstructure and properties of the resulting hybrid membranes were studied. The addition of silica improved the thermal stability, water uptake, and methanol permeability of the resulting nanocomposite membranes as compared with sulfonated polyimides. Addition of more than 3 wt% causes aggregation and decreased their performance. The high proton conductivity at different testing temperatures was observed on incorporation of 7 wt% into the sulfonated polyimides. It indicates that NP addition improves the proton conductivity of composite membranes.

Sadeghi et al. [30] investigated the gas separation properties of polyether-based polyurethane-silica nanocomposite membranes. Silica NPs were prepared through sol–gel method by hydrolysis of TEOS. Polyurethane and polyurethane-silica nanocomposite membranes were prepared by solution blending and casting method. The membranes were characterized using FTIR, TGA, and DSC analysis. Addition of 20 wt% of silica particles in polyurethane matrix showed reduction in CO_2

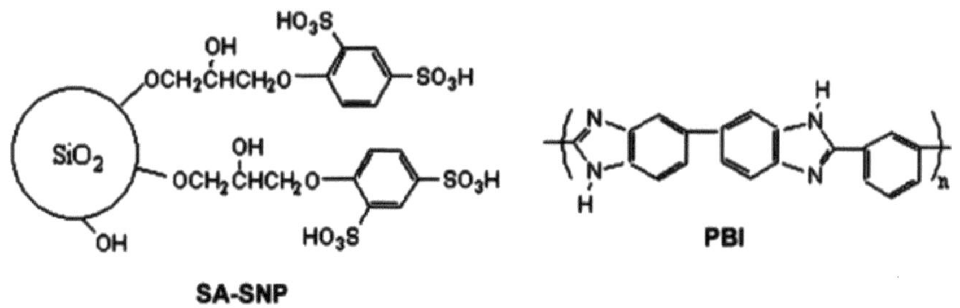

Fig. 13 Chemical structure of sulfonated silica NPs (SA-SNP) and PBI (reproduced from [28])

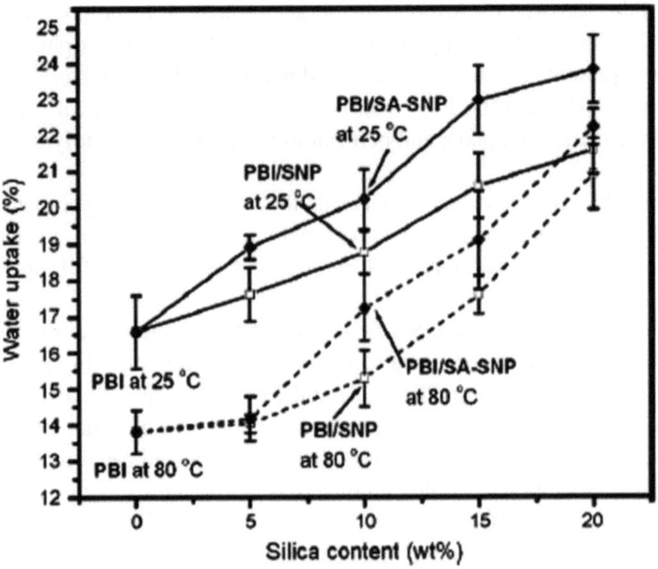

Fig. 14 Water uptake of the PBI nanocomposite membranes of (i) PBI, silica NPs, PBI/silica NPs and (ii) PBI, SA-SNP, PBI/SA-SNP at 2580 °C (reproduced from [28])

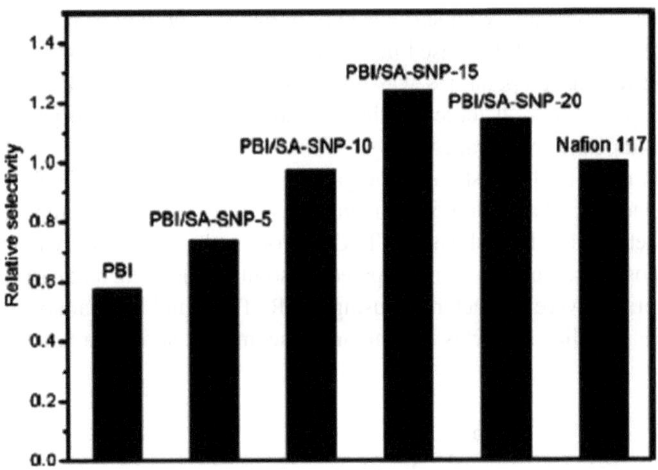

Fig. 15 The relative selectivity (over the Nafion 117 membrane) of membranes of PBI, SA-SNP, and PBI/SA-SNP (reproduced from [28])

permeability compared to pure polyurethane. Author also applied modified Higuchi model to predict the permeability and fairly good agreement was observed between experimental and predicted gas permeability values for these membranes.

3 Conclusion

The synthesis of organic-silica nanocomposite membranes is explained and their corresponding structure and properties are discussed. The chemical structure modification and grating of silica NPs alter the technical property of the prepared nanocomposite membranes. Many attempts had been made by several researchers to overcome the drawbacks of organic-silica-based membranes. The role of membrane structure, particle size, particle distribution, and surface properties in membrane performance has also been clearly explained.

References

1. Yin J, Deng B (2015) Polymer-matrix nanocomposite membranes for water treatment. J Membr Sci 479:256–275
2. Khoonsap S, Supanchaiyamat N, Hunt AJ et al (2015) Improving water selectivity of poly (vinyl alcohol) (PVA)—fumed silica (FS) nanocomposite membranes by grafting of poly (2-hydroxyethyl methacrylate) (PHEMA) on fumed silica particles. Chem Eng Sci 122: 373–383
3. Tian M, Yi-NingWang RW et al (2016) Synthesis and characterization of thin film nanocomposite forward osmosis membranes supported by silica nanoparticle incorporated nanofibrous substrate. Desalination. doi:10.1016/j.desal.2016.04.003
4. Leung PK, Xu Q, Zhao TS et al (2013) Preparation of silica nanocomposite anion-exchange membranes with low vanadium-ion crossover for vanadium redox flow batteries. Electrochim Acta 105:584–592
5. Akbari A, Yegani R, Pourabbas Behzad (2016) Synthesis of high dispersible hydrophilic poly (ethylene glycol)/vinyl silane grafted silica nanoparticles to fabricate protein repellent polyethylene nanocomposite. Eur Polym J 81:86–97
6. Nadargi D, Gurav J, Marioni M et al (2015) Methyltrimethoxysilane (MTMS)-based silica–iron oxide superhydrophobic nanocomposites. J Colloid Interface Sci 459:123–126
7. Takahashi S, Paul DR (2006) Gas permeation in poly(ether imide) nanocomposite membranes based on surface-treated silica. Part 1: without chemical coupling to matrix. Polym 47:7519–7534
8. Takahashi S, Paul DR (2006) Gas permeation in poly(ether imide) nanocomposite membranes based on surface-treated silica. Part 2: with chemical coupling to matrix. Polym 47:7535–7547
9. Maity S, Singha S, Jana T (2015) Low acid leaching PEM for fuel cell based on polybenzimidazole nanocomposites with protic ionic liquid modified silica. Polym 66:76–85
10. Devrim Y, Devrim H (2015) PEM fuel cell short stack performances of silica doped nanocomposite membranes. Int J Hydrogen Energy 40:7870–7878
11. Ghadimi A, Mohammadi T, Kasiri N (2015) Gas permeation, sorption and diffusion through PEBA/SiO_2 nanocomposite membranes (chemical surface modification of nanoparticles). Int J Hydrogen Energy 40:9723–9732
12. Roelofs KS, Hirth T, Schiestel T (2010) Sulfonated poly(ether ether ketone)-based silica nanocomposite membranes for direct ethanol fuel cells. J Membr Sci 346:215–226
13. Zeng J, Shen PK, Shanfu Lu et al (2012) Correlation between proton conductivity, thermal stability and structural symmetries in novel HPW-meso-silica nanocomposite membranes and their performance in direct methanol fuel cells. J Membr Sci 397:92–101

14. Naghsh M, Sadeghi M, Moheb A et al (2012) Separation of ethylene/ethane and propylene/propane by cellulose acetate–silicanano composite membranes. J Membr Sci 423:97–106
15. Huang J, Zhang K, Wang K et al (2012) Fabrication of polyethersulfone-mesoporous silica nanocomposite ultrafiltration membranes with antifouling properties. J Membr Sci 423: 362–370
16. Treekamol Y, Schieda M, Robitaille L et al (2014) Nafion®/ODF-silica composite membranes for medium temperature proton exchange membrane fuel cells. J Power Sour 246:950–959
17. Tancharernrat T, Rempel GL, Prasassarakich Pattarapan (2014) Preparation of styrene butadiene copolymer–silica nanocomposites via differential micro emulsion polymerization and NR/SBR–SiO_2 membranes for pervaporation of water–ethanol mixtures. Chem Eng J 258:290–300
18. Torabi B, Ameri E (2016) Methyl acetate production by coupled esterification-reaction process using synthesized cross-linked PVA/silica nanocomposite membranes. Chem Eng J 288:461–472
19. Huang K-Y, Weng C-J, Huang L-T et al (2010) Systematically comparative studies on the preparation and physical properties of PMMA–silica mesocomposite and nanocomposite membranes. Micropor Mesopor Mater 131:192–203
20. Liu T, An Q-F, Wang X-S et al (2014) Preparation and properties of PEC nanocomposite membranes with carboxymethyl cellulose and modified silica. Carbohyd Polym 106:403–409
21. Li Q, Hui Y, Feiyang W et al (2016) Fabrication of semi-aromatic polyamide/spherical mesoporous silica nanocomposite reverse osmosis membrane with superior permeability. Appl Surf Sci 363:338–345
22. Seck S, Magana S, Prebe A et al (2015) PVDF-HFP/silica-SH nanocomposite synthesis for PEMFC membranes through simultaneous one-step solegel reaction and reactive extrusion. Mater Chem Phys 163:54–62
23. Ahsani M, Yegani R (2015) Study on the fouling behavior of silica nanocomposite modified polypropylene membrane in purification of collagen protein. Chem Eng Res Des 102: 261–273
24. Zhuang G-L, Wey M-Y, Tseng H-H (2015) The density and crystallinity properties of PPO-silica mixed-matrix membranes produced via the in situ sol-gel method for H_2/CO_2 separation. II: effect of thermal annealing treatment. Chem Eng Res Des 104:319–332
25. Nambi Krishnan N, Henkensmeier D, Jang Jong Hyun et al (2011) Sulfonated poly (ether sulfone)-based silica nanocomposite membranes for high temperature polymer electrolyte fuel cell applications. Int J Hydrogen Energy 36:7152–7161
26. Salarizadeh P, Javanbakht M, Abdollahi M et al (2013) Preparation, characterization and properties of proton exchange nanocomposite membranes based on poly(vinyl alcohol) and poly(sulfonic acid)-grafted silica nanoparticles. Int J Hydrogen Energy 38:5473–5479
27. Shahabadi R, Abdollahi M, Sharif A (2015) Preparation, characterization and properties of polymer electrolyte nanocomposite membranes containing silica nanoparticles modified via surface-initiated atom transfer radical polymerization. Int J Hydrogen Energy 40:3749–3761
28. Suryani Y-LL (2009) Preparation and properties of nanocomposite membranes of polybenzimidazole/sulfonated silica nanoparticles for proton exchange membranes. J Membr Sci 332:121–128
29. Liu D, Geng L, Yuqin F et al (2011) Novel nanocomposite membranes based on sulfonated mesoporous silica nanoparticles modified sulfonated polyimides for direct methanol fuel cells. J Membr Sci 366:251–257
30. Sadeghia M, Semsarzadeh MA, Barikani M et al (2011) Gas separation properties of polyether-based polyurethane–silica nanocomposite membranes. J Membr Sci 376:188–195

Chapter 4
Organic/Zeolites Nanocomposite Membranes

Eliana B. Souto, Patrícia Severino, Patrícia Hissae Yassue-Cordeiro, Romilda Fernandez Felisbino, Eliezer Ladeia Gomes and Classius Ferreira da Silva

Abstract There are different types of fuel cells, but Polymer Electrolyte Membrane (PEM) Fuel Cells are one of the most promising because they are low-temperature fuel cells. The solid polymer electrolyte has to present transport selectivity, besides the ion conductivity; thus in case of using in fuel cells, the membrane must let the hydrogen ions pass easily and block the passage of fuel (methanol or hydrogen) as

---

E.B. Souto (✉)
Department of Pharmaceutical Technology, Faculty of Pharmacy, University of Coimbra, Pólo das Ciências da Saúde - Azinhaga de Santa Comba, 3000-548 Coimbra, Portugal
e-mail: ebsouto@ff.uc.pt; souto.eliana@gmail.com

P. Severino
Laboratory of Nanotechnology and Nanomedicine, Institute of Technology and Research, University of Tiradentes, Av. Murilo Dantas, 300, Aracaju, SE CEP 49010-390, Brazil
e-mail: pattypharma@gmail.com

P.H. Yassue-Cordeiro
Department of Chemical Engineering, Laboratório de Catálise, Cinética e Reatores Químicos (LCCRQ), State University of Maringá, Avenida Colombo 5790 - Bloco D-90, Maringá CEP 87020-900, Brazil
e-mail: patricia.yassue@gmail.com

R.F. Felisbino · C.F. da Silva
Biotechnology Laboratory of Natural Products, Institute of Environmental Sciences, Chemical and Pharmaceutical, Federal University of São Paulo, Rua São Nicolau, 210, Diadema, SP CEP 09913-030, Brazil
e-mail: fernandez.romilda@unifesp.br

C.F. da Silva
e-mail: cfsilva@unifesp.br

E.L. Gomes · C.F. da Silva
Department of Chemical Engineering, Institute of Environmental Sciences, Chemical and Pharmaceutical, Federal University of São Paulo, Rua Prof. Artur Riedel, 275, Jd. Eldorado, Diadema CEP 09972-270, Brazil
e-mail: eliezer.ladeia@unifesp.br

E.B. Souto
REQUIMTE/LAQV, Group of Pharmaceutical Technology, Faculty of Pharmacy, University of Coimbra, Coimbra, Portugal

Inamuddin et al. (eds.), *Organic–Inorganic Composite Polymer Electrolyte Membranes*, DOI 10.1007/978-3-319-52739-0_4

well as oxidant molecules (oxygen) which have to be kept separated from each other. In addition to these properties, list other important properties that PEMs must show for high performance: low electronic conductivity, low water transport through diffusion and electro-osmosis, oxidative and hydrolytic stability, good mechanical stability in both dry and hydrated states, low cost, and capability for fabrication into membrane electrode assembly (MEA). This chapter is divided into seven sections. The first section presents some statistical data of publications concerning fuel cell, PEM fuel cells (PEMFC) and PEMFC with zeolites. The second section exhibits some concepts about zeolites types, structure, properties and industrials applications. The third section presents the role of the zeolite properties on the performance of the PEMFC. The fourth section describes the main technique used for producing zeolite/polymer nanocomposite membrane for PEMFCs. The two following sections outline the state of the art of using the zeolite for PEMFC applications, being the fifth and sixth sections dedicated to the synthetic and natural polymers, respectively.

Keywords Fuel cells · Zeolites · Nanocomposites · Polymorphs · Nanostructures

Abbreviations

APTS	3-Aminopropyltrimethoxysilane
DMFC	Direct methanol fuel cells
IZA-SC	International Zeolite Association Structure Commission
MEA	Membrane electrode assembly
MPD	Maximum power density
OCV	Open circuit voltage
PEEK	Poly(ether ether ketone) or nafion
PEM	Polymer electrolyte membrane
PEMFC	Polymer electrolyte membrane fuel cells
PES	Polyethersulfone
PFSA	Perfluorosulfonic acid
PTCS	Phenyltrichrolosilane
PTFE	Polytetrafluoroethylene
SPAES	Poly(arylene ether sulfone)
S-PEKES	Sulfonated poly(ether ketone ether sulfone)
SPK-ZrP	Sulfonated poly(ether ketone)-zeolite-zirconium hydrogen phosphate
SSM	Stainless steel mesh

1 Introduction

The number of academic papers about fuel cells has increased significantly in recent years. Figure 1 shows the number of scientific articles resulting from searches conducted on the Scopus® that is the most outstanding abstract and citation database of peer-reviewed literature. The search field chosen to obtain the data was article title, abstract, and keywords. We also restricted the search to the scientific journals and conference proceedings in the English language between 1996 and 2015, i.e., during the last 20 years. Moreover, we improved our search by using "*" as a wildcard to find plural and inflected forms of words; then the first search was Fuel Cell* which could be "fuel cell" or "fuel cells." In the last 20 years, the number of papers jumped from 712 in 1996 to 7810 in 2015, i.e., an increasing of more than ten times (Fig. 1). The total number of papers was 85607. By refining the search, it was included another term to the search field that was "Fuel Cell* and PEM*". The term PEM refers to the Polymer Electrolyte Membrane, and the "*" was used to increase to find the plural word (PEMs) or even inflected words like PEMFC (Polymer Electrolyte Membrane Fuel Cell). This search presented the same tendency, and the numbers jumped from 19 in 1996 to 1590 in 2015, i.e., an increasing of more than eighty times. On the other hand, the total number of papers in this refining search was 17,871; that is about 21% of the whole papers about Fuel Cells.

The inset Figure shows the last search that was "Fuel Cell* and PEM* and Zeolite*". The number of papers was null between 1996 and 2000; the first paper was just found in 2001. Moreover, there was one record with this search in 2011 that was nine papers. It is important to mention that the USA, China, and Japan, together, have about half of the papers for the first (Fuel Cell) and second search (Fuel Cell and PEM). On the other hand, Italy and South Korea got the first position for the third search when the term zeolite was added. Finally, USA and China got the second and third place, respectively.

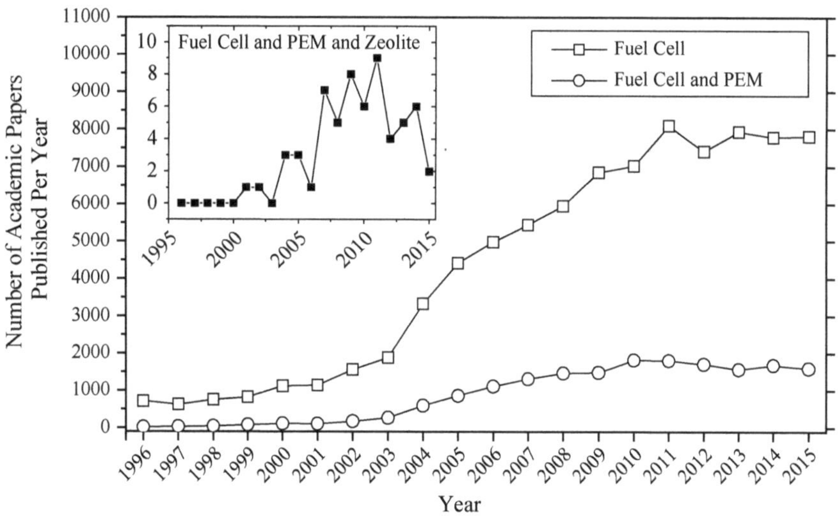

Fig. 1 Number of academic papers published per year according to Scopus database

It is known that fuel cell (PEMFC) works well when hydrogen is used as fuel, producing only electricity and water. However, as it has been presented earlier, it is difficult to store hydrogen, be it as compressed gas, cryogenic liquid, metal hydride, or carbon nanotubes [1]. The construction of a cell that directly uses methanol as organic fuel, instead of previously converting it to hydrogen through industrial reforming, is a very promising alternative compared to the trivial solution. Methanol oxidation potential (E° = 1.21 V) is comparable to hydrogen potential (E° = 1.23 V) and methanol volume energy density is 50% higher [2], which consists as a great incentive for researching the so-called DMFCs (Direct Methanol Fuel Cells).

The matrix of PEM membranes is usually composed of solid polymer, known as Nafion, which is an efficient insulating for free electrons, a good proton conductor (1×10^{-1} S cm^{-1}, saturated in water, room temperature) [3], and adaptable and resistant to most chemical products. However, Nafion loses its proton conducting properties when submitted to temperatures over 100 °C, which is the temperature regime that provides the best cathode kinetics. Another significant disadvantage to be overcome is that Nafion membranes used in these DMFCs allow methanol crossover from the anode to cathode. This crossover results in two main effects: first, methanol chemical energy is reduced when it percolates the membrane, drastically reducing its efficiency when used as fuel. Second, the two simultaneous cathode electrochemical reactions directly compete for cathode catalytic sites, reducing the global efficiency of the cell. Currently, methanol crossover limits feed concentration from 0.5 to 1.0 mol L^{-1} of methanol. By increasing methanol feed concentration, it would be possible to enhance methanol oxidation kinetics [4]. In such context, there is a quest for high performance and low-cost polymeric materials that enable high-ionic conductivity and low-methanol crossover.

There has been a significant contribution of research in the quest for incorporating inorganic materials to the polymeric membrane to overcome such difficulties. This organic/inorganic composite material must have the double function of blocking most of the methanol crossover and keeping high-proton conductivity [5]. Some inorganic materials have already been combined with conductive polymers, such as calcium phosphate [6], montmorillonite [7, 8], zeolites [9] and TiO_2 nanoparticles [10], in order to make use of their particular properties.

Zeolites, with their structural properties, shape selectivity and framework composition, are attractive inorganic fillers for DMFC membranes. Such inorganic microporous materials may be incorporated into the polymer matrices producing organic/inorganic composite membranes with the desired properties. Polymer-zeolite membranes have shown good performance in reverse osmosis processes and PEMs due to the synergy of the system [11]. Zeolite has evaluated as filler for polymers electrolyte membrane fuel cell to reduce the methanol permeability and enhance the thermal stability.

2 Basic Concepts About Zeolites

The term "zeolite" was originally used to designate a family of natural minerals which had ion exchange properties and reversible water desorption. This last property originated the word zeolite, which is derived from the Greek words: *zeo* = "to boil" and *lithos* = "stone" [12].

Zeolites are crystalline aluminosilicates with well-defined structures formed by the combination of three-dimensional tetrahedral TO_4 (SiO_4 or AlO_4^-), linked by oxygen atoms to form subunits and vast networks of identical blocks. Aluminum or silicon atoms (T-atoms) occupy the center of the tetrahedron, and the oxygen atoms occupy the vértices [12, 13]. Zeolites are microporous materials that typically present a high specific surface area (>350 m^2/g), variable diameter of micropores (2 and 12 Å), high-thermal stability (200 and 1000 °C) and high-chemical stability [12, 14].

The tetrahedrons may be connected to form a structure containing regular and uniform channels and cavities, with dimensions of the order of a few angstroms, which allow the passage of a particular size and shape molecules. This characteristic has given rise to the name "molecular sieves," which applies to zeolites and other similar structure materials like pillared clays and carbon polymers. More recently, the name zeolite was limited to only aluminosilicates, using the name Molecular Sieves when there are other structural atoms.

Zeolites can be natural, found directly in nature, or synthesized in the laboratory or industrially. Natural zeolites were initially considered as impurities of volcanic or basaltic rocks, limiting their commercial use. Only later, starting at 50s was discovered large sedimentary deposits, such as those located in the Western of the United States, which allowed their exploitation and marketing [15].

There are about 40 naturally occurring zeolites, which formed through slow hydrothermal reactions of volcanic lava deposited over thousands of years, under the influence of high temperatures in the presence of water or saline and alkaline solutions. The first attempts to obtain synthetic zeolites in laboratory imitate the natural conditions (high temperature and pressure and an extended period of synthesis). Mild conditions and shorter synthesis times were only obtained after using reagents that were more reactive. The synthetic zeolites obtained in the laboratory have several advantages over the natural, such as high-purity and well-defined chemical composition, because their composition did not vary according to the deposit of origin that had been extracted.

As the aluminum oxidation number is +3 and it is coordinated by four oxygen atoms, the tetrahedral $(AlO_4)^-$ induce the formation of negative charges in the zeolite structure. This negative charge is counterbalanced by a cation, called compensation cation, which is located in the zeolite cavities and channels without belonging to the crystalline structure [12].

In natural zeolites, the compensation cations can be sodium, potassium, calcium, magnesium, or barium. In synthetic zeolites, these can be Na^+, K^+, H^+, NH_4^+, La^{+3}, and several organic cations used as structure-directing agents from the synthesis of zeolite or ion exchange processes. It can be performed an isomorphic substitution of Si and Al by other atoms such as Ge, P, Ga, and B. In addition to the cations, water molecules can also be contained within the channels and cavities of the zeolites, giving properties of ion exchange and reversible dehydration without disruption of the framework.

The structural formula of a zeolite can be written as (Eq. 1) [13]:

$$M_{x/n}\left[(\mathrm{AlO_2})_x(\mathrm{SiO_2})_y\right]\cdot w\mathrm{H_2O} \tag{1}$$

where M is the compensation cation with valence n, w is the number of water molecules, and the ratio y/x usually has values of 1–5 depending on the structure. The sum $(x + y)$ is the total number of tetrahedra per unit cell.

According to the Atlas of Zeolite Framework Types (6th Edition, 2007), 176 distinct types of zeolite framework have been approved by Commission of the International Zeolite Association (IZA-SC). To the present day, 231 structures have been approved by IZA-SC [16]. Such structures can be visualized at the database (http://www.iza-structure.org/databases/) of zeolite that provides structural information on all of the zeolite framework types that have been approved by the Structure Commission of the International Zeolite Association (IZA-SC).

3 Polymer-Zeolite Composite Membranes: The Role of the Zeolite

The use of zeolites as an inorganic component in DMFC composite membranes reflects some important aspects that must be observed during the preparation of membrane [17–19]. Hereafter, a brief discussion on the parameters related to zeolites that directly influence the polymer-zeolite performance will be presented.

3.1 Influence of Si/Al Ratio

Zeolites are aluminum silicates on which Si/Al molar ratio intensely modifies the microporous structure properties. Each structural aluminum atom generates a negative charge on the solid structure that must be compensated by an exchangeable cation. Thus, the concentration of ion exchange sites is proportional to the content of aluminum in the framework. These compensating charge cations in aqueous or ammoniacal medium suffer solvation, which is directly related to proton conductivity. Since the purpose is the maximization of this conductivity, the lower the Si/Al ratio, the higher the hydrophilic character of zeolite, which permits preferential adsorption of water or ammonia and, as a result, preferred proton transference.

It is important to consider that the distance among the exchange sites is directly related to the Si/Al ratio: more the aluminum content, the greater is the proximity among them in this framework and, therefore, to its compensating charge cations. The mobility provided by the compensating charge cations is responsible for the ionic conductivity exhibited by zeolites.

3.2 *Proton Mobility in Zeolites*

It has been observed in several works that solvated zeolites which contain NH_3 and H_2O molecules enabled faster proton transport and lesser activation energy compared to unsolvated zeolites. It is believed that protons are moving along the channels where solvated molecules are encapsulated within these channels, according to *Grotthuss* transport mechanism [20, 21]. In hydrated zeolites, the proton moves from oxonium ion to the neighboring H_2O molecule that is present in the zeolite channel. This process occurs through tunneling with H_2O molecule reorientation to accommodate proton movement. *Grotthuss* proton transport mechanism is the principal mechanism observed in PFSA (perfluorosulfonic acid) and liquid electrolytes.

Simon et al. [22], in their study on the construction of zeolite-based sensors, observed through the use of impedance measurements that the presence of NH_3 in the pores influences the proton mobility through H-Beta zeolite channels, which was adequately observed by conductivity measurements. However, by using the Na-β zeolite, it was found that the conductivity is not affected by the presence of NH_3. The authors concluded that NH_3 predominantly influences the mobility of Brønsted protons. β-zeolite (or BEA structure), used in this study, is formed by the intergrowth of two polymorphs (A and B) closely related to each other. Polymorphs consist of a three-way system of channels delimited by 12-membered rings which are interconnected, forming channels. There is a high incidence of stacking faults in this zeolite, due to successive interconnection on the plane [001] creating an intersection of two systems of linear channels, according to Fig. 2.

Research on NH_4-chabazite and NH_4-clinoptilolite zeolites demonstrates that proton transport occurred via NH_4^+ cations with conducting properties present in ammoniated zeolites. Krever et al. [23] proposed the *vehicle transport mechanism*, where the proton is bonded to the "vehicle" solvate, i.e., H_2O or NH_3, generating H_3O^+ or NH_4^+ species carriers. "Uncharged" vehicles move in the opposite direction, and the conduction of the proton is determined by the diffusion rate of the vehicle solvate. For clarity, Fig. 3 shows the representation of the CHA structure which presents selectivity shape and large diameter of micropores (0.6 nm $< d_p < 0.8$ nm) and its framework composition is $[Ca^{2+}{}_6(H_2O)_{40}][Al_{12}Si_{24}O_{72}]$ (CHA structure) [16]. These characteristics allow adsorption of molecules with an average kinetic diameter and also provide increased mobility of proton due to the structural flexibility of the framework.

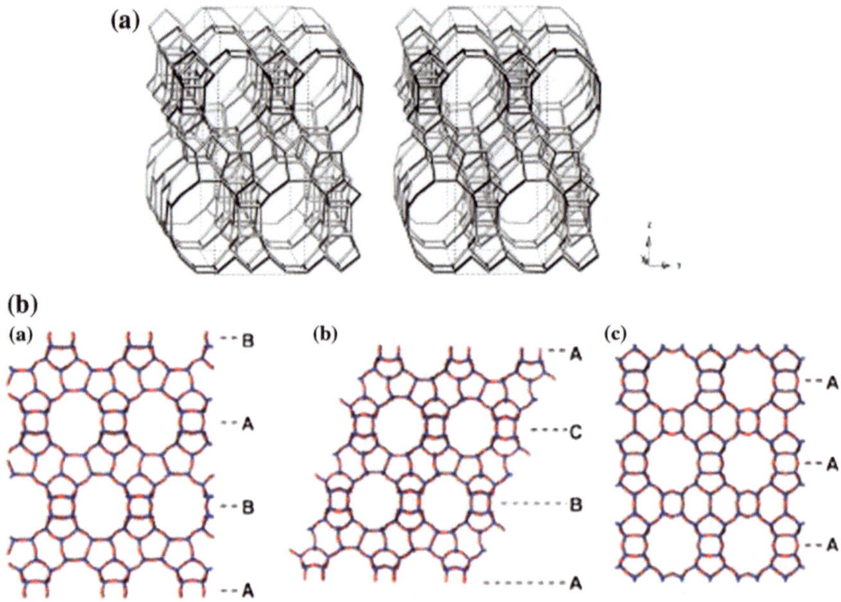

Fig. 2 **a** Beta zeolite structure. **b** Polymorphs *A*, *B* and *C*

Fig. 3 Scheme for CHA structure

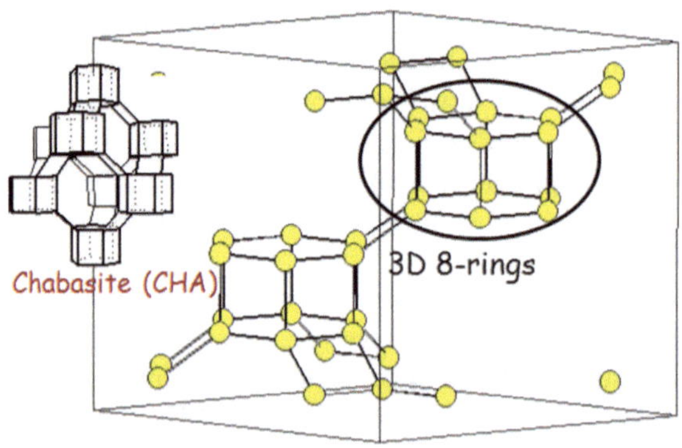

Franke and Simon reported that the proton transport in solvated H-ZSM-5 zeolite (Si/Al $\leq$ 40) might present both kinds of mechanisms [24]: *Grotthuss* type and vehicle transport. *Grotthuss transport mechanism* is predominant in temperatures under 120 °C, whereas *vehicle transport mechanism* occurs at higher temperatures.

Experiments and calculations based on theoretical models demonstrate that the activation energy range for conducting ions in zeolites is between 40 and 130 kJ mol^{-1}, which is lower than the deprotonation energy (1300 kJ mol^{-1}). The data suggest that the proton transport in zeolites in mild temperatures follow

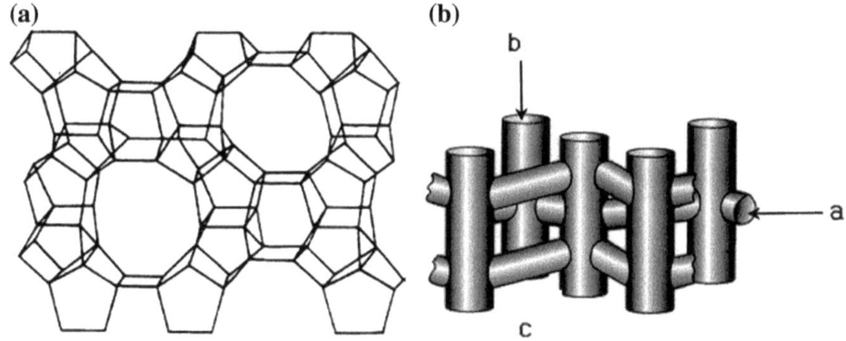

Fig. 4 Scheme for MFI structure (**a**) and interconnected channel system (**b**)

different mechanisms that are related to the zeolite framework composition and the metal cation present in the cavities [25].

Franke and Simon have studied the proton mobility in H-ZSM-5 zeolite and found that the proton movement in zeolites with adsorbed water and ammonia vapor showed proton movement activation energy dependent on the temperature [26]. The H-ZSM-5 zeolite shows medium-sized micropore diameter (0.5 nm < d_p < 0.6 nm), as shown in Fig. 4. ZSM-5 zeolite (MFI structure), whose composition is $Ca_2\ [(AlO_2)_4(SiO_2)_8]\cdot H_2O$ presents ion exchange capacity, a structure that allows adsorption of molecules with an average kinetic diameter and framework aluminum content that makes it moderately hydrophilic. H-ZSM-5 has Brønsted acidity similar to sulfuric acid due to the presence of H^+ protons in its framework.

Franke and Simon also found that zeolites in conditions of low temperature and high ammonia and water loads have their surface covered by a condensed phase of solvate molecules weakly bonded (called solvent molecule chains) among adjacent Brønsted sites, which permit *Grotthuss* proton type transport [26]. The presence of a significant amount of adsorbed solvated molecules may induce the formation of solvent complexes attached to acid protons of Brønsted sites. Thus, the proton transfer takes place with a relatively small activation energy (Fig. 5a). By gradually elevating the temperature, a growing desorption of the adsorbed weakly bonded molecules takes place, dismounting the complex solvate chains and opening larger space among them. As a consequence, the proton transport activation energy among groups becomes higher, resulting in more difficulties for the occurrence of the *Grotthuss*-type mechanism.

Franke and Simon observed that the continuous increasing of temperature leads to complete desorption of the weakly bonded molecules, and the proton transport may occur only by *vehicle* mechanism. NH_4^+ cations or H_3O^+ group enable proton transport that moves among neighboring Brønsted sites [26]. However, when comparing proton mobility, the movement of these vehicle species presents lower activation energy [27]. In mild to higher temperatures, proton mobility increases

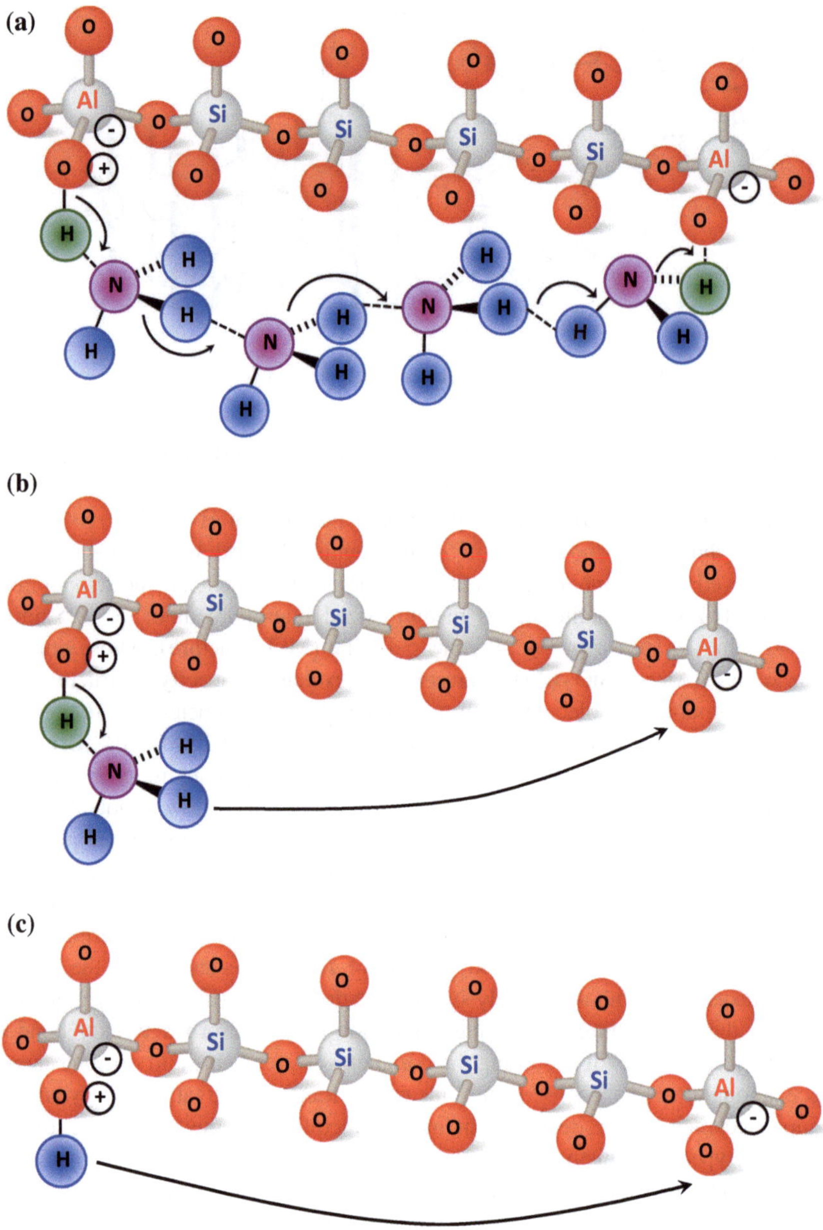

Fig. 5 **a** *Grotthuss* transport mechanism in low temperatures; **b** vehicle transport mechanism: NH_4^+ works as proton transport vehicle; **c** submitted to high temperatures, the elimination of adsorbed molecules presents the same observed conductivity for unsolvated zeolite

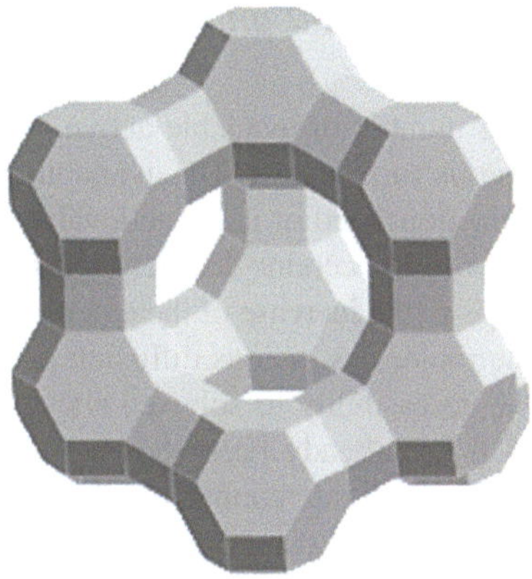

Fig. 6 Scheme for X and Y zeolites (FAU structure)

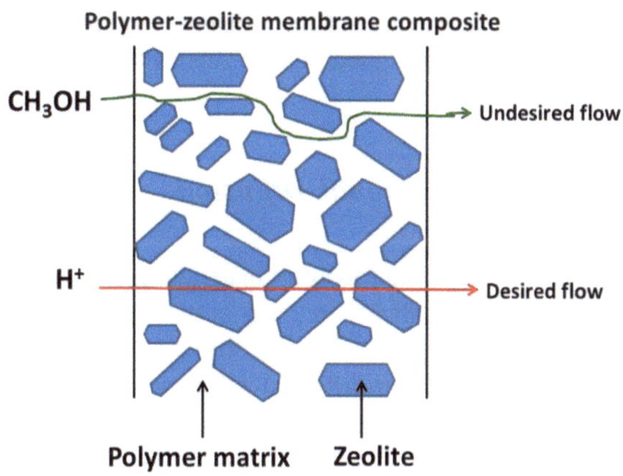

Fig. 7 Methanol and H^+ diffusion by using a polymer matrix containing zeolite

and, as a consequence, zeolite proton conductivity increases as well (Fig. 5b). The complete elimination of solvates in high temperatures leads to almost constant conductance and this conductance, as well as the activation energy, gets close to the values observed in the unsolvated zeolite (Fig. 5c) [27].

These differences found in proton mobility mechanisms may explain that typical zeolites, such as H-ZSM-5 (Fig. 4), HY (Fig. 6), and mordenite presented ionic conductivity no higher than 10^{-4} S cm^{-1}. However, some hydrated zeolites, such as Sn–Mordenite, present high-ionic conductivity closer to the one shown by PFSA polymer, around 0.1 S cm^{-1} [28]. Now Fig. 7 illustrates the FAU structure, with large-sized micropore diameter (0.6 nm < d_p < 0.8 nm), suitable for shape selectivity of larger molecules. Depending on the Si/Al ratio, i.e., on the composition of its framework, the zeolite will be named Y or X. The zeolite with FAU structure and Si/Al ratio in the range from 1.0 to 1.5 is called X. The zeolite with Si/Al ratio above 1.5 is named Y, and its composition is $Na_{86}[(AlO_2)_{86}(SiO_2)_{106}]\cdot H_2O$. Such characteristics enable them to adsorb molecules which have average kinetic diameter and hydrophilic ion exchange.

3.3 Internal and External Surface Area

The number of available active sites in a zeolite is proportional to its high microporous surface area. The availability and accessibility of the active sites of the solid (functionalized or nonfunctionalized) directly influence proton transport. It must be taken into account the fact that the reduction in the size of the zeolite crystallites leads to the increase of the external surface area. This contributes to the reduction of the molecules or ions diffusion path and as a consequence, there is an increase of the accessibility to solvated sites.

It has already been shown that large zeolite crystallites form composite membranes with interconnecting defects between the zeolite and the polymer due to the small external contact surface area [11]. Such defects may transform the membrane into a barrier for proton conduction and an open door for methanol diffusion. It would be possible to produce membranes with good zeolite-polymer compatibility making use of the size of functionalized zeolite nanocrystals, such as the sulfonic acid [9], promoting good contact due to the high external surface area.

3.4 Configurational Diffusion

It is known that microporous materials have high resistance to chemical species diffusion. Since micropores are approximately of the same size of the kinetic diameter of the molecules, the diffusion coefficient will be between 10^{-14} and 10^{-5} cm^2 s^{-1}, i.e., the diffusion will depend strongly on this parameter [27]. Due to its small size, the proton will not suffer much effect of the constraint of the mean free path internally available in the zeolites. However, partial blocking of the pores may constraint proton conductivity in functionalized surfaces.

3.5 Crystallite Size [17, 18]

The crystallite size of the zeolite used in the construction of the membrane is critical to DMFC performance. Crystallite size and surface area are deeply connected. As said before, large crystallites do not enable good connectivity to polymer matrix [29–33]. A good polymer-zeolite membrane depends on the incorporation technique of the zeolite nanocrystallites to the composite membrane because the agglomeration of these nanocrystallites, when removed from the solution, must be avoided so that good dispersion must be kept during simultaneous synthesis of the polymer [18, 34].

3.6 Functionalization of Zeolite Surface

A different way to partially solve the problems related to the proton conductivity is the surface functionalization with organosilanes or by modifying functional groups attached to polymer backbone chain. Functional groups attached to zeolite surface, such as sulfonic acid, considerably enhance proton conductivity. The handling of the inorganic surface by functionalization or incorporation of heteropolyacids in the molecular sieve pores provides higher proton conductivity [18]. There is lots of space for innovation and development in this field of research.

3.7 Selectivity, Proton Conductivity, and Permeability

Hereafter, some important aspects for understanding DMFC membranes will be discussed. The selectivity of the organic/inorganic composite membrane (β, in S s cm^{-3}) is defined as a relation between proton conductivity (σ, in S cm^{-1}) and the membrane methanol permeability (P, in cm^2 s^{-1}), that is, $\beta = \sigma/P$ [35, 36].

When developing a DMFC membrane, high selectivity is sought, resulting in high-proton conductivity and low permeability (Fig. 7). In the past, this objective was thought only by obtaining high-proton conductivity. However, while the current membrane of a standard Nafion fuel cell presents very high conductivity, its selectivity is compromised due to high-methanol permeability. If another membrane shows highest selectivity, albeit lowest conductivity, its conductance could be enhanced by making a thinner membrane.

Composites with zeolite particles in hydrophobic polymer have been under study to make it possible to discover membranes with higher selectivity. As it is shown in Fig. 8, membrane components were chosen in a way that protons are transported through the dispersing phase and polymer matrix. If the dispersed material is impermeable to methanol, protons will follow a more direct and shorter path compared to methanol [35].

Another important aspect that must be taken into consideration when preparing polymer-zeolite membranes is the choice of polymers to be employed. Initially, the polymer matrix used to develop these polymer-zeolite membranes was PTFE (polytetrafluoroethylene), inert and nonconductive [37]. Conductivity would be attributed solely to zeolite phase. PTFE-zeolite X or PTFE-zeolite Y composites presented better conductivity compared to similar PTFE-NaA, PTFE–Na–ZSM-5, and PTFE-Na-mordenite membranes. PTFE-LiY membrane reached conductivity of 2×10^{-3} S cm^{-1} (348 K), and PTFE-Na-mordenite was stable at 623 K. PFSA-zeolite membrane (PFSA: perfluorosulfonic acid) is one of the most studied zeolite composite membranes.

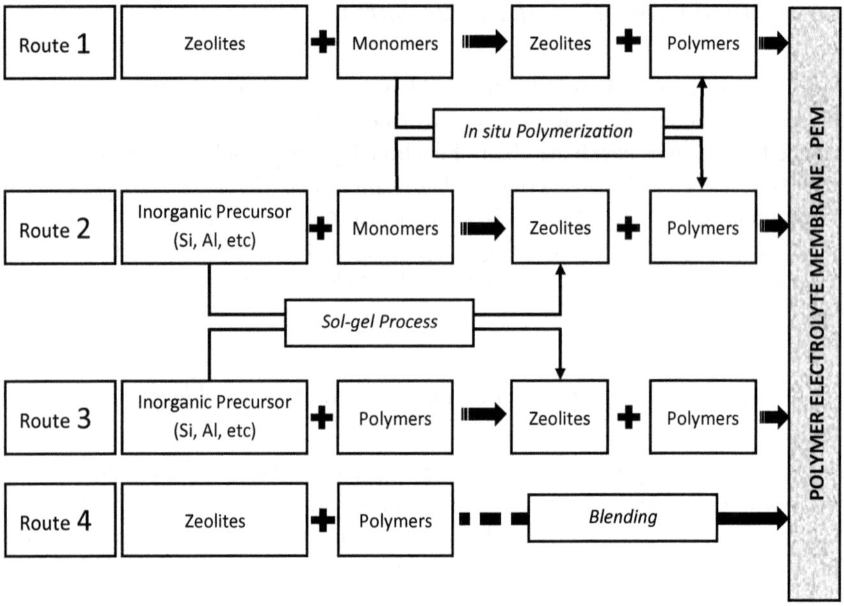

Fig. 8 Schematic routes for the preparation of polymers electrolyte membranes

4 Techniques for Producing Organic/Zeolite Nanocomposite Membranes

To better understand the different routes for producing organic/inorganic nanocomposite membranes, it is important to think about zeolite as being crystalline inorganic polymer that is formed by repeated tetrahedral units. In this sense, it is possible to combine at four raw materials for producing organic/zeolite nanocomposite membranes:

(1) organic monomers;
(2) "inorganic monomers" (AlO_4 and SiO_4 tetrahedra) of zeolite structure. In truth, they are inorganic precursors that are alkoxide precursors $M(OR)_n$, where M is Si or Al in the zeolite structure and R is an alkyl group;
(3) organic polymers;
(4) inorganic polymers (zeolites).

All the above raw materials can be combined according to Fig. 8.

When the monomers are provided to the composite membrane preparation, it is imperative to conduct the polymerization during composite membrane formation. Thus, classical techniques like radical or addition polymerization are employed to yield the organic polymer, as well as, the sol–gel process for the inorganic polymer (zeolite). The combination of those raw materials allows to design four routes according to Fig. 8.

The Route 4 (Fig. 8) would be the simplest route to produce the organic–inorganic hybrid PEM using both organic and inorganic polymers. In this route, the zeolite can be dispersed into the organic polymer that was previously melted (melt blending) or was solubilized in an appropriate solvent (liquid-state blending). As both polymers are blended, this could also be called Blending Process. Needless to say that the melt blending is a more environmentally friendly process because no organic solvent is used. The interaction between zeolite and organic matrix polymer impact on the performance of nanocomposite membrane because the dispersion of zeolite may lead to undesirable agglomeration, then sometimes surface modification techniques are carried out to enhance the interaction between the organic and inorganic phases. Such surface modification may be irradiation grafting or functionalization of zeolite. Additionally, in another version, the zeolite is previously synthesized on a support surface and then polymer solution is casted on this support.

On the opposite side, the Route 2 would be the most complicated because both the raw materials are monomers (organic monomers and inorganic precursors monomers), such route will require the inorganic polymerization (sol–gel process) to yield the inorganic polymer (zeolite) and the classical polymerization for the organic polymer. That is why this route could not be found in the literature. There are also intermediates routes when just the organic polymer is formed during the PEM preparation (Route 1), or just the inorganic polymer (zeolite) is formed (Route 3).

However, there are also variations of Routes 3 and 4 which are called doping or infiltration technique. In this technique, the preformed membrane is swelled in a solvent to enlarge the pore before the doping or the infiltration of the zeolite nanoparticles or inorganic precursors. Then the composite is cured, and covalent bonds are carried out inside the polymer matrix. The curing process can be carried out by heat, radiation, or chemical grafting of the inorganic precursor. The leaching of the zeolite from the membrane is one of the challenges to be overcome.

Finally, layer-by-layer is one of the newest techniques to produce self-assembled hybrid membrane. The organic and inorganic components are organized in a nanoscale by electrostatic interactions. Every oppositely charged component is deposited alternately on each other.

The casting or recasting techniques are included in all routes. The casting consists in using solubilizing the organic polymer in an appropriated solvent and then the dispersion is cast in molds like Petri dish (in a lab version) before drying the solvent. On the other hand, the recasting technique starts with the commercial solution of the polymer; the solvent of this solution is evaporated to get a dry residue which is solubilized again in an appropriate solvent and concentration usually different from those of the commercial one.

5 Synthetic Polymers/Zeolite Nanocomposite Membranes for PEMFCs

The Route 2 has not been found in literature because once both monomers are used, synthesis become more difficult.

5.1 *Route 1: Zeolite + Organic Monomers*

Nur et al. synthesized zeolite ZSM-5 and compared two silylation processes of zeolite. One process was carried out by reaction of ZSM-5 with 3-mercaptopropyltrimethoxysilane, this sample was labeled as ZSM-5-SH [38]. This zeolite was followed by grafting polymerization of styrene and divinylbenzene, and afterward, the sulfonation reaction was performed by pouring the acetyl sulfate solution into the polystyrene-zeolite composite. The obtained zeolite was labeled as ZSM-5-SH-PS-SO_3H. The ZSM-5-SH was also oxidized with hydrogen peroxide to obtain propylsulfonic acid functionalized zeolite, which was labeled as ZSM-5-SH-SO_3H. The other silylation process was carried out by reaction of ZSM-5 with phenyltrichrolosilane (PTCS) and then was sulfonated with sulfuric acid, the zeolites were denominated ZSM-5-PTCS and ZSM-5-PTCS-SO_3H, respectively. It was observed that the higher the amount of sulfonic groups in the sample, the greater the sample hydrophilicity, and consequently, the higher the water uptake capacity and ion exchange. The water molecules may interact with the polymeric matrix by hydrophilic interaction with –SO_3H and the ZSM-5 zeolite may retain water molecules at the channels and cavities especially at high temperatures. The ZSM-5-SH-PS-SO_3H sample had the greatest amount of the functional group on the ZSM-5 surface (65 wt%), that is, a highest carrier (proton) concentration of sulfonic acid groups, consequently, this sample showed the highest proton conductivity, followed by ZSM-5-SH-SO_3H > ZSM-5-PTCS-SO_3H samples.

5.2 *Route 3: Inorganic Precursor + Organic Polymer*

Nafion/acid functionalized zeolite Beta (NAFB) nanocomposite membranes were prepared by Chen et al. using in situ hydrothermal crystallization technique. A precursor solution of organically functionalized zeolite beta was prepared, and the crystallization of the zeolite was conducted in situ in the swollen Nafion 115 membrane. According to Chen et al., the micrometer-sized particles usually have low-proton conductivity, but the performance can be improved by using nanometer-sized particles once those smaller particles are crucial to the compatibility between the inorganic filler and Nafion [39]. Moreover, the compatibility has a significant effect on the proton conductivity and methanol permeability of the membranes. Acid functionalized zeolite nanocrystals were produced to minimize the loss of proton conductivity and reduce the methanol permeability. Such nanocrystals presented aggregation during the membrane casting. To circumvent the problem, Chen et al. developed an in situ crystallization process to fabricate nanocomposite membranes [39]. Firstly, the Nafion membrane was impregnated in the precursor solution for the organically functionalized Beta synthesis, followed by the hydrothermal treatment of the impregnated membrane in the synthesis solution to grow organic functionalized Beta nanocrystals inside the Nafion membrane. The hydrophilic ionic clusters of the Nafion membrane could be considered one nanoreactor for the formation of the zeolite. Finally, concentrated sulfuric acid was used to remove the template in the zeolite pores and also sulfonate the organic functional groups. They prepared Nafion/acid functionalized zeolite Beta using the technique as mentioned above, and the zeolite loading was 5%. They compared the composite membrane with the commercial Nafion membrane, although the proton conductivity was similar, the composite membrane presented lower methanol permeability (40% reduction). The composite membrane was tested in a Direct Methanol Fuel Cells (DMFC) with two methanol concentrations (1 and 5 M). The composite membrane presented higher open-circuit voltage (OCV) and maximum power density than the commercial membrane; the differences were even bigger for the 5 M methanol.

5.3 *Route 4: Zeolite + Organic Polymer*

Han et al. prepared microfabricated zeolite proton conducting membrane and structured zeolite-PFSA composite membrane [19]. The first consists of an array of self-supported zeolite micromembranes fabricated on a silicon wafer. The second were built on stainless steel mesh (SSM) with 110 μm diameter holes and Nafion resin was casted on the zeolite-coated SSM. In both methods, they previously synthesized the zeolite on the support material (silicon wafer or SSM). The Nafion solution was casted just to the SSM systems and compared to the silicon wafer with zeolite. The zeolite in a general way acts as an electric insulating, and was used to retain water in the structured zeolite-PFSA composite membrane, improving membrane thermomechanical properties and enhanced PEMFC performance. The purpose of the second study by the same research group was to improve the properties of confined PFSA-zeolite structured composite membrane using bi-layered zeolites [40]. The zeolite served as a protective barrier against corrosion for the stainless support. Interactions between zeolite surface and PFSA polymer chain resulted in higher glass transition temperature and increased proton conductivity. Despite the better mechanical properties, composite membranes presented better performance than the Nafion membrane, an order of magnitude higher MPD (maximum power density) was obtained for hydrogen PEMFC and a substancial increase was observed for passive DMFC.

Han et al. studied supported and self-supporting ZSM-5 membranes (Si/Al = 25) [41]. The supported ZSM-5 membrane was synthesized by seeding as well as by secondary regrowth method on cellulose paper (the polymer). The self-supporting ZSM-5 membrane was prepared by microfabrication method that consists of an array of micromembranes on a silicon substrate. To evaluate the performance of the PEMFC, they previously coated the anode and cathode electrodes with the catalyst and Nafion solution was used to glue every of the membranes mentioned above, as a "membrane sandwich." Due to the contact between the electrode, catalyst and membrane layers in the assembly, the self-supporting H-ZSM-5 membrane showed the values: an open-circuit voltage (OCV) (0.77 V), supported H-ZSM-5 (0.90 V) and Nafion 117 (0.98).

Reference [42] evaluated the effect of the Si/Al ratio of H-ZSM-5 zeolite incorporated in various loadings (1, 3, and 5 wt%) into Nafion membranes. The researchers found that the lowest methanol permeability was obtained using composite membranes with 5 wt% of zeolite with Si/Al ratio 25. This result may be due the induration of a polymer layer around the aggregate zeolite particles. This membrane also showed a high power density and stable performance in time. The highest proton conductivity was obtained using 1 wt% of zeolites with Si/Al ratio 50.

Changkhamchom and Sirivat synthesized composite ZSM-5/sulfonated poly (ether ketone ether sulfone) (S-PEKES) membranes [43]. The ZSM-5 zeolites with different Si/Al ratios (23, 50, 80, and 180) were tested as the inorganic fillers at various volume percentages (2–10% v/v). They observed that the Si/Al ratio directly influence the increase of the sample hydrophilicity, which is inversely proportional to the proton conductivity. In other words, the more is the hydrophilic character of the membrane, the higher is the amount of water molecules retained and better is the protons transport capacity. All composite membranes showed lower methanol permeability than the pristine S-PEKES and Nafion 117 membranes. The methanol permeability decreases with the increasing of Si/Al ratio and with the amount of zeolite added to the membrane. As the methanol molecules are preferentially adsorbed and trapped within the zeolites, the trapped molecules induce the delay in the methanol diffusion, considerably decreasing methanol permeability.

Yu et al. used 1,3-propane sultone (5, 10, and 15%) to functionalize ZSM-5 zeolites and then incorporated them to the sulfonated poly(arylene ether sulfone) (SPAES) copolymers using the blending technique [44]. As reported by Nur et al. [38], the proton conductivity of the functionalized zeolite increases due to the high amount of sulfonic acid groups. However, the amount of zeolite was found to directly influence this parameter. The addition of 5 wt% of sulfonated zeolite resulted in the increase of proton conductivity up to 0.030 cm^{-1} (120 °C and 50% RH) compared to the pristine membrane (0.022 S cm^{-1}). This result was attributed to the dispersion of zeolite in the membrane and generation of the hydrophilic channel networks by the sulfonic acid groups, increasing the movement of protons and resulting in high conductivity proton. However, increasing the amount of sulfonated zeolite (10 or 15 wt%), sulfonated zeolite aggregated and acted as a barrier for proton conducting channels and networks, decreasing proton conductivity.

Sancho et al. studied four different materials, namely, NaA zeolite, mordenite, umbite, and ETS-10 [45, 46]. The materials were prepared by pressing (7 ton/5 min), to obtain pellets of 13 mm of diameter, using the polymer PVDF (10 wt%) as a binder, and not as material to composite membrane production. Nafion membrane showed the best results of conductivity up to 100 °C, and after, progressively declined due to degradation process. However, zeolite-polymer composites were shown to be more stable at high temperature. NaA zeolite increased the proton conductivity up to 120 °C and ETS-10 up to 150 °C. The ionic conductivity of the zeolites increases with temperature mainly due to high ion mobility and the large quantity of adsorbed water inside the pores. ETS-10 pellet presented values in the same order of magnitude as Nafion membrane (40 °C/70 RH). ETS-10 (1.97 mS s cm^{-3}) also showed the best selectivity, with the same order of magnitude as Nafion membrane (5.84 mS s cm^{-3}) at 40 °C.

In another work, the functionalized zeolite was previously synthesized and mixed with the Nafion solution, then was cast, and the solvent was evaporated [18]. Zeolite loading was evaluated in the 2.5–20 wt% range, demonstrating that zeolite nanocrystals were homogenously dispersed within the polymer matrix for 2.5 and 5.0 wt%. But the dispersion was heterogeneous for 10, and 20 wt% zeolite membrane was overloaded and brittle. The selectivity decreased when the temperature increased for all the membranes. The selectivity was evaluated ex situ and in situ (in a Membrane Electrode Assemblies). The results ex situ showed that incorporation of 10 wt% of zeolite impaired the selectivity, although the 5 wt% showed better results than 10 wt% of zeolite. Moreover, similar results were verified for in situ configuration when 2.5 and 5.0 wt% were tested, and the first presented best results.

Li et al. prepared Nafion/Zeolite A composite membranes. Zeolite NaA has low Si/Al ratio which offers a hydrophilic character and preferential adsorption of water; thus, proton transfer can be improved [17]. Moreover, zeolite NaA could exclude methanol from the zeolite channels which could provide low-methanol permeability. The poor interfacial compatibility between the organic polymer matrix and zeolite crystals promotes pinholes between them, which favor the methanol transport; therefore, the functionalization of the zeolite surface can be carried out to overcome the poor compatibility between zeolite and the organic polymer. The organic polymeric chains were grafted onto the inorganic zeolite surface using 3-aminopropyltrimethoxysilane (APTS) as silane coupling agent. Functionalization could improve the interface compatibility between zeolite crystals and Nafion, resulting in a decrease of methanol permeability, which could be attributed to the inhibition of pinhole formation. On the other hand, a noticeable reduction in the proton conductivity of functionalized composite membrane has also been verified. The decrease was attributed to the alkaline nature of NH_2-groups of the APTS, a large molecule that is unable to get into the NaA porous; thus modifying only the external surface of the zeolite.

On the other hand, Tripathi et al. studied sulfonated poly (ether ether ketone) (SPK)-zeolite-zirconium hydrogen phosphate (ZrP) nanocomposite PEM prepared by in situ infiltration and precipitation [47]. The authors had already verified promising results with SPK-ZrP composite membranes. First, SPK membranes doped with surfactant (cetylpyridinium chloride) were prepared. The surfactant was used to form pores. Secondly, the surfactant was leached out with HCl solution, and fine zeolite particles were infiltrated in the swelled SPK membrane for developing SPK-zeolite composite, and finally ZrP was precipitated in the SPK-zeolite composite membrane. Although the thermal, mechanical strengths, oxidative and dimensional stabilities, these nanocomposite membranes offered no significant advantages over SPK or Nafion (N117) membrane for proton conductivity at 30 °C. However, SPK-zeolite-ZrP nanocomposite membranes showed improvement over SPK or N117 membrane at 70 °C due to their high selectivity index, which indicates that nanocomposite PEMs was a suitable candidate for DMFC applications at moderate temperatures (between 60 and 150 °C).

Micro and nano-sized NaX zeolites were prepared by Cui et al. using the sol–gel process. Then, the sodium ion was exchanged by ammonium to form the NH_4X zeolite [48]. According to the authors, the ammonium ions would establish a chemical equilibrium with protons and ammonia, which could prevent the methanol permeability and supply further protons and increase the ionic conductivity. Zeolite (5 and 10 wt%) and Nafion solution were blended, and the composite membranes were prepared by the recasting method. Concerning the zeolite loading, the membranes with 10 wt% loading of zeolite presented microcracking and segregation for both nano and micro-sized, but with 5 wt% loading the membranes showed uniform distribution. They also verified that the selectivity of NH_4X/Nafion micro-sized 5 wt% composite membrane was more than twice that of pristine Nafion. Moreover, DMFC cell tests at 60 °C showed that peak power density for this NH_4X/Nafion was threefold higher than the value for pristine Nafion membrane.

Kongkachuichay et al. incorporated two types of natural zeolite (Analcime and Faujasite) into Nafion membranes using the blending technique [49]. The water uptake, ion exchange capacity, and proton conductivity were higher for composite membranes than for Nafion membranes. Moreover, the authors also found that Analcime/Nafion nanocomposite membranes exhibited better properties than Faujasite. Based on this results, Intaraprasit and Kongkachuichay selected Analcime to study the effect of sulfonation on poly(ether ether ketone) (PEEK or Nafion) [50]. First, sulfonation of PEEK was carried out using concentrated sulfuric acid and different zeolite content was evaluated in nanocomposite membranes prepared by blending technique. The increasing amount of added Analcime decreased the ion exchange capacity and water uptake of the membranes. The proton conductivity was enhanced up to 10 wt% of Anacilme, but the agglomeration of zeolite particles led a decrease above this value, being even lower than that of the SPEEK. The improvement of conductivity was attributed to the hydrophilicity and protons inside the connected channels of Analcime.

Devrim and Albostan evaluated the preparation of Nafion/Beta zeolite composite membranes with different zeolite loading [51]. The authors also used the recasting method and the maximum zeolite loading was 12.5% because the membrane cracked above this value. The proton conductivity increased as the temperature increased. Moreover, the growth of zeolite loading with the decrease of proton conductivity may be attributed to barrier properties of nonhomogenously dispersed zeolite particles. The Nafion/zeolite membrane with 10 wt% zeolite loading showed the best performance for single cell PEMFC tests, although all the Nafion/zeolite membrane performed better than Nafion membrane.

Mecheri et al. compared the recasting method to the hot pressing method with different faujasite zeolite loading [52]. For the hot pressing method, membrane and zeolite were hot pressed at 140 °C under a pressure of 227 kg/cm^2 for 40 s. The water uptake and the proton conductivity of nanocomposite membranes increased with increasing zeolite loading. The recasting method presented higher values than hot pressing. The decrease of the membranes porosity could explain the lower water uptake of hot pressing. The surface composition of the zeolite, as well as the introduction of porosity at the polymer/filler interface could account for the enhancement of the water mobility degree in the membrane.

Patet et al. used the atom transfer radical polymerization to synthesize poly (vinylidene fluoride-co-chlorotrifluoroethylene)-g-poly(styrene sulfonic acid) (graft copolymer) with 47 wt% of poly(styrene sulfonic acid) and posteriorly blended with zeolite type A (3–10 wt%) [53]. The authors found that the zeolite particles strongly interact with the sulfonic acid groups of the graft copolymer matrix. Both the proton conductivities and water uptake decreased with the increase of both zeolites (3A and 5A).

Zeolites beta crystals with various SiO_2/Al_2O_3 ratios were added to SPES-SPEEK blend membranes [54]. Sulfonation of poly (ether ether ketone) (PEEK) and polyethersulfone (PES) was carried out to enhance the proton conductivity of the polymers. The polyethersulfone was used to improve the thermohydrolytic stability of the membrane. The proton conductivity of SPEEK was enhanced by the addition of zeolite beta. Moreover, the thermohydrolytic stability was improved by blending poly-ether-sulfone (PES). The SiO_2/Al_2O_3 ratio of 50 at 10 wt% loading presented the best conductivity results for zeolite Beta/SPEEK composite membranes. On the other hand, SPES/SPEEK membrane was more hydrodynamically stable and also performed better than pristine SPEEK membranes which excessively swell.

Chabazite and Clinoptilolite were used as fillers to Nafion composite membranes. Zeolites presented uniform distribution up to 40 vol%, and although more brittle than Nafion, the composite membranes showed reasonable flexibility below 40 vol% when hydrated. Zeolitic fillers could provide notable changes of conductivity and selectivity with respect to Nafion membranes without fillers [9].

6 Natural Polymers/Zeolite Nanocomposite Membranes for PEMFCs

Natural polymers have shown great promise in the production of devices for generating and storing energy at low cost. Natural polymers are cost-effective and eco-friendly derived from renewable sources and become a promising replacement for synthetic polymers. There are a wide variety of natural polymers present in nature; the main polymer employed for a zeolites incorporation for methanol fuel cell is chitosan [55].

Chitosan has been a promising source because it is abundant in the environment [56], it has hydrophilic character, can be use in high temperature and low humidity environments, it has low-methanol permeability, and certain functional groups that allow chemical modification to adjust its properties [11, 57]. Furthermore, chitosan has high-proton conductivity [11, 58].

Chitosan can be applied in fuel cell electrodes based polymer and in biofuel cells [59]. It is a natural polymer derived from chitin and it is composed of randomly distributed beta (1–4)-linked D-glucosamine (deacetylated unit) and *N*-acetyl-D-glucosamine. It is insoluble in water and in most of the organic solvents and alkali. However, it is soluble in diluted organic acids such as acetic acid, formic acid, and lactic acid. Chitosan contains three different polar functional groups, namely hydroxyl (–OH), primary amine ($–NH_2$), and ether (C–O–C) groups, and due to the presence of these functional groups, it has a high water attraction ability [56].

Several studies describe the employment of chitosan for the production of hydrogen from methanol [11, 60–62].

The production of natural polymers nanocomposites is based on casting method. This method is based on the solubilisation of chitosan in an aqueous acetic acid solution while stirring by 80 °C. After complete solubilization of the polymer, the desired zeolite in added and homogeneously stirred. Then, bubbles need to be removed, followed by plating and dry at ambient temperature. After drying, the membrane should be crosslinked with H_2SO_4, and subsequently washed with buffer solution and finally, dried [62].

The compatibility between the polymer and inorganic filler surface is a key issue to determine the ownership of the final performance and membrane. To improve the interfacial morphology of the hybrid membrane, a transition will be created between the organic and inorganic phases, mitigate or eliminate nonselective voids [61]. Considering the high selectivity, low cost, and environmentally friendly, as well as easy to manufacture, chitosan offers great promise for the development of Nanocomposites Natural Polymers.

Wang et al. developed chitosan membrane incorporating zeolite beta particles (300–800 nm) [62]. A homogeneous matrix with size larger than 800 nm was shown to be difficult to be produced. The best membrane was obtained by employing zeolite beta-2. To evaluate the performance of membrane, it is important to know the water and methanol crossover and proton conductivity. The authors observed that the increase of zeolite in the membranes decreased the water and methanol uptake. This fact can be explained by the high hydrophobic character of zeolite compared to chitosan.

In another study, Wang et al. developed a chitosan membrane incorporating zeolites (3A, 4A, 5A, 13X, mordenite, and H-ZSM-5) [11]. The diffusion resistance of methanol and methanol permeability were evaluated. The results were dependent on the type of zeolite incorporated in the matrix which influenced the efficiency.

The application of modified chitosan membrane as the electrolyte in fuel cells is aimed for improved chemical, mechanical, and thermal stability, also contributing to the proton conductivity. Transport of protons in the chitosan membrane can occur by two mechanisms, namely, *Grotthuss* mechanism and electrostatic interactions. The *Grotthuss* mechanisms protons can be transferred jumping from functional group to another. In this type of low-temperature ionic conductor, migration of protons occurs mainly by OH^-, which allows jumping through the protons in water molecules together. The electrostatic interactions occur between the amine group (NH_3^+) of the chitosan and sulfate ions (SO_4^{2-}) of crosslinking (crosslinking in the case of sulfuric acid). In this mechanism, protons are transferred along the chains through ionic interactions, facilitated by reducing the energy barrier for transport of protons through ion (SO_4^{2-}) [63].

7 Conclusions

Some promising possibilities related to the use of DMFC composite membranes have been discussed employing a polymer-zeolite system. ZSM-5 and Beta zeolites are produced at extremely low cost in oil refining industry and would not represent extra costs to the production of membranes that include inorganic components. The versatility in the use of zeolites in systems with various temperatures enables the handling of the proton transport mechanism, either of *Grotthuss* type or of vehicle type. In lower operation temperature, the materials used in the fuel cell construction would be less thermally and mechanically demanding, thus making possible a simpler and cheaper technology. Thus, the control of the ionic conductivity properties would lie mainly on the chosen polymers.

The clear advantages in the use of zeolites and the vast knowledge in their application in other technological areas indicate their potential use in these energy generation systems. However, there are inherent disadvantages when using a microporous system, which constraints the transport to the interior of the inorganic structure that must be creatively overcome. Configurational diffusion and shape selectivity must be handled, avoiding the undesired methanol crossover and allowing the proton conductivity, enabling the construction of the cell. The surface functionalization of the zeolites needs to be more carefully studied, but it configures as a good alternative to make zeolite and other inorganic solids viable.

Acknowledgements This work was financed through the projects M-ERA-NET/0004/2015-PAIRED and UID/QUI/50006/2013, receiving support from the Portuguese Science and Technology Foundation, Ministry of Science and Education (FCT/MEC) through national funds, and co-financed by FEDER, under the Partnership Agreement PT2020.

References

1. Appleby AJ (1999) The electrochemical engine for vehicles. Sci Am 281(1):6
2. Dyer CK (1999) Replacing the battery in portable electronics. Sci Am 281:88–93
3. Zawodzinski TA, Springer TE, Uribe F, Gottesfeld S (1993) Characterization of polymer electrolytes for fuel cell applications. Solid State Ionics 60(1–3):199–211
4. Narayanan S, Valdez T, Rohatgi N, Chun W, Hoover G, Halpert G (eds) (1999) Recent advances in direct methanol fuel cells. In: The fourteenth annual conference on battery conference on applications and advances, 12–15 Jan
5. Antonucci V, Arico A, Modica E, Creti P, Staiti P, Antonucci P (2001) Polymer-silica composite membranes for direct methanol fuel cells. Stud Surf Sci Catal 140:37–45
6. Park Y-S, Yamazaki Y (2005) Low methanol permeable and high proton-conducting Nafion/calcium phosphate composite membrane for DMFC. Solid State Ionics 176(11):1079–1089
7. Song M-K, Park S-B, Kim Y-T, Kim K-H, Min S-K, Rhee H-W (2004) Characterization of polymer-layered silicate nanocomposite membranes for direct methanol fuel cells. Electrochim Acta 50(2):639–643
8. Silva R, Passerini S, Pozio A (2005) Solution-cast Nafion®/montmorillonite composite membrane with low methanol permeability. Electrochim Acta 50(13):2639–2645
9. Tricoli V, Nannetti F (2003) Zeolite-Nafion composites as ion conducting membrane materials. Electrochim Acta 48(18):2625–2633
10. Baglio V, Arico A, Di Blasi A, Antonucci P, Nannetti F, Tricoli V et al (2005) Zeolite-based composite membranes for high temperature direct methanol fuel cells. J Appl Electrochem 35 (2):207–212
11. Wang J, Zheng X, Wu H, Zheng B, Jiang Z, Hao X et al (2008) Effect of zeolites on chitosan/zeolite hybrid membranes for direct methanol fuel cell. J Power Sour 178(1):9–19
12. Pace GG, Montes A, Rodriguez G (2000) Zeolitas: características, propiedades y aplicación industriales: editorial Innovación Tecnológica, Facultad de Ingeniería, UCV
13. Breck DW (1974) Zeolite molecular sieves: structure, chemistry, and use. Wiley, 771 p
14. Inglezakis V, Hadjiandreou K, Loizidou M, Grigoropoulou H (2001) Pretreatment of natural clinoptilolite in a laboratory-scale ion exchange packed bed. Water Res 35(9):2161–2166
15. Jacobs P, Flanigen E, Jansen J, van Bekkum H (2001) Introduction to zeolite science and practice. Elsevier
16. Baerlocher C, McCusker LB, Olson DH (2007) Atlas of zeolite framework types. Elsevier
17. Li X, Roberts EP, Holmes SM, Zholobenko V (2007) Functionalized zeolite A–nafion composite membranes for direct methanol fuel cells. Solid State Ionics 178(19):1248–1255
18. Holmberg BA, Wang X, Yan Y (2008) Nanocomposite fuel cell membranes based on Nafion and acid functionalized zeolite beta nanocrystals. J Membr Sci 320(1):86–92
19. Han W, Kwan SM, Yeung KL (2012) Zeolite applications in fuel cells: water management and proton conductivity. Chem Eng J 187:367–371
20. Chen P, Schönebaum S, Simons T, Rauch D, Dietrich M, Moos R et al (2015) Correlating the integral sensing properties of zeolites with molecular processes by combining broadband impedance and DRIFT spectroscopy—a new approach for bridging the scales. Sensors 15 (11):28915–28941
21. Yeung KL, Han W (2014) Zeolites and mesoporous materials in fuel cell applications. Catal Today 236:182–205

22. Simon U, Flesch U, Maunz W, Müller R, Plog C (1998) The effect of NH 3 on the ionic conductivity of dehydrated zeolites Nabeta and Hbeta. Microporous Mesoporous Mater 21(1):111–116
23. Kreuer KD, Rabenau A, Weppner W (1982) Vehicle mechanism, a new model for the interpretation of the conductivity of fast proton conductors. Angew Chem, Int Ed Engl 21 (3):208–209
24. Franke M, Simon U (1999) Proton mobility in H-ZSM5 studied by impedance spectroscopy. Solid State Ionics 118(3):311–316
25. Saad KB, Hamzaoui H, Mohamed M (2007) Ionic conductivity of metallic cations encapsulated in zeolite Y and mordenite. Mater Sci Eng, B 139(2):226–231
26. Franke ME, Simon U (2004) Solvate-supported proton transport in zeolites. Chem Phys Chem 5(4):465–472
27. Ruthven DM, Post MF (2001) Diffusion in zeolite molecular sieves. Stud Surf Sci Catal 137:525–577
28. Knudsen N, Andersen EK, Andersen IK, Skou E (1989) Tin-mordenites, syntheses and ionic conductivity. Solid State Ionics 35(1–2):51–55
29. Li Y, Guan H-M, Chung T-S, Kulprathipanja S (2006) Effects of novel silane modification of zeolite surface on polymer chain rigidification and partial pore blockage in polyethersulfone (PES)–zeolite A mixed matrix membranes. J Membr Sci 275(1):17–28
30. Duval JM, Kemperman A, Folkers B, Mulder M, Desgrandchamps G, Smolders C (1994) Preparation of zeolite filled glassy polymer membranes. J Appl Polym Sci 54(4):409–418
31. Metın D, Tihminlioğlu F, Balköse D, Ülkü S (2004) The effect of interfacial interactions on the mechanical properties of polypropylene/natural zeolite composites. Compos A Appl Sci Manuf 35(1):23–32
32. Alberti G, Casciola M (2003) Composite membranes for medium-temperature PEM fuel cells. Annu Rev Mater Res 33(1):129–154
33. Boom J, Pünt I, Zwijnenberg H, De Boer R, Bargeman D, Smolders C et al (1998) Transport through zeolite filled polymeric membranes. J Membr Sci 138(2):237–258
34. Mahajan R, Koros WJ (2002) Mixed matrix membrane materials with glassy polymers. Part 1. Polym Eng Sci 42(7):1420–1431
35. Libby B, Smyrl W, Cussler E (2003) Polymer-zeolite composite membranes for direct methanol fuel cells. AIChE J 49(4):991–1001
36. Freeman BD (1999) Basis of permeability/selectivity tradeoff relations in polymeric gas separation membranes. Macromolecules 32(2):375–380
37. Yahiro H, Konda Y, Okada G (2003) Conductivity of zeolite/poly(tetrafluoroethylene) composite membrane in the presence of water vapor. Phys Chem Chem Phys 5(3):620–623
38. Nur H, Kee GL, Hamdan H, Mahlia TMI, Efendi J, Metselaar HSC (2012) Organosulfonic acid functionalized zeolite ZSM-5 as temperature tolerant proton conducting material. Int J Hydrogen Energy 37(17):12513–12521
39. Chen Z, Holmberg B, Li W, Wang X, Deng W, Munoz R et al (2006) Nafion/zeolite nanocomposite membrane by in situ crystallization for a direct methanol fuel cell. Chem Mater 18(24):5669–5675
40. Han W, Cheung CT, Poon HY, Yeung KL (2012) A new structured composite membrane for fuel cell applications. Catal Today 193(1):194–199
41. Han W, Kwan SM, Yeung KL (2010) Zeolite proton conducting membrane for micro fuel cell applications. Top Catal 53(19–20):1394–1400
42. Yildirim MH, Curos AR, Motuzas J, Julbe A, Stamatialis DF, Wessling M (2009) Nafion®/H-ZSM-5 composite membranes with superior performance for direct methanol fuel cells. J Membr Sci 338(1):75–83
43. Changkhamchom S, Sirivat A (2014) High proton conductivity ZSM-5/sulfonated poly(ether ketone ether sulfone) (S-PEKES) composite proton exchange membrane for using in direct methanol fuel cell. Solid State Ionics 263:161–166

44. Yu DM, Yoon YJ, Kim T-H, Lee JY, Hong YT (2013) Sulfonated poly(arylene ether sulfone)/sulfonated zeolite composite membrane for high temperature proton exchange membrane fuel cells. Solid State Ionics 233:55–61
45. Sancho T, Soler J, Pina M (2007) Conductivity in zeolite–polymer composite membranes for PEMFCs. J Power Sour 169(1):92–97
46. Sancho T, Lemus J, Urbiztondo M, Soler J, Pina M (2008) Zeolites and zeotype materials as efficient barriers for methanol cross-over in DMFCs. Microporous Mesoporous Mater 115 (1):206–213
47. Tripathi BP, Kumar M, Shahi VK (2009) Highly stable proton conducting nanocomposite polymer electrolyte membrane (PEM) prepared by pore modifications: an extremely low methanol permeable PEM. J Membr Sci 327(1):145–154
48. Cui Y, Baker AP, Xu X, Xiang Y, Wang L, Lavorgna M et al (2015) Enhancement of Nafion based membranes for direct methanol fuel cell applications through the inclusion of ammonium-X zeolite fillers. J Power Sour 294:369–376
49. Kongkachuichay P, Pimprom S (2010) Nafion/Analcime and Nafion/Faujasite composite membranes for polymer electrolyte membrane fuel cells. Chem Eng Res Des 88(4):496–500
50. Intaraprasit N, Kongkachuichay P (2011) Preparation and properties of sulfonated poly(ether ether ketone)/Analcime composite membrane for a proton exchange membrane fuel cell (PEMFC). J Taiwan Inst Chem Eng 42(1):190–195
51. Devrim Y, Albostan A (2015) Enhancement of PEM fuel cell performance at higher temperatures and lower humidities by high performance membrane electrode assembly based on Nafion/zeolite membrane. Int J Hydrogen Energy 40(44):15328–15335
52. Mecheri B, Felice V, Zhang Z, D'Epifanio A, Licoccia S, Tavares AC (2012) DSC and DVS investigation of water mobility in nafion/zeolite composite membranes for fuel cell applications. J Phys Chem C 116(39):20820–20829
53. Patel R, Park JT, Lee WS, Kim JH, Min BR (2009) Composite polymer electrolyte membranes comprising P (VDF-co-CTFE)-g-PSSA graft copolymer and zeolite for fuel cell applications. Polym Adv Technol 20(12):1146–1151
54. Şengül E, Erdener H, Akay RG, Yücel H, Bac N, Eroğlu İ (2009) Effects of sulfonated polyether-etherketone (SPEEK) and composite membranes on the proton exchange membrane fuel cell (PEMFC) performance. Int J Hydrogen Energy 34(10):4645–4652
55. Ma J, Sahai Y (2013) Chitosan biopolymer for fuel cell applications. Carbohyd Polym 92 (2):955–975
56. Feketeföldi B, Cermenek B, Spirk C, Schenk A, Grimmer C, Bodner M et al (2016) Chitosan-based anion exchange membranes for direct ethanol fuel cells. J Membr Sci Technol
57. Djelad A, Morsli A, Robitzer M, Bengueddach A, Di Renzo F, Quignard F (2016) Sorption of Cu (II) ions on chitosan-zeolite X composites: impact of gelling and drying conditions. Molecules 21(1):109
58. Muthumeenal A, Neelakandan S, Kanagaraj P, Nagendran A (2016) Synthesis and properties of novel proton exchange membranes based on sulfonated polyethersulfone and *N*-phthaloyl chitosan blends for DMFC applications. Renew Energy 86:922–929
59. Karimi A, Othman A, Uzunoglu A, Stanciu L, Andreescu S (2015) Graphene based enzymatic bioelectrodes and biofuel cells. Nanoscale 7(16):6909–6923
60. Yuan W, Wu H, Zheng B, Zheng X, Jiang Z, Hao X et al (2007) Sorbitol-plasticized chitosan/zeolite hybrid membrane for direct methanol fuel cell. J Power Sour 172(2):604–612
61. Wu H, Zheng B, Zheng X, Wang J, Yuan W, Jiang Z (2007) Surface-modified Y zeolite-filled chitosan membrane for direct methanol fuel cell. J Power Sour 173(2):842–852
62. Wang Y, Yang D, Zheng X, Jiang Z, Li J (2008) Zeolite beta-filled chitosan membrane with low methanol permeability for direct methanol fuel cell. J Power Sour 183(2):454–463
63. Wang Z, Yan Y (2016) Zeolite thin films and membranes: from fundamental to applications. Zeolites in sustainable chemistry. Springer, pp 435–472

Chapter 5
Composite Membranes Based on Heteropolyacids and Their Applications in Fuel Cells

Ebrahim Abouzari-lotf, Mohamed Mahmoud Nasef, Masoumeh Zakeri, Arshad Ahmad and Adnan Ripin

Abstract Heteropolyacids (HPAs) are a class of inorganic materials that have been widely used as additives to enhance the performance of fuel cell membranes, recently. This chapter covers the use of HPAs in the preparation of proton exchange membranes (PEM) for polymer electrolyte membrane fuel cells (PEMFCs). The fundamental aspects of HPAs and their corresponding salts in addition to various structural configurations such as Keggin, Wells–Dawson, and Lacunar are discussed. The use of HPAs for preparation of membranes for high-temperature PEMFC and direct methanol fuel cell (DMFC) based on the immobilization on various substrates including perfluorinated sulfonic acids (PFSAs), aromatic hydrocarbons, poly(vinyl alcohol) (PVA), and polybenzimidazole (PBI) are reviewed. The research challenges that need to be addressed to bring the new composite membranes to practical application are also discussed.

E. Abouzari-lotf · M.M. Nasef (✉) · M. Zakeri · A. Ahmad · A. Ripin
Advanced Materials Research Group, Center for Hydrogen Energy,
Institute of Future Energy, Universiti Teknologi Malaysia, Kuala Lumpur, Malaysia
e-mail: mahmoudeithar@cheme.utm.my

E. Abouzari-lotf
e-mail: ebrahim@utm.my

M. Zakeri
e-mail: ms.zakeri@gmail.com

A. Ahmad
e-mail: arshad@cheme.utm.my

A. Ripin
e-mail: adnan@cheme.utm.my

A. Ahmad · A. Ripin
Department of Chemical Engineering, Universiti Teknologi Malaysia,
Johor Bahru 81310, Malaysia

M.M. Nasef
Malaysia–Japan International Institute of Technology, Universiti Teknologi Malaysia,
Kuala Lumpur, Malaysia

Inamuddin et al. (eds.), *Organic–Inorganic Composite Polymer Electrolyte Membranes*, DOI 10.1007/978-3-319-52739-0_5

Keywords Proton exchange membranes · Heteropolyacids · High-temperature PEMFC · DMFC

List of abbreviations

CS-HEC	Chitosan-hydroxy ethyl cellulose
CsPW	Cs hydrogen salts of phosphotungstic acid
HPA	Heteropolyacid
LbL	Layer by layer
mGO	Graphene oxide modified with 3-aminopropyl-triethoxysilane
MMT	K10 montmorillonite
MO	Metal oxide
MOR	Methanol oxidation reaction
ORR	Oxygen reduction reaction
PBI	Polybenzimidazole
PDDA	Poly(diallyl dimethyl ammonium chloride)
PEFC	Polymer electrolyte fuel cell
PEMFC	Polymer electrolyte fuel cell
PFSA	Perfluorosulfonic acid
PMA	Molybdophosphoric acid or phosphomolybdic acid, $H_3PMo_{12}O_{40}$
POM	Polyoxometalate
Ppy	Polypyrrole
PVA	Poly(vinyl alcohol)
PWA	Phosphotungstic acid, $H_3PW_{12}O_{40}$
QDPSU	Quaternary diazabicyclo-octane polysulfone
rGO	Reduced graphene oxide
SiMA	Silicomolybdic acid, $H_4SiMo_{12}O_{40}$
SiWA	Silicotungestico acid $H_4SiW_{12}O_{40}$
SPAEK	Sulfonated poly(aryl ether ketone)
SPBN	Sulfonated polynorbornene
SPEEK	Sulfonated poly(ether ether ketone)s
SPS	Sulfonated polystyrene
ZrP	Zirconium phosphate

1 Introduction

The interest in developing highly stable and cost-effective new proton exchange membranes for polymer electrolyte fuel cells (PEMFCs) has been ever growing to promote the commercialization for this type of fuel cell [1–12]. This is to overcome the limitation of perfluorosulfonic acid (PFSA) membranes, which rely on water in proton conduction that prevent the operation of PEMFCs above 80 °C [13–15]. Higher fuel cell operation is desired to enhance reaction kinetics of electrochemical reactions, increase CO tolerance of electrodes and reduce or eliminate water management system [16–18].

Various approaches have been used to develop alternative membranes including modification of commercially available PFSA membranes, chemical synthesis of sulfonate hydrocarbon membranes, and formation of composite membranes [19]. Particularly, composite ionic membranes composed of organic–inorganic components have received great attention in the past decade because of their ability to be operated under wide ranges of temperatures including much higher ones than membranes made of the pure polymers [20]. Moreover, the incorporation of inorganic fillers provides a number of advantages ranging from improvement in the mechanical properties and water management of the membrane to inhibition of the fuel crossover by increasing the tortuosity of transport pathways. These composite membranes can be formed by: (1) introducing nanosized hygroscopic inorganic fillers such as SiO_2, ZrO_2, and TiO_2, (2) doping of basic substrates with nonvolatile proton conducting solvents such as phosphoric acid, N-heterocycles, heteropolyacids (HPAs), and ionic liquids, and (3) incorporation of solid proton conductors such as phosphate salts of zirconium (ZrP) and HPAs. Various synthesis routes can be used to introduce one phase into the other one. A review on various methods of introducing each phase in organic/inorganic composite membrane is available [21].

Among composite membranes, those doped with or containing solid proton conductors such as HPAs are promising candidates for PEMFC [22, 23, 21]. This is because HPAs display the highest proton conductivity among inorganic solids near ambient temperatures. They also have distinct discrete ionic structures including heteropoly anions and counter cations (e.g., H^+, H_3O^+, and $H_5O_2^+$) and therefore exhibit high-proton mobility. Moreover, the presence of bounded water molecule in HPAs makes them interesting for the development of moderate temperature PEMFC since proton conduction is independent of external humidification [2, 23].

The objective of this chapter is to review the progress of developments in regard to composite ionic membranes based on HPAs. A special attention is given to the types and structures of HPAs and their use in synthesis of nanocomposite polymeric materials with ion conducting functional groups for various types of PEMFC applications. The challenges remaining for further research to achieve important breakthroughs are also discussed.

2 Heteropolyacids Types and Structures

Polyoxometalates (POMs) are group of chemicals that have attracted an increasing attention due to the diversity and selectivity of their properties making them suitable for wide number of applications [24]. HPAs are complex proton acids incorporating polyoxometalate anions (heteropolyanions) having metal-oxygen octahedral as basic structural units [25, 26]. Keggin and Wells–Dawson are two important categories of the HPAs. In 1933, Keggin used X-ray diffraction technique to determinate the $H_3[PW_{12}O_{40}]5H_2O$ structure and provided a clear description of the bonds between the WO_6 octahedra of the molecule [27]. The Wells–Dawson structure was theoretically described by Wells in 1945 and experimentally confirmed in 1953 [28].

HPAs are subsets of metal oxides distinguished by central heteroatoms, which are surrounded by a number of metal-oxygen octahedra. There are strong bonds between the atoms supporting the polyhedra structures and connect them with the heteroatoms. The metal in HPAs is usually tungsten or molybdenum and less usually vanadium or uranium. The heteropoly anions having metal-oxygen octahedral as the basic structural units make up the primary structure of the HPA as shown in the Keggin anion (Fig. 1).

The Keggin HPAs, which is the most studied, represented by the formula $X^{x+}M_{12}O_{40}{}^{x-8}$, contain 12 addenda atoms and one heteroatom, where *X* is the heteroatom (central atom, commonly P^{5+}, Si^{4+}, or B^{3+}), *x* is its oxidation state, and *M* is the addenda atom (metal ion, commonly W and Mo). It is composed of a central tetrahedron (XO_4), which is surrounded by 12 edge- and corner-sharing metal oxide octahedra (MO_6). The octahedra are arranged in four M_3O_{13} groups, and each group has three octahedra sharing edges. They are linked through strong oxygen bonds, which also connect the central tetrahedron [29]. A typical FTIR-ATR spectrum of Keggin-type phosphotungstic acid (PWA) showing four different W–O bonds and characteristic bands at 1080 (P–O_a), 983 (W = O_d), 893 (W–O_b–W), and 797 (W–O_c–W) cm^{-1} is shown in Fig. 2.

Due to the complexity, other polyoxometalates including Wells–Dawson heteropolyanions $X_2{}^{x+}M_{18}O_{62}{}^{2x-16}$, Keggin and Dawson lacunary anions,

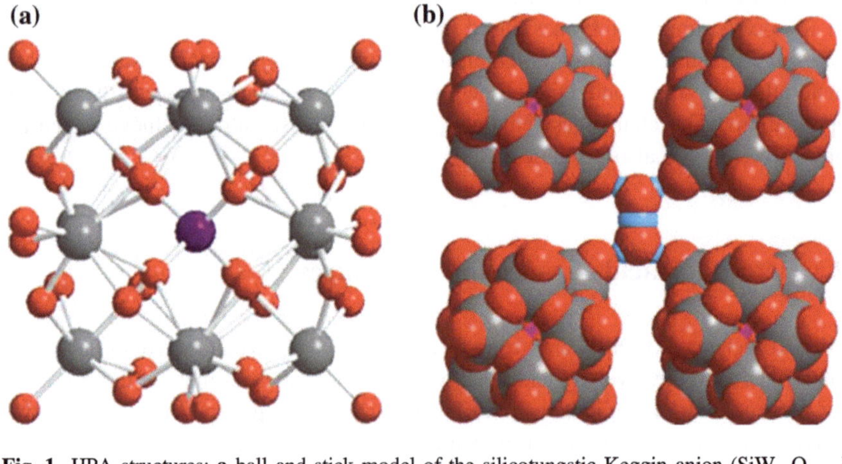

Fig. 1 HPA structures: **a** ball and stick model of the silicotungstic Keggin anion ($SiW_{12}O_{40-4}$) and **b** space filling model of the secondary crystalline structure of Keggin HPA with base-centered cubic arranged cavities containing $H_5O_2{}^+$ cations (*red* O, *gray* W, *blue* H, *purple* P, or Si)

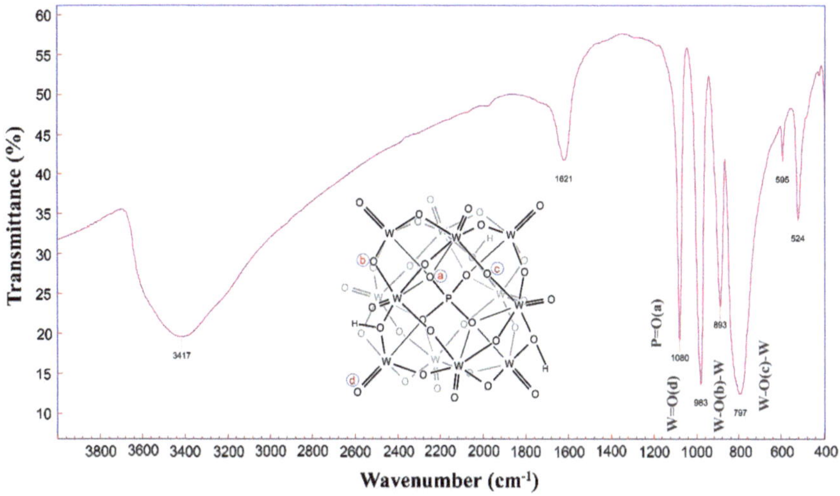

Fig. 2 FTIR-ATR spectrum of Keggin-type PWA showing four characteristic bands

$XM_{11}O_{39}^{x-12}$ and $X_2M_{17}O_{61}^{2x-20}$, and transition metal complexes have not been fully examined.

In addition to metals, HPAs are interestingly having multiple active sites in their structure including protons and oxygens. Functionally, protons act as Brønsted acids to promote acid-catalyzed process including proton transports. Besides, some oxygen atoms on the surface of HPA (especially oxygens located on the lacunary sites of lacunary POM anions with a high negative charge) are basic enough to react with protons and hydronium ions (protonate) and can practically behave as active sites in base-catalyzed reactions. On the other hand, the metals of HPAs are active sites in all oxidative reactions. The details of various catalytic activities of HPAs were reviewed [30–32].

Although Dawson HPAs were used in various applications including proton transport materials [33], majority of the patents and investigations are based on the applications of the Keggin-type HPAs and their salts. This primarily includes $H_3PMo_{12}O_{40}$ (PMA), $H_3PW_{12}O_{40}$ (PWA), $H_4SiMo_{12}O_{40}$ (SiMA), and $H_4SiW_{12}O_{40}$ (SiWA). Solid HPAs commonly form ionic crystals composing of anions (which is heteropolyanions), cationic counterions (H^+, H_3O^+, $H_5O_2^+$, etc.), and hydration water. The HPAs are noticeably crystallizing with a large number of water molecules as in PWA and PMA, both of which have 29 water molecules compared to 30 for SiWA [34, 35]. The status of water is important for assessing HPA role as a proton transfer material. Some HPAs loss water very easily even at room temperature where as others can retain water tightly even at high temperature.

In HPA crystals, the Keggin anions are quite mobile. Due to structural flexibility and mobility of HPAs, not only water but also a diversity of polar molecules can enter and leave HPA crystals that is important when using HPAs in fuel cell applications. It should be noted that proton conductivity of solids is generally related to their acid-based catalytic activities [36]. HPAs display strong acidity, good solubility in polar solvents such as water, lower alcohols, ketones, ethers and esters, in addition to good thermal stability and reversible redox behavior. One of the most important properties of HPAs is the ability to be chemically adjusted or tuned by simple modification of their structures. The acidity of the HPAs is dependent upon the total charge of the anion and the heteroatom in their structure [37–39]. The strong acidity level means fast proton mobility, which results in a high catalytic activity [29]. The acidity strength of crystalline HPAs decreases in the series PW > SiW $\geq$ PMo > SiMo, which is identical to that in solutions [40].

Due to their versatile properties, HPAs have been used in an extensive range of applications, including chemical analysis, biochemistry, ion selective membranes, sensors, and electrochemical energy devices [41, 42]. Particularly, HPAs have been useful constituent of polymer electrolyte fuel cells (PEFCs) and their corresponding fuel cells using liquid fuels such as methanol or ethanol. Intensive work has been reported on membranes incorporating HPAs for fuel cells, specifically, saturated HPAs and unsaturated (defective) or lacunary HPAs were evaluated for fuel cell applications [43]. Figure 3 compares the structure of saturated HPA of $[SiW_{12}O_{40}]^{4-}$ in a hydrated form of $H_4[SiW_{12}O_{40}]22H_2O$, which is also known as 12-HSiW and one of the corresponding lacunary structures $[SiW_{11}O_{39}]^{8-}$ in a hydrated form of $H_8[SiW_{11}O_{39}]26H_2O$ known as 11-HSiW. As shown, the number of acidic hydrogens and water molecules is higher in lacunary HPAs. Interestingly, lacunary HPAs can be simply prepared by hydrolysis of the parent HPAs with a base under controlled conditions of temperature and ionic strength. Removal of one, two, or three metal oxide (MO) units from the parent HPAs results in a mono-, di-, or tri-lacunary POMs. A general reaction scheme for the preparation of various lacunary HPAs is shown Fig. 4.

Lacunary HPAs could be further modified by bonding with functional silanes that can subsequently crosslinked or polymerized to achieve more stable HPAs.

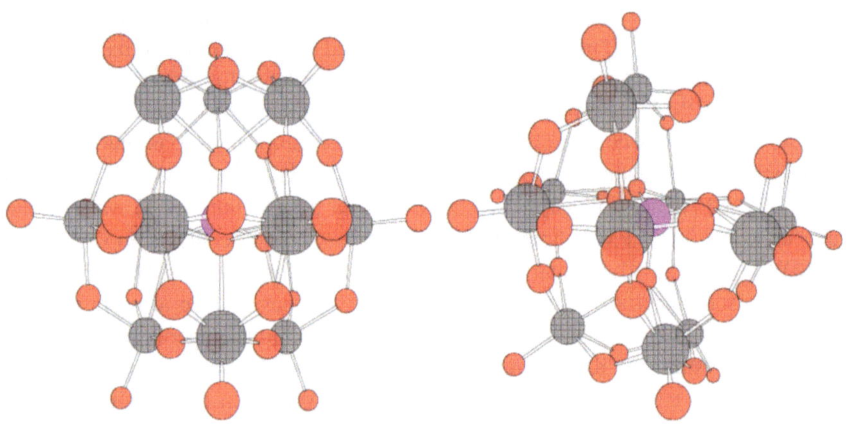

Fig. 3 Comparison of HPA and corresponding mono-lacunary structures

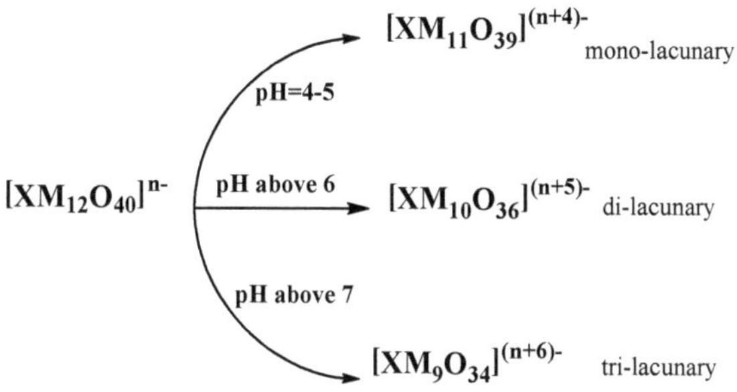

Fig. 4 General pathways for preparation of various lacunary HPAs

Such HPAs offer high ionic conductivities prompting their use in fuel cells. For example, indium-substituted HPAs with Keggin $H_4[In(H_2O)PW_{11}O_{39}]11H_2O$ and $H_5[In(H_2O)SiW_{11}O_{39}]8H_2O$, structures were reported to be solid high-proton conductors with a conductivity of 2.60×10^{-4} S cm^{-1} and 5.25×10^{-4} S cm^{-1} (at 18 °C and 80% relative humidity), respectively. The conductivity of these materials also increased with the rise of the temperature [44].

The conductivity performance of HPAs varies widely depending upon the type of the HPAs' structure. Particularly, the change in the constitutional elements of polyanion can lead to the difference in the degree of hydration and proton mobility [45]. Typically, $H_5[SiW_{11}VO_{40}]15H_2O$ shows an excellent conductivity of 7.93×10^{-3} S cm^{-1} at 15 °C and 50% RH [38]. In addition, the size of the HPAs influences considerably their performances specially when they are incorporated in the composite membranes [46].

Generally, HPAs have high thermal stability, in the order of PW > SiW > PMo > SiMo (Keggin types) that decompose at 465, 445, 375, and 350 °C, respectively. On the other hand, the influence of the heteroatom on the thermal stability is negligible [47]. The redox activity of HPAs promotes their application as redox bi-functional catalysts. With increasing electronegativity of the addenda atom, the reduction potential of the HPA decreases, while increases as the electronegativity of the counter cations or the heteroatoms increases, but it decreases as the electronegativity of the addenda atom increases [28]. The oxidation potential decreases in the Keggin structures as PMo > SiMo $\gg$ PW > SiW.

HPAs have been used in an extensive range of applications, including chemical analysis, biochemistry, ion selective membranes, sensors, and electrochemical energy devices [41, 42]. Most generally, they are used as inorganic catalysts. Depending on their composition, HPAs can be effective catalysts for both acid catalysis and redox catalysis [47]. One of the important specific characteristics of the HPAs is that they can be tuned by adding different elements to their structures, allowing for the design of HPAs for special reactions. The high selectivity, thermal stability as well as hydrolytic stability, long lifetimes in solution, high oxidation potential, and noncorrosive nature of many HPAs make them economically and environmentally attractive as a natural 'green catalysts' [48]. In addition, the list of proteins or enzymes that can interact with HPA is long which supports the use of HPAs in a number of biochemical and biomedical applications [49]. Moreover, due to antiviral and antitumor activity of HPAs, these find use in biomedicine applications [50, 51].

The appealing applications of HPAs as dopants and fillers in ionic composite membranes are indicated by the large number of patents and publications appeared recently [2, 52, 53, 37, 54, 55, 56]. This was motivated by their high ionic conductivity, thermal stability, very strong Brønsted acidity approaching the superacid region (more acidic than sulfuric acid and Nafion), and their capability to undergo redox processes under the mild conditions. Such membranes also found uses in ion selective electrodes, gas detection apparatuses, solid-state electrochromic devices, and solid or liquid electrolytes in electrolytic cells [57, 58, 59, 31, 32]. A great deal of work has been focused on the self-immobilization of POMs especially the Keggin type into the polymers matrices such as polypyrrole, polythiophene, and polyaniline to increase the performance of hybrid membranes.

3 HPAs and Proton Transport in Fuel Cells

Due to the dissociated protons combined with the anions having good mobility, HPAs are excellent proton conductors and promising solid electrolytes. The interest in HPAs began after reporting remarkably high conductivities of 2×10^{-1} S cm^{-1} at 25 °C, and low activation energies of 15.5 and 13.7 kJ/mol for crystals of $H_3[Mo_{12}PO_{40}]29H_2O$ and $H_3[W_{12}PO_{40}]29H_2O$ [60], respectively. However, upon application for fuel cell, keeping the samples in their fully hydrated form without decomposition is difficult since HPAs dissolve in the water that formed in a fuel cell during the operation [61]. Highly concentrated aqueous solution of HPA was reported for the use as a fuel cell electrolyte, which resulted in a comparable performance with PEMFC [62, 63]. Although stable cell performance for HPA electrolyte was reported for 300 h, however, the high crossover results in voltage loss of fuel cell.

HPAs with high stability in water have been obtained upon immobilization on clay [64], aluminum phosphate [53], and silica supports [65]. It was reported that the addition of aluminum phosphate in SPEEK/PWA remarkably increased water retention and effectively reduced leaching of PWA from the composite membrane [53]. On the other hand, the proton conductivity of mesoporous silica is as low as 10^{-6}–10^{-4} S cm^{-1}. However, its remarkable structural order, large surface area and pore volumes, availability and simple functionalization make it an ideal porous framework for HPA based proton conductors [47, 66, 67].

Typically, an inorganic glass composite membrane containing a mixture of PWA and PMAPWA/PMA–P_2O_5–SiO_2 resulted in very highly conductivity values of 1.014 S cm^{-1} at 30 °C and 85% RH [68] and 1.01×10^{-1} S cm^{-1} at 85 °C under 85% RH for a mesoporous-structured PWA–P_2O_5–SiO_2 glass [69]. The fuel cell performance of these inorganic materials was 35–42 mW cm^{-2} in H_2/O_2 at 30% RH and 30 °C [68, 69]. However, they cannot be used as electrolytes in fuel cells in their powdered form. Therefore, composite membranes with a polymer binder have been developed for fuel cell application.

Several approaches to incorporate HPA into fuel cell membranes include fabricating unsupported HPA pellets [61], doping polymer electrolytes [70], and sol–gel methods [71]. Among these, HPAs have been mainly used as dopants in composite proton exchange membranes. It was suggested that specific interactions between HPAs and polymers could have a significant influence on the fuel cell performance at elevated temperatures and thus they have received considerable attention [71]. More details of various preparation methods can be found elsewhere [72–75]. Five different procedures were commonly used to produce such composite membranes as schematized in Fig. 5. They include mixing of HPA with a polymer in a solution form followed by casting [76, 77] or impregnation of HPA in a porous substrate [78, 70]. HPAs were supported on various substrates including silica. When preparing the membrane from a polymer solution, the incorporation of HPAs supported on silica within a polymeric membrane can be achieved by two different ways: (i) pre-formation of HPA–silica particles followed by direct mixing with a polymer solution [65, 79] and (ii) in situ preparation of the inorganic phase within the matrix during film formation, e.g., via the sol–gel process of an alkoxide precursor in the presence of PWA [80–82]. Particle size and distribution heterogeneity are two main concerns when using the first method. However, the sol–gel method is normally lead to better distribution, more stability and possible control for the size

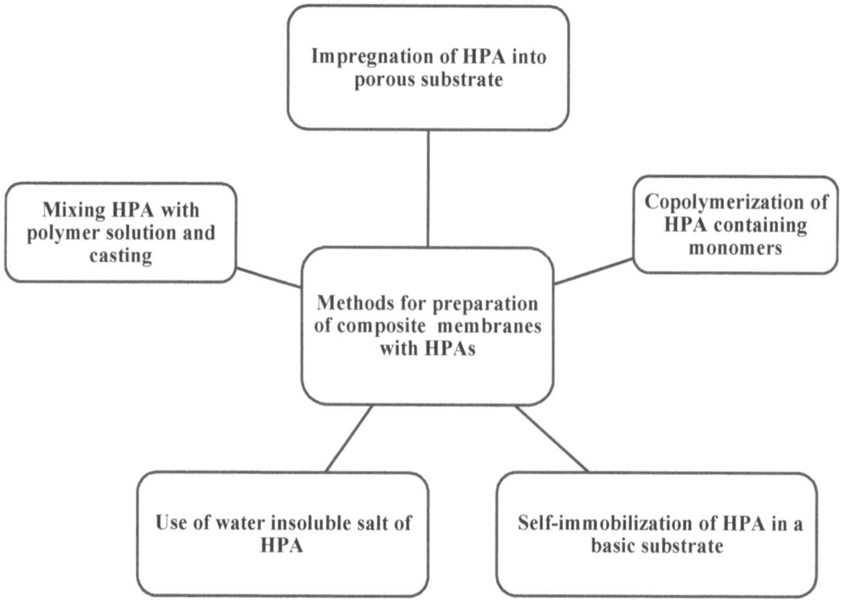

Fig. 5 Various methods for preparation of composite proton exchange membranes containing HPAs

of the particles by variation of the preparation parameters such as pH, temperature, and concentration of starting components [83, 84].

HPA supported on the polymers and porous fillers are likely subjected to leaching out of HPA during the fuel cell operation, despite the improvement took place in other properties including proton conductivities [85]. Such limitation is due to low loading level, loose interaction with solid organic substrate and possible agglomeration leading to the formation of large particles or clusters in the polymer substrate during the casting process [86, 87]. To overcome this problem, a third effective method involving copolymerization of HPA containing monomers or covalent bonding of HPA to a polymer backbone was proposed [87, 88, 89, 43]. The covalent bonding of dopant to the polymer backbone minimizes leaching and enhances the immobilization level of HPAs. Nevertheless, this synthetic procedure is costly and, therefore, the application of this method was limited to few examples [52].

The forth method involves a self-immobilization of HPAs onto basic polymer backbones typically in a fiber form [2, 52] or adding another filler with high specific surface area such as reduced graphene oxide (rGO) [54] or rGO covalently modified with 3-aminopropyl-triethoxysilane (mGO) [90]. For example, amounts of up to 50 wt% of PWA could be simply immobilized onto nylon nanofibers by self-immobilization. The fifth proposed method for solving the leaching problem of HPAs was to use water-insoluble salt of HPAs such as $Cs_xH_{3-x}PW_{12}O_{40}$ [91]. In this regard, HPA salts such as CsH_2PO_4 have been intensively investigated because of their desirable high operating temperature (230–260 °C) [92, 93] and high-protonic conductivity ($\sigma = 2.2 \times 10^{-2}$ S cm^{-1} at 240 °C) [94]. Such advantages explain the reason for applying the compositions of CsH_2PO_4 with oxyanions [95], cations [96], and inorganic/organic scaffolds [97, 98] for improving proton conductivity, mechanical properties, and thermal stability at high temperatures in addition to reducing methanol permeability.

4 HPAs in PEM Fuel Cell

HPAs have been used extensively in the PEMFC as additive to the catalyst layer and membrane. As additive to the catalyst layer, HPAs have been used in both cathode and anode mainly to address the CO-poisoning by enhancing the tolerance. Particularly, it was reported that the addition of PMA and corresponding vanadium substituted analogs into Pt results in a remarkable performance improvements when the fuel cell was fed with reformed hydrogen containing 100 ppm CO [99]. It was concluded that PMA acts as an oxidizing agent for the reaction between CO and water in the presence of Pt, Pd, and Au [100, 22]. Besides, the addition of HPAs in the electrode is likely to improve the catalytic activity towards oxygen reduction reaction (ORR) and methanol oxidation reaction (MOR) [101, 55, 102]. However, HPA applications in the catalyst layers of electrodes are beyond the scope of this chapter and more details can be found elsewhere [22].

PFSAs are the choice materials for the membranes for PEMFC applications. However, these membranes are not gas tight enough especially at temperature above 80 °C [20, 5, 103]. Several low cost alternatives such as poly(vinyl alcohol) (PVA), sulfonated poly(ether ether ketone)s (SPEEK), and sulfonated polystyrene (SPS) have been evaluated for PEMFC applications. These membranes exhibited good chemical stability and low fuel crossover. However, their proton conductivity is significantly lower than PFSAs. Typically, the proton conductivity of PVA and SPEEK are in the ranges of 10^{-14}–10^{-10} S cm^{-1} [22, 104] and 10^{-4}–10^{-3} S cm^{-1} [83], respectively. HPAs have been introduced to various low cost PFSA alternatives to enhance the proton conductivity. This led to an increase in the conductivity up to the range of 10^{-3}–10^{-1} S cm^{-1} [83, 105, 106]. Interestingly, such improvement in the conductivity was accompanied by an increase in the thermal stability of membranes.

5 HPAs in High-Temperature and Low-Humidity PEMFC

The main limitation of PFSA family of membranes is their dependence on level of hydration to achieve high conductivity, and this poses a problem when operating at temperatures higher than 80 °C as the proton conductivity of the membranes decreased significantly. The variation of hydration level undermines the permeability of H_2 and O_2 due to the involvement of the formed hydrated ionic clusters in the gas permeation mechanism [107, 108]. On the other hand, the retention and management of water within these types of membranes is challenging, costly and in most cases affect the PEMFC performance. Therefore, there is a strong demand for the PEMs to work at lower relative humidity and/or under anhydrous conditions. However, acceleration of electrochemical reactions and simplification of water management system of fuel cells under lower humidity and/or anhydrous conditions is challenged by dramatic losses of proton conductivity of PFSA membranes [109, 110, 5]. Typically, Nafion 112 losses an order of magnitude of its proton conductivity at around 60% RH, which increases two orders of magnitude with the reduction of RH to 25%.

Various materials and designing strategies were proposed to prepare alternative proton conductors for high-temperature fuel cell operation under dry conditions. The use of other media of proton conductions or less-volatile liquids such as phosphoric and phosphonic acids [1, 111] and amine heterocycles [112] have been widely considered. However, in most cases, the proton conductivity is not comparable to PFSAs and there is a risk of liquid electrolytes leaching. On the other hand, addition of an inorganic material with hydrophilic properties such as HPAs was proposed to maintain high water content levels at elevated temperatures or under dry conditions. There is a linear relationship between water content of the membrane and its conductivity at lower RH%.

A lot of work has been reported on incorporation of different HPAs in PFSA membranes [76, 113, 114] and other hydrocarbon membranes [83, 115, 105, 106, 116, 117] for PEMFCs application. It was shown that the activation energy for proton transport (E_a) decreased significantly for PFSA membranes upon increasing the amount of HPAs under lower RH% [118]. For example, the addition of 5 wt% of PWA lowered E_a by 40% at 60% RH compared to pristine membrane. Initial investigation on various HPAs such as PWA, SiWA, PMA, and SiMA indicated that the additives containing molybdenum were less stable in the membrane environment than those containing tungsten [76]. Upon operating in fuel cell for a few hours, the amount of molybdenum-based additive present in the cathode catalyst layer increased. It was suggested that molybdates migrated into the catalyst layers, where they had a detrimental effect on the performance by undergoing parasitic redox reactions on the surfaces of carbon and platinum, resulting in an increase in the activation overpotential. Table 1 summarizes some of the efforts to introduce HPAs to membranes used in the PEMFC.

Initial studies reported the limiting changes in the stability of HPA in hydrated conditions and the strong influence of humidity on the proton conductivities and diffusional problems of the membranes. The high water solubility of HPAs triggers leaching problem and consequently high drop in a performance. Two different approaches of using alternative heteropolysalts instead of their acid and supported HPAs were widely used to avoid the leaching problem in high-temperature PEMs. More stable HPA containing membranes were achieved using heteropolysalts as fillers in membranes for PEMFCs. In this regard, larger cations such as Cs^+, NH_4^+, Rb^+, and Tl^+ [133] as well as organic cations such as 4,4′- and 2,2′-bipyridines [134] were combined with HPAs to produce heteropoly salts. Precipitation from aqueous solutions was normally used to achieve such partially substituted heteropolysalts [135, 136]. On the other hand, it was shown that a solid-state reaction involving mechano-chemical treatments using a high-energy ball mill is more efficient to prepare heteropolysalts and normally lead to more conductive heteropolysalts [137, 138], and subsequent composite membranes [115]. Additional investigation revealed that the reduction in the particle size facilitates the conductivity due to enhancement in the surface to volume ratio, which permits more efficient proton hopping and increase in conductivity [23]. For example, reducing a filler size from 1–2 mm to 30–50 nm resulted in 35% enhancement in proton conductivity [139].

In the second approach, HPAs or their corresponding salts were supported on different materials including SiO_2 [86, 121, 84, 122, 77, 140, 123], ZrO_2 [84, 120, 77], WO_2 [77], and TiO_2 [84, 77] followed by immobilization in polymeric membranes [86]. PWA was found to be the only heteropolyacid that has been found suitable for these hybrid membranes, and it was concluded that SiO_2 is the best supporting material since there is a strong interaction between the $H_3PW_{12}O_{40}$, Nafion®, and the SiO_2 molecules [121]. The proton conductivity and the fuel cell performance at 100% RH of the composite membrane close to those of pristine membranes. However, the merits of HPA became evident upon increasing the temperature or reducing the RH%. Improved performance under dry conditions was

Table 1 Properties of some HPA containing membranes for proton exchange membrane fuel cell application

Polymer substrate	HPA (loading level)	Thickness	Proton conductivity ($S\ cm^{-1}$)	Condition	Fuel cell performance[a] ($mA\ cm^{-2}$)	Refs.
Nafion®	NaPWA	175	2.3×10^{-2}	100 °C, 100% RH	136	[119]
	NaPWA	165	1.5×10^{-2}		140	
	NaPMA	160	1.3×10^{-2}		145	
Nafion®	PWA (25 wt%)	–	6×10^{-2}	80 °C, 75% RH	–	[76]
			1.5×10^{-2}	120 °C, 35% RH	–	
	SiWA (25 wt%)	–	8.5×10^{-2}	80 °C, 75% RH	–	
			1.0×10^{-2}	120 °C, 35% RH	–	
	PMA (25 wt%)	–	–	–	–	
	SiMA (25 wt%)	–	8×10^{-2}	80 °C, 75% RH	–	
			1.0×10^{-2}	120 °C, 35% RH	–	
	PWA-ZrO_2 (30 wt%)	50–75	6×10^{-3}	120 °C, 25% RH	–	[120]
	PWA-SiO_2	70 ± 5	1.19×10^{-3}	90 °C, 40% RH	82	[121]
	PWA-SiO_2 (10 wt%)	120	0.01	Conductivity:100 °C, 40% RH	540	[122]
				Fuel cell: 110 °C, 100% RH		
	PWA/ SiO_2 (14.3 wt%)	120	2.67×10^{-2}	110 °C, 70% RH	540	[77]
	PWA/meso-SiO_2	–	0.07	120 °C, 25 RH%	510	[123]
	PWA/mGO (1 wt%)	40	0.0092	80 °C, 20% RH	841	[90]
Chitosan	CsPWA (2 wt%)	30	0.028	25 °C	410	[124]
	CsPWA (2 wt%)	30	0.018	25 °C, low RH (T_{gas} = 25 °C)	350	[125]
	CsPWA (2 wt%)	3	0.014	25 °C, low RH (T_{gas} = 25 °C)	550	[56]

(continued)

Table 1 (continued)

Polymer substrate	HPA (loading level)	Thickness	Proton conductivity (S cm^{-1})	Condition	Fuel cell performance[a] (mA cm^{-2})	Refs.
SPEEK	CsPWA (20 wt%)	84	2.06×10^{-3}	80 °C, 80 RH%	245	[115]
	CsSiWA (20 wt%)	75	2.25×10^{-3}		247	
	CsPWA/SiO_2 (2 wt%)	30–40	6.2×10^{-3}	100 °C, 90% RH		[83]
	$H_4PW_{11}VO_{40}{\cdot}8H_2O$ (70 wt%)	31	10^{-2}	30 °C, 100% RH	–	[116]
	$H_4PW_{11}VO_{40}{\cdot}8H_2O$/rGO (70 wt%/ 2 wt%)	134	6.43×10^{-2}	65 °C, 65% RH	–	[54]
	PWA (10 wt%)/aluminum phosphate (3 wt%)	52	6.8×10^{-4}	30 °C, 100% RH	–	[53]
SPSU	SiO_2 (2 wt%)/PMA 20 wt%	50	Same to Nafion 112	30–80 °C, 100% RH	–	[126]
PVA	$H_5[SiW_{11}VO_{40}]15H_2O$ (70 wt%)	33	10^{-4}	30 °C, 100% RH	–	[116]
	PMA (33.3 wt%)		0.0101	–	–	[127]
PBI	PWA (30 wt%)/H_3PO_4	–	0.036	140 °C, 20% RH	–	[128]
	SiWA (30 wt%)/H_3PO_4	–	0.046	200 °C, 5% RH	–	[128]
	SiWA-SiO_2 (30 wt%)/H_3PO_4	50–80	0.206	150 °C	–	[129]
	$Cs_{2.5}H_{0.5}PMo_{12}O_{40}$	30	0.00004	150 °C, 0% RH		[130]
	$Cs_{2.5}H_{0.5}PMo_{12}O_{40}/H_3PO_4$	30	0.04	Conductivity: 8.4% RH, 180 °C Fuel cell: 150 °C, 0% RH	800	
	CsPOMo (30 wt%)/H_3PO_4	50 ± 3	0.12	150 °C, 0% RH	630	[131]
	CsPOW (30 wt%)/H_3PO_4		0.1		590	
	CsSiOMo (30 wt%)/H_3PO_4		0.051		480	
	CsSiOW (30 wt%)/H_3PO_4		0.057		510	

(continued)

Table 1 (continued)

Polymer substrate	HPA (loading level)	Thickness	Proton conductivity (S cm^{-1})	Condition	Fuel cell performance[a] (mA cm^{-2})	Refs.
	CsPWA/H_3PO_4		1.91×10^{-2}	160 °C, 0% RH	84	[98]
	CsSiWA/H_3PO_4		1.71×10^{-2}		–	
PTFE porous	CsPOMo/QDPSU[b]/H_3PO_4	30	0.04	150 °C, 0% RH	240	[132]
Nylon[c]	HPA (51 wt%)	67	0.08	Conductivity: 80 °C, 80% RH	930	[2]
				Fuel cell: 60 °C, 40% RH		

[a]Maximum power density (otherwise stated)
[b]Quaternary diazabicyclo-octane polysulfone
[c]Sandwiched between two Nafion layers

also achieved for membranes containing $Cs_xH_{3-x}PWO_{40}$ immobilized on TiO_2 [141] and CeO_2 [142]. Nevertheless, immobilization of HPA salts and their stability on the inorganic supports were much better than pure HPAs [141]. Table 1 illustrates some of the results on HPA salts loading level and properties of the obtained composite membranes.

Commercially available polybenzimidazole (PBI) has been the most extensively studied and used for the PEMFC under high-temperature. Particularly, PBI membranes impregnated with phosphoric acid have been studied as polymer electrolytes in PEMFCs for two decades and were reasonably successful with excellent thermo-chemical stability and good conductivity [143, 144]. Several organic modifiers including ZrP [7] and PWA [145] have been incorporated into PBI to improve the performance of the PBI/H_3PO_4 membrane.

PWA and SiWA [128, 146], PMA [147], PWA/SiO_2 [148], and SiWA/SiO_2 [145] were initially used to fabricate PBI composite membranes. Initial results indicated that the conductivities of the composite membranes containing 20 and 30 wt% of PWA at 140 °C and that of SiWA at 200 °C were lower than that of pristine PBI membranes under the same conditions [128]. It was reported that the SiO_2 support provided a stable structure and membranes were thermally stable up to 400 °C. The conductivity values of 1.5×10^{-3} S cm^{-1} at 150 °C for PWA/SiO_2/PBI and 2.23×10^{-3} S cm^{-1} at 160 °C for SiWA/SiO_2/PBI were reported. In a separate study, 37.5 wt% of SiWA-SiO_2/mPBI showed excellent proton conductivity of 1.32×10^{-3} S cm^{-1} at room temperature [149]. Despite the good conductivity values of the composite membranes, there are few reports on fuel cell performance. Membranes composed of PWA/PBI or SiWA/PBI doped with phosphoric acid showed high conductivity compared to neat PBI membranes in high-temperature PEMFC [150]. At 150 °C, the conductivity of the 40% SiWA/PBI composite membrane was 0.1774 S cm^{-1}. In a single cell test, for example, 40 wt% SiWA/PBI exhibits only around 10–15 mV loss in cell potential due to IR drop which is around 75 mV at 500 mA cm^{-2} for PBI. The best fuel cell performance was achieved with a 40 wt% SiWA/PBI composite membrane at 120 ° C. However, at higher temperatures, contrary to the PBI membrane, the performance of the SiWA/PBI composite membrane declined in fuel cell. It has been suggested that the PWA/PBI composite membranes were suitable for the temperature limit of 120 °C.

CsHPA was used extensively for PEMFC due to its insolubility in water. Typically, PBI/$Cs_{2.5}H_{0.5}PMo_{12}O_{40}$/$H_3PO_4$ composite membrane exhibited conductivity as high as 0.15×10^{-2} S cm^{-1} and in a fuel cell gave a high power density of 0.7 W cm^2 (at atmospheric pressure and 150 °C with H_2/O_2) [130]. Higher proton conductivities of 1.91×10^{-2} S cm^{-1} and 1.71×10^{-2} S cm^{-1} were achieved at 160 °C under anhydrous conditions for mechano chemically synthesized CsHPA composites of PBI-50$H_3PW_{12}O_{40}$·50$CsHSO_4$ with 87 wt% phosphoric acid doping level and PBI-50$H_4SiW_{12}O_{40}$·50$CsHSO_4$ with 82 wt% doping level, respectively [98]. Nevertheless, these composite membranes show better conductivity with comparable single cell results, despite the lower phosphoric acid doping level compared to PBI.

Comparative conductivity, stability and fuel cell performance of PBI composite membranes with four different cesium salts of CsPOMo, CsPOW, CsSiOMo, and CsSiOW were evaluated [131]. The composite membranes loaded with H_3PO_4 showed higher conductivities than that of the phosphoric acid loaded PBI membrane. The conductivity increased with an increase in the percentage of inorganic component in the composite up to 30%. CsPOMo/PBI/H_3PO_4 exhibited a conductivity of 0.12 S cm^{-1} under anhydrous conditions at 150 °C. The membrane having P form of the CsHPA demonstrated higher conductivity than those containing Si atom, although the mechanical strength was inferior. The performance of the fuel cell with these composite membranes was better than that with a phosphoric acid-doped PBI membrane under the same conditions. The CsPOMo gave the best power density of around 0.6 W cm^2 with oxygen at atmospheric pressure and 150 °C.

6 HPAs in DMFC

DMFCs is an alternative cell to the hydrogen PEMFC, which avoids the limitations associated with the use of H_2 as a fuel. In order to be competitive, the DMFC must be reasonably cheap and capable of delivering high power densities. However, the performance of DMFCs is limited by the catalyst and membranes. The slow kinetics of the anode reaction and poisoning of the platinum catalyst together with the high methanol crossover are among main concerns. Typically, PFSA membranes showed exceptional proton conductivity upon employing in DMFCs but excess methanol permeability causes fuel loss, which lowers energy efficiency and fuel cell performance. The use of diluted methanol fuel (<4 molar of methanol), which supposed to minimize excessive fuel permeability is likely to reduce the energy density of fuel cell significantly [151]. Particularly, these drawbacks reduce the energy density to as low as 2000 Wh kg^{-1} and yield low operating voltage compared to PEMFC [152]. The ultimate solution to high methanol crossover, which is a key barrier for the development of DMFC, is the utilization of membranes with low methanol permeability that have potential for operation at higher temperatures. Introducing nanoparticles into polymers substrates is an effective tool to suppress methanol crossover, but is often accompanied by the partial loss of proton conductivity because protons and methanol are transported through the same pathway. HPAs having very strong Brønsted acidity approaching to the level of the superacids (more acidic than 100% sulfuric acid and Nafion) and fast reversible redox transformations are interesting candidates for preparation of composite membranes for DMFC [153]. HPAs have been immobilized in various proton conducting membranes to improve methanol barrier property and/or proton conductivity [154, 155]. DMFC performance based on the PWA—meso-silica nanocomposite membrane and low Pt catalyst loading (1 mg cm^2)—was found close to the advanced DMFCs (100–200 mW cm^2) [156].

Majority of reported studies of HPAs based composite membranes were dedicated to PFSA based membranes including Nafion. Generally, various HPAs were used as dopants that are usually sonicated with liquid Nafion in a high boiling point solvent before casting. Typically, stable molybdophosphoric acid (PMA)-impregnated membranes were reported to be prepared either by exposing cast membranes to solvent vapor at room temperature for 24 h and annealing at 150 °C [157] or by imbedding in polymer films using mixed solvents [158]. The obtained membranes showed noticeable enhancement in the conductivity [157] and improvement in the methanol rejection [158] compared to commercial Nafion.

Operation at high temperatures significantly enhances the kinetics of methanol oxidation. It was reported that the addition of inorganic hygroscopic materials to Nafion extends the operating temperature range of DMFC [159, 160, 140]. Nafion–silica composite membranes doped with phosphotungstic and silicotungstic acids were investigated at 145 °C and improvements in the electrochemical characteristics at high current densities were reported [140]. For instance, a maximum power density of 400 mW cm^{-2} was obtained at 145 °C in the presence of oxygen feed, whereas the maximum power density in the presence of air feed was around 250 mW cm^{-2} [140]. In another study, the power densities of 33 mW cm^{-2} at 80 °C, 39 mW cm^{-2} at 160 °C, and 44 mW cm^{-2} at 200 °C were reported [160]. Such improved performance was correlated to the strong interaction of inorganic fillers with water at high temperature. Table 2 illustrates additional examples of HPAs

Table 2 PWA loading level, thickness, conductivities, and methanol permeability of various DMFC membranes

Sample	Thickness (μm)	Proton conductivity (S cm^{-1})	MeOH permeability ($\times 10^{-8}$ cm^2 s^{-1})	DMFC performance[a] (mW cm^{-2})	Refs.
SPBN[b]	–	6.1×10^{-2} at 100% RH, 30 °C	93.9	63 at 80 °C, (1 M methanol)	[161]
SPEEK/PWA	–	1.7×10^{-2} at 100% RH, 100 °C	–	–	[162]
SPEEK/MMT[c]-SiWA (30 wt%)	–	43.8×10^{-2} at 100% RH, 30 °C	35.0	–	[64]
Nylon[d] (51.12)	95	3.59	16.3	127.1 at 60 °C, (5 M methanol)	[52]
40%-*smp*CTS/PWA (52.4)	–	29 at 80 °C	0.47	16 at 80 °C, (2 M methanol)	[10]
PPy-layered SPEES/PWA	–	6.0 at 25 °C	21	–	[163]
PES/PVP-PWA	–	78 at 80 °C	165	132 (2 M methanol)	[11]
Nafion/PPO/PMA	100 ± 5	3.41×10^{-2} at room temperature	44.1	Higher OCV than Nafion	[158]
Nafion-silica/PWA	80	–	–	250 at 145 °C	[140]
Nafion-silica/ PWA	–	0.024	25	70 at 80 °C	[164]
N212/PDDA-PWA[e]	51	–	–		[165]

(continued)

Table 2 (continued)

Sample	Thickness (μm)	Proton conductivity (S cm^{-1})	MeOH permeability ($\times 10^{-8}$ cm^2 s^{-1})	DMFC performance[a] (mW cm^{-2})	Refs.
SPAEK–COOH/polycation chitosan (CTS)-PWA	50	0.24 at 80 °C	100	–	[166]
Pore-filled PVDF with PWA	–	0.098 at 80 °C	90	–	[167]
Pore-filled PVDF with Si-PWA	60	1.01×10^{-2} at 60 °C	–	21.6 at 25 °C	[168]
PVA–ZrP–CsSiWA	180	13 at 100 °C, 50% RH	200	2.1	[169]
PVA immobilized Si-PWA	80	0.0105×10^{-2} at 50 °C	16	29.6 at 35 °C	[170]
PVA/PAA–CsPMA	250 ± 20	3.7	11	36 at 25 °C	[171]
PVA/PAA–CsPWA	250 ± 20	3	6.7	30 at 25 °C	[171]
PVA/PAA–CsSiWA	250 ± 20	1.8	1.38	39 at 25 °C	[171]
CsPW–Nafion	50	0.05	15.4	30.6 at 25 °C, (2 M methanol)	[172]
CsPW–Chitosan	–	75 at 80 °C	156	–	[173]
CsPW–SPEEK	80–100	1.3×10^{-1} at 80 °C 100% RH	47	–	[174]
PWA-MC/Imidazole		31.6 at 80 °C, room temperature	166	–	[154]
PVA–ZrP–SiWA	250	1×10^{-2} at 60 °C, 60% RH	60	–	[175]
PVA-PWA-SiO_2	–	0.004–0.017	1–10	–	[176]
PVA/PMA	–	2.05	154	–	[177]
PEG/SiO_2-PWA	150-400	5.04×10^{-3}	121	–	[178]

[a]Maximum power density (otherwise stated)
[b]Sulfonated polynorbornene membranes
[c]K10 montmorillonite
[d]Sandwiched between two Nafion layers
[e]Four bilayers of poly(diallyl dimethyl ammonium chloride) and PWA

composite membranes for DMFC applications at high temperature as well as normal operation temperature.

PWA in SPEEKs and poly(1,4-phenylene ether ether sulfone) stabilized by surface modification with polypyrrole (Ppy) using an in situ polymerization method showed a reduction in the methanol crossover without significant leaching in PWA [163, 179]. In addition, the water uptake, swelling ratio, and the tensile strength of the surface-coated membranes were improved with the increase in the thickness of Ppy layer. In a similar approach, PWA was microcapsulated into imidazole and incorporated into SPEEK membrane matrix prior to evaluation in DMFC and revealed improved DMFC performance [154].

Pore-filling of polymer substrates with HPAs was also reported to achieve an improved performance of DMFC [168, 167]. Typically, pores of PVDF substrates filled by sulfonated poly(aryl ether ketone sulfone) membranes blended with phosphotungstic acid resulted in a membrane with comparable conductivity and lower methanol permeability than Nafion [167]. The layer-by-layer (LbL) self-assembly is another method to introduce a thin multilayer film on a substrate by sequential electrostatic adsorption between the negatively and positively charged polyelectrolytes. A 50% reduction in the methanol permeability was reported for most of the LbL membranes. Typically, bilayers of chitosan and PWA onto sulfonated poly (aryl ether ketone) (SPAEK) substrate resulted in a membrane with high-proton conductivity values up to 0.24 S cm^{-1} at 80 °C and extremely low water swelling ratio and methanol permeability [166].

Inorganic PEM such as PWA immobilized into mesoporous silica matrix such as MCM-41 (Si-PWA) was assembled and used in DMFC [180, 181]. The PWA/MCM-41 electrolyte has a high-proton conductivity of 4.5×10^{-2} S cm^{-1} at 150 °C with an activation energy of ca. 8 kJ mol^{-1}. Single fuel cell tests showed a peak power density of 95 mW cm^{-2} at 100 °C and 100% RH with H_2/O_2 system and 90 mW cm^{-2} in methanol/O_2 at 150 °C and an extremely low RH of 0.67%. This sub-micron to nano-sized Si-PWA immobilized onto PVA crosslinked glutaraldehyde [170] and porous PVDF [168] were also reported. Lower methanol permeability than Nafion and a maximum power density of around 44 mW cm^{-2} were obtained at 60 °C.

An interesting class of materials for suppressing methanol permeability is the polymers containing basic groups including amine, amide, imine, and imidazole [182–184]. One of the key membranes of interest in DMFC is PBI based materials such as commercialized Celtec-V membrane, which is a blend of PBI and poly (vinylphosphonic acid) [185]. Although some encouraging results were reported [186], the conductivity was low and needed to be improved. PWA was self-immobilized on various substrates with basic groups to reduce methanol permeability [52, 10]. Since the HPA is immobilized, the leaching risk is minimized and the conductivity could be simply tuned by controlling the level of PWA. Upon using micro-pores chitosan as a substrate, comparable performance and excellent methanol-blocking performance of 60% lower than that of Nafion 212 membrane was achieved [10].

Recently, it was reported that self-immobilization of high level of PWA on electrospun nylon nanofiberous results in a highly selective methanol barrier with undetectable leaching of PWA [52]. Upon assembling in a sandwiched proton conducting membranes (Fig. 6) with Nafion outer layers, superior methanol barrier properties (viz. $P = 3.59 \times 10^{-8}$ cm^2 s^{-1}) and selectivity mounting to more than 20 times higher than Nafion 115 coupled with proton conductivities reaching 58.6×10^{-3} mS cm^{-1} at 30 °C were achieved. When tested in DMFC single cell, the performance of hybrid membrane was far better than Nafion 115 especially at higher methanol concentrations. For example, at 2 M methanol feed, the composite membrane had an improvement in power peak by 49.6% over Nafion 115. An increase in the power peak by 113.3% was obtained with 5 M methanol feed and

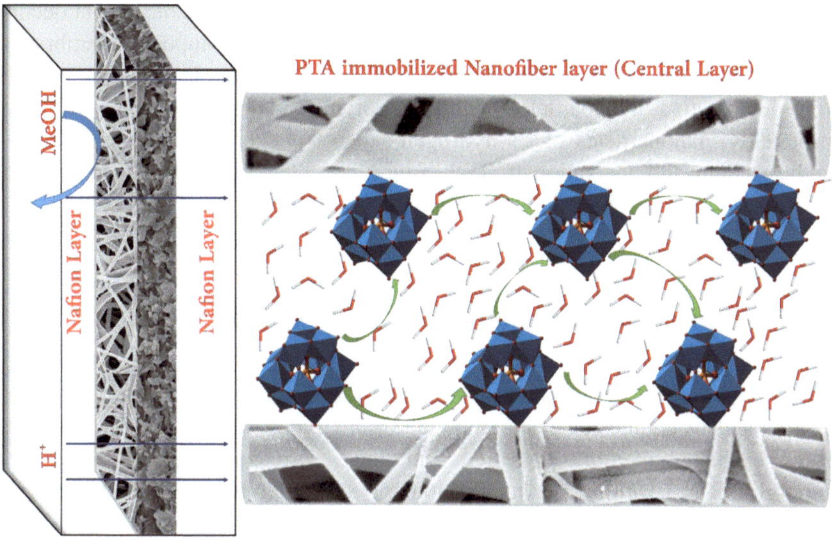

Fig. 6 Three-layered membranes composed of self-immobilized PWA on electrospun nylon and two outer Nafion layers [52]

the maximum power density of 127.1 mW cm^{-2} that was observed for 3L membrane compared to 59.6 mW cm^{-2} for Nafion 115. In addition, the OCV, which is directly related to the methanol crossover, was improved to 693 and 675 mV giving about 41 and 78 mV higher than that of Nafion 115 in 2 and 5 M methanol, respectively. Self-immobilization of PWA on PES/PVP-PWA also resulted in comparable performance of 132 mW cm^{-2} to that of Nafion 212 at 80 °C [187, 11].

Cesium phosphotungstate salt enhances the proton conductivity and reduces methanol permeability of various host substrates including SPEEK [174] chitosan-hydroxy ethyl cellulose (CS-HEC) [188], chitosan [173], PVA [169] and PVA/PAA [171], etc. Cesium phosphotungstate salt has excellent conductive capability and increased stability in aqueous media. PVA/PAA membranes with various Cs salt of heteropolyacids were assembled for DMFC [171]. The blend of PVA and PAA was crosslinked with glutaraldehyde, and the Cs salts of different HPAs, including PMA, PWA, and silicotungstic acid (SiWA), were incorporated into the polymer network to form PVA/PAA–CsPMA, PVA/PAA–CsPWA, and PVA/PAA–CsSiWA. A dense network formation was achieved through crosslinking with glutaraldehyde, which led to an order of magnitude decrease in the methanol permeability compared to Nafion 115 membrane. The hybrid membrane containing CsSiWA exhibited a very low methanol permeability of 1.4×10^{-8} cm^2 s^{-1}) compared to other membranes. Better performance was achieved when sulfosuccinic acid used as a crosslinker possibly due to the presence of sulfonic acid in the structure of crosslinker [189]. Ultralow methanol permeability resulted to a power density of 150 mW cm^{-2} at a current density of 500 mA cm^{-2}.

To avoid any heterogeneity in the membrane preparation and consequent decline in both proton conductivity and mechanical properties of composite membranes, which may cause by the relatively large particle sizes, an in situ synthesis of proton-conductive nanoparticles within a polymer matrix was reported [172]. Particularly, nanoparticles of Cs hydrogen salts of phosphotungstic acid (CsPW) synthesized in situ in Nafion (CsPW–Nafion) resulted in a 101.3% increase of maximum power density relative to pristine Nafion in DMFC [172].

7 Concluding Remarks and Future Perspectives

Various types of HPAs in a pure or modified forms and as an additive have been incorporated in composite membranes in various types of PEMFC. In composite proton exchange membranes, HPAs and theirs corresponding salts have been immobilized into various inert and ionomeric substrates to reduce fuel crossover and enhance the proton transport and fuel cell performance especially at high temperatures and in dry conditions. In addition, HPAs were evaluated in DMFC to suppress methanol crossover. Among various types of HPAs, phosphotungstic acid and corresponding cesium salts have been used widely due to their long-term stability in fuel cell.

Effective immobilization of HPAs onto polymeric substrates is still challenging due to the highly solubility of such fillers in water. Using of corresponding salts of HPAs, covalent attachments to the polymer backbones and immobilization on the porous silica or fibrous structured amid-type backbones have been proposed to overcome the HPAs leaching during PEMFC or DMFC operation. Although some preliminarily reports indicate a stable immobilization of HPAs in the composite membranes, the long-term stability under dynamic fuel cell conditions has been not provided in many cases. Particularly, the PEMFC performance under higher temperature and low relative humidity is lacking and this aspect is needed to be investigated comprehensively.

Acknowledgements Authors wish to acknowledge the financial support from the Malaysian Ministry of Higher Education under the LRGS program (vote no. #4L817) and Universiti Teknologi Malaysia (vote no. #00D20).

References

1. Abouzari-Lotf E, Ghassemi H, Mehdipour-Ataei S, Shockravi A (2016) Phosphonated polyimides: enhancement of proton conductivity at high temperatures and low humidity. J Membr Sci 516:74–82. doi:10.1016/j.memsci.2016.06.009
2. Abouzari-lotf E, Jacob MV, Ghassemi H, Ahmad A, Nasef MM, Zakeri M, Mehdipour-Ataei S (2016) Enhancement of fuel cell performance with less-water dependent

composite membranes having polyoxometalate anchored nanofibrous interlayer. J Power Sources 326:482–489. doi:10.1016/j.jpowsour.2016.07.027
3. Aili D, Zhang J, Dalsgaard Jakobsen MT, Zhu H, Yang T, Liu J, Forsyth M, Pan C, Jensen JO, Cleemann LN, Jiang SP, Li Q (2016) Exceptional durability enhancement of PA/PBI based polymer electrolyte membrane fuel cells for high temperature operation at 200 °C. J Mater Chem A 4(11):4019–4024. doi:10.1039/C6TA01562J
4. He C, Han KF, Yu JH, Zhu H, Wang ZM (2016) Novel anti-oxidative membranes based on sulfide-containing polybenzimidazole for high temperature proton exchange membrane fuel cells. Eur Polymer J 74:168–179. doi:10.1016/j.eurpolymj.2015.11.024
5. Nasef MM (2014) Radiation-grafted membranes for polymer electrolyte fuel cells: current trends and future directions. Chem Rev 114(24):12278–12329. doi:10.1021/cr4005499
6. Nasef MM, Fujigaya T, Abouzari-Lotf E, Nakashima N (2016a) Electrospinning of poly (vinylpyrrodine) template for formation of ZrO_2 nanoclusters for enhancing properties of composite proton conducting membranes. Int J Polym Mater. doi:10.1080/00914037.2016.1201829
7. Nasef MM, Fujigaya T, Abouzari-Lotf E, Nakashima N, Yang Z (2016) Enhancement of performance of pyridine modified polybenzimidazole fuel cell membranes using zirconium oxide nanoclusters and optimized phosphoric acid doping level. Int J Hydrogen Energy 41 (16):6842–6854. doi:10.1016/j.ijhydene.2016.03.022
8. Tanaka M (2016) Development of ion conductive nanofibers for polymer electrolyte fuel cells. Polym J 48(1):51–58. doi:10.1038/pj.2015.76
9. Wu H-W (2016) A review of recent development: transport and performance modeling of PEM fuel cells. Appl Energy 165:81–106. doi:10.1016/j.apenergy.2015.12.075
10. Wu Q, Wang H, Lu S, Xu X, Liang D, Xiang Y (2016) Novel methanol-blocking proton exchange membrane achieved via self-anchoring phosphotungstic acid into chitosan membrane with submicro-pores. J Membr Sci 500:203–210. doi:10.1016/j.memsci.2015.11.019
11. Xu X, Wang H, Lu S, Peng S, Xiang Y (2016) A phosphotungstic acid self-anchored hybrid proton exchange membrane for direct methanol fuel cells. RSC Adv 6(49):43049–43055. doi:10.1039/C6RA07318B
12. Ye Y, Wu X, Yao Z, Wu L, Cai Z, Wang L, Ma X, Chen Q-H, Zhang Z, Xiang S (2016) Metal-organic frameworks with a large breathing effect to host hydroxyl compounds for high anhydrous proton conductivity over a wide temperature range from subzero to 125 [degree] C. J Mater Chem A 4(11):4062–4070. doi:10.1039/C5TA10765B
13. Hickner MA, Ghassemi H, Kim YS, Einsla BR, McGrath JE (2004) Alternative polymer systems for proton exchange membranes (PEMs). Chem Rev 104(10):4587–4612
14. Kim YS, Lee K-S (2015) Fuel cell membrane characterizations. Polym Rev 55(2):330–370. doi:10.1080/15583724.2015.1011275
15. Zhang L, Chae S-R, Hendren Z, Park J-S, Wiesner MR (2012) Recent advances in proton exchange membranes for fuel cell applications. Chem Eng J 204–206:87–97. doi:10.1016/j.cej.2012.07.103
16. Li J, Wang Z, Li J, Pan M, Tang H (2014) Nanostructure-based proton exchange membrane for fuel cell applications at high temperature. J Nanosci Nanotechnol 14(2):1181–1193. doi:10.1166/jnn.2014.8744
17. Subianto S (2014) Recent advances in polybenzimidazole/phosphoric acid membranes for high-temperature fuel cells. Polym Int 63(7):1134–1144. doi:10.1002/pi.4708
18. Zhang J, Xie Z, Zhang J, Tang Y, Song C, Navessin T, Shi Z, Song D, Wang H, Wilkinson DP, Liu ZS, Holdcroft S (2006) High temperature PEM fuel cells. J Power Sources 160(2):872–891. doi:10.1016/j.jpowsour.2006.05.034
19. Nasef H, Saidi H, Uthman H, Sithambaranathan P (2013) Advances in membranes for high temperature polymer electrolyte membrane fuel cells. In: Inamuddin (ed) Advanced functional polymers and composites: materials devices and allied applications, vol 1. Nova Science Publishers Inc., USA, pp 1–44

20. Alberti G, Casciola M (2003) Composite membranes for medium-temperature PEM fuel cells. Annu Rev Mater Res 33(1):129–154. doi:10.1146/annurev.matsci.33.022702.154702
21. Tripathi BP, Shahi VK (2011) Organic–inorganic nanocomposite polymer electrolyte membranes for fuel cell applications. Prog Polym Sci 36(7):945–979. doi:10.1016/j.progpolymsci.2010.12.005
22. Kourasi M, Wills RGA, Shah AA, Walsh FC (2014) Heteropolyacids for fuel cell applications. Electrochim Acta 127:454–466. doi:10.1016/j.electacta.2014.02.006
23. Sachdeva S, Turner JA, Horan JL, Herring AM (2011) The use of heteropoly acids in proton exchange fuel cells. Struct Bond 141:115–168
24. Katsoulis DE (1998) A Survey of applications of polyoxometalates. Chem Rev 98(1):359–388. doi:10.1021/cr960398a
25. Maksimov GM (1995) Advances in the synthesis of polyoxometalates and in the study of heteropolyacids. Russ Chem Rev 64(5):445
26. Pope MT, Müller A (1991) Polyoxometalate chemistry: an old field with new dimensions in several disciplines. Angew Chem Int Ed Engl 30(1):34–48. doi:10.1002/anie.199100341
27. Calligaris M (1985) Structural inorganic chemistry by A.F. Wells. Acta Crystallogr Sect B 41(3):208. doi:10.1107/S0108768185001987
28. Baker LCW, Glick DC (1998) Present general status of understanding of heteropoly electrolytes and a tracing of some major highlights in the history of their elucidation. Chem Rev 98(1):3–50. doi:10.1021/cr9603921
29. Kozhevnikov IV (1998) Catalysis by heteropoly acids and multicomponent polyoxometalates in liquid-phase reactions. Chem Rev 98(1):171–198
30. Song Y-F, Tsunashima R (2012) Recent advances on polyoxometalate-based molecular and composite materials. Chem Soc Rev 41(22):7384–7402. doi:10.1039/C2CS35143A
31. Walsh JJ, Bond AM, Forster RJ, Keyes TE (2016) Hybrid polyoxometalate materials for photo(electro-) chemical applications. Coord Chem Rev 306(Part 1):217–234. doi:10.1016/j.ccr.2015.06.016
32. Wang S-S, Yang G-Y (2015) Recent advances in polyoxometalate-catalyzed reactions. Chem Rev 115(11):4893–4962. doi:10.1021/cr500390v
33. Wu X, Wu W, Qian X, Wu Q, Yan W (2015) Proton-conducting materials based on heteropoly acid and matrixes. J Non-Cryst Solids 426:88–91. doi:10.1016/j.jnoncrysol.2015.06.010
34. Pope MT, Müller A (2001) Polyoxometalate chemistry from topology via self-assembly to applications. Kluwer Academic Publishers, New York
35. Singh RP, Khatri RP, Dubois J, Gaur SS, Abe M (1990) Synthesis of dodecamolybdoantimonate(V) salts containing the Keggin structure. J Chem Soc Dalton Trans (3):947–951. doi:10.1039/DT9900000947
36. Pandey G (1994) Zeolite, clay, and heteropoly acid in organic reactions. Von Y. Izumi, K. Urabe und M. Onaka. VCH verlagsgesellschaft, weinheim, 1992. 166 S., geb. 128.00 DM. —ISBN 3-527-29011-7. Angewandte Chemie 106(21):2316–2317. doi:10.1002/ange.19941062133
37. Cai H, Huang T, Wu Q, Yan W (2016) Synthesis and conduction mechanism of high proton conductor $H_6SiW_{10}V_2O_{40}{\cdot}14H_2O$. Funct Mater Lett 09(03):1650048. doi:10.1142/S179360471650048X
38. Huang T, Tian N, Wu Q, Yan Y, Yan W (2015) Synthesis, crystal structure and conductive mechanism of solid high-proton conductor tungstovanadosilicic heteropoly acid. Mater Chem Phys 165:34–38. doi:10.1016/j.matchemphys.2015.08.026
39. Jansen RJJ, van Veldhuizen HM, Schwegler MA, van Bekkum H (1994) Recent (1987-1993) developments in heteropolyacid catalysts in acid catalyzed reactions and oxidation catalysis. Recl Trav Chim Pays-Bas 113(3):115–135. doi:10.1002/recl.19941130302
40. Misono M (1987) Heterogeneous catalysis by heteropoly compounds of molybdenum and tungsten. Catal Rev 29(2–3):269–321. doi:10.1080/01614948708078072

41. Coronado E, Gómez-García CJ (1998) Polyoxometalate-based molecular materials. Chem Rev 98(1):273–296. doi:10.1021/cr970471c
42. Cronin L, Muller A (2012) From serendipity to design of polyoxometalates at the nanoscale, aesthetic beauty and applications. Chem Soc Rev 41(22):7333–7334. doi:10.1039/C2CS90087D
43. Vernon DR, Meng F, Dec SF, Williamson DL, Turner JA, Herring AM (2005) Synthesis, characterization, and conductivity measurements of hybrid membranes containing a mono-lacunary heteropolyacid for PEM fuel cell applications. J Power Sources 139(1–2):141–151. doi:10.1016/j.jpowsour.2004.07.027
44. Wu X, Huang T, Wu Q, Xu L (2015) Synthesis and conductive performance of indium-substituted ternary heteropoly acids with Keggin structures. Dalton Trans 45(1):271–275. doi:10.1039/c5dt02541a
45. Tong X, Tian N, Wu W, Zhu W, Wu Q, Cao F, Yan W, Yaroslavtsev AB (2013) Preparation and electrochemical performance of tungstovanadophosphoric heteropoly acid and its hybrid materials. J Phys Chem C 117(7):3258–3263. doi:10.1021/jp3102223
46. Ramani V, Kunz HR, Fenton JM (2005) Effect of particle size reduction on the conductivity of Nafion®/phosphotungstic acid composite membranes. J Membr Sci 266(1–2):110–114. doi:10.1016/j.memsci.2005.05.019
47. Athens GL, Ein-Eli Y, Chmelka BF (2007) Acid-functionalized mesostructured aluminosilica for hydrophilic proton conduction membranes. Adv Mater 19(18):2580–2587. doi:10.1002/adma.200602781
48. Misono M (2009) Recent progress in the practical applications of heteropolyacid and perovskite catalysts: catalytic technology for the sustainable society. Catal Today 144(3–4):285–291. doi:10.1016/j.cattod.2008.10.054
49. Jelikić-Stankov M, Uskoković-Marković S, Holclajtner-Antunović I, Todorović M, Djurdjević P (2007) Compounds of Mo, V and W in biochemistry and their biomedical activity. J Trace Elem Med Biol 21(1):8–16. doi:10.1016/j.jtemb.2006.11.004
50. Crans DC (1994) Enzyme interactions with labile oxovanadates and other polyoxometalates. Comments Inorg Chem 16(1–2):35–76. doi:10.1080/02603599408035851
51. Rhule JT, Hill CL, Judd DA, Schinazi RF (1998) Polyoxometalates in Medicine. Chem Rev 98(1):327–358. doi:10.1021/cr960396q
52. Abouzari-Lotf E, Nasef MM, Ghassemi H, Zakeri M, Ahmad A, Abdollahi Y (2015) Improved methanol barrier property of nafion hybrid membrane by incorporating nanofibrous interlayer self-immobilized with high level of phosphotungstic acid. ACS Appl Mater Interfaces 7(31):17008–17015. doi:10.1021/acsami.5b02268
53. Banerjee S, Kar KK (2016) Synergistic effect of aluminium phosphate and tungstophosphoric acid on the physicochemical properties of sulfonated poly ether ether ketone nanocomposite membrane. J Appl Polym Sci 133(5). doi:10.1002/app.42952
54. Cai H, Lian X, Wu Q, Cao F, Yan W (2016) A novel SPEEK/PW11V/rGO hybrid film for proton conduction. J Non-Cryst Solids 447:202–206. doi:10.1016/j.jnoncrysol.2016.04.034
55. Renzi M, Mignini P, Giuli G, Marassi R, Nobili F (2016) Rotating disk electrode study of $Pt/Cs_3HPMo_{11}VO_{40}$ composite catalysts for performing and durable PEM fuel cells. Int J Hydrogen Energy 41(26):11163–11173. doi:10.1016/j.ijhydene.2016.04.194
56. Santamaria M, Pecoraro CM, Di Franco F, Di Quarto F, Gatto I, Saccà A (2016) Improvement in the performance of low temperature H_2–O_2 fuel cell with chitosan–phosphotungstic acid composite membranes. Int J Hydrogen Energy 41(11):5389–5395. doi:10.1016/j.ijhydene.2016.01.133
57. Bijelic A, Rompel A (2015) The use of polyoxometalates in protein crystallography—an attempt to widen a well-known bottleneck. Coord Chem Rev 299:22–38. doi:10.1016/j.ccr.2015.03.018
58. Herrmann S, Ritchie C, Streb C (2015) Polyoxometalate—conductive polymer composites for energy conversion, energy storage and nanostructured sensors. Dalton Trans 44(16):7092–7104. doi:10.1039/C4DT03763D

59. Jameel U, Zhu M, Chen X, Tong Z (2016) Recent progress of synthesis and applications in polyoxometalate and nanogold hybrid materials. J Mater Sci 51(5):2181–2198. doi:10.1007/s10853-015-9503-1
60. Nakamura O, Kodama T, Ogino I, Miyake Y (1979) High-conductivity solid proton conductors: dodecamolybdophosphoric acid and dodecatungstophosphoric acid crystals. Chem Lett 8(1):17–18. doi:10.1246/cl.1979.17
61. Nakamura O, Ogino I (1982) Electrical conductivities of some hydrates of dodecamolybdophosphoric acid and dodecatungstophosphoric acid and their mixed crystals. Mater Res Bull 17(2):231–234. doi:10.1016/0025-5408(82)90150-7
62. Creeth A (2013) PEM fuel cells can now rival diesel engines on performance. Fuel Cells Bull 2013(7):12–14. doi:10.1016/S1464-2859(13)70270-0
63. Giordano N, Staiti P, Hocevar S, Aricò AS (1996) High performance fuel cell based on phosphotungstic acid as proton conducting electrolyte. Electrochim Acta 41(3):397–403. doi:10.1016/0013-4686(95)00316-9
64. Mohtar SS, Ismail AF, Matsuura T (2011) Preparation and characterization of SPEEK/MMT-STA composite membrane for DMFC application. J Membr Sci 371(1–2):10–19. doi:10.1016/j.memsci.2011.01.009
65. Staiti P, Freni S, Hocevar S (1999) Synthesis and characterization of proton-conducting materials containing dodecatungstophosphoric and dodecatungstosilic acid supported on silica. J Power Sources 79(2):250–255. doi:10.1016/S0378-7753(99)00177-9
66. Colomer MT (2006) Nanoporous anatase thin films as fast proton-conducting materials. Adv Mater 18(3):371–374. doi:10.1002/adma.200500689
67. Xiong L, Nogami M (2006) Proton-conducting ordered mesostructured silica monoliths. Chem Lett 35(8):972–973. doi:10.1246/cl.2006.972
68. Celik SU, Akbey U, Graf R, Bozkurt A, Spiess HW (2008) Anhydrous proton-conducting properties of triazole-phosphonic acid copolymers: a combined study with MAS NMR. Phys Chem Chem Phys 10(39):6058–6066. doi:10.1039/B807659F
69. Uma T, Nogami M (2007) Structural and transport properties of mixed phosphotungstic acid/phosphomolybdic acid/SiO_2 glass membranes for H_2/O_2 fuel cells. Chem Mater 19 (15):3604–3610. doi:10.1021/cm070567f
70. Malhotra S, Datta R (1997) Membrane-supported nonvolatile acidic electrolytes allow higher temperature operation of proton-exchange membrane fuel cells. J Electrochem Soc 144(2):L23–L26
71. Honma I, Nakajima H, Nishikawa O, Sugimoto T, Nomura S (2003) Family of high-temperature polymer-electrolyte membranes synthesized from amphiphilic nanostructured macromolecules. J Electrochem Soc 150(5):A616–A619. doi:10.1149/1.1566018
72. Honma I, Nakajima H, Nishikawa O, Sugimoto T, Nomura S (2003) Organic/inorganic nano-composites for high temperature proton conducting polymer electrolytes. Solid State Ionics 162–163:237–245. doi:10.1016/S0167-2738(03)00260-1
73. Honma I, Nakajima H, Nomura S (2002) High temperature proton conducting hybrid polymer electrolyte membranes. Solid State Ionics 154–155:707–712. doi:10.1016/S0167-2738(02)00431-9
74. Honma I, Takeda Y, Bae JM (1999) Protonic conducting properties of sol–gel derived organic/inorganic nanocomposite membranes doped with acidic functional molecules. Solid State Ionics 120(1–4):255–264. doi:10.1016/S0167-2738(98)00562-1
75. Lavrenčič Štangar U, Grošelj N, Orel B, Schmitz A, Colomban P (2001) Proton-conducting sol–gel hybrids containing heteropoly acids. Solid State Ionics 145(1–4):109–118. doi:10.1016/S0167-2738(01)00920-1
76. Ramani V, Kunz HR, Fenton JM (2004) Investigation of Nafion®/HPA composite membranes for high temperature/low relative humidity PEMFC operation. J Membrane Sci 232(1–2):31–44. doi:10.1016/j.memsci.2003.11.016
77. Shao Z-G, Xu H, Li M, Hsing IM (2006) Hybrid nafion–inorganic oxides membrane doped with heteropolyacids for high temperature operation of proton exchange membrane fuel cell. Solid State Ionics 177(7–8):779–785. doi:10.1016/j.ssi.2005.12.035

78. Lu JL, Fang QH, Li SL, Jiang SP (2013) A novel phosphotungstic acid impregnated meso-nafion multilayer membrane for proton exchange membrane fuel cells. J Membrane Sci 427:101–107. doi:10.1016/j.memsci.2012.09.041
79. Zaidi SMJ, Ahmad MI (2006) Novel SPEEK/heteropolyacids loaded MCM-41 composite membranes for fuel cell applications. J Membr Sci 279(1–2):548–557. doi:10.1016/j.memsci.2005.12.048
80. Aparicio M, Mosa J, Etienne M, Durán A (2005) Proton-conducting methacrylate-silica sol–gel membranes containing tungstophosphoric acid. J Power Sources 145(2):231–236. doi:10.1016/j.jpowsour.2005.01.071
81. Lavrenčič Štangar U, Grošelj N, Orel B, Schmitz A, Colomban P (2001) Proton-conducting sol–gel hybrids containing heteropoly acids. Solid State Ionics 145(1–4):109–118. doi:10.1016/S0167-2738(01)00920-1
82. Nakajima H, Nomura S, Sugimoto T, Nishikawa S, Honma I (2002) High temperature proton conducting organic/inorganic nanohybrids for polymer electrolyte membrane. J Electrochem Soc 149(8):A953–A959. doi:10.1149/1.1485080
83. Colicchio I, Wen F, Keul H, Simon U, Moeller M (2009) Sulfonated poly(ether ether ketone)-silica membranes doped with phosphotungstic acid. Morphology and proton conductivity. J Membr Sci 326(1):45–57. doi:10.1016/j.memsci.2008.09.008
84. Ramani V, Kunz HR, Fenton JM (2006) Metal dioxide supported heteropolyacid/Nafion® composite membranes for elevated temperature/low relative humidity PEFC operation. J Membr Sci 279(1–2):506–512. doi:10.1016/j.memsci.2005.12.044
85. Herring AM (2006) Inorganic-polymer composite membranes for proton exchange membrane fuel cells. J Macromol Sci C 46(3):245–296. doi:10.1080/00222340600796322
86. Bose AB, Gopu S, Li W (2014) Enhancement of proton exchange membrane fuel cells performance at elevated temperatures and lower humidities by incorporating immobilized phosphotungstic acid in electrodes. J Power Sources 263:217–222. doi:10.1016/j.jpowsour.2014.04.043
87. Horan JL, Genupur A, Ren H, Sikora BJ, Kuo M-C, Meng F, Dec SF, Haugen GM, Yandrasits MA, Hamrock SJ, Frey MH, Herring AM (2009) Copolymerization of divinylsilyl-11-silicotungstic acid with butyl acrylate and hexanediol diacrylate: synthesis of a highly proton-conductive membrane for fuel-cell applications. ChemSusChem 2 (3):226–229. doi:10.1002/cssc.200800237
88. Horan JL, Lingutla A, Ren H, Kuo M-C, Sachdeva S, Yang Y, Seifert S, Greenlee LF, Yandrasits MA, Hamrock SJ, Frey MH, Herring AM (2013) Fast proton conduction facilitated by minimum water in a series of divinylsilyl-11-silicotungstic acid-co-butyl acrylate-co-hexanediol diacrylate polymers. J Phys Chem C 118(1):135–144. doi:10.1021/jp4089657
89. Motz AR, Horan JL, Kuo M-C, Herring AM (2015) Water uptake in novel, water-stable, heteropoly acid films. ECS Trans 69(17):587–590. doi:10.1149/06917.0587ecst
90. Kim Y, Ketpang K, Jaritphun S, Park JS, Shanmugam S (2015) A polyoxometalate coupled graphene oxide-Nafion composite membrane for fuel cells operating at low relative humidity. J Mater Chem A 3(15):8148–8155. doi:10.1039/c5ta00182j
91. Dsoke S, Kolary-Zurowska A, Zurowski A, Mignini P, Kulesza PJ, Marassi R (2011) Rotating disk electrode study of Cs 2.5H 0.5PW 12O 40 as mesoporous support for Pt nanoparticles for PEM fuel cells electrodes. J Power Sources 196(24):10591–10600. doi:10.1016/j.jpowsour.2011.09.010
92. Haile SM, Liu H, Secco RA (2003) High-temperature behavior of CsH_2PO_4 under both ambient and high pressure conditions. Chem Mater 15(3):727–736. doi:10.1021/cm020138b
93. Kim G, Griffin JM, Blanc F, Haile SM, Grey CP (2015) Characterization of the dynamics in the protonic conductor CsH_2PO_4 by ^{17}O solid-state NMR spectroscopy and first-principles calculations: correlating phosphate and protonic motion. J Am Chem Soc 137(11):3867–3876. doi:10.1021/jacs.5b00280

94. Haile SM, Chisholm CRI, Sasaki K, Boysen DA, Uda T (2007) Solid acid proton conductors: from laboratory curiosities to fuel cell electrolytes. Faraday Discuss 134:17–39. doi:10.1039/B604311A
95. Chisholm CRI, Haile SM (2000) Superprotonic behavior of $Cs_2(HSO_4)(H_2PO_4)$—a new solid acid in the $CsHSO_4$–CsH_2PO_4 system. Solid State Ionics 136–137:229–241. doi:10.1016/S0167-2738(00)00315-5
96. Ikeda A, Kitchaev DA, Haile SM (2014) Phase behavior and superprotonic conductivity in the Cs1-xRbxH_2PO_4 and Cs1-xKxH_2PO_4 systems. J Mater Chem A 2(1):204–214. doi:10.1039/C3TA13889E
97. Oh S-Y, Kawamura G, Muto H, Matsuda A (2012) Mechanochemical synthesis of proton conductive composites derived from cesium dihydrogen phosphate and guanine. Solid State Ionics 225:223–227. doi:10.1016/j.ssi.2012.02.058
98. Oh S-Y, Yoshida T, Kawamura G, Muto H, Sakai M, Matsuda A (2010) Inorganic-organic composite electrolytes consisting of polybenzimidazole and Cs-substituted heteropoly acids and their application for medium temperature fuel cells. J Mater Chem 20(30):6359–6366. doi:10.1039/C0JM00318B
99. Stanis RJ, Kuo MC, Turner JA, Herring AM (2008) Use of W, Mo, and v substituted heteropolyacids for CO mitigation in PEMFCs. J Electrochem Soc 155(2):B155–B162. doi:10.1149/1.2815917
100. Kim WB, Voitl T, Rodriguez-Rivera GJ, Evans ST, Dumesic JA (2005) Preferential oxidation of CO in H_2 by aqueous polyoxometalates over metal catalysts. Angew Chem Int Ed 44(5):778–782. doi:10.1002/anie.200461601
101. Baker PS, Bonville LJ, Kunz HR (2014) Performance of silicotungstic acid modified platinum cathodes at high temperature and low relative humidity for PEMFCs. J Electrochem Soc 161(12):F1307–F1313. doi:10.1149/2.1101412jes
102. Stanis RJ, Kuo M-C, Rickett AJ, Turner JA, Herring AM (2008) Investigation into the activity of heteropolyacids towards the oxygen reduction reaction on PEMFC cathodes. Electrochim Acta 53(28):8277–8286. doi:10.1016/j.electacta.2008.06.052
103. Rhee HW, Ghil LJ (2012) Polymer nanocomposites in fuel cells. In: Advances in polymer nanocomposites: types and applications. Elsevier Inc., pp 433–471
104. Thanganathan U (2010) Proton-conducting PVA/PMA hybrid membranes for fuel cell applications. World Acad Sci Eng Technol 71:403–406
105. Shanmugam S, Viswanathan B, Varadarajan TK (2006) Synthesis and characterization of silicotungstic acid based organic-inorganic nanocomposite membrane. J Membr Sci 275(1–2):105–109. doi:10.1016/j.memsci.2005.09.009
106. Thanganathan U, Kumar S, Kishimoto A, Kimura K (2012) Synthesis of organic/inorganic hybrid composite membranes and their structural and conductivity properties. Mater Lett 72:81–87. doi:10.1016/j.matlet.2011.12.066
107. Broka K, Ekdunge P (1997) Oxygen and hydrogen permeation properties and water uptake of Nafion® 117 membrane and recast film for PEM fuel cell. J Appl Electrochem 27(2):117–124
108. Yılmaztürk S, Ercan N, Deligöz H (2012) Influence of LbL surface modification on oxygen cross-over in self-assembled thin composite membranes. Appl Surf Sci 258(7):3139–3146. doi:10.1016/j.apsusc.2011.11.051
109. Ketpang K, Shanmugam S, Suwanboon C, Chanunpanich N, Lee D (2015) Efficient water management of composite membranes operated in polymer electrolyte membrane fuel cells under low relative humidity. J Membr Sci 493:285–298. doi:10.1016/j.memsci.2015.06.055
110. Li Q, He R, Jensen JO, Bjerrum NJ (2003) Approaches and recent development of polymer electrolyte membranes for fuel cells operating above 100 °C. Chem Mater 15(26):4896–4915. doi:10.1021/cm0310519
111. Abouzari-Lotf E, Ghassemi H, Shockravi A, Zawodzinski T, Schiraldi D (2011) Phosphonated Poly(arylene ether)s as potential high temperature proton conducting materials. Polymer 52(21):4709–4717. doi:10.1016/j.polymer.2011.08.020

112. Amiinu IS, Li W, Wang G, Tu Z, Tang H, Pan M, Zhang H (2015) Toward anhydrous proton conductivity based on imidazole functionalized mesoporous silica/nafion composite membranes. Electrochim Acta 160:185–194. doi:10.1016/j.electacta.2015.02.070
113. Tazi B, Savadogo O (2000) Parameters of PEM fuel-cells based on new membranes fabricated from Nafion, silicotungstic acid and thiophene. Electrochim Acta 45(25–26):4329–4339. doi:10.1016/S0013-4686(00)00536-3
114. Tian H, Savadogo O (2006) Silicotungstic acid Nafion composite membrane for proton-exchange membrane fuel cell operation at high temperature. J New Mater Electrochem Syst 9(1):61–71
115. Oh SY, Yoshida T, Kawamura G, Muto H, Sakai M, Matsuda A (2010) Proton conductivity and fuel cell property of composite electrolyte consisting of Cs-substituted heteropoly acids and sulfonated poly(ether-ether ketone). J Power Sources 195(18):5822–5828. doi:10.1016/j.jpowsour.2010.01.063
116. Tian N, Wu X, Yang B, Wu Q, Cao F, Yan W, Yaroslavtsev AB (2015) Proton-conductive membranes based on vanadium substituted heteropoly acids with Keggin structure and polymers. J Appl Polym Sci 132(27). doi:10.1002/app.42204
117. Zaidi SMJ, Mikhailenko SD, Robertson GP, Guiver MD, Kaliaguine S (2000) Proton conducting composite membranes from polyether ether ketone and heteropolyacids for fuel cell applications. J Membr Sci 173(1):17–34. doi:10.1016/S0376-7388(00)00345-8
118. Meng F, Aieta NV, Dec SF, Horan JL, Williamson D, Frey MH, Pham P, Turner JA, Yandrasits MA, Hamrock SJ, Herring AM (2007) Structural and transport effects of doping perfluorosulfonic acid polymers with the heteropoly acids, $H_3PW_{12}O_{40}$ or $H_4SiW_{12}O_{40}$. Electrochim Acta 53(3):1372–1378. doi:10.1016/j.electacta.2007.06.047
119. Tazi B, Savadogo O (2001) Effect of various heteropolyacids (HPAs) on the characteristics of Nafion®—HPAS membranes and their H_2/O_2 polymer electrolyte fuel cell parameters. J New Mater Electrochem Syst 4(3):187–196
120. Saccà A, Carbone A, Pedicini R, Marrony M, Barrera R, Elomaa M, Passalacqua E (2008) Phosphotungstic acid supported on a nanopowdered ZrO_2 as a filler in nafion-based membranes for polymer electrolyte fuel cells. Fuel Cells 8(3–4):225–235. doi:10.1002/fuce.200800009
121. Mahreni A, Mohamad AB, Kadhum AAH, Daud WRW, Iyuke SE (2009) Nafion/silicon oxide/phosphotungstic acid nanocomposite membrane with enhanced proton conductivity. J Membr Sci 327(1–2):32–40. doi:10.1016/j.memsci.2008.10.048
122. Shao Z-G, Joghee P, Hsing IM (2004) Preparation and characterization of hybrid Nafion–silica membrane doped with phosphotungstic acid for high temperature operation of proton exchange membrane fuel cells. J Membr Sci 229(1–2):43–51. doi:10.1016/j.memsci.2003.09.014
123. Yan X-M, Mei P, Mi Y, Gao L, Qin S (2009) Proton exchange membrane with hydrophilic capillaries for elevated temperature PEM fuel cells. Electrochem Commun 11(1):71–74. doi:10.1016/j.elecom.2008.10.040
124. Pecoraro CM, Santamaria M, Bocchetta P, Di Quarto F (2015) Influence of synthesis conditions on the performance of chitosan–heteropolyacid complexes as membranes for low temperature H_2–O_2 fuel cell. Int J Hydrogen Energy 40(42):14616–14626. doi:10.1016/j.ijhydene.2015.06.083
125. Santamaria M, Pecoraro CM, Di Quarto F, Bocchetta P (2015) Chitosan-phosphotungstic acid complex as membranes for low temperature H_2-O_2 fuel cell. J Power Sources 276:189–194. doi:10.1016/j.jpowsour.2014.11.147
126. Martínez-Morlanes MJ, Martos AM, Várez A, Levenfeld B (2015) Synthesis and characterization of novel hybrid polysulfone/silica membranes doped with phosphomolybdic acid for fuel cell applications. J Membr Sci 492:371–379. doi:10.1016/j.memsci.2015.05.031
127. Anis A, Banthia AK, Mondal S, Thakur AK (2006) Synthesis and characterization of hybrid proton conducting membranes of poly(vinyl alcohol) and phosphomolybdic acid. Chin J Polym Sci 24(5):449–456. doi:10.1142/S0256767906001515 (English Edition)

128. He R, Li Q, Xiao G, Bjerrum NJ (2003) Proton conductivity of phosphoric acid doped polybenzimidazole and its composites with inorganic proton conductors. J Membr Sci 226 (1–2):169–184. doi:10.1016/j.memsci.2003.09.002
129. Lee JW, Kim K, Khan SB, Han P, Seo J, Jang W, Han H (2014) Synthesis, characterization, and thermal and proton conductivity evaluation of 2,5-polybenzimidazole composite membranes. J Nanomater 2014. doi:10.1155/2014/460232
130. Li M-Q, Shao Z-G, Scott K (2008) A high conductivity $Cs_{2.5}H_{0.5}PMo_{12}O_{40}$/polybenzimidazole (PBI)/H_3PO_4 composite membrane for proton-exchange membrane fuel cells operating at high temperature. J Power Sources 183(1):69–75. doi:10.1016/j.jpowsour.2008.04.093
131. Xu C, Wu X, Wang X, Mamlouk M, Scott K (2011) Composite membranes of polybenzimidazole and caesium-salts-of- heteropolyacids for intermediate temperature fuel cells. J Mater Chem 21(16):6014–6019. doi:10.1039/c1jm10093a
132. Xu C, Wang X, Wu X, Cao Y, Scott K (2012) A composite membrane of caesium salt of heteropolyacids/quaternary diazabicyclo-octane polysulfone with poly (tetrafluoroethylene) for intermediate temperature fuel cells. Membranes 2(3):384
133. Safronova EY, Osipov AK, Baranchikov AE, Yaroslavtsev AB (2015) Proton conductivity of $MxH_{3\text{-}x}PX_{12}O_{40}$ and $MxH_{4\text{-}x}SiX_{12}O_{40}$ (M = Rb, Cs; X = W, Mo) acid salts of heteropolyacids. Inorg Mater 51(11):1157–1162. doi:10.1134/S0020168515110102
134. Dey C, Kundu T, Aiyappa HB, Banerjee R (2015) Phosphate enriched polyoxometalate based ionic salts for proton conduction. RSC Adv 5(3):2333–2337. doi:10.1039/C4RA07598F
135. Narasimharao K, Brown DR, Lee AF, Newman AD, Siril PF, Tavener SJ, Wilson K (2007) Structure-activity relations in Cs-doped heteropolyacid catalysts for biodiesel production. J Catal 248(2):226–234. doi:10.1016/j.jcat.2007.02.016
136. Okuhara T, Watanabe H, Nishimura T, Inumaru K, Misono M (2000) Microstructure of cesium hydrogen salts of 12-tungstophosphoric acid relevant to novel acid catalysis. Chem Mater 12(8):2230–2238. doi:10.1021/cm9907561
137. Daiko Y, Takagi H, Katagiri K, Muto H, Sakai M, Matsuda A (2008) Mechanochemically synthesized cesium-ion-substituted phosphotungstic acid using several types of cesium-containing salts. Solid State Ionics 179(21–26):1174–1177. doi:10.1016/j.ssi.2008.02.004
138. Matsuda A, Kikuchi T, Katagiri K, Daiko Y, Muto H, Sakai M (2007) Mechanochemical synthesis of proton conductive cesium hydrogen salts of 12-tungstophosphoric acid and their composites. Solid State Ionics 178(7–10):723–727. doi:10.1016/j.ssi.2007.02.033
139. Ramani V, Kunz HR, Fenton JM (2005) Stabilized composite membranes and membrane electrode assemblies for elevated temperature/low relative humidity PEFC operation. J Power Sources 152:182–188. doi:10.1016/j.jpowsour.2005.03.135
140. Staiti P, Aricò AS, Baglio V, Lufrano F, Passalacqua E, Antonucci V (2001) Hybrid Nafion–silica membranes doped with heteropolyacids for application in direct methanol fuel cells. Solid State Ionics 145(1–4):101–107. doi:10.1016/S0167-2738(01)00919-5
141. Wang L, Yi BL, Zhang HM, Xing DM (2007) $Cs_{2.5}H_{0.5}PWO_{40}/SiO_2$ as addition self-humidifying composite membrane for proton exchange membrane fuel cells. Electrochim Acta 52(17):5479–5483. doi:10.1016/j.electacta.2007.03.007
142. Zhao D, Yi BL, Zhang HM, Yu HM, Wang L, Ma YW, Xing DM (2009) Cesium substituted 12-tungstophosphoric ($CsxH_{3-x}PW_{12}O_{40}$) loaded on ceria-degradation mitigation in polymer electrolyte membranes. J Power Sources 190(2):301–306. doi:10.1016/j.jpowsour.2008.12.133
143. Carollo A, Quartarone E, Tomasi C, Mustarelli P, Belotti F, Magistris A, Maestroni F, Parachini M, Garlaschelli L, Righetti PP (2006) Developments of new proton conducting membranes based on different polybenzimidazole structures for fuel cells applications. J Power Sources 160(1):175–180. doi:10.1016/j.jpowsour.2006.01.081

144. Li Q, Jensen JO, Savinell RF, Bjerrum NJ (2009) High temperature proton exchange membranes based on polybenzimidazoles for fuel cells. Prog Polym Sci 34(5):449–477. doi:10.1016/j.progpolymsci.2008.12.003
145. Staiti P, Minutoli M, Hocevar S (2000) Membranes based on phosphotungstic acid and polybenzimidazole for fuel cell application. J Power Sources 90(2):231–235. doi:10.1016/S0378-7753(00)00401-8
146. Verma A, Scott K (2008) Development of high-temperature PEMFC based on heteropolyacids and polybenzimidazole. J Solid State Electrochem 14(2):213–219. doi:10.1007/s10008-008-0678-0
147. Asensio JA, Borrós S, Gómez-Romero P (2003) Enhanced conductivity in polyanion-containing polybenzimidazoles. Improved materials for proton-exchange membranes and PEM fuel cells. Electrochem Commun 5(11):967–972. doi:10.1016/j.elecom.2003.09.007
148. Staiti P (2001) Proton conductive membranes based on silicotungstic acid/silica and polybenzimidazole. Mater Lett 47(4–5):241–246. doi:10.1016/S0167-577X(00)00241-X
149. Xie X, Mao Z, Xu J, Zhen Y (2002) A hybrid membrane of modified polybenzimidazole and heteropoly acid for direct methanol fuel cell. In: PowerCon 2002–2002 international conference on power system technology, Proceedings, pp 169–172
150. Zhu X, Liu Y, Zhu L (2009) Polymer composites for high-temperature proton-exchange membrane fuel cells. Polymer membranes for fuel cells. Springer, US, pp 159–184
151. Li X, Faghri A (2013) Review and advances of direct methanol fuel cells (DMFCs) Part I: design, fabrication, and testing with high concentration methanol solutions. J Power Sources 226:223–240. doi:10.1016/j.jpowsour.2012.10.061
152. Maiyalagan T (2009) Silicotungstic acid stabilized Pt-Ru nanoparticles supported on carbon nanofibers electrodes for methanol oxidation. Int J Hydrogen Energy 34(7):2874–2879. doi:10.1016/j.ijhydene.2009.01.069
153. Cavani F (1998) Heteropolycompound-based catalysts: a blend of acid and oxidizing properties. Catal Today 41(1–3):73–86. doi:10.1016/S0920-5861(98)00039-X
154. Wu H, Shen X, Cao Y, Li Z, Jiang Z (2014) Composite proton conductive membranes composed of sulfonated poly(ether ether ketone) and phosphotungstic acid-loaded Imidazole microcapsules as acid reservoirs. J Membrane Sci 451:74–84. doi:10.1016/j.memsci.2013.09.058
155. Xiang Y, Yang M, Zhang J, Lan F, Lu S (2011) Phosphotungstic acid (HPW) molecules anchored in the bulk of nafion as methanol-blocking membrane for direct methanol fuel cells. J Membrane Sci 368(1–2):241–245. doi:10.1016/j.memsci.2010.11.049
156. Thomas SC, Ren X, Gottesfeld S, Zelenay P (2002) Direct methanol fuel cells: progress in cell performance and cathode research. Electrochim Acta 47(22–23):3741–3748. doi:10.1016/S0013-4686(02)00344-4
157. Dimitrova P, Friedrich K, Stimming U, Vogt B (2002) Modified Nafion®-based membranes for use in direct methanol fuel cells. Solid State Ionics 150(1):115–122
158. Sauk J, Byun J, Kim H (2005) Composite Nafion/polyphenylene oxide (PPO) membranes with phosphomolybdic acid (PMA) for direct methanol fuel cells. J Power Sources 143(1–2):136–141. doi:10.1016/j.jpowsour.2004.11.030
159. Aricò AS, Baglio V, Di Blasi A, Creti P, Antonucci PL, Antonucci V (2003) Influence of the acid–base characteristics of inorganic fillers on the high temperature performance of composite membranes in direct methanol fuel cells. Solid State Ionics 161(3–4):251–265. doi:10.1016/S0167-2738(03)00283-2
160. Kim H-J, Shul Y-G, Han H (2006) Sulfonic-functionalized heteropolyacid–silica nanoparticles for high temperature operation of a direct methanol fuel cell. J Power Sources 158(1):137–142. doi:10.1016/j.jpowsour.2005.10.001
161. He XH, Zheng Y, Yao HL, Chen YW, Chen DF (2014) Hybrid network sulfonated polynorbornene/silica membranes with enhanced proton conductivity by doped phosphotungstic acid. Fuel Cells 14(1):26–34. doi:10.1002/fuce.201300142

162. Ponce ML, Prado L, Ruffmann B, Richau K, Mohr R, Nunes SP (2003) Reduction of methanol permeability in polyetherketone–heteropolyacid membranes. J Membr Sci 217(1–2):5–15. doi:10.1016/S0376-7388(02)00309-5
163. Neelakandan S, Kanagaraj P, Sabarathinam RM, Nagendran A (2015) Polypyrrole layered SPEES/TPA proton exchange membrane for direct methanol fuel cells. Appl Surf Sci 359:272–279. doi:10.1016/j.apsusc.2015.10.122
164. Xu W, Lu T, Liu C, Xing W (2005) Low methanol permeable composite Nafion/silica/PWA membranes for low temperature direct methanol fuel cells. Electrochim Acta 50(16–17):3280–3285. doi:10.1016/j.electacta.2004.12.014
165. Yang M, Lu S, Lu J, Jiang SP, Xiang Y (2010) Layer-by-layer self-assembly of PDDA/PWA-Nafion composite membranes for direct methanol fuel cells. Chem Commun 46(9):1434–1436. doi:10.1039/b912779h
166. Zhao C, Lin H, Cui Z, Li X, Na H, Xing W (2009) Highly conductive, methanol resistant fuel cell membranes fabricated by layer-by-layer self-assembly of inorganic heteropolyacid. J Power Sources 194(1):168–174. doi:10.1016/j.jpowsour.2009.05.021
167. Xu L, Xu J, Liu M, Han H, Ni H, Ma L, Wang Z (2015) Fabrication of sulfonated poly(aryl ether ketone sulfone) membranes blended with phosphotungstic acid and microporous poly (vinylidene fluoride) as a depository for direct-methanol fuel cells. Int J Hydrogen Energy 40 (22):7182–7191. doi:10.1016/j.ijhydene.2015.02.139
168. Pandey J, Shukla A (2014) PVDF supported silica immobilized phosphotungstic acid membrane for DMFC application. Solid State Ionics 262:811–814. doi:10.1016/j.ssi.2013.10.029
169. Helen M, Viswanathan B, Srinivasa Murthy S (2006) Fabrication and properties of hybrid membranes based on salts of heteropolyacid, zirconium phosphate and polyvinyl alcohol. J Power Sources 163(1):433–439. doi:10.1016/j.jpowsour.2006.09.041
170. Pandey J, Mir FQ, Shukla A (2014) Synthesis of silica immobilized phosphotungstic acid (Si-PWA)-poly(vinyl alcohol) (PVA) composite ion-exchange membrane for direct methanol fuel cell. Int J Hydrogen Energy 39(17):9473–9481. doi:10.1016/j.ijhydene.2014.03.237
171. Helen M, Viswanathan B, Murthy SS (2010) Poly(vinyl alcohol)–polyacrylamide blends with cesium salts of heteropolyacid as a polymer electrolyte for direct methanol fuel cell applications. J Appl Polym Sci 116(6):3437–3447. doi:10.1002/app.31940
172. Rao S, Xiu R, Si J, Lu S, Yang M, Xiang Y (2014) In situ synthesis of nanocomposite membranes: comprehensive improvement strategy for direct methanol fuel cells. ChemSusChem 7(3):822–828. doi:10.1002/cssc.201301060
173. Xiao Y, Xiang Y, Xiu R, Lu S (2013) Development of cesium phosphotungstate salt and chitosan composite membrane for direct methanol fuel cells. Carbohydr Polym 98(1):233–240. doi:10.1016/j.carbpol.2013.06.017
174. Doğan H, Inan TY, Unveren E, Kaya M (2010) Effect of cesium salt of tungstophosphoric acid (Cs-TPA) on the properties of sulfonated polyether ether ketone (SPEEK) composite membranes for fuel cell applications. Int J Hydrogen Energy 35(15):7784–7795. doi:10.1016/j.ijhydene.2010.05.045
175. Helen M, Viswanathan B, Murthy SS (2007) Synthesis and characterization of composite membranes based on α-zirconium phosphate and silicotungstic acid. J Membr Sci 292(1–2):98–105. doi:10.1016/j.memsci.2007.01.018
176. Xu W, Liu C, Xue X, Su Y, Lv Y, Xing W, Lu T (2004) New proton exchange membranes based on poly (vinyl alcohol) for DMFCs. Solid State Ionics 171(1–2):121–127. doi:10.1016/j.ssi.2004.04.009
177. Anis A, Banthia AK, Bandyopadhyay S (2008) Synthesis and characterization of polyvinyl alcohol copolymer/phosphomolybdic acid-based crosslinked composite polymer electrolyte membranes. J Power Sources 179(1):69–80. doi:10.1016/j.jpowsour.2007.12.041
178. Lin CW, Thangamuthu R, Chang PH (2005) PWA-doped PEG/SiO_2 proton-conducting hybrid membranes for fuel cell applications. J Membr Sci 254(1–2):197–205. doi:10.1016/j.memsci.2005.01.007

179. Xu D, Zhang G, Zhang N, Li H, Zhang Y, Shao K, Han M, Lew CM, Na H (2010) Surface modification of heteropoly acid/SPEEK membranes by polypyrrole with a sandwich structure for direct methanol fuel cells. J Mater Chem 20(41):9239–9245. doi:10.1039/C0JM02167A
180. Lu S, Wang D, Jiang SP, Xiang Y, Lu J, Zeng J (2010) HPW/MCM-41 phosphotungstic acid/mesoporous silica composites as novel proton-exchange membranes for elevated-temperature fuel cells. Adv Mater 22(9):971–976. doi:10.1002/adma.200903091
181. Zeng J, Shen PK, Lu S, Xiang Y, Li L, De Marco R, Jiang SP (2012) Correlation between proton conductivity, thermal stability and structural symmetries in novel HPW-meso-silica nanocomposite membranes and their performance in direct methanol fuel cells. J Membr Sci 397–398:92–101. doi:10.1016/j.memsci.2012.01.018
182. Argun AA, Ashcraft JN, Hammond PT (2008) Highly conductive, methanol resistant polyelectrolyte multilayers. Adv Mater 20(8):1539–1543. doi:10.1002/adma.200703205
183. Ni HJ, Pei Y, Zhang QT, Zhao J, Wang XX, Huang MY (2015) Preparation and performances of SPEEK/PANI/PWA composite proton exchange membranes. Xiandai Huagong/Mod Chem Ind 35(11):72–76. doi:10.16606/j.cnki.issn0253-4320.2015.11.017
184. Roziere J, Jones DJ (2003) Non-fluorinated polymer materials for proton exchange membrane fuel cells. Ann Rev Mater Res 33:503–555. doi:10.1146/annurev.matsci.33.022702.154657
185. Lobato J, Cañizares P, Rodrigo MA, Linares JJ, López-Vizcaíno R (2008) Performance of a vapor-fed polybenzimidazole (PBI)-based direct methanol fuel cell. Energy Fuels 22 (5):3335–3345. doi:10.1021/ef8001839
186. Gubler L, Kramer D, Belack J, Ünsal Ö, Schmidt TJ, Scherer GG (2007) Celtec-V: a Polybenzimidazole-based membrane for the direct methanol fuel cell. J Electrochem Soc 154 (9):B981–B987. doi:10.1149/1.2754078
187. Lu S, Xu X, Zhang J, Peng S, Liang D, Wang H, Xiang Y (2014) A self-anchored phosphotungstic acid hybrid proton exchange membrane achieved via one-step synthesis. Adv Energy Mater 4(17). doi:10.1002/aenm.201400842
188. Mohanapriya S, Bhat SD, Sahu AK, Pitchumani S, Sridhar P, Shukla AK (2009) A new mixed-matrix membrane for DMFCs. Energy Environ Sci 2(11):1210–1216. doi:10.1039/b909451b
189. Bhat SD, Sahu AK, Jalajakshi A, Pitchumani S, Sridhar P, George C, Banerjee A, Chandrakumar N, Shukla AK (2010) PVA–SSA–HPA Mixed-matrix-membrane electrolytes for DMFCs. J Electrochem Soc 157(10):B1403–B1412. doi:10.1149/1.3465653

[illegible]

Chapter 6
Organic/Montmorillonite Nanocomposite Membranes

Palaniappan Sathish Kumar, Sathyamangalam Munusamy Senthil, Samir Kumar Pal and Rathanasamy Rajasekar

Abstract In this chapter, organic/montmorillonite nanocomposite membrane and membrane fabrication techniques are discussed. The fabrication technique, properties of the fabricated membranes, and performance are explained in detail and compared. With the addition of clay addition, important parameters which affect the membrane performance, such as crystallinity, porous structure, hydrophobicity/hydrophilicity, membrane charge, and surface roughness were analyzed. Despite the fact that extensive knowledge exist on membrane pore structure after clay incorporation including its surface properties and cross-section morphology by selection of appropriate fabrication methods, there is still a challenge to produce reliable membranes with antifouling properties, thermal resistance, chemical resistance, high-mechanical strength with high flux and selectivity. To ensure progress in polymer–clay membrane performance, further improvements are needed for common membrane fabrication techniques, such as solution casting, phase inversion, and interfacial polymerization. At the same time, the potential of novel fabrication techniques such as electro spinning and track-etching are also assessed. A comprehensive understanding between structure-surface properties and performance is a key for further development and progress in organic/montmorillonite membrane technology.

P. Sathish Kumar (✉) · S.K. Pal
Department of Mining Engineering, Indian Institute of Technology Kharagpur, Kharagpur 721302, West Bengal, India
e-mail: sathishcad91@gmail.com

S.K. Pal
e-mail: pal.samir09@gmail.com

S.M. Senthil · R. Rajasekar
Department of Mechanical Engineering, Kongu Engineering College, Erode 638052, Tamil Nadu, India
e-mail: senthil.awaits@gmail.com

R. Rajasekar
e-mail: rajasekar.cr@gmail.com

Inamuddin et al. (eds.), *Organic–Inorganic Composite Polymer Electrolyte Membranes*, DOI 10.1007/978-3-319-52739-0_6

Abbreviations

Al_2O_3	Aluminum oxide
AFM	Atomic force microscopy
AC	Applied current
AMPS	2-acrylamide-2-methyl-1-propane sulfonic acid
CA	Cellulose acetate
CH_4	Methane
CS	Chitosan
CNTs	Carbon nanotubes
DCM	Dichloro methane
DMF	Dimethyl formamide
DC	Direct current
DSC	Differential scanning calorimeter
DMA	Dynamic mechanical analysis
DMDOC	Dimethyl dioctadecylammonium chloride
EDS	Energy dispersive X-ray spectroscopy
FESEM	Field emission scanning electron microscopy
FPSM	Free-path spacing measurement
FTIR	Fourier transforms infrared
HFP	Hexafluoro propylene
HNT	Halloysite nanotubes
I	Current
LiCl	Lithium chloride
MF	Microfiltration
MS	Mass spectrometer
MMMs	Mixed matrix membranes
MEA	Membrane electrode assembly
Na^+	Sodium
NaAlg	Alginate
NMP	*N*-methyl-2-pyrrolidone
OMMT	Organically modified montmorillonite
OPBI	Organically modified polybenzimidazole
PVP	Polyvinyl pyrrolidone
PEG	Polyethylene glycol
PVDF	Polyvinylidene fluoride
PAA	Poly (amic acid)
PVC	Poly (vinyl chloride)
PS	Polystyrene
PVA	Poly (vinyl alcohol)
PE	Polyethylene
PP	Polypropylene
PES	Poly (ether sulfones)
PEM	Proton exchange membrane
PEMFC	Proton exchange membrane fuel cells

PPO	Poly (2,6-dimethyl-1,4-phenylene oxide)
PV	Photovoltaic
PBI	Polybenzimidazole
PPA	Polyphthalamide
PINMs	Polymer-inorganic nanocomposite membranes
PLA	Poly (lactic acid)
PSf	Polysulfone
PSM	Particle size measurement
PDM	Particle density measurement
PAN	Polyacrylonitrile
PDMS	Polydimethylsiloxane
PTPE	Poly tetra fluoro ethylene
PFSI	Per fluorinated ionomers
RO	Reverse osmosis
SO_3H	Sulfonic acid
SDS	Sodium dodecyl sulfate
SiO_2	Silicon dioxide
SEM	Scanning electron microscopy
SHI	Swift heavy ions
SAXS	Small angle X-ray scattering
SPEEK	Sulfonated poly (ether ether ketone)
SA	Sulfanilic acid
SHMP	Sodium hexametaphosphate
SMMT/SPSU-BP	Sulfonated montmorillonite/sulfonated poly (biphenyl ether sulfone)
TAP	Triallyl phosphate
TiO_2	Titanium dioxide
TEM	Transmission electron microscopy
TGA	Thermogravimetric analysis
UF	Ultra-filtration
V	Voltage
WAXD	Wide angle X-ray diffraction
XRD	X-ray diffraction
ZrO_2	Zirconium dioxide

1 Introduction

The membranes play a key role in membrane-based processes and determine the economic efficiency of the technologies. In this chapter, the recent development of polymeric-montmorillonite membrane materials and membrane preparation methods with focus on structure–property relationships will be discussed. This chapter will provide a reference to the researchers and manufacturers working on fabrication of organic/montmorillonite nanocomposite membranes and materials.

2 Membrane Fabrication Methods

The technique for polymer-montmorillonite membrane fabrication depends on a choice of polymer, type of clay, and desired structure of the membrane. The most commonly used techniques for preparation of polymeric membranes include phase inversion, interfacial polymerization, stretching, track-etching, and electro spinning.

2.1 *Phase Inversion*

Phase inversion can be described as a demixing process whereby the initially homogeneous polymer solution is transformed in a controlled manner from a liquid to a solid state. This transformation can be accomplished in several ways as mentioned below:

Immersion precipitation: In this, the polymer solution is immersed in a non-solvent coagulation bath (typically water). Demixing and precipitation occurs due to the exchange of solvent (from polymer solution) and nonsolvent (from coagulation bath). Thermally induced phase separation: This method is based on the phenomenon that the solvent quality usually decreases when the temperature is decreased. After demixing is induced, the solvent is removed by extraction, evaporation or freeze drying. Evaporation-induced phase separation: In this, the polymer solution is made in a solvent or in a mixture of a volatile nonsolvent and the solvent is allowed to evaporate, leading to precipitation or demixing/precipitation. This technique is also known as a solution casting method. Vapor-induced phase separation: In this, the polymer solution is exposed to an atmosphere containing a nonsolvent (typically water). Absorption of nonsolvent causes demixing/precipitation. However, among these techniques, immersion precipitation and thermally induced phase separation are the most commonly used methods in the fabrication of polymeric membranes of various morphologies.

2.2 *Immersion Precipitation*

Immersion precipitation is a process where a polymer solution is cast on a suitable support, then immersed in a coagulation bath containing a nonsolvent where an exchange of solvent and nonsolvent takes place and the membrane is formed. The solvent diffused into the coagulation bath (at a flux = J2), whereas the nonsolvent diffused into the cast film (at a flux = J1). After a certain time, the exchange of solvent and nonsolvent proceeded until the solution becomes thermodynamically unstable. A solid polymeric film with an asymmetric structure finally formed. Usually at J2 ≫ J1 "skin" ultrafiltration (UF) membranes with pore size of 10–300 Å were obtained, while at J2 = J1 mainly microfiltration (MF) membranes with pore size of 0.2–0.5 μm were fabricated.

For membrane technologies, the development of the first high-flux anisotropic cellulose acetate (CA) reverse osmosis (RO) membranes via immersion precipitation was one of the most critical breakthroughs in desalination. Today, extensive knowledge exists on how to "tailor" the membrane's pore structure including its cross-section morphology by the proper selection of polymer, solvents and nonsolvents, additives, precipitation time, bath temperature, and other parameters during immersion precipitation. For example, different casting conditions and post treatments were proposed to improve the water flux and salt rejection of the CA membranes.

Apart from the chemical nature of a casting polymer, the concentration of the polymer is very important in membrane fabrication via immersion precipitation. Increasing polymer concentration in the casting solution produces membranes with low porosity and pore size. In this case, the macro-void formation is suppressed and the tendency to form sponge-like structures is enhanced. The UF membranes are obtained within a range of polymer concentration of 12–20 wt%, whereas RO membranes are typically prepared from casting solutions with polymer concentrations ≥20 wt%. Selection of solvent/nonsolvent system also strongly affects morphology and properties of casted membranes. The low miscibility of polymer in the solvent leads to fabrication of a nonporous membrane, while more porous membranes are obtained at higher miscibility of polymer. Generally, aprotic solvents (where no hydrogen atoms able to contribute to hydrogen bonding) are preferred for membrane casting. An aprotic polar solvent, such as *N*-methyl-2-pyrrolidone, dimethyl formamide, dimethyl acetamide or dimethyl sulfoxide is preferable for rapid precipitation (instantaneous demixing) upon immersion in the nonsolvent (water) and this produces anisotropic membranes of a high porosity.

To improve the membrane morphology and properties, various inorganic salts [such as lithium chloride (LiCl)] and high-molecular weight organic [such as polyvinyl pyrrolidone (PVP) or poly (ethylene glycol) (PEG)] additives were used. An additive can function as a pore former, solution viscosity enhancer or the phase inversion process accelerator. For example, the effect of LiCl addition in the membrane formation was investigated. Some researchers found that LiCl addition in the polyvinylidene fluoride (PVDF)/dimethylacetamide dope increases flux of the casted membranes at low LiCl concentration of 2.5 wt%, but it suppressed macro-void formation at a high concentration of 7.5% LiCl and resulted in a decrease of the membrane permeation flux. The similar results were obtained for poly (amic acid) (PAA) casting solutions in *N*-methyl-2-pyrrolidone. It was found that by increasing LiCl concentration in the PAA/N-methyl-2-pyrrolidone system, the solution viscosity can be raised to the point where macro-void formation is hindered and development of a finely porous structure is favored. The above observations were believed to be associated with the change of the thermodynamic and kinetic properties of the polymer dope system before and after LiCl addition. It was shown that LiCl addition increased the dope's thermodynamic instability in reaction with water, which facilitated a rapid phase demixing and resulted in macro-void formation (thermodynamic effect). On the other hand, LiCl has strong interactions with the polymer and solvent, which was supported by the significant increase in viscosity of LiCl added casting solutions. The strong interactions among the components of the casting solution tended to delay the dope precipitation (the kinetic effect), which partially offset the thermodynamic impact of LiCl addition. As a result, the size of the macrovoids in the fabricated membranes is reduced at high loading of LiCl dosage in the casting solutions.

Recently, the use of inorganic nanoparticles as additives to polymeric membranes has begun to attract wide interest due to the improved membrane properties, including increased strength and modulus, which result from the strong interfacial interactions. It should be mentioned however, that one of the limiting factors for incorporation of nanoparticles into polymeric membranes is high aggregation of nanoparticles that results in a low dispensability in the casting solution. Furthermore, careful control and monitoring of the nanoparticles released from the modified membranes are necessary to minimize potential (eco) toxicity effects.

2.3 Evaporation-Induced Phase Separation

Evaporation-induced phase separation is a facile technique to prepare membranes for various applications. At the first stage, a sufficiently viscous polymer solution is prepared in a solvent (or in binary/ternary mixture of solvents) plus a nonsolvent. The prepared polymer solution is casted on a flat porous substrate using a doctor blade technique. When the volatile solvent evaporates from the casted solution, a thin polymer film is formed on the porous support. The morphology of the solution casted films can be controlled by using solvents with different boiling points. Researchers developed PVDF, polyvinyl chloride (PVC), polystyrene (PS), and polyvinyl alcohol (PVA) microporous membranes using different organic solvents and studied the effect of different solvents on the surface morphology and pore size/shape. In addition, preparation of microporous PS films was done by using PEG as pore former. The pore size of the membranes was maintained in the range of 5–12 μm by varying PS/PEG concentrations and using PEG of different molecular weights. Silicon rubber microporous membranes using this technique were also developed. The pore size and the pore structure of the membranes were tuned by varying casting temperature and the concentration of liquid paraffin.

3 Montmorillonite-Based Nanocomposites Membranes

Polymer nanocomposites membranes based on the impermeable lamellar fillers such as montmorillonites have received considerable attention since they may lead to enhanced barrier properties with far less filler content than conventional composite membranes. As organic active membranes, these membranes are able to increase the time lag but can drastically reduce the steady-state gas flux. Certain equations are also already proposed to well adapt the membranes containing exclusively exfoliated structures and become less accurate as the dispersion of the silicate sheets is more complex, i.e., when different types of structures from exfoliated ones to intercalated coexist. However, such a kind of morphology is often observed, particularly for an intermediate and large clay content which justifies that additional models have been further proposed to consider polydisperse flakes. These models take into account a distribution of the impermeable flakes dimensions considering variations of width and constant thickness. The use of organoclay as fillers in comparison with other types of fillers like zeolites, carbon nanotubes (CNTs), carbon molecular sieves, silica, and metal organic frameworks is attractive because of the low cost and availability of the material. In addition, the ionic interactions between the organoclay platelets are weak contributing to the ease of delamination.

Pervaporation, a membrane process for separating liquid mixture has emerged as an energy-efficient alternative to distillation and other separation methods. It has been widely used in solvent dehydration due to its highly selective, economical, energy-efficient, and eco-friendly characteristics. Much attention has been directed for the membrane fabrication addressing the issues on separation efficiency and development of the skin layer. Separators used in commercial lithium-ion batteries are nearly exclusively semi-crystalline polyolefin membranes, including polyethylene (PE), polypropylene (PP), and their blends such as PE-PP. Although PE-based separators have good mechanical properties, chemical stability and acceptable cost, they cause safety problems in lithium-polymer batteries at temperatures above 140 °C. Some of the membrane researchers had already synthesized modified poly (ether sulfones) (PES)-clay composites and reported significant changes to the membrane skin layer and sublayer with increasing clay concentration that resulted in increased membrane permeability. They have also reported an increase in the pesticide retention capacity, but no significant changes in pore size. These results indicate that polymer–clay composite membranes have improved performance toward water treatment compared to unmodified membranes. Similar results were obtained for clay nanoparticle-polymer composite membranes.

Polymer-clay hybrid membranes have received greater attention for pervaporation dehydration of organic solvent due to their hydrophilic, thermal, and mechanical properties. Several techniques have been proposed in the literature to introduce hydrophilic groups on the hydrophobic polymeric hybrid membrane. Amongst these, proton conducting membranes for fuel cell application using radiation technique are most promising. Incorporation of high surface area, nanometer-size filler and nucleating agents like CNTs, organically modified layered silicate, montmorillonite and graphene into the polymer matrix can dramatically alter their crystallization behavior, mechanical strength, thermal, and gas barrier properties. Recently, polymeric hybrid membrane becomes one of the key materials for energy conversion technologies as it facilitates improvement in gas permeability, UF, MF, and proton exchange membrane (PEM) for fuel cell application with respect to its pristine polymer.

PEM-fuel cells (PEM-FCs) have wide applications that include road vehicles and residential heat/power supplies. Nafion®, the fluorinated membrane from DuPont, and similar membranes commercialized by Dow and Asahi have been intensively used in fuel cells as PEMs. These membranes combine mechanical strength and chemical/thermal stability with high proton conductivity, methanol permeability, and a large-scale cost of production. Perfluorinated sulfonic acid resin (such as Nafion®) is the most common PEM and must be kept well hydrated to retain acceptable proton conductivity. Loss of water content in the PEM leads to a sharp decrease in the proton conductivity and to shrinkage of the membrane. The latter results in mechanical degradation of the membrane catalyst interface. Incorporation of sodium (Na^+) montmorillonite into PVA matrix leads to pervaporation dehydration of aqueous mixtures of isopropanol and 1,4-dioxane which results in enhanced separation factor. In addition, intercalation of alginate (NaAlg) into partial Na^+ montmorillonite galleries and the selectivity to water was increased; this result is ascribed to the spatial distribution, rearrangement of intercalating polymer chains and interfacial interaction between the clay layers and polymers. Organically modified montmorillonite (OMMT) has an improved capability of dispersing, gelling, adsorption, and nanocomposite formation in organic systems. Styrene sulfonic moieties were bonded with a clay surface via plasma activation and the Nafion®/clay-SO_3H composite membrane was studied in comparison with a commercial Nafion® membrane. Montmorillonite was also modified by dodecylamine to prepare a Nafion® composite membrane.

In recent years, there has been an intensive research effort about the development of alternative membranes with potentially lower costs and better processability. Many polymers, such as sulfonated PES, sulfonated poly (arylene ether sulfone) and various other polymeric systems have been briefly described already. However, the methanol permeability is still relatively high. Reduction of methanol permeability was investigated by modifying the membrane surface and the size of the proton transport channels, developing new types of polyelectrolytes and introducing tortuous pathways. Tortuous pathways can be introduced by dispersing inorganic fillers, such as silica, heteropolyacid, zeolite, zirconium phosphate and montmorillonite within the polymer. Montmorillonite is a type of layered silicate composed of silica tetrahedral and alumina octahedral sheets. To render layered silicates miscible with polymer matrices, one must convert the normally hydrophilic silicate surface to an organophilic one. Organically modified montmorillonite has been obtained by ion-exchange reactions with cationic surfactants, such as alkyl ammonium or alkyl phosphonium cations. Nonfluorinated membranes based on the poly (2,6-dimethyl-1,4-phenylene oxide) (PPO) have been presented as materials with excellent chemical/thermal stability, mechanical strength, and good processability. Their poor proton conductivity was improved by the sulfonation process. Sulfonation is a powerful and versatile process that can be used to simultaneously render these polymers proton conductive as well as hydrophilic in nature. Like other sulfonated aromatic main-chain polymers, these sulfonated PPOs require certain acidic groups to achieve high proton conductivity. However, high loading of acidic groups induces excessive water swelling and methanol diffusion. Sulfonated PPO membranes have been used for gas separation and reverse osmosis due to their good permeation properties, as well as thermal and chemical stabilities.

For instance, polyamide/clay membranes were studied for photovoltaic (PV) of aqueous ethanol mixtures. The organoclay was prepared by ion-exchange reaction between the montmorillonite silicate layers and sodium dodecyl sulfate (SDS) as intercalating agent. The permeation rate was found to decrease and the separation factor to increase upon the insertion of 2 wt% SDS nanoclay in the membranes. PVA-based mixed matrix membranes loaded with 5 and 10 wt% of Na^+ montmorillonite were used for dehydration of aqueous mixtures of isopropanol and 1,4-dioxane. The water selectivity was found to be increase for PVA/Na^+ montmorillonite composite membranes; whereas the corresponding flux values were somewhat lower than the reference one. Crosslinked montmorillonite modified NaAlg membranes led to similar results. Intercalated nanocomposites were formed either by melt blending with OMMT or in situ polymerization within Na^+ montmorillonite. The barrier properties were studied for water vapor and dichloromethane (DCM) as an organic solvent. The sorption and the zero concentration diffusion coefficients were evaluated for both water and DCM vapors. The water sorption increases with the increase in montmorillonite content, particularly for the microcomposites containing the unmodified Na^+ montmorillonite. Some authors studied the dehydration of water and alcohol mixtures by vapor permeation through PVA/clay nanocomposite membranes and found that permeation rate was decreased, but selectivity increased, as the clay content increased. It has been shown that polybenzimidazole (PBI)-based PEMs can be operated for well over 20,000 h under steady-state conditions (stationary power plants) and are durable under start/stop cycling operation, thermal cycling, and load cycling. These methods can be classified into four main categories: (a) introduction of fillers (i.e., small organic or inorganic molecules) into the polymer matrix; (b) chemical crosslinking of the polymer; (c) alternative processing techniques—conventional imbibing, Polyphthalamide (PPA) process (sol–gel transition), thermal annealing, control of doping level and (d) novel polymer chemistries. In most of the papers on PBI nanocomposites membranes, the efforts have been made on improving the properties of PBI by low clay loading. Mixed matrix PVC membranes, with maghnite H clay as filler has also been previously investigated. Polymer-inorganic nanocomposite membranes (PINMs), also called mixed matrix membranes (MMMs), which were formed by inorganic particles uniformly dispersed in a polymer matrix, have received much attention in the field of gas separation, pervaporation, and UF. These MMMs present an interesting approach to improve the separation properties of polymer membranes because they possess properties of both organic and inorganic membranes, such as good permeability, selectivity, mechanical strength, thermal, and chemical stability. The structure and performance of nanocomposite membranes are generally a function of the physical and chemical properties of the polymer matrix and nanoparticles as well as the method of nanoparticle incorporation. Depending on how many dimensions in the nanometer range, inorganic particles dispersed in the PINMs can be distinguished as isodimensional nanoparticles, nanotubes or nanowires, and clays when the dimensions in the nanometer range are three, two and one respectively. Most of the recent studies on PINMs have focused on introduction of isodimensional nanoparticles, such as TiO_2, aluminum oxide (Al_2O_3), zirconium dioxide (ZrO_2), and silicon dioxide (SiO_2) nanoparticles for better understanding of membrane skin properties (thickness, overall porosity, and surface porosity) and the macro-void morphology of the support layer. The PINMs were prepared by blending polymer solutions with nanoparticles (ex situ) or with their precursors (in situ). CNTs are often employed as filler materials to prepare porous polymer nanocomposite hydrophobic membranes or gas separation membranes. However, the inorganic ingredients are liable to agglomerate in the membranes because of the huge difference between the polymer and inorganic materials in their properties and strong agglomeration of the nanofillers. The agglomeration of nanofillers in the membrane weakens the mechanical strength of the membrane.

Incorporation of nanoclay can be considered as an effective method to improve the mechanical strength of porous membrane supports. Some authors studied the barrier properties of polyamide/montmorillonite nanocomposite membranes and found that nanocomposite membranes exhibited superior barrier properties to gas and water in comparison to neat polyamide film. In addition, nanocomposite films had a permeability transition at low silicate content of 2.5 wt%. Up to 2.5 wt% of the Cloisite 30B (organophilic montmorillonite) nanofiller, the water permeability decreases with the increase in the nanofiller volume fraction. Researchers proved the particular barrier property through intercalation of clay in the poly (lactic acid) (PLA) matrix and found that both gas permeabilities of the prepared nanocomposite membranes decreased with the increase of clay content.

Polysulfone (PSf) membranes have been widely used as MF and UF membranes in many industrial fields because of their low cost, superior film forming ability, good mechanical and anti-compaction properties, strong chemical, and thermal stabilities as well as resistance outstanding acidic and alkaline solutions. However, their hydrophobic nature that often results in severe membrane fouling and decline of permeability has been a barrier for their application of water treatment. Organophilic montmorillonite was commonly used as additives in the PSf/clay nanocomposites. Motivational researchers prepared porous membranes and dense films with polysulfone solutions in *N*-methyl-2-pyrrolidone (NMP) containing different amounts of different types of clay (unmodified or organically modified) without any other additives. They found that Cloisite Na and Cloisite 93A formed microaggregates, Cloisite 30B yielded nanostructures composed of both single sheets and well-ordered multilayer silicate clusters. The addition of Cloisite 30B to the casting solution influenced the phase-separation process in the coagulation bath. Cloisite 30B was also found to improve the wettability and mechanical properties of dense films. In many cases, PSf/montmorillonite nanocomposite membranes were prepared by the combination of the solution dispersion method and the final step of the wet-phase inversion method. The Na^+ montmorillonite intercalated and exfoliated in the PSf matrix. Thermal stability and hydrophilicity of nanocomposite membranes were improved by the incorporation of montmorillonite. In addition, clay minerals are organophobic in nature, therefore organic modification is required to provide a microchemical environment for intercalation with hydrophobic polymers. Intergallery distance of unmodified montmorillonites such as Na^+ montmorillonite is relatively small and therefore, affects the distribution state of montmorillonite in the nanocomposites. The optimum organoclay contents in the nanocomposites were commonly 3 wt% of the polymers. Additives, such as PEG, PVP, and LiCl were often needed in the preparation of membrane as the pore forming agent and PSf membrane prepared without additives had lower permeability. It is necessary to study the effect of clay dosage on the structure and performance with the incorporation of additives. The use of PVP and PEG particles has been reported to improve both membrane permeability and fouling resistance by many researchers. Advances in nanomaterial synthesis have resulted in a new generation of research on colloidal nanoparticle-polymer composite membranes and the use of nanoparticle additives is hypothesized to improve membrane properties such as mechanical strength and performance characteristics including water flux and fouling resistance. The use of clay additives during the fabrication of polymer composite membranes has been reported for fuel cell, gas separation, and water treatment applications.

Koonaphapdeelert et al. [1] related the membrane morphological and structural changes of polymeric membranes utilized in contactor system to the hydrophobicity as well as the intrinsic membrane material properties. They later tried to overcome the inadequacies of polymeric membranes by fabricating high hydrophobic ceramic membranes. The utilization of inorganic membranes was due to their intrinsic high thermal and chemical stabilities and ability to achieve much higher gas fluxes in comparison to the polymeric membranes. The fabricated membranes were examined at high temperature carbon dioxide (CO_2) stripping process via membrane contactor. Higher and more stable performance compared to the available commercial polymeric membranes were also achieved. However, the preparation of ceramic membranes is complex, time consuming and expensive. In addition, low processability and flexibility that together with the other shortcomings prevent ceramic membranes from being used widely [2, 3]. Many factors influencing the wettability resistance and morphological changes of the polymeric membranes have been reported in the literature including operating temperature and pressure conditions, polymer properties and membrane surface characteristics, such as pore size, porosity and roughness [4, 5]. Sharma and Chellam [6] conducted a research on membrane nanofiltration application, where the effects of operating temperature on the morphology and structure of two commercially available thin film composite membranes were investigated. Temperature increase effect was found to enlarge the size of the network pores while simultaneously decreasing pore density. As a result, the water contaminants removal was decreased considerably. The contactor membranologists have simply reported that increasing absorbent temperature reduces the absorption performance of the membranes employed in gas/liquid contactors [7], but the phenomena are still not thoroughly investigated.

Rezaei-DashtArzhandi et al. [8] explained that clay particles embedded into the original PVDF membranes results in improved thermal stability. The prepared membranes immersed in water for 7 days @ 80 °C, dried and characterized. The field emission scanning electron microscopy (FESEM) images revealed that large pores became larger and small pores became smaller. High partial pore wetting resulted in 31 and 13% reduction in performance for neat PVDF and 5 wt% montmorillonite filled membranes. In addition, due to flux reduction at higher temperatures, membrane gas performance, porosity, contact angle, and wetting resistance were reduced. It has been suggested that impregnation of polymeric membranes by hydrophobic inorganic particles can be an effective method to stabilize the performance and properties of neat PVDF and montmorillonite filled PVDF membranes.

Picard et al. [9] studied the barrier properties of polyamide 6-montmorillonite membranes by varying the kinetic diameter and interaction capacity of different diffusing molecules. Tortuosity effect explained that relative permeability is independent on the different diffusing molecule. The experimental relative permeability using different geometrical model was also discussed. In connection to this, crystalline morphology of the polymer depends on the nanocomposite composition and permeation properties. Polydispersity of the impermeable filler shapes and presence of surfactant at the inorganic surface are necessary to model the transport properties in high nanoclay loadings.

Ismail et al. [10] explained that flat sheet asymmetric MMMs from PES and Cloisite 15A were characterized by using wide angle X-ray diffraction (WAXD), transmission electron microscopy (TEM), tensile test, and pure gas permeation measurement. The effects of dispersion of silicate layers on the properties and performance of MMMs for CO_2/methane (CH_4) separation were also investigated by qualitative and quantitative measurements. High degree of intercalation and exfoliation of silicate layers were identified by particle size measurement (PSM), particle density measurement (PDM) and free-path spacing measurement (FPSM) with dispersed single silicate layers and formation of several tactoid classes for PES/Cloisite 15A.

Hwang et al. [11] explained that PVDF composite membranes with various nanoclays showed enhanced mechanical strength (Fig. 1) without affecting the membrane porosity (around 80%) compared to pristine PVDF membrane. The samples were prepared using phase inversion method by controlling the retention time. High porous morphologies were confirmed using scanning electron microscopy (SEM) analysis as shown in Fig. 2. X-ray diffraction (XRD) results confirmed that the clay particles were well dispersed in PVDF and partially intercalated in PVDF/Na^+ nanocomposite membrane as depicted in Fig. 3.

Jana et al. [12] fabricated nanochannels in PVDF-co-hexafluoro propylene (HFP) films by bombarding swift heavy ions (SHI), creating latent tracks in the ion passage and finally chemical etching of the amorphous track. In addition, grafting

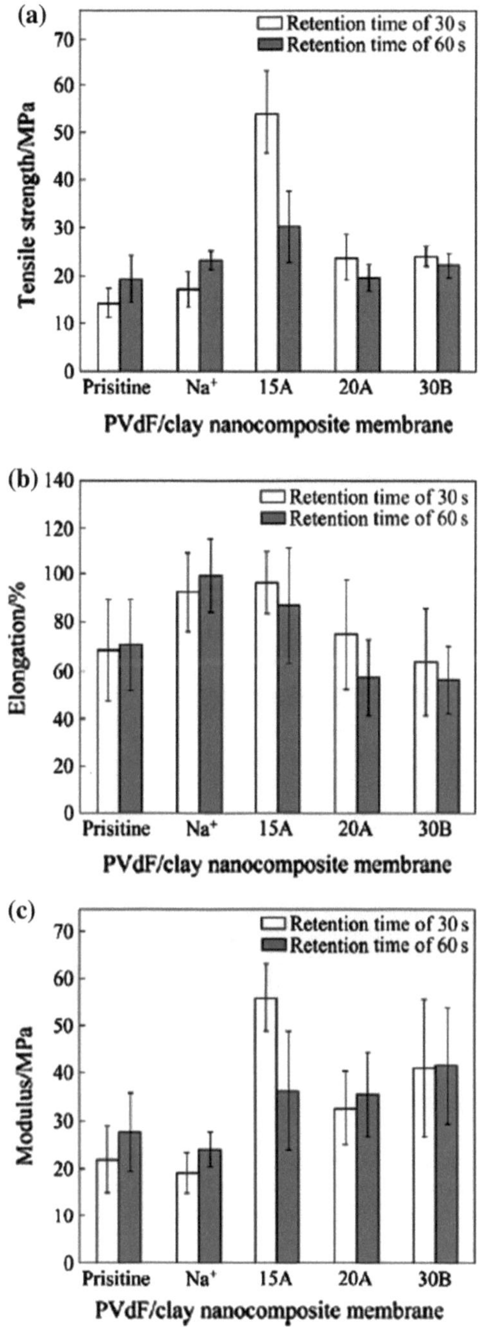

Fig. 1 Mechanical properties of PVDF and PVDF/clay nanocomposite membranes prepared with retention time of 30 and 60 s (reproduced from [11])

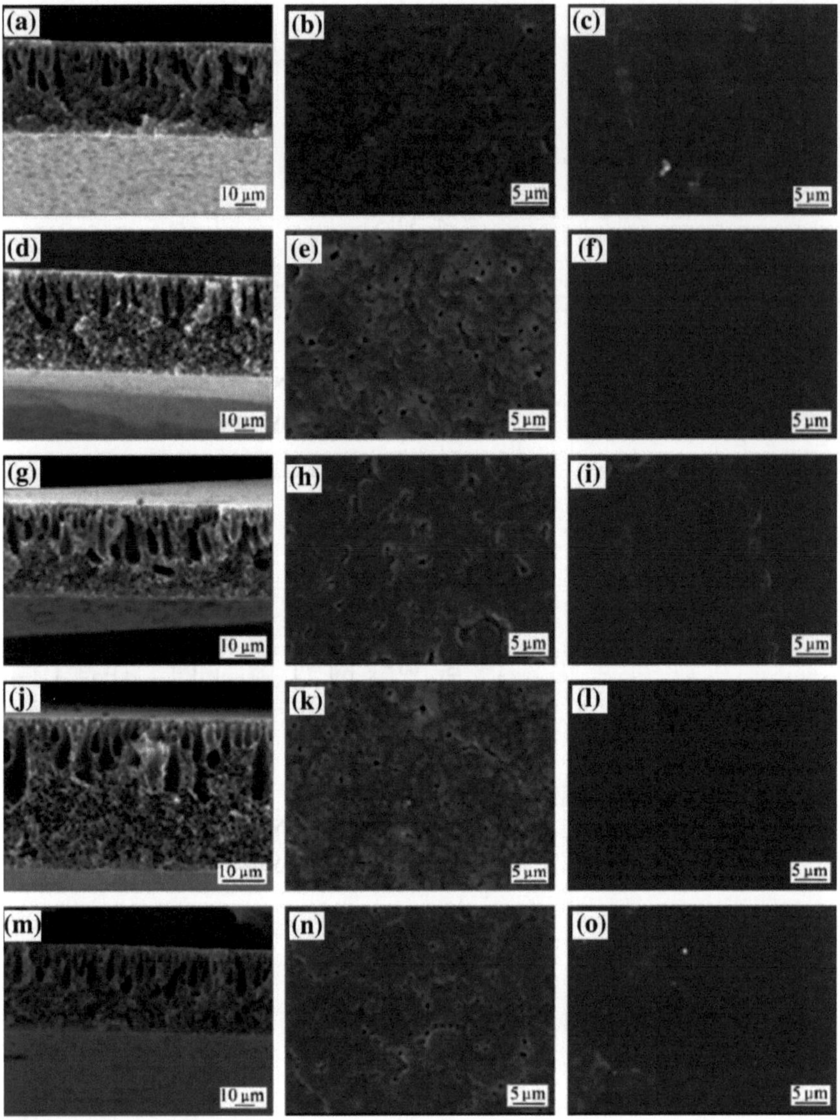

Fig. 2 SEM images of porous PVDF and PVDF nanocomposite membranes prepared with retention time of 30 s: **a**, **b**, **c** PVDF; **d**, **e**, **f** PVDF/Na^+; **g**, **h**, **i** PVDF/15A; **j**, **k**, **l** PVDF/20A; **m**, **n**, **o** PVDF/30B (reproduced from [11])

with polystyrene using free radicals and sulfonation was also carried out. The dimension of nanochannel was varied by dispersing organically modified nanoclay into the polymer matrix. Efficiency was studied through direct current (DC) conductivity of the bulk film, whereas current (I)–voltage (V) characteristics exhibits strong fluence dependency and shows superior conduction in nanohybrid.

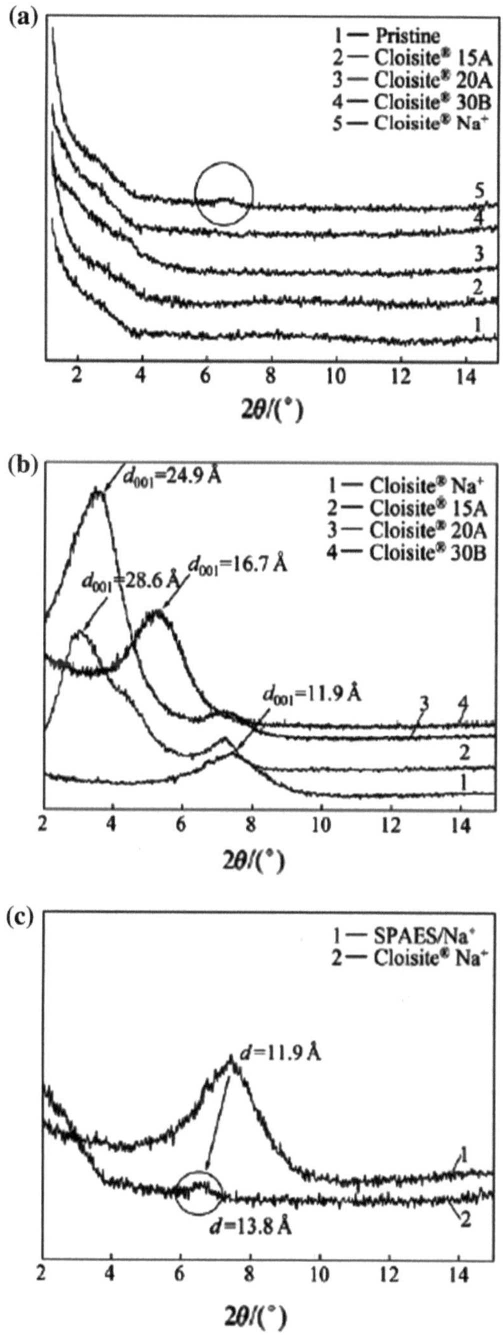

Fig. 3 XRD patterns of PVDF/clay composite membranes containing different clay species and original clay: **a** PVDF/clay nanocomposite membranes; **b** clay particles used in this study; **c** diffusion peak shift in PVDF/Na^+ nanocomposite membrane (reproduced from [11])

Table 1 Comparison of proton conductivity (k^m), activation energy (E_a) and methanol permeability (P) of HFP-g-s and NH-g-s with standard Nafion® 117 membrane (reproduced from [12])

Membrane	k^m ($\times 10^{-2}$ S cm^{-1})	E_a (kJ mol^{-1})	P ($\times 10^{-7}$ cm^2 s^{-1})
HFP-g-s	5.5	6.36	2.91
NH-g-s	6.2	5.79	1.84
Nafion® 117	9.56	6.52	13.10

Table 1 lists the data of proton conductivity, activation energy and methanol conductivity of Nafion® 117-based membrane.

Shubha et al. [13] proposed a new approach to fabricate electrospun PVDF-co-HFP membrane with nanoclay particles to enhance the ionic conductivity, mechanical properties, and electrochemical properties. XRD, differential scanning calorimeter (DSC), FESEM and thermogravimetric analysis (TGA) were used to analyze the developed polymer gel electrolytes. The PEGs showed better compatibility with lithium electrodes. The charge–discharge properties and cyclic performance have also been evaluated at room temperature.

Li et al. [14] prepared nanocomposite fibers of nylon-6, OMMT and Cloisite 30B by adopting electrospinning method. Dispersion of OMMT in base polymer was done using melt extrusion in a twin screw extruder prior to electrospinning. The diameter of the prepared fibers with 15% composite solution ranged from 70 to 140 nm. The effect of OMMT onto the membrane was also investigated and thus clay layers were exfoliated and oriented along the fiber direction. The crystallinity and crystalline size increased for the nanocomposite fibrous mats. The addition of OMMT has strong influence on the mechanical properties of the electrospun fibrous mats, whereas small fibers exhibited higher Young's modulus.

Alonso et al. [15] minimized the adverse effects on ionic conductivity of the prepared Nafion®-clay nanocomposite membranes, the clay particles were H^+ exchanged prior to mixing. Small angle X-ray scattering (SAXS) were performed to analyze the preferential orientation which is induced by the platy nature of clay nanoparticles. The prepared films are much stiffer to withstand high temperature compared to pure Nafion®, displayed dramatic reduction in methanol permeability and maintained high levels of proton conductivity. This research work was big challenge to obtain superior thermomechanical, electromechanical and barrier properties of the nanocomposite membranes for direct methanol fuel cell applications.

Singha and Jana [16] took away the big challenge of studying the chemistry of organic–inorganic interface on the final properties of organically modified polybenzimidazole (OPBI)/clay nanocomposites PEM for fuel cell. Lot of research had been done in the wide range to analyze the influence of type and quality of nanoclay on the micro properties of polymer nanocomposites. Influence of the different surfactants was analyzed using WAXD, TEM analysis, TGA, and dynamic mechanical analysis (DMA). Proton conductivity data depicted higher proton conductivity and decreased activation energy barrier for the proton hopping.

Kim et al. [17] prepared proton conducting membranes together with Nafion® by using functionalized montmorillonite as organic fillers. The organic fillers prepared by using simple one-step reaction showed higher ion-exchange capacity and thermal stability than two-step method. The composites membranes with grafted montmorillonite showed increased water uptake by maintaining comparable ionic conductivity relative to pristine Nafion® membrane, enhanced performance without external humidification of reactant gases. For example, current density of 0.5 V of membrane-electrode assembly (MEA) composite membrane fabricated containing montmorillonite with long chain organic sulfonic acid was twice that of montmorillonite with short chain and thrice that of pristine Nafion® 212. Figure 4 explains the WAXD pattern and TEM image of various loading of montmorillonite onto the Nafion®-based composite membranes. In addition, Fig. 5 illustrates the water uptake in percentage for 5 wt% loading of inorganics onto the Nafion® membrane.

Gosalawit et al. [18] developed Krytox–montmorillonite–Nafion® nanocomposite membrane for direct methanol fuel cell applications. The good compatibility was observed whereas, exfoliated and homogeneity were confirmed through XRD and SEM. Atomic force microscopy (AFM) confirmed the successful dispersion of

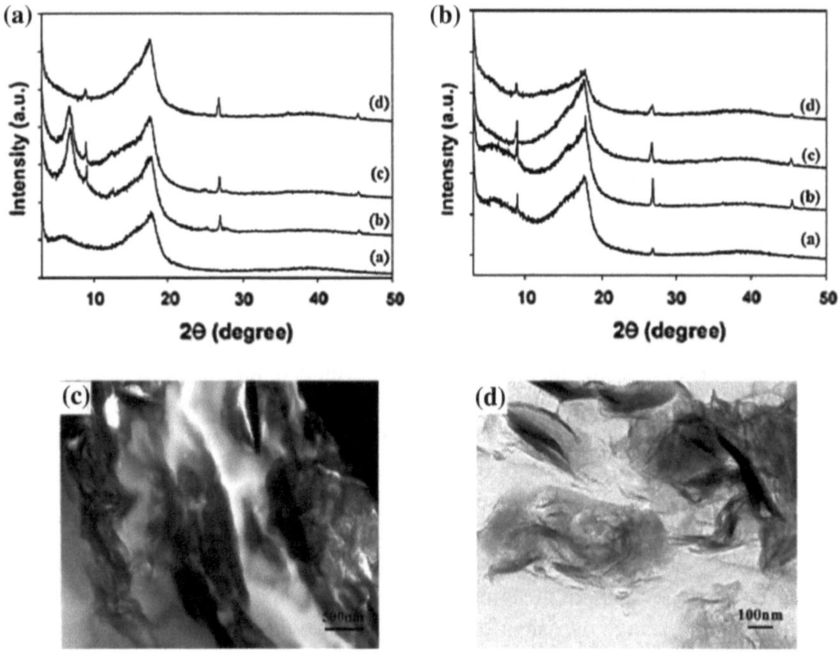

Fig. 4 **a** WXRD patterns: *a* recast Nafion® membrane, *b* 5 wt% H^+ montmorillonite/Nafion®, *c* HSO_3-R-montmorillonite/Nafion®, and *d* HSO_3-RSR-montmorillonite/Nafion® composite membranes, **b** WXRD of composite membrane with different HSO_3-RSR-montmorillonite contents: *a* 1 wt%, *b* 3 wt%, *c* 5 wt% and *d* 10 wt% TEM image of 5 wt% HSO_3-R-montmorillonite/Nafion® (**c**) and 5 wt% HSO_3-RSR-montmorillonite/Nafion® (**d**) composite membranes (reproduced from [17])

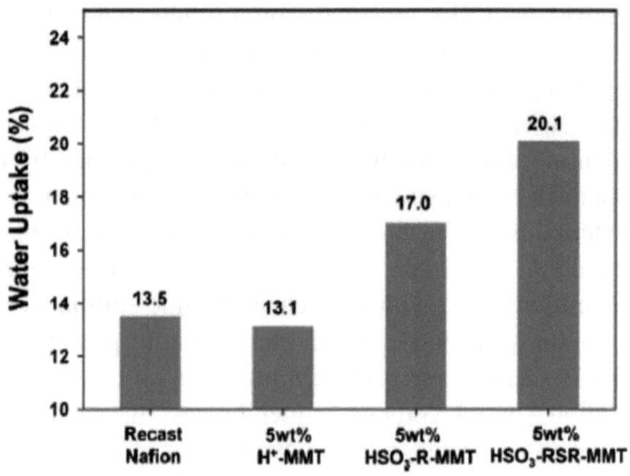

Fig. 5 Water uptake (%) of recast Nafion® and composite membranes containing 5 wt% inorganics (reproduced from [17])

montmorillonite clay all over the nanocomposite membrane. TGA exhibited improvement in water retention and thermal resistance compared to Nafion® membrane. Arrhenius plot suggested lower activation energy than Grotthus mechanism.

Anadao et al. [19] combined solution dispersion and immersion step of the wet-phase inversion method to prepare PSf/montmorillonite nanocomposite membrane. As a whole, 0.5 and 3.0 mass% montmorillonite was added and their morphology, thermal, mechanical, and hydrophilic properties were examined and compared with pure PSf membrane. Small angle XRD, TEM, SEM images depicted the intercalated structure, exfoliated structure, and good dispersion of the clay particles. Indeed, tensile or thermal properties and water contact angle measurements were high compared to pristine PSf.

Dogan et al. [20] modified montmorillonite with 2-acrylamido-2-methyl-1-propanesulfonic acid (AMPS), dimethyl dioctadecylammonium chloride (DMDOC) and sulfanilic acid (SA). Sulfonated poly (ether ether ketone) (SPEEK)-based OMMT composite membranes were developed for fuel cell applications. TGA results revealed improvement in thermal resistance compared to pristine membrane. Particularly, DMDOC-montmorillonite incorporation onto the polymer increased the proton conductivity of the composite membrane.

PVDF/nanoclay membranes were solution casted and phase inverted using different concentrations of OMMT. Rajabi et al. [21] characterized XRD, water contact angle, SEM, and AFM for the above prepared membranes and their performances were evaluated in terms of pure water flux and fouling parameters. Thinner skin layer, high porosity, formation of intercalated structure, enhanced pure water flux, antifouling properties, and flux recovery were compared to nascent PVDF membrane. The author strongly suggested that PVDF/montmorillonite membrane offered remarkable reusability and durability against biofouling.

Wu et al. [22] developed and modified Nafion® membranes with montmorillonite for direct ethanol fuel cells rather than methanol fuel cells. Different types of clay, such as Ca-montmorillonite, Na-montmorillonite, K-montmorillonite,

Mg-montmorillonite, H-montmorillonite, and Nafion® solution via casting method was used to prepare organic/inorganic composite membranes. Addition of montmorillonite decreases the ethanol conductivity and proton conductivity depending on the inserted amount and type of montmorillonites.

Tran et al. [23] prepared PSf/montmorillonite MMMs and investigated the feasibility of immobilization. Phase inversion method was used to develop pristine PSf and PSf/montmorillonite membranes, whereas various surface techniques were used to characterize the prepared membranes. The addition of montmorillonite enhanced the hydrophilicity and promoted the formation of porous structure which resulted in greater permeance but lower rejection compared to PSf membrane. Image analysis confirmed the hydrophobic interactions between the montmorillonite and protein molecules. This finds major applications of clay filler polymers in selectively removing protein from wine.

Wet spinning method was used to prepare porous PVDF-hydrophobic montmorillonite mixed matrix membranes by Rezaei et al. [24] to use in membrane gas absorption process. Finger like and sponge-like structures with increase in cross-section resulted in higher permeability and lower mass transfer resistance compared to control. Significant improvements in entry pressure, porosity and surface hydrophobicity with the addition of filler had been observed. Membrane with 5 wt% montmorillonite presented higher performance and improved CO_2 absorption.

Nesic et al. [25] synthesized the environmental application membrane containing chitosan (CS) and 10–50% montmorillonite. Fourier transforms infrared spectroscopy (FTIR), SEM, and TGA were used to characterize the prepared membrane. The results of adsorption kinetics showed enhanced adsorption capacity with the increase in clay content. Kinetic models were developed and compared with the experimental data which fitted and followed by intra-particle diffusion. At different concentrations of dye solution, Langmuir and Freundlich adsorption isotherms were applied to experimental equilibrium data. The competency of prepared membrane for Bezactiv Orange adsorption was also presented.

Xing et al. [26] explained that OMMT normally improved the permeability, water retention and proton conductivity of PEM for fuel cells. In this study, sulfonated montmorillonite/sulfonated poly (biphenyl ether sulfone)/poly tetra fluoro ethylene (SMMT/SPSU-BP/PTFE) composite membranes were prepared for fuel cells and their thermal stability was tested using TGA-mass spectrometer (MS) technique. XRD characterization was done to identify the structure of SMMT and it was found completely exfoliated. The properties of ion-exchange capacity, water uptake, swelling ratio, proton conductivity, and mechanical strength of the composite membranes were also investigated and found improvement in their properties. This promising composite membranes were used in PEM for medium temperature PEMFCs applications.

Yang [27] developed proton conducting composite membrane comprising of PVA with proton sources like modified montmorillonite fillers was fabricated using solution casting method. TGA, DMA, XRD, SEM, FTIR, applied current (AC) impedance method and methanol permeability measurement experiments were carried out to investigate the properties of these blend composite membranes. The direct methanol fuel cell with this composite polymer membrane exhibited good electrochemical performance. These composite membranes have potential future applications in direct methanol fuel cells.

Sulfonated PPO with various degrees of sulfonation was prepared by Hasani-Sadrabadi et al. [28]. Solvent casting method was used to mix the solution with modified montmorillonite. Membranes without montmorillonite and 40% sulfonation exhibited ion-exchange capacity, water uptake, and proton conductivity as 2.59 mequiv g^{-1}, 21% and 0.0182 S cm^{-1}, respectively. FTIR analysis for the absorption bands and XRD to find out exfoliated structures of clay (Fig. 6) in polymer matrices were performed. In addition, 27% sulfonated PPO with 2 wt% montmorillonite loading showed 63,500 membranes (40,500 for Nafion® 117) and higher power density (125 mW cm^{-1}) than Nafion® 117 for single cell direct methanol fuel cell in a 5 M methanol feed. The comparison of membranes for various direct methanol fuel cell applications is presented in Table 2.

Dehkordi et al. [29] combined solution dispersion and phase inversion methods to develop asymmetric nanocomposite membrane based on CA and modified OMMT. Properties and performance of the prepared nanocomposite membranes were investigated using XRD, FTIR, SEM, TGA, contact angle, porosity measurement, tensile strength, and humid acid adsorption. UF experiments were carried out to analyze the pure water flux and humic acid rejection range. Addition of small content of OMMT drastically influenced the properties of the prepared membranes.

Choudhari et al. [30] incorporated Na^+ montmorillonite into quaternized CS to prepare novel polymer-based composite membranes. The samples were characterized by using FTIR, WAXD, and TGA. Membrane swelling effect and pervaporation dehydration of isopropanol were studied by varying Na^+ montmorillonite content. The main highlight of this research was to propose Arrhenius activation

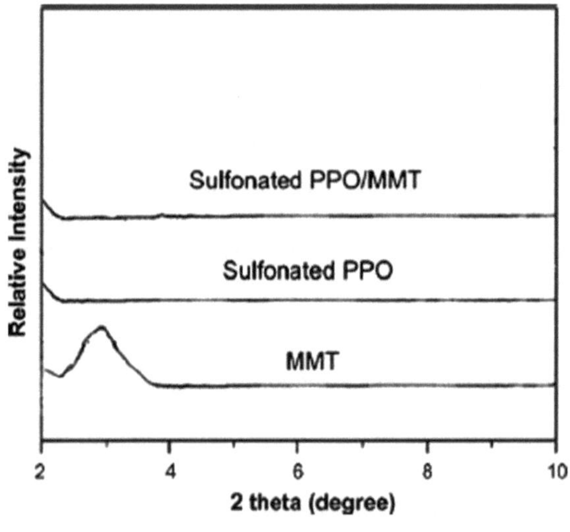

Fig. 6 The XRD pattern of montmorillonite, sulfonated PPO and sulfonated PPO/montmorillonite composite (reproduced from [28])

Table 2 Comparison of parameters for various direct methanol fuel cell membranes (reproduced from [28])

Proton exchange membrane	Proton conductivity (S cm^{-1})[a,b]	Methanol permeability (cm^2 s^{-1}) ($\times 10^6$)[b]	Membrane selectivity
Sulfonated PPO (27% of sulfonation)	0.0131	0.55	23,400
Sulfonated PPO (27% of sulfonation) +2% montmorillonite	0.0108	0.17	63,500
Nafion® 117	0.081	2.00	40,500
Nafion® + 5% SiO_2	0.270[c]	4.17	48,000
Nafion® + 5% montmorillonite	0.092	1.63	56,500
Sulfonated polystyrene and sulfonated PPO blend	0.034	2.35	14,500
Crosslinked PVA/poly (acrylic acid)/ silica hybrid	0.012	0.21	57,200
Crosslinked PVA using sulfosuccinic acid Tetra-fluoroethylene with poly (styrene sulfonic acid)	0.015	0.33	45,500
Pall R1010 (36 μm)	0.08	6	13,350
Pall R4010 (63 μm)	0.072	4.2	17,500

[a]Along the plane of the membrane
[b]Measured at room temperature
[c]Conductivity measured at 20 °C

parameters. The positive heat of sorption values was obtained to suggest that Henry's mode of sorption was predominant in the process.

Jaafar et al. [31] investigated the effect of clay (2.5 wt% cloisite 15A) loading on the physical properties of SPEEK membrane with 63% degree of sulfonation. In addition to this, various amounts of triallyl phosphate (TAP) were added and their effect on the physical and performance properties of the membranes were investigated. FESEM and XRD displayed the assistance of TAP and complete exfoliation of clay nanoparticles. Increase in clay content tends to enhanced swelling behavior and barrier properties. The prepared nanocomposite membrane served as a promising alternative proton exchange membrane for direct methanol fuel cell applications compared to commercial Nafion® 112.

Aouinti et al. [32] studied the separation properties of simple and cheap mixed PVC matrix membranes. Several types of nanocor clay particles, maghnite H, maghnite H^+, wyoming, and kaolin were added to prepare hybrid PVC membranes. The amount and types of clay incorporated onto the membranes drastically modified the transport properties. It can also be tuned easily either as barrier materials or as toluene selective membranes with strongly enhanced flux compared to pure PVC membranes.

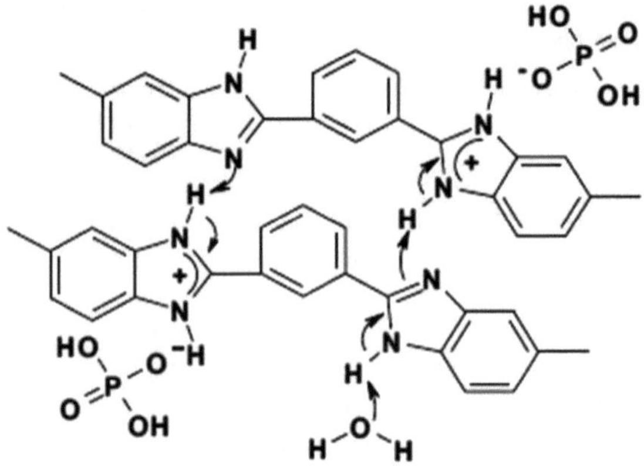

Fig. 7 Proton conductivity mechanism of PBI/polyamide membrane (reproduced from [33])

Ublekov et al. [33] used easy and efficient procedure to prepare p-PBI phosphoric acid doped membranes containing high levels of protonated montmorillonite-H. Compared to celtec-P membrane, i.e., commercial product of fuel cell GmbH, the prepared membrane showed 100% enhancement in proton conductivity and mechanical properties. Figure 7 depicts the proton conductivity mechanism for PBI/polyamide membrane. The major implementation leads to trapping of water and phosphoric acid in the montmorillonite-H channels.

Slavutsky et al. [34] developed and characterized the films based on gum exudates from Brea tree and nanoclay particles especially for food applications. The addition of montmorillonite onto the Brea gum-based films enhanced the functional properties of the same whereas increase in surface energy caused reduction in foam forming. Casting method was adopted to develop the films and in line with those physicochemical properties, FTIR, XRD, optical properties, mechanical properties, such as tensile strength and Young's modulus, water resistance and gas barrier properties of the prepared Brea gum/montmorillonite films were investigated. Incorporation of 5% montmorillonite improved the water resistance, water and gas barrier, and mechanical properties of the films. Nanoclay addition reduced the gas permeability and water isotherms, i.e., water uptake of Brea gum film.

Gao et al. [35] prepared novel polymer-inorganic hybrid membrane for the dehydration of water-acetone mixture by adding potassium (K^+) montmorillonite into CS. Morphology, chemical, and physical structure analysis showed that montmorillonite uniformly distributed onto the CS matrix which enhanced the thermal stability. Finally, 10 wt% of K^+ montmorillonite showed 8 times enhancement in their results compared to pristine CS membrane. A plausible model for water transportation was proposed and finds potential application for pervaporation. Figures 8, 9, 10, 11 and 12 display the surface morphology, cross-section morphology, FTIR, XRD and mass swelling of pure CS, CS/polyacrylonitrile (PAN), K^+ montmorillonite, and CS-K^+ montmorillonite-X/PAN hybrid membrane (Figs. 9, 10 and 11).

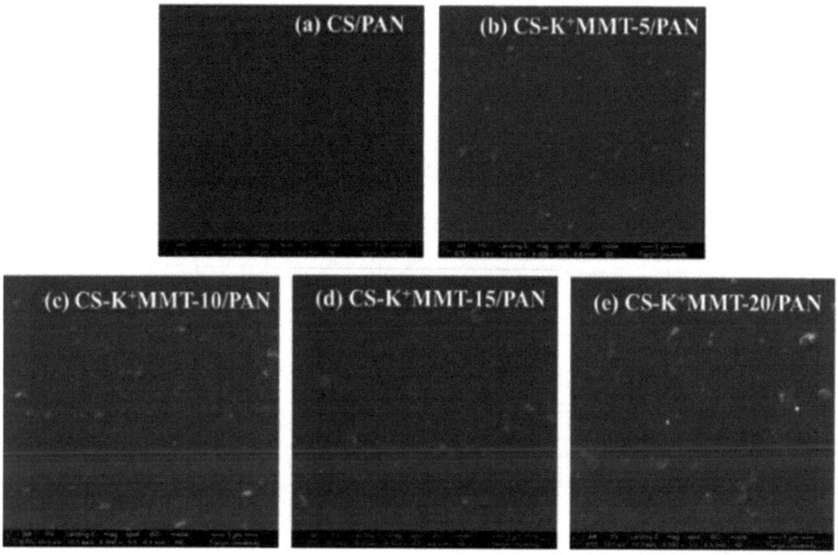

Fig. 8 The surface morphology of CS/PAN and CS-K^+ montmorillonite-X/PAN hybrid membrane (reproduced from [35])

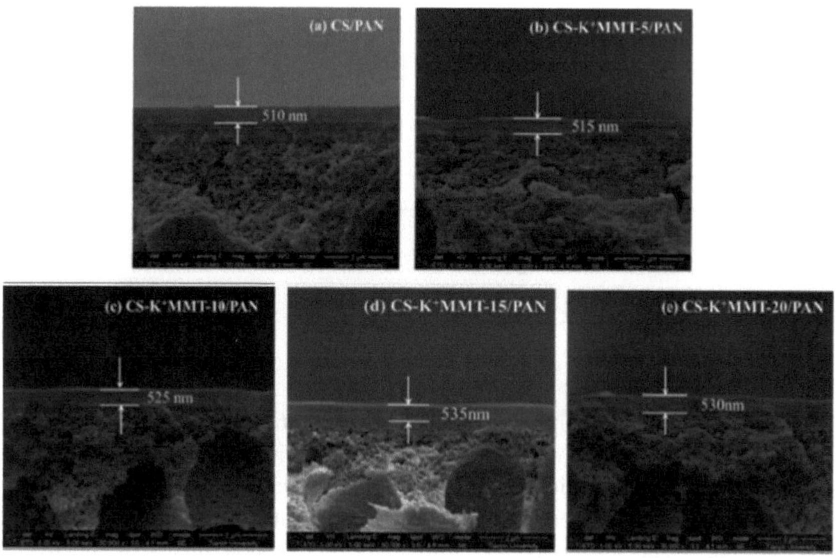

Fig. 9 The cross-section morphology of CS/PAN and CS-K^+ montmorillonite-X/PAN hybrid membrane (reproduced from [35])

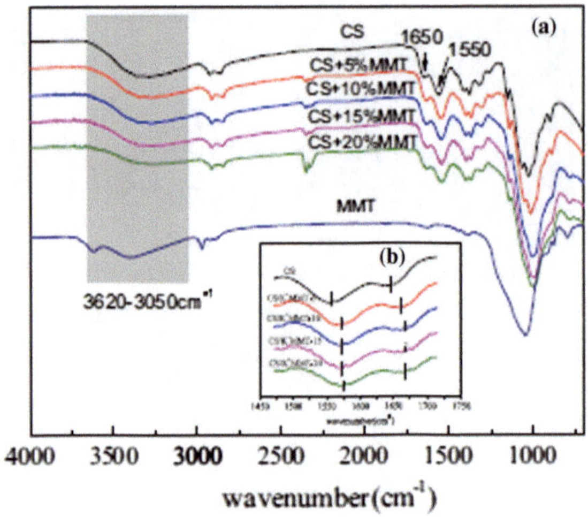

Fig. 10 FT-IR spectra of K^+ montmorillonite, CS-K^+ montmorillonite-X hybrid membrane (reproduced from [35])

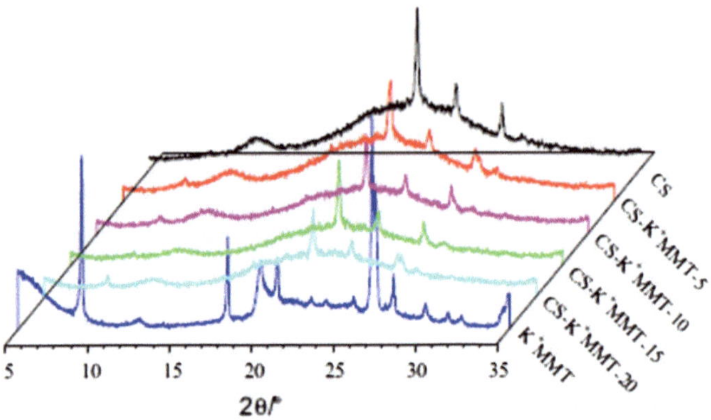

Fig. 11 XRD patterns of pure CS and CS-K^+ montmorillonite-X hybrid membrane (reproduced from [35])

Bhat Santosh Kumar and Aminabhavi Tejraj [36] developed novel hybrid composite membranes comprising Na^+ montmorillonite, NaAlg, and crosslinked glutaraldehyde. PV separation performance has been investigated and their effects of intercalated Na^+ montmorillonite were studied. Addition of Na^+ montmorillonite increases the water selectivity and lowers the permeation fluxes of the hybrid composite membranes compared to pristine NaAlg membrane.

Garg et al. [37] used in situ crosslinking of V-polydimethylsiloxane (PDMS) resin in the presence of 1–10 wt% clay content to prepare PDMS/clay nanocomposite membranes. The influence of layered silicate on pervaporation characteristics

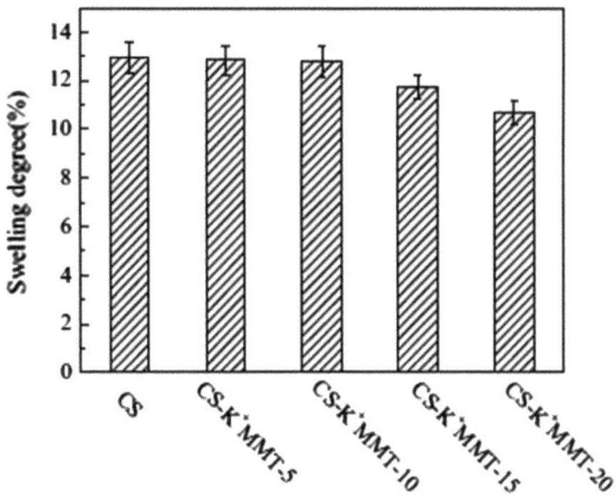

Fig. 12 Mass swelling degree of pure CS membrane and CS-K^+ montmorillonite-X hybrid membranes (reproduced from [35])

was also studied. Structural, mechanical, and thermal characterization of two clays with polar and nonpolar surfactants was done by using FTIR, tensile testing system and TGA. Uniform dispersion of nanoclay was observed using surface characterization. The total flux observed was lower for the prepared membranes compared to pristine PDMS membrane. It was concluded that polymer nanocomposite membranes could be an alternative way for tuning between permeation flux and selectivity to enhance thermal and mechanical properties.

Samberan et al. [38] prepared nanocomposite polyelectrolytes based on SPEEK and OMMT via solution intercalation technique. In addition, Na^+ montmorillonite and Cloisite 15A also incorporated. Transport properties, proton conductivity and methanol permeability were evaluated. Presence of montmorillonite detoriated the activation energy needed for proton conductivity. Optimum concentration of 3 wt% exhibited high selectivity and power density at the elevated methanol concentrations and exhibited higher open I–V compared to control.

Ma et al. [39, 40] adopted phase inversion method to prepare flat sheet asymmetric PSf/clay nanocomposite membranes using various clays. The morphology and structure of prepared membranes were determined using SEM, TEM and XRD. The results showed that clay had a good dispersion in the matrix and performances were evaluated in terms of pure water flux (Fig. 13), protein rejection (Fig. 13), porosity, contact angle, tensile strength (Fig. 14), and elongation at break (Fig. 14). Addition of clay additive enhanced the ratio of large pore in the skin layer, porosity and decreased the mechanical properties. The clay content was varied from 0–6 wt% to investigate the above mentioned properties. The author also used nonsolvent phase inversion method to develop PSf membrane and PSf/clay membrane. In this, the tensile strength attained its maximum of 5.34 MPa at 3 wt% clay loading. Normally, clay had high stability in the nanocomposite membranes.

Figure 15 explains the cross-section SEM image of flat sheet asymmetric PSf/clay nanocomposite membranes prepared using phase inversion method with different clay dosage.

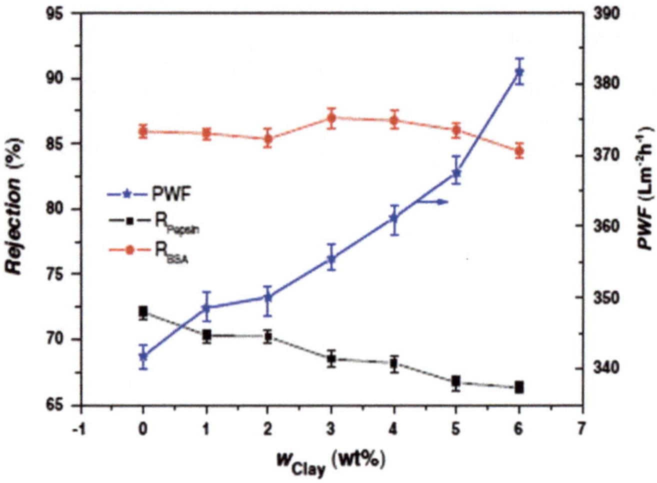

Fig. 13 Effect of w_{clay} on the pure water flux and protein rejection of PSf/clay nanocomposite membranes (reproduced from [39])

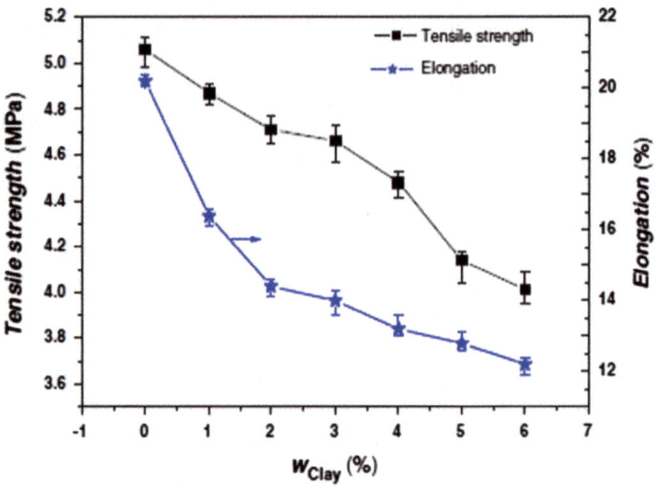

Fig. 14 Tensile strength and elongation at break of PSf/clay nanocomposite membranes with various w_{clay} (reproduced from [39])

Mierzwa et al. [41] studied the effect of clay nanoparticles on the morphology and performance of PES UF membranes by using casting solution additive. Pre water permeability has no correlation with the membrane thickness, porosity, and contact angle. SEM study determines the membrane surface porosity, surface pore size, and internal structures. The membranes with 2% clay and 1% clay + 1% sodium hexametaphosphate (SHMP) had a greater potential for fouling as

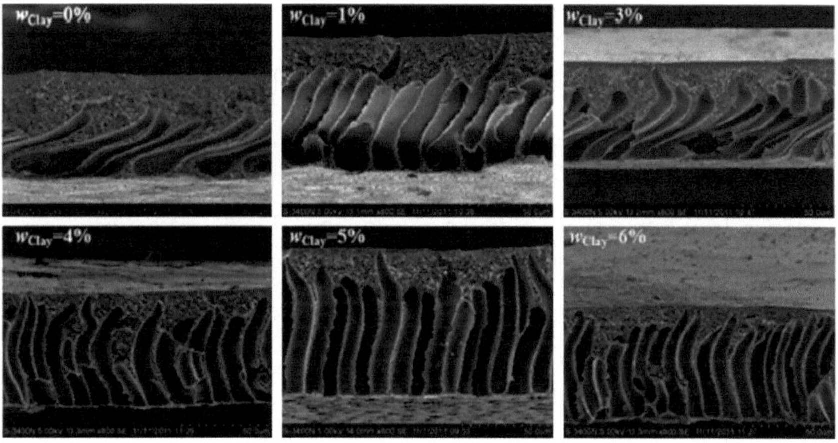

Fig. 15 Cross-section SEM images of PSf/clay nanocomposite membranes with different w_{clay} (reproduced from [40])

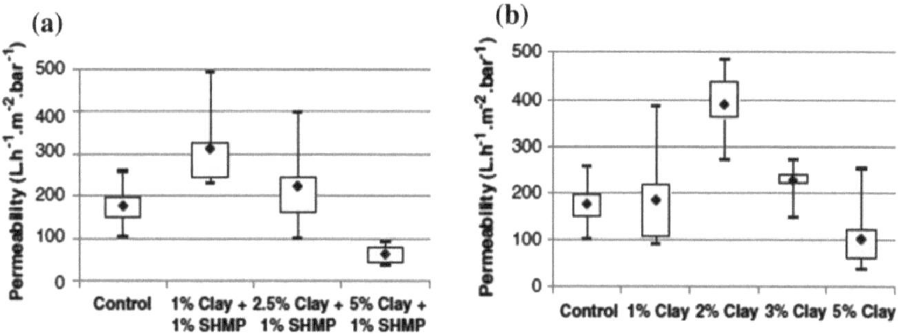

Fig. 16 Dead-end ultrapure water permeability: **a** Control membrane and membranes with 1% SHMP and clay nanoparticles (1–5%); **b** control membrane and membranes with clay nanoparticles (1–5%) (reproduced from [41])

compared to the control. The dead-end ultrapure water permeability for the control membrane was examined (Fig. 16). The effect of clay nanoparticles on membrane thickness, porosity, contact angle and dead-end permeability are tabulated in Table 3. Similar calculations were done and the Pearson's correlations between membrane permeability and different membrane properties (Table 4) were presented.

Lalia et al. [42] reviewed the structure, properties, and performance relationships of membrane fabrication. The polymeric membrane fabrication techniques for pressure-driven membrane process and membrane distillation were explained in detail. Crystallinity of the polymer membrane, porous structure, membrane charge, surface roughness, hydrophobicity, and hydrophilicity were also analyzed. Cross-section morphology, antifouling properties, chemical resistance,

Table 3 Effect of clay nanoparticles on membrane thickness, porosity, contact angle and dead-end permeability (reproduced from [41])

Membrane	Thickness (μm)	Porosity (%)	Contact angle (degree)	Average permeability ($L\ h^{-1}\ m^{-2}\ bar^{-1}$)
Control	57 ± 1	68 ± 1	60 ± 2	176 ± 56
1% clay + SHMP	58 ± 1	66 ± 2	64 ± 2	312 ± 89
2.5% clay + SHMP	58 ± 1	66 ± 2	67 ± 3	223 ± 91
5% clay + SHMP	56 ± 1	65 ± 2	69 ± 2	63 ± 19
1% clay	57 ± 2	68 ± 3	67 ± 6	185 ± 130
2% clay	58 ± 1	70 ± 2	67 ± 3	389 ± 71
3% clay	59 ± 1	73 ± 4	68 ± 3	227 ± 37
5% clay	57 ± 2	67 ± 4	70 ± 2	102 ± 70

Table 4 Pearson's correlations between membrane permeability and different membrane properties (reproduced from [41])

Membrane properties	Pearson's correlation with membrane permeability
Permeability ($L\ h^{-1}\ m^{-2}\ bar^{-1}$)	1.00
Thickness (μm)	0.61
Porosity (%)	0.36
Contact angle (degrees)	−0.21

high-mechanical strength with high flux, and selectivity are the most important properties of membrane fabrication. In addition, the two common membrane fabrication techniques such as phase inversion and interfacial polymerization were discussed. At the same time, electrospinning and track-etching methods were also reviewed. Membrane technology for water desalination clearly explains the structure-surface properties and performances.

Silva et al. [43] developed solution-cast membranes by immersing per fluorinated ionomers (PFSI) in dimethylformamide (DMF) solvent. Modified membranes were prepared by using 1 and 4 wt% montmorillonite salts onto it. Measurements of water uptake, solubility, and methanol permeation of the casted membranes were carried out. Addition of low quantity of silicate did not alter the conductivity but marked a decrease of methanol permeability. A simple model was proposed to enhance the tangential conductivity with decreasing thickness of Nafion® and recast membranes.

Mura et al. in 2007 [44] in this study fabricated Nafion® composite membranes with using montmorillonite and TiO_2 as fillers and DMF as the casting solvent. Addition of 1 wt% fillers to the ionomer prior to heat treatment showed increase in water content of the recast membranes. The tangential conductivity increases with increasing relative humidity of the environment. If the temperatures are low, fuel cell tests showed best performances.

De silva et al. [45] prepared CS membranes reinforced with 2–15 wt% of halloysite nanotubes (HNTs) using solution casting method. The 5 wt% of HNT yielded highest Young's modulus and tensile strength. TGA demonstrated the enhanced thermal stability of the prepared CS/HNT membranes.

Prasanth et al. [46] examined electrochemical characteristics of lithium-ion rechargeable batteries containing hybrid organic–inorganic polymer gel electrolytes based on polymer-clay nanocomposite microporous membranes. Solution casting method was used to develop PVDF-clay membrane. The morphology was examined by using FESEM, intercalation/exfoliation through XRD and thermal properties by means of DSC and TGA. The crystallinity of the membrane was decreased with increasing nanoclay addition. Addition of 2 wt% nanoclay onto the matrix enhanced the ionic conductivity, delivers high charge–discharge capacity and showed stable cycle performance.

Nanocomposite membranes with PVDF polymer and clay (Cloisite 15A) were prepared using phase inversion method by Lai et al. [47]. FTIR, TGA, tensile testing, SEM and EDS were to characterize the material. Water permeation test showed reduced water flux at low concentration of nanoclay. The abrasion resistance was high for the prepared composite membrane compared to neat PVDF membrane. Nanoclay of 1 wt% onto the membrane has greater tensile strength, high toughness, and small agglomerated particle size compared to high loading of nanoclay. The proper optimization of clay content resulted improvement in most of their technical properties.

4 Conclusion

In this chapter, synthesis of organic–inorganic polymer-montmorillonite nanocomposite membranes, their structure, surface properties, and performance are discussed. It appears that upto date remarkable progress has been made in the area of the fabrications of membranes. However, there is still a challenge to produce reliable membranes with antifouling properties, better mechanical strength, high resistance to chlorine attack, and minimal thickness of the membrane barrier layer to provide a high flux. To ensure progress in these fields, more efforts are needed for further improvement of common membrane fabrication methods as well as the development of new fabrication techniques. Membrane performance (flux, rejection, and fouling) is strongly influenced by membrane chemical composition and specific features of porous structure and the membrane surface, including hydrophobic/hydrophilic properties, membrane charge (zeta potential), surface roughness, pore size, and pore size distribution. Despite numerous studies attempting to relate physicochemical properties of membranes to solute rejection, the role of membrane structure, and surface properties in membrane performance is still not thoroughly understood. A systematic work is still needed to clarify, for example the role that membrane morphology plays in trans-membrane transport, the effect of surface roughness on membrane flux during membrane treatment of complex feed solutions with organic and inorganic species, the influence of membrane surface charge on membrane pore structure at various pH or for studying foulant-membrane interactions under severe fouling conditions. Identification of these parameters and others which can effectively be used to predict relationships among membrane fabrication, structure, surface properties, and performance could be very meaningful for further development of organic–inorganic montmorillonite membrane.

References

1. Koonaphapdeelert S, Wu Z, Li K (2009) Carbon dioxide stripping in ceramic hollow fiber membrane contactors. Chem Eng Sci 64:1–8
2. Smid J, Avci CG, Gunay V et al (1996) Preparation and characterization of microporous ceramic hollow fibre membranes. J Membr Sci 112:85–90
3. Way JD, Roberts DL (1992) Hollow fiber inorganic membranes for gas separations. Sep Sci Technol 27:29–41
4. Lv Y, Yu X, Tu ST et al (2010) Wetting of polypropylene hollow fiber membrane contactors. J Membr Sci 362:444–452
5. Mavroudi M, Kaldis SP, Sakellaropoulos GP (2006) A study of mass transfer resistance in membrane gas–liquid contacting processes. J Membr Sci 272:103–115
6. Sharma RR, Chellam S (2005) Temperature effects on the morphology of porous thin film composite nanofiltration membranes. Environ Sci Technol 39:5022–5030
7. Lu JG, Zheng YF, Cheng MD (2008) Wetting mechanism in mass transfer process of hydrophobic membrane gas absorption. J Membr Sci 308:180–190
8. Rezaei-Dasht Arzhandi M, Ismail AF, Ghanbari M et al (2016) An investigation of temperature effects on the properties and CO_2 absorption performance of porous PVDF/montmorillonite mixed matrix membranes. J Nat Gas Sci Eng 31:515–524
9. Picard E, Vermogen A, Gerard JF et al (2007) Barrier properties of nylon 6-montmorillonite nanocomposite membranes prepared by melt blending: influence of the clay content and dispersion state Consequences on modeling. J Membr Sci 292:133–144
10. Ismail NM, Ismail AF, Mustafa A et al (2015) Qualitative and quantitative analysis of intercalated and exfoliated silicate layers in asymmetric polyethersulfone/cloisite 15A mixed matrix membrane for CO_2/CH_4 separation. Chem Eng J 268:371–383
11. Hwang HY, Kim DJ, Kim HJ et al (2011) Effect of nanoclay in properties of porous PVDF membranes. Trans Nonferrous Metals Soc China 21:141–147
12. Jana KK, Srivastava A, Parkash O et al (2016) Nanoclay and swift heavy ions induced piezoelectric and conducting nanochannel based polymeric membrane for fuel cell. J Power Sources 301:338–347
13. Shubha N, Prasanth R, Hoon HH et al (2013) Dual phase polymer gel electrolyte based on non-woven poly(vinylidenefluorideco-hexafluoropropylene)-layered clay nanocomposite fibrous membranes for lithium ion batteries. Mater Res Bull 48:526–537
14. Li L, Bellan LM, Craighead HG et al (2006) Formation and properties of nylon-6 and nylon-6/montmorillonite composite nanofibers. Polymer 47:6208–6217
15. Alonso RH, Estevez L, Lian H et al (2009) Nafion-clay nanocomposite membranes: morphology and properties. Polymer 50:2402–2410
16. Singha S, Jana T (2016) Influence of interfacial interactions on the properties of polybenzimidazole/clay nanocomposite electrolyte membrane. Polymer 98:20–31

17. Kim Y, Choi Y, Kim HK et al (2010) New sulfonic acid moiety grafted on montmorillonite as filler of organic–inorganic composite membrane for non-humidified proton-exchange membrane fuel cells. J Power Sources 195:4653–4659
18. Gosalawit R, Chirachanchai S, Shishatskiy Sergey et al (2007) Krytox–Montmorillonite–Nafion® nanocomposite membrane for effective methanol crossover reduction in DMFCs. Solid State Ionics 178:1627–1635
19. Anadao P, Sato LF, Wiebeck H et al (2010) Montmorillonite as a component of polysulfone nanocomposite membranes. Appl Clay Sci 48:127–132
20. Dogan H, Inan TY, Koral M et al (2011) Organo-montmorillonites and sulfonated PEEK nanocomposite membranes for fuel cell applications. Appl Clay Sci 52:285–294
21. Rajabia Hamid, Ghaemic Negin, Madaenia Sayed S et al (2014) Nanoclay embedded mixed matrix PVDF nanocomposite membrane: Preparation, characterization and biofouling resistance. Appl Surf Sci 313:207–214
22. Wu XW, Wu N, Shi CQ et al (2016) Proton conductive montmorillonite-Nafion composite membranes for direct ethanol fuel cells. Appl Surf Sci
23. Tran ATT, Patterson DA, James BJ (2012) Investigating the feasibility of using polysulfone–montmorillonite composite membranes for protein adsorption. J Food Eng 112:38–49
24. Rezaei M, Ismaila AF, Hashemifard SA et al (2014) Preparation and characterization of PVDF-montmorillonite mixed matrix hollow fiber membrane for gas–liquid contacting process. Chem Eng Res Des 92:2449–2460
25. Nesic AR, Velickovic SJ, Antonovic DG (2012) Characterization of chitosan/montmorillonite membranes as adsorbents for Bezactiv Orange V-3R dye. J Hazard Mater 209–210:256–263
26. Xing D, He G, Hou Z et al (2011) Preparation and characterization of a modified montmorillonite/sulfonated polyphenylether sulfone/PTFE composite membrane. Int J Hydrogen Energy 36:2177–2183
27. Yang CC (2011) Fabrication and characterization of poly(vinyl alcohol)/montmorillonite/poly (styrene sulfonic acid) proton-conducting composite membranes for direct methanol fuel cells. Int J Hydrogen Energy 36:4419–4431
28. Hasani-Sadrabadi MM, Emami SH, Moaddel H (2008) Preparation and characterization of nanocomposite membranes made of poly(2,6-dimethyl-1,4-phenylene oxide) and montmorillonite for direct methanol fuel cells. J Power Sources 183:551–556
29. Dehkordi FS, Pakizeh M, Namvar-Mahboub M (2015) Properties and ultrafiltration efficiency of cellulose acetate/organically modified Mt (CA/OMMt) nanocomposite membrane for humic acid removal. Appl Clay Sci 105–106:178–185
30. Choudhari Santosh K, Kariduraganavar Mahadevappa Y (2009) Development of novel composite membranes using quaternized chitosan and Na^+-MMT clay for the pervaporation dehydration of isopropanol. J Colloid Interface Sci 338:111–120
31. Jaafar J, Ismaila AF, Matsuuraa T et al (2011) Performance of SPEEK based polymer-nanoclay inorganic membrane for DMFC. J Membr Sci 382:202–211
32. Aouinti L, Roizard D, Hu GH et al (2009) Investigation of pervaporation hybrid polyvinylchloride membranes for the separation of toluene-n-heptane mixtures—case of clays as filler. Desalination 241:174–181
33. Ublekov F, Penchev H, Georgiev V et al (2014) Protonated montmorillonite as a highly effective proton-conductivity enhancer in p-PBI membranes for PEM fuel cells. Mater Lett 135:5–7
34. Slavutsky AM, Bertuzzi MA, Armada M et al (2014) Preparation and characterization of montmorillonite/brea gum nanocomposites films. Food Hydrocolloids 35:270–278
35. Gao C, Zhang M, Jiang Z et al (2015) Preparation of a highly water-selective membrane for dehydration of acetone by incorporating potassium montmorillonite to construct ionized water channel. Chem Eng Sci 135:461–471
36. Bhat Santosh Kumar D, Aminabhavi Tejraj M et al (2006) Novel sodium alginate–Na^+ MMT hybrid composite membranes for pervaporation dehydration of isopropanol, 1,4-dioxane and tetrahydrofuran. Sep Purif Technol 51:85–94

37. Garg P, Singh RP, Choudhary V (2011) Pervaporation separation of organic azeotrope using poly(dimethyl siloxane)/clay nanocomposite membranes. Sep Purif Technol 80:435–444
38. Samberan MF, Hasani-Sadrabadi MM, Ghaffarian SR et al (2013) Investigation of the effects of AMPS-modified nanoclay on fuel cell performance of sulfonated aromatic proton exchange membranes. Int J Hydrogen Energy 38:14076–14084
39. Ma Y, Shi F, Wang Z et al (2012) Preparation and characterization of PSf/clay nanocomposite membranes with PEG 400 as a pore forming additive. Desalination 286:131–137
40. Ma Y, Shi F, Zhao W et al (2012) Preparation and characterization of PSf/clay nanocomposite membranes with LiCl as a pore forming additive. Desalination 303:39–47
41. Mierzwa JC, Arieta V, Verlage M et al (2013) Effect of clay nanoparticles on the structure and performance of polyethersulfone ultrafiltration membranes. Desalination 314:147–158
42. Lalia BS, Kochkodan V, Hashaikeh R et al (2013) A review on membrane fabrication: structure, properties and performance relationship. Desalination 326:77–95
43. Silva RF, Passerini S, Pozio A (2005) Solution-cast Nafion®/montmorillonite composite membrane with low methanol permeability. Electrochim Acta 50:2639–2645
44. Mura F, Silva RF, Pozio A (2007) Study on the conductivity of recast Nafion®/montmorillonite and Nafion®/TiO_2 composite membranes. Electrochim Acta 52:5824–5828
45. De Silva RT, Pooria Pasbakhsh, Goh KL et al (2013) Physico-chemical characterisation of chitosan/halloysite composite membranes. Polym Testing 32:265–271
46. Prasanth R, Shubha N, Hng HH et al (2013) Effect of nano-clay on ionic conductivity and electrochemical properties of poly(vinylidene fluoride) based nanocomposite porous polymer membranes and their application as polymer electrolyte in lithium ion batteries. Eur Polymer J 49:307–318
47. Lai CY, Groth A, Gray S et al (2014) Preparation and characterization of poly(vinylidene fluoride)/nanoclay nanocomposite flat sheet membranes for abrasion resistance. Water Res 57:56–66

Chapter 7
Electrospun Nanocomposite Materials for Polymer Electrolyte Membrane Methanol Fuel Cells

Nuha Awang, Juhana Jaafar, Ahmad Fauzi Ismail, Takeshi Matsuura, Mohd Hafiz Dzarfan Othman and Mukhlis A. Rahman

Abstract Recently, the demands of modern society on energy have become enormous. This consequently rises up significant concerns on the ecological and environmentally friendly energy conversion to power up applications from portable devices up to stationary power plant. Direct methanol fuel cell (DMFC) is one of the zero-pollution energy supply fuel cells that has gained much attention for their high efficiency and high power density yet compact in size. In order to ensure a continuously high performance power output from DMFC, a promising proton exchange membrane (PEM) with high proton conductivity and low methanol permeability is desirable. As one of the most promising and versatile fabrication methods for one-dimensional microstructure nanomaterials composed of organic and inorganic components prepared as randomly arranged continuous nanofibrous

N. Awang · J. Jaafar · A.F. Ismail (✉) · M.H.D. Othman · M.A. Rahman
Advanced Membrane Technology Research Centre (AMTEC), Universiti Teknologi Malaysia, 81310 Johor Bahru, Johor, Malaysia
e-mail: afauzi@utm.my

N. Awang
e-mail: nuhaawang@yahoo.com

J. Jaafar
e-mail: juhana@petroleum.utm.my

M.H.D. Othman
e-mail: hafiz@petroleum.utm.my

M.A. Rahman
e-mail: mukhlis@petroleum.utm.my

N. Awang · J. Jaafar · A.F. Ismail · T. Matsuura · M.H.D. Othman · M.A. Rahman
Faculty of Chemical and Energy Engineering, Universiti Teknologi Malaysia, 81310 Johor Bahru, Johor, Malaysia
e-mail: matsuura@eng.uottawa.ca

T. Matsuura
Department of Chemical Engineering, University of Ottawa, 161 Louis Pasteur Street, Ottawa, ON KIN 6N5, Canada

Inamuddin et al. (eds.), *Organic–Inorganic Composite Polymer Electrolyte Membranes*, DOI 10.1007/978-3-319-52739-0_7

mats, electrospinning has been widely investigated to fabricate PEM applied in DMFC because of their dimensional, directional, and compositional flexibility. In this chapter, the application of electrospun nanofibers from organic, inorganic, and composite organic–inorganic is reviewed in details. Particular progresses with the use of electrospun nanofibers to improve fuel cell performance in terms of power density, ionic conductivity, interfacial resistance, and chemical stability, as well as mechanical strength are emphasized and discussed. The meaningful critical review could contribute to further enhance the development and evolution of fuel cells as one of the advanced energy conversion systems.

List of Abbreviations

DMFC	Direct methanol fuel cell
PEM	Proton exchange membrane
MEA	Membrane electrode essembly
OCV	Open circuit voltage
CO	Carbon monoxide
PVA	Poly(vinyl alcohol)
SiO_2	.
SPEEK	Sulfonated poly(ether ether ketone)
TPA	Tungstophosphoric acid
MCM-41	Mesoporous material composite
OMB	Sulfonated mesoporous benzene-silica
BI	Benzimidazole
TiO_2	Titanium dioxide
ZrO_2	Zirconium dioxide
MMT	Montmorillonite
GO	Graphene oxide
SPTA	Sulfonated polytriazole
SPPO	Sulfonated polyphenylene oxide
PAA	Perfluorosulfonc acid polymer

1 Introduction

Direct methanol fuel cell (DMFC) is a type of fuel cell which uses methanol as a fuel. Polyelectrolyte membrane (PEM) is sandwiched between two catalytic electrodes which are made of gas diffusion layer and catalyst layer. Methanol oxidation and oxygen reduction occur at anode and cathode respectively. The PEM conducts protons from anode to cathode, while other compounds diffusion is blocked. The stacking of the electrodes and membranes is called membrane electrode assembly (MEA) [1]. As compared to gaseous and solid fuels, liquid fuels are more preferable in portable devices and mobile applications. Hence, researches have been carried out to study a suitable liquid fuel. As a result, methanol has proved as the most suitable fuel for it is environmental friendly and can be readily manufactured by usual process from many raw materials. In addition, unlike other fuels, methanol is relatively hydrogen dense (4/6 atoms in methanol are hydrogen). Besides, oxygen in methanol is more reactive than plain hydrocarbon and it can be fed directly into a fuel cell [2].

Another advantage of DMFC is that it provides required run time and power to the portable electronic devices. Lithium ion batteries applied in an operating laptop, for example, when charged to 100% capacity, it can lose 35% of their energy capacity permanently over 12 months if exposed to 40 °C. The DMFC can potentially compete with advanced lithium ion batteries in terms of gravimetric energy density of the energy density (measured in watt-hours per liter) of the new fuel cells is 385 Wh/L, which is superior to lithium ion batteries of 270 Wh/L [3].

Several important features required for a high performance DMFC are: (1) high proton conductivity, (2) low methanol permeability, (3) low electronic conductivity, (4) acceptable water uptake (hydrolytic stability), (5) good thermal stability as well as mechanical stability and (6) low cost. Among those properties, methanol permeability and proton conductivity were the most critical characteristics for PEMs in DMFC application.

2 Methanol Crossover and Low Proton Conductivity

High methanol crossover and low proton conductivity are two technical challenges encountered by PEM in DMFC application. Methanol crossover occurs when methanol permeates through membrane and dragged along with hydrated protons from anode to cathode. There are three transport mechanisms that describe methanol crossover [4] which are (a) convection by the hydraulic pressure, (b) electroosmotic drag by proton transport and (c) diffusion by methanol concentration gradient. The diffusive mode of methanol transport dominates when the cell is idle while electroosmotic drag dominates when the cell is operating [5].

Methanol crossover causes low power output during operation of DMFC due to the chemical oxidation of methanol at the cathode (with the support of catalyst). The chemical oxidation brings several problems such as

(a) Electrode depolarization.
(b) A critical water accumulation on the cathode (water being produced by methanol oxidation).
(c) Mixed potential, resulting in the open-circuit voltage (OCV) of the DMFC below 0.8 V.
(d) Consumption of O_2.
(e) Cathode catalyst poisoning by CO (an intermediate of methanol oxidation).

All the problems mentioned above are not only limiting O_2 access to catalyst at cathode but also lowering the fuel cell efficiency [6]. Unfortunately, methanol crossover can be reduced when water uptake is reduced. This can simultaneously sacrifice the proton conductivity [7]. Even though Nafion® is considered as an established membrane, its high methanol crossover and high dependence on humidity still limit its commercialization. Since the highest proton conductivity happens at 100% humidity, the permeation of water in membranes not only allows water but also dragged methanol along the flow. The mixture of water and methanol can be referred as mixed potential.

New membranes with the ability to reduced water transport (by electroosmotic drag or diffusion) and methanol permeability together with high stabilities and conductivity are required for DMFCs. The membrane complexity and resistivity were improved using diluted methanol as the fuel to reduce energy density. Both water uptake and of the membrane rely on the ionic group concentration (usually sulphonic acid). However, a high density of ionic groups can also cause excessive membrane swelling, which deteriorates membrane durability and mechanical stability. Hence, the fundamental properties for PEMs (proton conductivity, IEC, and water uptake) need to be optimized under the operating environment. The development of organic–inorganic nanocomposite PEMs seems to be attractive in solving the issues mentioned above [8]. Table 1 presents the polymer electrolyte membranes developed for the DMFC. Composite membranes have been other alternatives to be synthesized to reduce methanol crossover and improve proton conductivity for DMFCs such as inorganic–organic material [17], composite with organic, composite with inorganic [18], nanocomposite [19], and multilayer composite [20].

Owing to the improvements in the mechanical, thermal, and barrier properties of the pure polymer, layered silicates-polymer nanocomposite has the potential to provide a solution to overcome the drawbacks of polymeric membranes. These improvements result from nano-dispersion of layered silicates throughout polymer matrix. The combination of the advantages from base materials for instance the flexibility and process ability of polymers, and the selectivity and thermal stability of the inorganic nanofillers, may result in the above-mentioned properties [21].

The overall membrane characteristic of a PEM for DMFC is always related by the ratio of proton conductivity to methanol permeability. Hence, introducing inorganic nanofillers can affect the membrane cell performance in two ways: (1) the complete exfoliation morphology allows more cations to available and mobile for conduction, (2) the uniform nanosized distribution of inorganic filler particles produces a winding diffusion pathway for methanol to migrate through the nanocomposite membrane [22].

Despite these positive progresses, the segmental flexibility of polymeric membranes is still restricted by their disability to discriminate small polar molecules notably methanol [23]. Therefore, polymeric materials are interesting material to be discovered as alternative approaches that can improve methanol permeability and proton conductivity better than that of Nafion®. One of the most promising

Table 1 Developed polymer electrolyte membranes for direct methanol fuel cells

Membranes	Abbreviations	Methanol permeability ($\times 10^{-6}$ cm^2 s^{-1})	Proton conductivity (mS cm^{-1})	References
Composite membranes				
Poly(vinyl alcohol) sulfosuccinic acid silica hybrid	PVA/SiO_2	0.08	10.00	[9]
Sulfonated poly(ether ether ketone) sub-layer and polyvinyl alcohol sub-layer	SPEEK/PVA	1.31	14.20	[10]
Sulfonated poly(ether ether ketone)/tungstophosphoric acid/mesoporous material composite	SPEEK/TPA/MCM-41	0.0057	2.75	[11]
Self-cross-linkable sulfonated poly(ether ether ketone)	SPEEK	2.52	40.50	[12]
Organic–inorganic hybrid	CS/silica	0.395	29.00	[13]
Cross-linked sulfonated poly(ether ether ketone)/silica hybrid	SPEEK/SiO_2	0.715	88.00	[14]
Sulfonated mesoporous benzene-silica incorporated poly (ether ether ketone) composite	SPEEK-OMB	0.50	79.00	[15]
Benzimidazole grafted poly(ether ether ketone and sulfonated poly(ether ether ketone) composite	SPEEK-PEEK-BI	1.61	28.00	[16]

aromatic polymers that has been extensively studied as PEM for DMFC application is sulfonated poly (ether ether ketone) (SPEEK).

3 Composite SPEEK

SPEEK has been discovered as highly potential alternatives for Nafion® owing to their excellent thermal and mechanical properties, high proton conductivities, and low cost. In addition, SPEEK can be easily functionalized for application as a good proton exchange membranes (PEMs) [24].

Hydrophobic poly (ether ether ketone) (PEEK) is the most studied as compared to aromatic polymers due to higher chemical and thermal stability and it also offers a good proton conductivity when sufficiently sulfonated. Nevertheless, the adverse effect of these membranes is that their, both chemical and mechanical properties deteriorate with increasing degree of sulfonation (DS) [25].

Materials with high degree of sulfonation (DS) usually show large swelling at high temperatures which lead to membrane dissolution in water. To curb the problem, cross-linked, composite, or hybrid polymers are being observed to enhance these properties. Currently, a series of studies in organic–inorganic composite membranes based on SPEEK were reported [25]. Figure 1 presents the classifications of PEMs.

The new unconventional membrane requires good mechanical properties, high thermal stability, and excellent chemical resistance can be operated at high temperature, reduce methanol crossover, and high proton conductivity [26]. The new unconventional membrane technology can be subdivided into the following three areas which are (a) direct copolymerization of ion containing monomers to form ion conducting polymers, (b) composite structures based on polymer/polymer

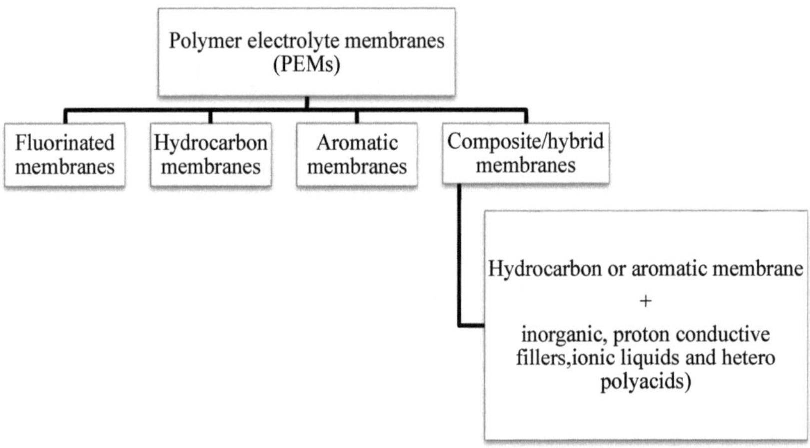

Fig. 1 Classifications of PEMs

composites or polymer/inorganic composites and (c) post-reactions on polymers to form ion conducting membranes [27]. Composite membranes are the most favorable subject to be studied in DMFC field as a highly potential material for PEM.

The introduction of nanoparticles into the SPEEK matrix is also an important method in PEM study. This method has three purposes: (a) improve proton conductivity, (b) enhance the mechanical properties of the composite membranes and (c) reduce methanol crossover [28]. Previous researches proved that functionalization of polymers, reduction of particle sizes, morphological properties as well as the role of fillers can bring some positive improvements in DMFC performance [29]. Figure 2 shows the modification of SPEEK as composite membranes for DMFC application.

Nanocomposites are the new type of composites that are particle-filled polymers for which at least one dimension of the dispersed particles is in the nanometer range. The types of nanocomposites can be differentiated according to number of dimensions of the dispersed particles [30].

Particles with three dimensions in the order of nanometers are typically isodimensional, such as spherical silica nanoparticles obtained by in situ sol-gel methods or by polymerization promoted directly from their surface the particles are including semiconductor nanoclusters [31]. When two dimensions are in the nanometer scale and the third is larger often related with nanotubes or whiskers which an elongated structure is formed [32].

There were many types of filler and materials which have been applied to develop a new composites membrane SPEEK. The effects on membrane performance during the modification of SPEEK membranes are summarized in Table 2.

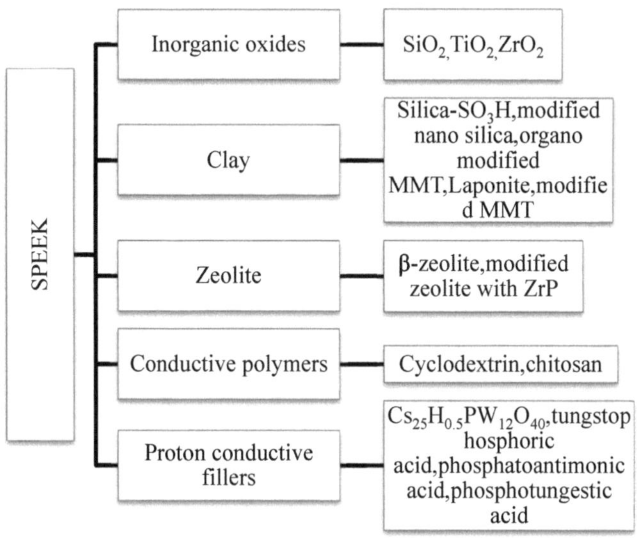

Fig. 2 The modification of SPEEK as composite membranes for DMFC application

Table 2 Summary of SPEEK nanocomposite PEM membranes

Membranes	Methanol permeability ($\times 10^{-6}$ cm^2 s^{-1})	Proton conductivity (mS cm^{-1})	Power density (mW cm^{-2})	Remarks	References
Inorganic oxides					
SPEEK/GO (graphene oxide)	0.132	29.4 at 65 °C	43.0	Reasonable proton conductivity and methanol permeability compared to pristine SPEEK. Performed well at high temperature 65–100 °C	[33]
SPEEK/polypyrrole/$AlPO_4$	–	0.30	–	Thermally stable up to 300 °C. However, the membrane has low proton conductivity	[34]
Clay					
Functionalized Bentonite (HSO_3-BEN)/SPEEK	–	21.00 at room temperature	140.00	Good proton conductivity with agglomeration and poor dispersion of fillers	[35]
SPEEK/(cloisite clay) CC/TAP (triaminopyrimidine)	0.52 at 60 °C	47.10 at 60 °C	54.93	Good proton and methanol permeability. However, the surface structure was half exfoliated	[36]
Charged surface modifying macromolecule (cSMM)/SPEEK	0.0019	10.82	–	Good proton conductivity and methanol permeability. cSMM was not well dispersed (two different layers were obviously observed)	[12]
Silicotungstic acid (STA)/montmorillonite (MMT)/SPEEK	0.35	6.08	–	Encorporating of MMT entrapped with STA into SPEEK enhanced proton conductivity and methanol permeability but presented poor mechanical property	[37]
Zeolite					
β-zeolite/SPEEK	–	40.00	0.03	The increasing of proton conductivity was observed. β-zeolite/SPEEK composite membranes was the best, among other composites	[38]

(continued)

Table 2 (continued)

Membranes	Methanol permeability ($\times 10^{-6}$ cm^2 s^{-1})	Proton conductivity (mS cm^{-1})	Power density (mW cm^{-2})	Remarks	References
Conductive polymers					
SPEEK/chitosan/MMT	0.03	4.66	–	Low methanol permeability but low proton conductivity	[39]
SPEEK/sulfonated cyclodextrin	9.23	35.00	29.52	Less exfoliated structure, high methanol permeability but possesses a good proton conductivity	[40]
Proton conductive fillers					
Cesium salt of tungstophosphoric acid (Cs-TPA)/SPEEK	0.47	13.00	–	Composite membranes were hydrolytically more stable than sPEEK70/Cs-TPA composite membranes. On the other hand, water vapor, methanol and hydrogen permeability values of SPEEK60 composite membranes were observed to be lower than that of Nafion	[41]
SPEEK/TPA/MCM-41	0.0057	2.75	–	Low methanol permeability was obtained. Proton conductivity value still lower than that of Nafion	[42]

The fabrication of composite polymer electrolyte membrane SPEEK with adding clay, inorganic oxide, conductive polymer, protons conductive fillers, and zeolite was considered to be the most promising technique in producing a high potential polymer electrolyte membrane (PEM) for DMFC application [43]. When the composite has the potential to be balanced between two important characteristics at polymer electrolyte membrane performance, methanol crossover and proton conductivity become very important elements. Thus, a coherent study in composite techniques in developing electrolyte membrane needs to be focused since these techniques have proven their effectiveness [44]. Among all techniques, the incorporation of clay as a function of reducing methanol crossover and stable to high temperature is interesting to be explored.

4 SPEEK-Clay Nanocomposite as PEM for DMFC

The layered silicates are synthetic minerals or natural consisting of ordered stacks of aluminosilicate layers with a high surface area and high aspect ratio. The current most popular layered silicates applied in preparation of polymer nanocomposites are clays [4]. Clay is known as a part of soil fraction with less than 2 μm of the particle size and the thickness of 1 nm. There are numerous types of clays with few differences in their properties, formula, and structure including exfoliation and swelling [45]. The clays with high ability of being exfoliated by monomers or polymer chains besides well distributed are suitable candidate for nanocomposite preparation [46]. These types of special characteristics can improve polymer properties drastically due to their interfacial interactions and high aspect ratio with polymer matrix [47]. Clay minerals are classified as one of the nanoplatelet-structured filler particles in polymer composite technology. Hectorite, saponite, and montmorillonite are the popular smectites which are widely applied in the polymer nanocomposites preparation (Table 3) [48].

The availability of montmorillonite (MMT) has made this clay as the most well-known material for polymer nanocomposite production. Besides, MMT has good properties in exfoliation chemistry, high surface reactivity, and high surface area [49]. The good aspect ratio of MMT which is in range of 1000 and surface area of about 750 m^2/g proved this type of clay has good dispersion without breaking of

Table 3 Chemical structure of common smectite clays

Smectites	Chemical formula
Montmorillonite (MMT)	$M_x(Al_{4-x}Mg_x)Si_8O_{20}(OH)_4$
Hectorite	$M_x(Mg_{6-x}Li_x)Si_8O_{20}(OH)_4$
Saponite	$M_x\ Mg_6\ (Si_{8-x}Al_x)O_{20}(OH)_4$

M Monovalent cation, *x* degree of cations isomorphous substitution in octahedral sheets [50]

layers. The breaking of clay layers into small plates occurs and the aspect ratio decreases to about 300–500 [50].

Mica, 2:1 pyrophillits, and 1:1 structures are the examples of other clay minerals that are not commonly applied in polymer nanocomposites. These clays not only lack exfoliation/intercalation characteristics but are also not easily separated into layers. The major factors which contribute to the difficulty in handling these clays include intercalation of polymer chains between layers or the hydration of interlayer and the higher charge density on the layers which are strongly stacked together [51].

The study of SPEEK-clay is important due to the special properties in layered silicate materials. Layered silicate, also known as nanofillers is able to improve membrane properties [52]. The important feature of polymer nanocomposite is better interfacial area as compared to traditional composites due to small size of fillers. The small amounts of fillers are able to create a significant improvement in polymer properties [53]. SPEEK-clay nanocomposites also provide a new group of polymeric composites with outstanding transport and mechanical properties. The biodegradability and electronic/ionic conductivity of these nanocomposites were improved by the addition of a small loading of clay [54]. Clay usually acts as a nucleating agent for crystallization of semi-crystalline polymers. Nevertheless, intercalating polymers become amorphous with clay loading [55].

Layered nanoporous materials consist of an intermediate structure between crystalline nanoporous frameworks (zeolites) and typical layered materials (smectite clays) [56]. Each layer comprises a porous network, while the gallery between layers provides the ability for exfoliation, intercalation, and pillaring [57]. Due to their high surface areas, high cation exchange capacities (1.25 meq g^{-1}), high barrier properties, and high surface reactivity, smectites which are a family of both hectorites and montmorillonite (MMT), are a valuable mineral class for industrial applications. These excellent properties also play an important role for nanocomposite applications notably for DMFC [53]. Even though there are many positive outcomes can be found from incorporating MMT into polymer matrix, there is an adverse effect of MMT due to chemical microstructure characteristic [53]. Since the silicate clays are lacking in affinity towards hydrophilic and hydrophobic polymers, a surface modification is needed to increase their compatibility with matrix or polymeric materials [53]. Fortunately, its good intercalation chemistry allows it to become compatible with various polymers and be chemically modified. When these modified particles are intercalated into polymeric matrices, they are known to increase the barrier and overall mechanical properties of the composite membrane [51].

The compatibility properties of MMT clay can be improved by modification of inorganic clays to organoclays [58]. The organophilicity from the surfactant layer allows inorganic clays to be dispersed in organic polymers [51]. In order to extend their application, organic modifier such as alkyl ammonium cations has been applied to reduce their organophilic characteristics. One of the organically modified MMT is cloisite15A®. The high aspect ratio (70–150) and high interlayer distance

Table 4 The typical dispersion states of a nanoclay within a polymer [60]

	Phase separated	Intercalated	Exfoliated/delaminated
Microstructures nanoclay within polymers	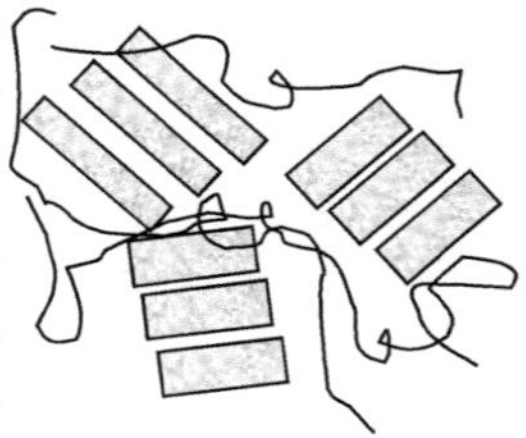		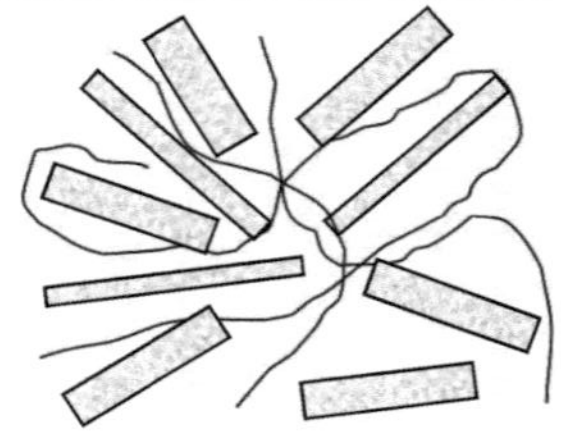
Explanations	• When unmodified clay interacted with polymer, the clay stacked together within the polymer matrix because the polymer is unable to intercalate with the clay • The phase separated nanocomposites is generally in range of micro composites	• The increasing of the interlayer spacing occurred when polymer chains are introduced into the clay interlayer. The periodic arrangement of clay still exists prior to the insertion of polymer chains • The electrostatic forces decrease due to the presence of polymer chains. However, the polymer did not completely dissolved • The intercalated arrangement is addressed as a well-ordered multilayered nanocomposite with high interference of polymer and filler or clay	• The separation of the layers and the good dispersion of fillers as a result of the introduction of polymer chains into the fillers galleries or clay galleries • The exfoliated structure is obtained when interlayer spacing more than 80–100 Å happened due to the insertion of polymer chains • The well dispersion of clay as fillers, high aspect ratio also obtained in this morphology. Lower clay content is applied to exfoliate the nanocomposite. Exfoliation improves most of properties due to well interaction between polymer and fillers

(~ 31.5 Å) features make cloisite 15A® as a promising material which contribute to high barrier properties and high ion mobility performance [52].

There are few reasons that may lead to the improvement of conductivity and barrier properties of the polymer electrolyte membrane (PEM) for example, state of the dispersion of filler particles, amount of filler loading, length to width ratio of the filler particles, and relative orientation of the filler particles [53]. Among the mentioned reasons, the effect of filler loading has been proved as an important parameter that has to be addressed as it determines whether the composite formed is exfoliated, normal, or intercalated composite and it affects the state of filler dispersion in polymer matrix [59].

5 Morphology Types and the Importance of Exfoliated Surface Structure on DMFC Performance

The morphology or polymer/clay nanocomposite structures are usually classified as intercalated or exfoliated according to the level of intercalation or exfoliation of polymer chains into the clay galleries. Several parameters such as polymer matrix, organic modifier, preparation processes, and clay nature are important factors that define the morphology of nanofibers [53]. The typical dispersions of nanoclays within polymers are explained in Table 4.

6 Preparation of Exfoliated Nanocomposite Membranes

Membrane structure can be either controlled by the degree of cross-linking of the polymer matrix or by the types of connection bonds between inorganic phases and the polymer in the nanocomposite material: van der Waals force, covalent bond, or hydrogen bonds [61–63]. Solution blending method was reported to produce polymer-inorganic nanocomposite membrane with strong hydrogen bonds or van der Waals force. This type of bonding causes an aggregation of inorganic ingredients in the membranes [63]. Meanwhile, by applying in situ polymerization method, functional group in inorganic nanoparticles is allowed to connect with polymer chains by covalent bonds. Nevertheless, aggregation of inorganic nanoparticles is still difficult to be avoided.

In order to exfoliate clay sheets in polymeric membrane, first, the polymer solution was obtained by the replacement of alcoholic co-solvent with an organic solvent such as dimethyl formamide (DMF) and dimethyl acetamide (DMA) and then the compatible polymer-inorganic nanocomposite solution could be obtained [64]. New preparation processes of polymer-inorganic-based PEM membranes have shown improvement in reducing the methanol crossover through the membrane. Nevertheless, they either suffer the increase of the membrane resistance or sacrifice the mechanical strength of the native commercial PEM [65]. Generally, four preparation methods that commonly applied are solution blending, melt blending, in situ polymerization, and sol-gel. The selection of suitable methods is very crucial in minimizing aggregation of nanofillers [66]. Table 5 shows the summary of polymer-clay nanocomposite, preparation methods, and morphological properties of the obtained products.

From Table 5, it can be concluded that intrinsic incompatibility between hydrophobic polymer chains and hydrophilic clay layers causes weak interfacial interaction due to poor clay dispersion. The situation prevents exfoliation to occur. Exfoliation is vitally important in improving properties. Modification of clay layers with hydrophobic agents is important in order to increase compatibility of clay and polymer chains. The method is known as surface modification. Even though the compatibility between two materials was improved, the reductions of surface

Table 5 Polymer-clay nanocomposite, preparation method and morphological properties

Polymer-clay nanocomposite	Preparation method	Power density (mW cm^{-2})	Remarks on morphology	References
SPEEK/Silica [the silica is generated in situ via the water free sol–gel process of polyethoxysiloxane (PEOS)]	Sol-gel	23.90	The formation of the organic-inorganic phase separation	[4]
Poly(vinyl alcohol/montmorillonite/poly (styrene sulfonic acid) (PVA/MMT/PSSA)	Melt blending	65.23	Unexfoliated. Uniform surface of membrane without phase separation	[14]
SPEEK/Charged surface modifying macromolecules. (cSMM)	Solution blending	–	Nodular structures. The batch processes are not suitable for mass production due to slow removal of solvents and fluctuation of mechanical electrochemical properties of each batch	[54]
Sulfonated polytriazole-clay (SPTA-clay) nanocomposites	In situ polymerization	–	Inorganic nanoparticles with functional groups can be connected with polymer chains by covalent bonds. However, the aggregation of inorganic fillers problems in the produced membranes is still remains	[67]

energy of clay layers and their surface polarity with polymer polarity occurred [68]. Hence, electrospinning as another approach towards producing an exfoliated PEM was discovered.

7 Electrospinning as a Membrane Morphological Modification Technique

Electrospinning is a very versatile and sophisticated technique in producing nanofibers. Electrospinning are found in few categories which are magneto-electrospinning, bubble electrospinning, and vibration electrospinning [69]. The electrospinning process had been first patent by Formhals in October 1934 [15]. The study on electrospinning in fuel cell is considered new because the process has been widely applied specifically in biomedical field. Hence, the electrospinning technique became one of the most interesting fabrication methods for the past few years [70].

In 1846, the research by Christian Friedrich Schonbein found electrostatic attraction of a liquid which enables to produce a highly nitrate cellulose. Charles Vernon Boys explained the theory in paper on nanofiber manufacture in 1887 [71]. John Zenely has published work on the behavior of fluid droplets at the end of the metal capillaries in 1914. The effort was the beginning of a study of mathematical model; the fluid behavior under electrostatic forces [72]. In addition, the work reported by Formhals and Zenely was specially applied to the polymer. Electrospinning was superior to other techniques such as dry spinning, melt spinning, and wet spinning. The electrospinning is capable to produce fibers of larger specific surface area, smaller diameter of 10–100 nm, and smaller pore sizes while others produce fibers in range of 5–500 μm [27].

Electrospinning method involves the use of high voltage electrostatic field to charge surface of the polymer solution droplet and induces the ejection of a liquid jet through spinneret [73]. At this point, the electrostatic force overcomes the surface tension of the droplet and the formation of the Taylor cone is obtained when the solution exiting the tip of spinneret and leading to the charged jet [74]. The formation of Taylor cone is proportional to the applied voltage, the voltage is kept increasing until the equilibrium condition is achieved between the electrostatic force and the surface tension (Fig. 3).

The electric field controlled the route of charge jet and the solidified electrospun fiber was formed on the collector [36, 75]. Setup for electrospinning is shown as Fig. 4.

There are few methods that usually applied in developing nanofibers; phase separation process, self-assembly process, template synthesis process, and drawing process. However, electrospinning process makes it more favorable to be applied for producing highly porous nanofibrous polymeric [76].

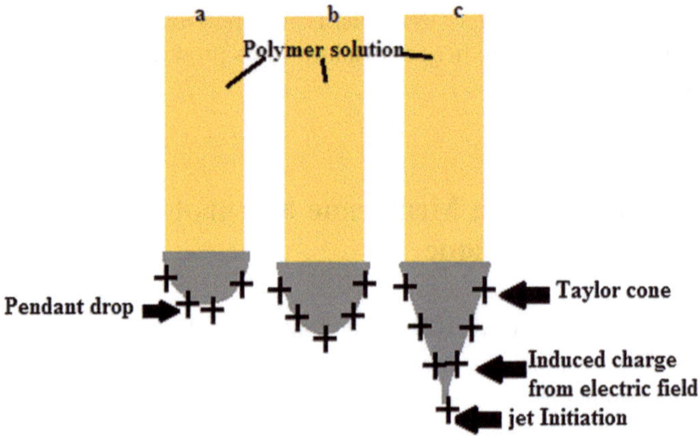

Fig. 3 Formation of Taylor cone

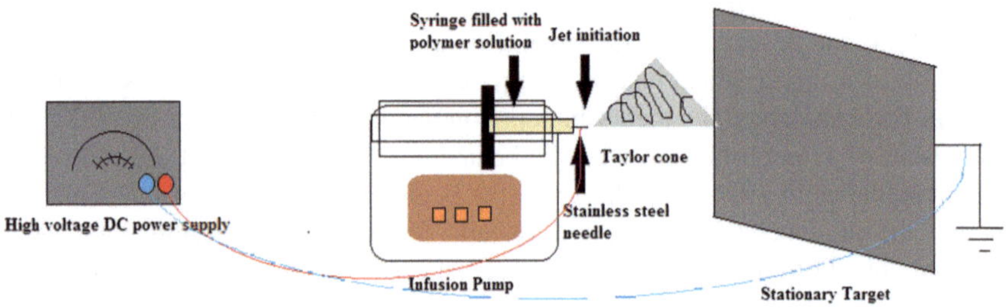

Fig. 4 Electrospinning setup

8 Electrospun Polymer-Based Nanofiber Membranes for DMFC Application

The research by Formhals and Zenely became the first step in exploring the potential of electrospun nanofibers from different types of polymers for fuel cell application [42]. Table 6 shows the summary of polymer-based nanofibers membranes for fuel cell application.

High proton conductivity from Nafion® has made Nafion® as a membrane that commercially applied as PEM [77]. A new approach in turning Nafion® into nanofibers has been studied by Dong in 2010. In the research, 0.1 wt% of poly (ethylene oxide) PEO was used as carrier polymer. A higher proton conductivity result was observed 1.5 S cm^{-1} as compared to bulk Nafion® film. The study also found the reduction in fiber diameter from 500 to 400 nm. As a result, the smaller fiber diameter improved proton conductivity of PEM.

Another important factor which affects the improvement of proton conductivity is ionic morphology. The orientation of electrospun nanofibers along fibers axis is better than that of non-electrospun fibers [78, 79]. Another research on electrospun fibers for fuel cell application has been done by Tamura and Kawakami in 2010 who synthesized sulfonated polyimide nanofibers for PEM. According to their observation, the orientation of electrospun nanofibers was well oriented. As compared to sulfonated polyimide which prepared via solvent-casting technique, the increasing amount of well-oriented electrospun nanofibers not only improved proton conductivity but also improved membrane stability.

Development of nanowire-based for fuel cell application was handled by Pan in 2008. Nafion/poly(vinyl pyrrolidone) (PVP) nanowires (NPNWs) were electropun. The use of PVP in this study was dedicated to the high molecular weight that possesses by PVP in order to successfully electrospin Nafion without any contribution on proton conductivity [80]. Since Nafion® is not soluble in most common solvents; electrospinning process for Nafion® is difficult to be handled. The non-soluble property will produce the formation of micelles that contribute to the decreasing in chain entanglement [81]. A research by Pan in 2008 found that using two conductive substrates (silicon) separated by a void gap, the transportation of protons to reach the cathode in NPNWs becomes easier. The smaller the diameter of fibers, the higher is the proton conductivity value. This condition happened because of the increase in the degree of the "texture" resulting from preferential orientation. The study also proved that the proton conductivity can be improved by adjusting the diameter of the NPNWs to less than 2.3 μm [81]. The others findings on electrospun fiber polymer-based membrane for fuel cell application are summarized in Table 6.

Table 6 Findings on electrospun fiber polymer-based membrane for fuel cell application

Based polymer	Filler/carrier polymer	Proton conductivity (mS cm^{-1})	References
Poly(vinyl alcohol)	Nafion®	22	[82]
Polyvinylidene fluoride	Nafion®	2	[83]
Sulfonated polyethersulfone	Nafion®	~85	[83]
Sulfonated random copolyimide	Sulfonated polyimide	Up to 370	[84]
Bromomethylated sulfonated polyphenylene oxide (BPPO)	Sulfonated poly (2,6-dimethyl-1,4-phenylene oxide) (SPPO)	30–80	[84]
3M perfluorosulfonc acid polymer	PEO	55	[85]
Nafion®	5 wt% PVA or PEO	8.7–16	[85]

(continued)

Table 6 (continued)

Based polymer	Filler/carrier polymer	Proton conductivity (mS cm^{-1})	References
3M perfluorosulfonc acid polymer	PAA	498	[85]
Polymerised Ionic Liquid	poly(MEBIm-BF4), PAA	7.1×10^{-4}	[85]
Sulfonated poly(ether ether ketone ketone)	None	37 (solvent DMF) 41 (solvent DMAc)	[60]
Sulfonated poly (arylene ether sulfone)	None	86 94	[60]
Sulfonated copolyimide	Sulfonated polyhedral oligomeric silsesquioxane (sPOSS)	~100	[86]
Polyvinylidene	Phosphotungstic acid (PWA, up to 12.8 wt%)	~0.4	[86]
Aquivion™	PEO (M_w 1×10^6)	66	[87]
Sulfonated Zro_2	PVP, poly (2-acrylamido-2-methylpropanesulfonic acid) (pAMPS)	240	[86]

Most of the previous studies were focusing on polymer reinforced composite proton exchange membranes. Nevertheless, the improvement of proton conductivity was proved as dominant rather than solving methanol crossover problem. It is vitally essential to curb the methanol crossover issue in order to improve a DMFC performance [88]. Sulfonated polyetherketones (PEEKK and PEEK) were durable under the fuel cell operating conditions for several thousand hours. The hydrophilic property is important because the presence of water molecules can facilitate the proton as well as increases the conductivity of the solid electrolytes which is crucial in fuel cell applications. While clay minerals were used as a barrier in polymer membrane to improve the methanol crossover and water permeability for direct methanol fuel cell (DMFC) [89]. Sulfonated poly (ether ether ketone) has currently been electrospun incorporated with SiO_2 as supported PEM in fuel cell application [90]. The main material was Nafion®. The SiO_2/SPEEK was impregnated in Nafion® solution to form a dense membrane. As a result, excellent proton conductivity was obtained. However, the method could not significantly reduce the methanol permeability.

The advantages of clay composite over other nanosized fillers had been proved by Bafna in 2004. The research found the surface area to volume ratio (A/V) for nanofibers two times higher than clay layer. This condition means that the nanofiber would have higher surface area of the filler exposed to the polymer matrix. Therefore, the nanofibers would have higher reinforcing ability as compared to the clay layers [91]. However, when oriented strongly in a particular direction in the composite, the ability was served to the clay layers to reinforce the composite biaxially. In preventing the transport of gases or fluid through a particular composite direction, a large amount of filler surface area is required and for this highly oriented filler, the area would be significantly larger in clay composite. This gave a reason to Bafna, 2004 in order to prepare a polyethylene-clay nanocomposite in their work. Depending on clay quantity and the dispersion state, clay can be a nucleating agent as well as obstacle to polymer mobility. However, electrospinning process helps to improve the ordering of polymer and elongation of polymer chains [92]. A research conducted by Jaafar et al. [30] revealed the improvement of PEM in studying SPEEK-cloisite 15A®. Without applying electrospinning process, the PEM was able to reduce methanol crossover problem as compared to Nafion 112® and pristine SPEEK itself. The combination of SPEEK and cloisite 15A® also prove to be one of the highest potential materials in producing good PEM. However, the studies on electrospun SPEEK-cloisite 15A® are limited. Therefore, a study on electrospinning of SPEEK-cloisite 15A® needs to be further explored [93].

9 Electrospinning Parameters

There are few parameters that control the electrospinning process. The parameters are classified as solution parameters, process parameters, and ambient parameters. Solution parameters include conductivity, viscosity, surface tension, and molecular weight. While process parameter, include distance of tip to collector, electric field, and flow rate. Every parameter significantly affects the fibers morphology [51]. In addition, ambient parameters which include the humidity and temperature of surroundings play an important part in defining fiber structures and morphologies [94]. Table 7 summarizes the effects of parameters on fiber morphologies.

Table 7 Electrospinning parameters and their effects on fiber morphology

Parameters	Effects on fiber morphology	References
Solution parameters		
Viscosity	The higher the viscosity the more beads appear on fibers	[46, 95–98]
Polymer concentration	The higher the concentration the larger the fibers diameters formed	[98–101]
Molecular weight of polymer	As the molecular weight increased, the number of beads and droplets reduced	[32, 102–104]
Conductivity	Fiber diameter decreased with decreasing in conductivity	[105–108]
Surface tension	Fibers morphology affected indirectly by surface tension. The surface tension causes the instability of jets	[108–111]

(continued)

Table 7 (continued)

Parameters	Effects on fiber morphology	References
Processing parameters		
Applied voltage	The increase of voltage decreased fiber diameter	[111, 112]
Distance between tip and collector	Too small and too large distance will generate beads and nonuniform fibers	[112, 113]
Feed rate/flow rate	Decrease of flow rate decreased the fiber diameters. Very high flow rate generated beads	[19, 114–117]
Ambient parameters		
Humidity	Porous fibers caused by high humidity	[118]
Temperature	Increase in temperature causes in decrease in fiber diameter	[119]

10 Future Directions and Conclusion

A previous study by Jaafar in 2009 has investigated SPEEK prepared using the mixture of fuming sulfuric acid as the sulfonating agent. The study showed the improvement of conductivity and methanol permeability as compared to commercial ion exchange membrane particularly for direct methanol fuel cell (DMFC) applications. Further study on SPEEK has been carried out by Jaafarin 2012 by incorporating 2,4,6-Triaminopyrimidine (TAP) as a compatibilizer to improve the compatibility between SPEEK and Cloisite 15A®. The intercalated SPEEK/Cloisite 15A®/TAP had successfully improved the membrane barrier properties due to the unique feature of Cloisite 15A® which contributed to the formation of a longer pathway towards methanol across the membrane. The methanol permeability also was significantly reduced [30].

According to the above significant improvement, the studies demonstrated that non-fluorinated polymer membranes such as SPEEK have the highest potential and viable strategy to overcome the difficulties of the currently used Nafion® membrane. However, swelling at high degree of sulfonation may reduce the performance of DMFC. There are several modifications that need to be studied to curb the swelling aspects in SPEEK such as

I. Offering bridging links to the reactive sulfonic groups via thermal activation and cross-linking of polymer chains by polyols [120].
II. Providing hydrophobic-hydrophilic block for SPEEK in different ratios [121].
III. Adjusting SPEEK by blending technique. The compatible blends occur between SPEEK-polyether sulfone (PES), SPEEK-polyether imide (PEI), and polybenzimidazole (PBI) [122–124].

Several studies have recommended the introduction of dispersive fillers to SPEEK to solve the methanol crossover problem in DMFC. Nanofillers, for instance silica and intercalated structure such as organic-montmorillonite (OMMT),

[125] have been applied in modifying SPEEK. Methanol crossover was reduced in these nanocomposite SPEEK membranes. Apart from these studies, it is very important to state that membrane morphology is very important in providing the path for methanol. Hence, exfoliated membrane pathway is an ideal membrane morphology which has the ability to create methanol path in solving the methanol crossover problem. One of the solutions is by fabricating the polymer membrane by electrospinning as it is one of the most suitable fabrication methods to reduce the size of filler, thus ease the filler to disperse in the polymer matrix without agglomeration [30].

References

1. Apanel G, Johnson E (2004) Direct methanol fuel cells—ready to go commercial. Fuel Cells Bull 2004:12–17
2. Ahmadian-Alam L, Kheirmand M, Mahdavi H (2016) Preparation, characterization and properties of PVDF-g-PAMPS/PMMA-co-PAMPS/silica nanoparticle as a new proton exchange nanocomposite membrane. Chem Eng J 284:1035–1048
3. Ayyaru S, Dharmalingam S (2011) Development of MFC using sulphonated polyether ether ketone (SPEEK) membrane for electricity generation from waste water. Bioresour Technol 102:11167–11171
4. Banerjee S, Kar KK (2016) Superior water retention, ionic conductivity and thermal stability of sulfonated poly ether ether ketone/polypyrrole/aluminum phosphate nanocomposite based polymer electrolyte membrane. J Environ Chem Eng 4:299–310
5. Chiou B Sen, Yee E, Wood D, Shey J, Glenn G, Orts W (2006) Effects of processing conditions on nanoclay dispersion in starch-clay nanocomposites. Cereal Chem 83:300–305
6. Cho E-B, Luu DX, Kim D (2010) Enhanced transport performance of sulfonated mesoporous benzene-silica incorporated poly(ether ether ketone) composite membranes for fuel cell application. J Membr Sci 351:58–64
7. Cui Y, Kundalwal SI, Kumar S (2016) Gas barrier performance of graphene/polymer nanocomposites. Carbon 98:313–333
8. Dicks AL (2012) Comprehensive renewable energy. In: Comprehensive renewable energy. Elsevier, Amsterdam, pp 203–245
9. Elabd YA, Napadensky E, Sloan JM, Crawford DM, Walker CW (2003) Triblock copolymer ionomer membranes. J Membr Sci 217:227–242
10. Fei S-T, Wood RM, Lee DK, Stone DA, Chang H-L, Allcock HR (2008) Inorganic–organic hybrid polymers with pendent sulfonated cyclic phosphazene side groups as potential proton conductive materials for direct methanol fuel cells. J Membr Sci 320:206–214
11. Fu R-Q, Julius D, Hong L, Lee J-Y (2008) PPO-based acid-base polymer blend membranes for direct methanol fuel cells. J Membr Sci 322:331–338
12. Hashemifard SA, Ismail AF, Matsuura T (2011) Effects of montmorillonite nano-clay fillers on PEI mixed matrix membrane for CO_2 removal. Chem Eng J 170:316–325
13. Ilbeygi H, Mayahi A, Ismail AF, Nasef MM, Jaafar J, Ghasemi M, Matsuura T, Zaidi SMJ (2014) Transport properties of SPEEK nanocomposite proton conducting membranes: optimization of additives content by response surface methodology. J Taiwan Inst Chem Eng 45:2265–2279
14. Baptista AC, Martins JI, Fortunato E, Martins R, Borges JP, Ferreira I (2011) Thin and flexible bio-batteries made of electrospun cellulose-based membranes. Biosens Bioelectron 26:2742–2745

15. Battirola LC, Schneider JF, Torriani ÍCL, Tremiliosi-Filho G, Rodrigues-Filho UP (2013) Improvement on direct ethanol fuel cell performance by using doped-Nafion® 117 membranes with Pt and Pt–Ru nanoparticles. Int J Hydrogen Energy 38:12060–12068
16. Yu DM, Yoon S, Kim T-H, Lee JY, Lee J, Hong YT (2013) Properties of sulfonated poly (arylene ether sulfone)/electrospun nonwoven polyacrylonitrile composite membrane for proton exchange membrane fuel cells. J Membr Sci 446:212–219
17. Hasani-Sadrabadi MM, Shabani I, Soleimani M, Moaddel H (2011) Novel nanofiber-based triple-layer proton exchange membranes for fuel cell applications. J Power Sour 196:4599–4603
18. Jin J, Hao R, He X, Li G (2015) Sulfonated poly(phenylsulfone)/fluorinated polybenzoxazole nanofiber composite membranes for proton exchange membrane fuel cells. Int J Hydrogen Energy 40:14421–14427
19. Kim JM, Joh H-I, Jo SM, Ahn DJ, Ha HY, Hong S-A, Kim S-K (2010) Preparation and characterization of Pt nanowire by electrospinning method for methanol oxidation. Electrochim Acta 55:4827–4835
20. Lee JY, Bashur CA, Goldstein AS, Schmidt CE (2009) Polypyrrole-coated electrospun PLGA nanofibers for neural tissue applications. Biomaterials 30:4325–4335
21. Liu HY, Liu LL, Yang CL, Li ZH, Xiao QZ, Lei GT, Ding YH (2014) A hard-template process to prepare three-dimensionally macroporous polymer electrolyte for lithium-ion batteries. Electrochim Acta 121:328–336
22. Chen Y, Guo R, Lee CH, Lee M, McGrath JE (2012) Partly fluorinated poly(arylene ether ketone sulfone) hydrophilic–hydrophobic multiblock copolymers for fuel cell membranes. Int J Hydrogen Energy 37:6132–6139
23. Jia C, Liu J, Yan C (2012) A multilayered membrane for vanadium redox flow battery. J Power Sources 203:190–194
24. Kang NR, Lee SY, Shin DW, Hwang DS, Lee KH, Cho DH, Kim JH, Lee YM (2016) Effect of end-group cross-linking on transport properties of sulfonated poly(phenylene sulfide nitrile)s for proton exchange membranes. J Power Sources 307:834–843
25. Yuan W, Zhou B, Deng J, Tang Y, Zhang Z, Li Z (2014) Overview on the developments of vapor-feed direct methanol fuel cells. Int J Hydrogen Energy 39:6689–6704
26. Ferreira SLC, Bruns RE, da Silva EGP, Dos Santos WNL, Quintella CM, David JM, de Andrade JB, Breitkreitz MC, Jardim ICSF, Neto BB (2007) Statistical designs and response surface techniques for the optimization of chromatographic systems. J Chromatogr A 1158:2–14
27. Alexandre M, Dubois P (2000) Polymer-layered silicate nanocomposites: preparation, properties and uses of a new class of materials. Mater Sci Eng R Rep 28:1–63
28. Baji A, Mai YW, Wong SC, Abtahi M, Chen P (2010) Electrospinning of polymer nanofibers: effects on oriented morphology, structures and tensile properties. Compos Sci Technol 70:703–718
29. Ma J, Sahai Y (2013) Chitosan biopolymer for fuel cell applications. Carbohydr Polym 92:955–975
30. Jaafar J, Ismail AF, Matsuura T (2009) Preparation and barrier properties of SPEEK/Cloisite 15A®/TAP nanocomposite membrane for DMFC application. J Membr Sci 345:119–127
31. Shao Z, Zhou W, Zhu Z (2012) Advanced synthesis of materials for intermediate-temperature solid oxide fuel cells. Prog Mater Sci 57:804–874
32. Ahmed FE, Lalia BS, Hashaikeh R (2015) A review on electrospinning for membrane fabrication: challenges and applications. Desalination 356:15–30
33. Luo CJ, Nangrejo M, Edirisinghe M (2010) A novel method of selecting solvents for polymer electrospinning. Polymer 51:1654–1662
34. Peponi L, Puglia D, Torre L, Valentini L, Kenny JM (2014) Processing of nanostructured polymers and advanced polymeric based nanocomposites. Mater Sci Eng R Rep 85:1–46
35. Wang SH, Lin HL (2014) Poly (vinylidene fluoride-co-hexafluoropropylene)/polybenzimidazole blend nanofiber supported Nafion membranes for direct methanol fuel cells. J Power Sour

36. Won J-H, Lee H-J, Lim J-M, Kim J-H, Hong YT, Lee S-Y (2014) Anomalous behavior of proton transport and dimensional stability of sulfonated poly(arylene ether sulfone) nonwoven/silicate composite proton exchange membrane with dual phase co-continuous morphology. J Membr Sci 450:235–241
37. Rezaei M, Ismail AF, Bakeri G, Hashemifard SA, Matsuura T (2015) Effect of general montmorillonite and Cloisite 15A on structural parameters and performance of mixed matrix membranes contactor for CO_2 absorption. Chem Eng J 260:875–885
38. Etmimi HM, Mallon PE, Sanderson RD (2013) Polymer/graphite nanocomposites: Effect of reducing the functional groups of graphite oxide on water barrier properties. Eur Polymer J 49:3460–3470
39. Wang Y, Yang D, Zheng X, Jiang Z, Li J (2008) Zeolite beta-filled chitosan membrane with low methanol permeability for direct methanol fuel cell. J Power Sources 183:454–463
40. Mohtar SS, Ismail AF, Matsuura T (2011) Preparation and characterization of SPEEK/MMT-STA composite membrane for DMFC application. J Membr Sci 371:10–19
41. Sasikala S, Meenakshi S, Bhat SD, Sahu AK (2014) Functionalized Bentonite clay-sPEEK based composite membranes for direct methanol fuel cells. Electrochim Acta 135:232–241
42. Wu H, Cao Y, Li Z, He G, Jiang Z (2015) Novel sulfonated poly (ether ether ketone)/ phosphonic acid-functionalized titania nanohybrid membrane by an in situ method for direct methanol fuel cells. J Power Sour 273:544–553
43. Yin Y, Xu T, He G, Jiang Z, Wu H (2015) Fabrication of sulfonated poly(ether ether ketone)-based hybrid proton-conducting membranes containing carboxyl or amino acid-functionalized titania by in situ sol–gel process. J Power Sour 276:271–278
44. Yin Y, Wang H, Cao L, Li Z, Li Z, Gang M, Wang C, Wu H, Jiang Z, Zhang P (2016) Sulfonated poly(ether ether ketone)-based hybrid membranes containing graphene oxide with acid-base pairs for direct methanol fuel cells. Electrochim Acta 203:178–188
45. Erdener H, Akay RG, Yücel H, Baç N, Eroğlu İ (2009) Effects of sulfonated polyether-etherketone (SPEEK) and composite membranes on the proton exchange membrane fuel cell (PEMFC) performance. Int J Hydrogen Energy 34:4645–4652
46. Yang T, Liu C (2011) SPEEK/sulfonated cyclodextrin blend membranes for direct methanol fuel cell. Int J Hydrogen Energy 36:5666–5674
47. Purwanto M, Atmaja L, Mohamed MA, Salleh MT, Jaafar J, Ismail AF, Widiastuti N (2016) Biopolymer-based electrolyte membranes from chitosan incorporated with montmorillonite-crosslinked GPTMS for direct methanol fuel cells. RSC Adv 6(3):2314–2322
48. Doğan H, Inan TY, Unveren E, Kaya M (2010) Effect of cesium salt of tungstophosphoric acid (Cs-TPA) on the properties of sulfonated polyether ether ketone (SPEEK) composite membranes for fuel cell applications. Int J Hydrogen Energy 35:7784–7795
49. Bello M, Zaidi SMJ, Rahman SU (2008) Proton and methanol transport behavior of SPEEK/TPA/MCM-41 composite membranes for fuel cell application. J Membr Sci 322:218–224
50. Colicchio I, Demco DE, Baias M, Keul H, Moeller M (2009) Influence of the silica content in SPEEK–silica membranes prepared from the sol–gel process of polyethoxysiloxane: Morphology and proton mobility. J Membr Sci 337:125–135
51. Yang CC, Chiu SJ, Kuo SC (2011) Preparation of poly(vinyl alcohol)/montmorillonite/poly (styrene sulfonic acid) composite membranes for hydrogen–oxygen polymer electrolyte fuel cells. Curr Appl Phys 11:S229–S237
52. Norddin MNAM, Ismail AF, Rana D, Matsuura T, Mustafa A, Tabe-Mohammadi A (2008) Characterization and performance of proton exchange membranes for direct methanol fuel cell: Blending of sulfonated poly(ether ether ketone) with charged surface modifying macromolecule. J Membr Sci 323:404–413
53. Huang YJ, Ye YS, Syu YJ, Hwang BJ, Chang FC (2012) Synthesis and characterization of sulfonated polytriazole-clay proton exchange membrane by in situ polymerization and click reaction for direct methanol fuel cells. J Power Sour 208:144–152

54. Baradie B, Dodelet J, Guay D (2000) Hybrid Nafion®-inorganic membrane with potential applications for polymer electrolyte fuel cells. J Electroanal Chem 489:101–105
55. Kwak S (2003) Nafion/mordenite hybrid membrane for high-temperature operation of polymer electrolyte membrane fuel cell. Solid State Ionics 160:309–315
56. Gang M, He G, Li Z, Cao K, Li Z, Yin Y, Wu H, Jiang Z (2016) Graphitic carbon nitride nanosheets/sulfonated poly(ether ether ketone) nanocomposite membrane for direct methanol fuel cell application. J Membr Sci 507:1–11
57. Li H, Zhang G, Wu J, Zhao C, Zhang Y, Shao K, Han M, Lin H, Zhu J, Na H (2010) A novel sulfonated poly(ether ether ketone) and cross-linked membranes for fuel cells. J Power Sour 195:6443–6449
58. Kim J, Kim B, Jung B (2002) Proton conductivities and methanol permeabilities of membranes made from partially sulfonated polystyrene-block-poly(ethylene-ran-butylene)-block-polystyrene copolymers. J Membr Sci 207:129–137
59. Paneri A, Heo Y, Ehlert G, Cottrill A, Sodano H, Pintauro P, Moghaddam S (2014) Proton selective ionic graphene-based membrane for high concentration direct methanol fuel cells. J Membr Sci 467:217–225
60. Tijing LD, Choi J-S, Lee S, Kim S-H, Shon HK (2014) Recent progress of membrane distillation using electrospun nanofibrous membrane. J Membr Sci 453:435–462
61. Krishnan NN, Henkensmeier D, Park Y-H, Jang J-H, Kwon T, Koo CM, Kim H-J, Han J, Nam S-W (2016) Blue membranes: sulfonated copper(II) phthalocyanine tetrasulfonic acid based composite membranes for DMFC and low relative humidity PEMFC. J Membr Sci 502:1–10
62. Anadão P, Sato LF, Wiebeck H, Valenzuela-Díaz FR (2010) Montmorillonite as a component of polysulfone nanocomposite membranes. Appl Clay Sci 48:127–132
63. Qiu J, Zhai M, Chen J, Wang Y, Peng J, Xu L, Li J, Wei G (2009) Performance of vanadium redox flow battery with a novel amphoteric ion exchange membrane synthesized by two-step grafting method. J Membr Sci 342:215–220
64. Ladewig BP, Knott RB, Martin DJ, Diniz da Costa JC, Lu GQ (2007) Nafion-MPMDMS nanocomposite membranes with low methanol permeability. Electrochem Commun 9:781–786
65. Balbaşı M, Gözütok B (2010) Poly(vinyl alcohol)-colloidal silica composite membranes for fuel cells. Synth Met 160:150–155
66. Ballengee JB, Pintauro PN (2011) Composite fuel cell membranes from dual-nanofiber electrospun mats. Macromolecules 44:7307–7314
67. Barbora L, Acharya S, Verma A (2009) Synthesis and ex-situ characterization of Nafion/TiO_2 composite membranes for direct ethanol fuel cell. In: Macromolecular symposia, vol 277, No. 1, pp 177–189. Wiley, New York
68. Baş D, Boyacı İH (2007) Modeling and optimization II: comparison of estimation capabilities of response surface methodology with artificial neural networks in a biochemical reaction. J Food Eng 78:846–854
69. Basavarajappa M, Draper T, Toth P, Ring TA, Miskovic S (2015) Numerical and experimental investigation of single phase flow characteristics in stirred tanks using Rushton turbine and flotation impeller. Miner Eng 83:156–167
70. Kim S, Lee H, Ahn D, Woong Park H, Chang T, Lee W (2013) Direct sulfonation and photocross linking of unsaturated poly(styrene-b-butadiene-b-styrene) for proton exchange membrane of direct methanol fuel cell. J Membr Sci 427:85–91
71. Sharma S, Pollet BG (2012) Support materials for PEMFC and DMFC electrocatalysts—a review. J Power Sour 208:96–119
72. Kahveci EE, Taymaz I (2014) Experimental investigation on water and heat management in a PEM fuel cell using response surface methodology. Int J Hydrogen Energy 39:10655–10663
73. Winters-Miner LA, Bolding PS, Hilbe JM, Goldstein M, Hill T, Nisbet R, Walton N, Miner GD, Winters R (2015) Practical predictive analytics and decisioning systems for

medicine. In: Practical predictive analytics and decisioning systems for medicine. Elsevier, Amsterdam, pp 757–794
74. Wolfe JE, Lind OT (2008) Influence of suspended clay on phosphorus uptake by periphyton. Hydrobiologia 610:211–222
75. Woo Y, Oh SY, Kang YS, Jung B (2003) Synthesis and characterization of sulfonated polyimide membranes for direct methanol fuel cell. J Membr Sci 220:31–45
76. Wu C-S, Lin F-Y, Chen C-Y, Chu PP (2006) A polyvinyl alcohol/p-sulfonate phenolic resin composite proton conducting membrane. J Power Sour 160:1204–1210
77. Wu HC, Li YY, Sakoda A (2010) Synthesis and hydrogen storage capacity of exfoliated turbostratic carbon nanofibers. Int J Hydrogen Energy 35:4123–4130
78. Wu X, Tong L (2013) Optical microfibers and nanofibers. Nanophotonics 2:407–428
79. Lin Z, Yao Y, Zhang X (2014) Electrospun nanofibers for design and fabrication of electrocatalysts and electrolyte membranes for fuel cells. In: Electrospun nanofibers for energy and environmental applications, pp 41–67. Springer, Berlin
80. Lucking AD, Pan L, Narayanan DL, Clifford CEB (2005) Effect of expanded graphite lattice in exfoliated graphite nanofibers on hydrogen storage. J Phys Chem B 109:12710–12717
81. Lue SJ, Pai Y-L, Shih C-M, Wu M-C, Lai S-M (2015) Novel bilayer well-aligned nafion/graphene oxide composite membranes prepared using spin coating method for direct liquid fuel cells. J Membr Sci 493:212–223
82. Lueking AD, Gutierrez HR, Fonseca DA, Dickey E (2007) Structural characterization of exfoliated graphite nanofibers. Carbon 45:751–759
83. Lueking AD, Pan L, Narayanan D, Clifford CEB (2005a) Exfoliated graphite nanofibers for hydrogen storage, pp 457–459. ACS Division of Fuel Chemistry (preprints)
84. Thiam HS, Daud WRW, Kamarudin SK, Mohamad AB, Kadhum AAH, Loh KS, Majlan EH (2013) Performance of direct methanol fuel cell with a palladium–silica nanofibre/nafion composite membrane. Energy Convers Manag 75:718–726
85. Thomassin J-M, Pagnoulle C, Caldarella G, Germain A, Jérôme R (2005) Impact of acid containing montmorillonite on the properties of Nafion® membranes. Polymer 46:11389–11395
86. Tong HW, Wang M (2013) A novel technique for the fabrication of 3D nanofibrous scaffolds using simultaneous positive voltage electrospinning and negative voltage electrospinning. Mater Lett 94:116–120
87. Touati N, Kaci M, Bruzaud S, Grohens Y (2011) The effects of reprocessing cycles on the structure and properties of isotactic polypropylene/cloisite 15A nanocomposites. Polym Degrad Stab 96:1064–1073
88. Tran NH, Dennis GR, Milev AS, Kannangara GSK, Williams P, Wilson MA, Lamb RN (2006) Dispersion of organically modified clays within n-alcohols. J Colloid Interface Sci 297:541–545
89. Tricoli V, Carretta N, Bartolozzi M (2000) A comparative investigation of proton and methanol transport in fluorinated ionomeric membranes. J Electrochem Soc 147(4):1286–1290
90. Xia X, Cai S, Hu J, Xie C (2006) Water absorption characteristics of novel Cu/LDPE nanocomposite for use in intrauterine devices. J Biomed Mater Res Part B Appl Biomat 79:345–352
91. Xie J, MacEwan MR, Schwartz AG, Xia Y (2010) Electrospun nanofibers for neural tissue engineering. Nanoscale 2:35–44
92. Xin Q, Wu H, Jiang Z, Li Y, Wang S, Li Q, Li X, Lu X, Cao X, Yang J (2014) SPEEK/amine-functionalized TiO_2 submicrospheres mixed matrix membranes for CO_2 separation. J Membr Sci 467:23–35
93. Xue S, Yin G, Cai K, Shao Y (2007) Permeabilities of methanol, ethanol and dimethyl ether in new composite membranes: a comparison with Nafion membranes. J Membr Sci 289(1):51–57
94. Yang L, Sun H, Wang S, Jiang L, Sun G (2012) Reversible and irreversible loss in performance in direct methanol fuel cells during freeze/thaw cycles. J Power Sour 219:193–198

95. Yang T (2008) Preliminary study of SPEEK/PVA blend membranes for DMFC applications. Int J Hydrogen Energy 33:6772–6779
96. Yang T (2009) Composite membrane of sulfonated poly(ether ether ketone) and sulfated poly(vinyl alcohol) for use in direct methanol fuel cells. J Membr Sci 342:221–226
97. Yang T, Xu Q, Wang Y, Lu B, Zhang P (2008) Primary study on double-layer membranes for direct methanol fuel cell. Int J Hydrogen Energy 33:6766–6771
98. Zaidi L, Bruzaud S, Bourmaud AP, Kaci M, Grohens Y (2010) Relationship between structure and rheological, mechanical and thermal properties of polylactide/cloisite 30B nanocomposites. J Appl Polym Sci 116:1357–1365
99. Zeng J, Haoqing H, Schaper A, Wendorff JH, Greiner A (2003) Poly-L-lactide nanofibers by electrospinning–Influence of solution viscosity and electrical conductivity on fiber diameter and fiber morphology. e-Polymers 3(1):102–110
100. Zhang S, He G, Gong X, Zhu X, Wu X, Sun X, Zhao X, Li H (2015) Electrospun nanofiber enhanced sulfonated poly (phthalazinone ether sulfone ketone) composite proton exchange membranes. J Membr Sci 493:58–65
101. Zhang Y-Q (1998) Natural silk fibroin as a support for enzyme immobilization. Biotechnol Adv 16:961–971
102. Bognitzki M, Czado W, Frese T, Schaper A, Hellwig M, Steinhart M, Greiner A, Wendorff JH (2001) Nanostructured fibers via electrospinning. Adv Mater 13:70–72
103. Cai Z, Yang G (2010) Research progress in electrospinning technique for biomedical materials. Sheng wu yi xue gong cheng xue za zhi = J Biomed Eng Shengwu yixue gongchengxue zazhi 27:1389–1392
104. Casper CL, Stephens JS, Tassi NG, Chase DB, Rabolt JF (2004) Controlling surface morphology of electrospun polystyrene fibers: effect of humidity and molecular weight in the electrospinning process. Macromolecules 37(2):573–578
105. Cavaliere S, Subianto S, Savych I, Jones DJ, Rozière J (2011) Electrospinning: designed architectures for energy conversion and storage devices. Energy Environ Sci 4(12):4761–4785
106. Chronakis IS (2005) Novel nanocomposites and nanoceramics based on polymer nanofibers using electrospinning process—a review. J Mater Process Technol 167:283–293
107. Cong Y, Liu S, Chen H (2013) Fabrication of conductive polypyrrole nanofibers by electrospinning. J Nanomater
108. Demir MM, Yilgor I, Yilgor EEA, Erman B (2002) Electrospinning of polyurethane fibers. Polymer 43(11):3303–3309
109. Frenot A, Chronakis IS (2003) Polymer nanofibers assembled by electrospinning. Curr Opin Colloid Interface Sci 8:64–75
110. Graeser M, Bognitzki M, Massa W, Pietzonka C, Greiner A, Wendorff JH (2007) Magnetically anisotropic cobalt and iron nanofibers via electrospinning. Adv Mater 19:4244–4247
111. Greiner A, Wendorff JH (2007) Electrospinning: a fascinating method for the preparation of ultrathin fibers. Angewandte Chemie—International Edition
112. Gupta P, Elkins C, Long TE, Wilkes GL (2005) Electrospinning of linear homopolymers of poly (methyl methacrylate): exploring relationships between fiber formation, viscosity, molecular weight and concentration in a good solvent. Polymer 46(13):4799–4810
113. Hohman MM, Shin M, Rutledge G, Brenner MP (2001) Electrospinning and electrically forced jets. I. Stability theory. Phys Fluids (1994–present) 13(8):2201–2220
114. Hutmacher DW, Dalton PD (2011) Melt electrospinning. Chemi Asian J
115. Inagaki M, Yang Y, Kang F (2012) Carbon nanofibers prepared via electrospinning. Adv Mater
116. Ji HM, Lee HW, Karim MR, Cheong IW, Bae EA, Kim TH, Islam MS, Ji BC, Yeum JH (2009) Electrospinning and characterization of medium-molecular-weight poly(vinyl alcohol)/high-molecular-weight poly(vinyl alcohol)/montmorillonite nanofibers. Colloid Polym Sci 287:751–758
117. Kang IK, Xing ZC, Han SJ, Shin YS (2011) Fabrication of biodegradable polyester nanocomposites by electrospinning for tissue engineering. J Nanomater

118. Kong L, Ziegler GR (2013) Quantitative relationship between electrospinning parameters and starch fiber diameter. Carbohydr Polym 92:1416–1422
119. Koski A, Yim K, Shivkumar S (2004) Effect of molecular weight on fibrous PVA produced by electrospinning. Mater Lett 58(3):493–497
120. Kostakova E, Meszaros L, Gregr J (2009) Composite nanofibers produced by modified needleless electrospinning. Mater Lett 63:2419–2422
121. Niu H, Lin T (2012) Fiber generators in needleless electrospinning. J Nanomater
122. Pham QP, Sharma U, Mikos AG (2006) Electrospinning of polymeric nanofibers for tissue engineering applications: a review. Tissue Eng 12:1197–1211
123. Park JH, Lee HW, Chae DK, Oh W, Yun JD, Deng Y, Yeum JH (2009) Electrospinning and characterization of poly(vinyl alcohol)/chitosan oligosaccharide/clay nanocomposite nanofibers in aqueous solutions. Colloid Polym Sci 287:943–950
124. Rafiei S, Maghsoodloo S, Noroozi B, Mottaghitalab V, Haghi AK (2013) Mathematical modeling in electrospinning process of nanofibers: a detailed review. Cellulose Chem Technol 47:323–338
125. Reneker DH, Yarin AL (2008) Electrospinning jets and polymer nanofibers. Polymer 49 (10):2387–2425

Chapter 8
A Basic Overview of Fuel Cells: Thermodynamics and Cell Efficiency

Narcis Duteanu, Adriana Balasoiu, Pritha Chatterjee and Makarand M. Ghangrekar

Abstract In the last century, there has been rapid urbanization leading to increased energy demand with an ever increasing load on nonrenewable resources and subsequent escalation of pollution. A viable solution to these two problems can be a power supply technology that is able to produce energy with minimum or zero pollutant emission into the environment. Fuel cells appear to be an eco-friendly power supply technology. Main advantage of fuel cell technology is represented by direct conversion of fuels into electrical energy, with zero emissions, when hydrogen is used as fuel. This article describes the basic overview of fuel cell technology in order to better understand the construction and also the working principle of this eco-friendly technology.

1 What Is a Fuel Cell?

Fuel cells (FC) appear as a *sci fi* system and represent a device able to convert the chemical energy of a fuel directly into electrical energy [1–3], and it is characterized by the continuous feed with active chemical species to undergo redox reactions [4]. In classical energy production technologies, fuel is burned to generate heat, which is further converted to steam in a steam engine. Produced steam is used to drive a

N. Duteanu (✉) · A. Balasoiu
CAICAM, Faculty of Industrial Chemistry and Environmental Engineering,
Politehnica University of Timisoara, 6, Pirvan Street, 300223 Timisoara, Romania
e-mail: narcis.duteanu@upt.ro

A. Balasoiu
e-mail: adriana.balasoiu@ymail.com

P. Chatterjee · M.M. Ghangrekar
Department of Civil Engineering, Indian Institute of Technology Kharagpur,
Kharagpur 721302, India
e-mail: pritha.besu@gmail.com

M.M. Ghangrekar
e-mail: ghangrekar@civil.iitkgp.ernet.in

Inamuddin et al. (eds.), *Organic–Inorganic Composite Polymer Electrolyte Membranes*, DOI 10.1007/978-3-319-52739-0_8

turbine which converts the thermal energy present in steam to mechanical energy. Finally, the mechanical energy is converted to electrical energy using a generator [1].

In a fuel cell, electricity is produced directly by burning of fuel, hence the intermediate steps of production of heat and mechanical energy can be avoided. This helps to evade the thermodynamic limitation of any engine. This thermodynamic limitation is known as Carnot efficiency which is defined as the upper limit on the efficiency that any classical thermodynamic engine can achieve during the conversion of heat into work [5]. Since the fuel cell system does not have to go through the Carnot cycle, hence its efficiency is expected to be higher than classical heat engines.

Combustion of fuels in conventional energy producing technologies has a huge environmental impact. However in a fuel cell, electrochemical conversion of the chemical energy, in presence of catalyst, present in the bonds of the fuel into electrical energy, produces power with minimal or zero pollution. In comparison with batteries, fuel cells are able to produce electricity as long as the fuel and also the oxidant are continuously replenished at anode and cathode, respectively [5, 6].

2 Fuel Cell Structure and Classification

A single unit cell represents core of FC which consists of two electrodes and one electrolyte layer. Anode (negative electrode) is placed in contact with electrolyte layer on one side and the cathode (positive electrode) is placed in contact on the other side of the electrolyte [5, 6]. A schematic representation of a classical H_2/O_2 fuel cell is presented in Fig. 1. Starting from a fuel cell basic unit, stacks are developed where individual core units are modularly combined in order to deliver the desired output current and voltage [6].

Fuel cells are able to process a wide variety of fuels, such as hydrogen, hydrocarbons, alcohols, natural gas, and derivatives. Burning of fossil fuels in conventional energy producing technologies has a high global warming potential due to emission of greenhouse gases and a high environmental impact due to use of nonrenewable fuels. Taking into account, the lower global warming potential by reducing greenhouse gases emissions, FC using hydrogen, are becoming more and more attractive. In all classical FC, during electricity production, anode must be fed continuously with the fuel, and simultaneously the cathode must be furnished continuously with an oxidant, preferably oxygen from air. Electricity generation from the FC is due to the spontaneous anodic and cathodic half-cell reactions [1, 4, 6, 7]. In a hydrogen fuel cell, at the anode catalyst layer, hydrogen undergoes an oxidation reaction producing protons and electrons. The generated protons are then conducted by the electrolyte membrane and delivered at the cathode, where they participate in reduction reaction along with oxygen and electrons, which traverse through the external electrical circuit producing water (Fig. 1).

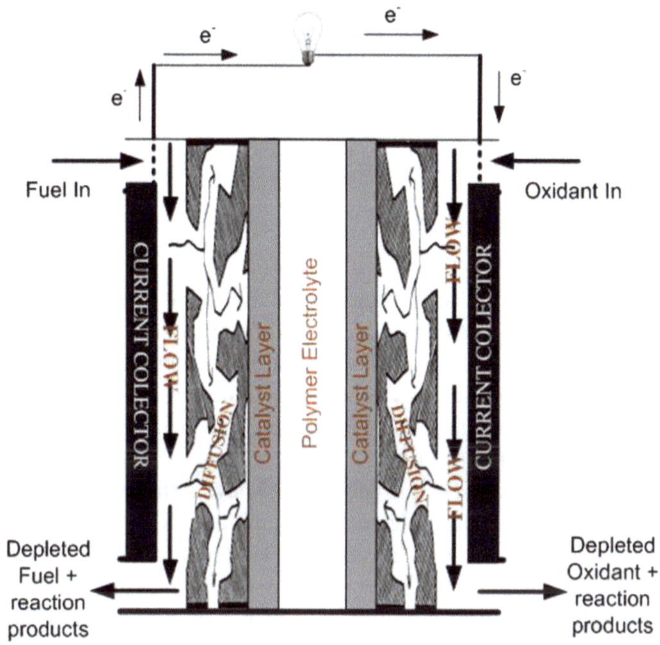

Fig. 1 Working principle of PEMFC

Fuel cells can be classified either based on operating temperature [4] or more commonly by the electrolyte used [1, 4, 6, 8]. Based on electrolyte used fuels cells are classified as follows [1, 4, 6, 8–12]:

- Proton Exchange Membrane Fuel Cells (PEMFC) or polymer electrolyte fuel cell:
  - H_2/O_2 PEMFC;
  - **D**irect **M**ethanol **F**uel **C**ells (**DMFC**);
  - **D**irect **E**thanol **F**uel **C**ells (**DEFC**);
  - **D**irect **F**ormic **A**cid **F**uel **C**ells (**DFAFC**);
  - **D**irect **B**orohydride **F**uel **C**ells (**DBFC**).
- **A**lkaline **F**uel **C**ells (**AFC**):
  - **P**roton **C**eramic **F**uel **C**ell (**PCFC**);
  - **D**irect **B**orohydride **F**uel **C**ell (**DBFC**);
  - **D**irect **A**lcohol **F**uel **C**ell (**DAFC**).
- **P**hosphoric **A**cid **F**uel **C**ell (**PAFC**);
- **M**olten **C**arbonate **F**uel **C**ells (**MCFC**);
- **S**olid **O**xide **F**uel **C**ells (**SOFC**);
- **D**irect **C**arbon **F**uel **C**ells (**DCFC**).

Based on operating temperature, fuel cells can be classified as:

- Low-operating temperature fuel cell (50–250 °C): PEMFC, AFC, PAFC;
- High-operating temperature fuel cell (650–1000 °C): MCFC, SOFC, DCFC.

3 Fuel Cell Construction

Polymer electrolyte fuel cells are constructed by using a Membrane Electrode Assembly (MEA). In all cases, MEA is formed from a central polymer electrolyte (membrane) and two electrodes on either side of the membrane [13–18]. By placing the two electrodes very close, it is expected to improve the cell performance by reducing the internal resistance of electrolyte. Schematically the core of any PEMFC can be represented as in Fig. 2.

To obtain a higher efficiency, it is necessary to have a strong contact between the catalyst layer and the polymer electrolyte membrane. This represents the most critical and important portion of a FC and is called the three phase interface, where the electrocatalytic reactions take place [4, 6, 16, 17, 19]. A catalyst particle can be active for the electrochemical reactions only if it is in contact with the reactants, electrolyte and also with the electrode [4, 20–23]. Contact of catalyst layer with electrolyte is necessary in order to be able to get the ionic reaction product from

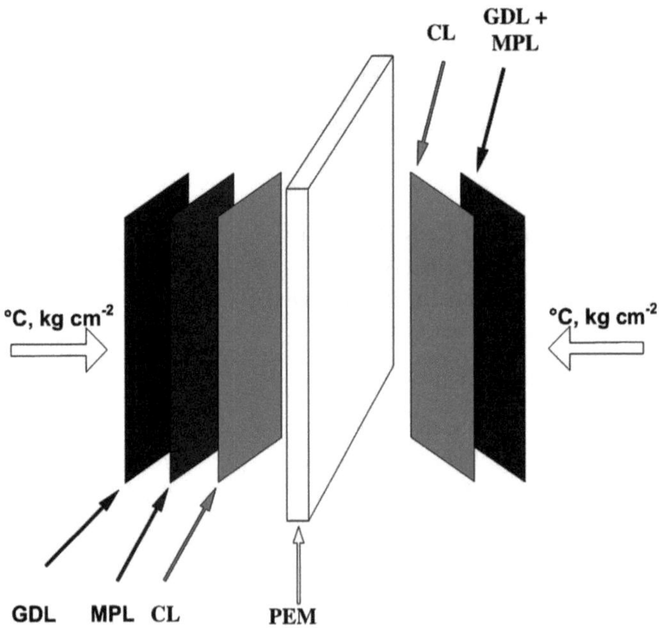

Fig. 2 Schematic representation of MEA production

production site and conduct them into the consumption place, and at the same time electrical contact with solid electrode is needed in order to retrieve electrons [24, 25]. Also to enhance reaction rate, sufficient catalyst should be present on the electrodes [4, 6]. This can be achieved either by applying a high catalyst loading rate or using a small quantity of electrolyte per cell [14, 26].

A practical method used to increase the electrocatalytic surface area of the electrode, along with reduction of catalyst loading, is by usage of 3D structured catalyst layers. Such 3D structures can be produced by using catalyst nanoparticles supported on the surface of some larger support particles. In actual stage of development carbon particles with sizes of around 10 nm, [13, 16, 27–31], carbon nanofibers [28, 29, 32–36] and graphene [37] are used as support particles.

Platinum is the most commonly used catalyst for hydrogen oxidation and oxygen reduction reactions (ORR) [13, 14, 20, 29–31, 36, 38–43]. However, platinum is expensive, and hence to reduce the cost of fuel cell without compromising with the performance, it is necessary to use nanoparticles of catalyst on support particles [1, 6, 22, 29, 33, 44–51], which increases the active surface area available for reactions. Increased surface area helps in improved reactant and products transport to and from bulk to the three phase interface [18, 25, 52, 53].

As can be seen from Fig. 2, electrode used in a fuel cell construction is a complex structure consisting of a gas diffusion layer, a microporous layer, and a catalyst layer [52]. One of the most important part of MEA is represented by the gas diffusion layer (GDL), which provides durability to the entire assembly and at the same time should ensure proper electrical contact between catalyst layer and the current collectors [52–55]. Presence of GDL increases contact resistance at the interface between the GDL and current collectors, which reduces energy conversion efficiency. At the same time, it is important to obtain a porous structure of GDL in order to be able to have continuous reactant and product flow to and from the catalyst layer to the bulk liquid [25, 53, 56]. Maximum power can be achieved only if the GDL present optimum hydrophobicity in order to remove produced water from the catalyst layer. Water retention is equivalent to blockage of internal channels, leading to a limitation of the supply of reactant and inefficient transport of reaction products, thus reducing the fuel cell efficiency [52–60]. Graphite paper consisting of woven carbon fibers is the most commonly used GDL (Fig. 3). This mesh of carbon fibers forms a series of interconnected channel like structure, which allows efficient transport of reactants and products. Hydrophobic layer on the graphite paper links the individual carbon fibers together.

The microporous layer (MPL), located between GDL and catalyst layer have the following features [61]:

- assures a physical microporous base for catalyst layer, which helps in transportation of products and reactants to and from the catalyst layer;
- prevents penetration of catalyst particles inside the GDL;
- plays an important role in electrode water management;
- reduces the contact resistance between catalyst layer and the GDL.

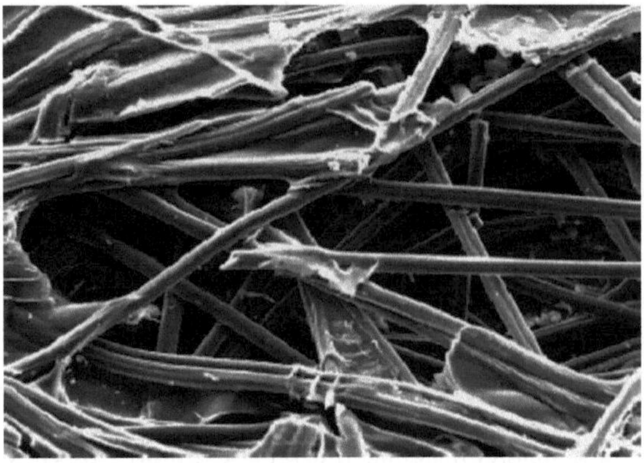

Fig. 3 SEM obtained for commercial carbon paper with 20% wet-proof at 500 X magnification

In short, MPL is responsible for rapid removal of reaction products, and reduction of water accumulation inside the electrode. In this way, it avoids blockage of catalyst layer with water molecule and reaction products.

Water management inside electrodes can be done by using hydrophobic/hydrophilic organic compounds (such as PTFE or Nafion 117) [20, 50, 52, 54, 55, 58, 62]. PTFE or Nafion should be used in an optimum loading because higher loading increases the internal resistance of the MPL and limits porosity (number and diameter) thus causing mass transfer limitations inside the electrode [20, 58, 63, 64]. All these secondary effects in turn reduces the efficiency of fuel cells [20, 55, 58–60, 62].

To improve fuel cell efficiency, internal resistance of the cell should be reduced. Internal resistance of MPL is dependent on the MPL thickness and the internal resistance of a gas diffusion electrode decreases when the MPL thickness is reduced [20, 54, 55, 59, 60]. However, excessive reduction of MPL or GDL thickness will cause faulty water management and corrosion of catalyst layer [18, 25, 53, 65, 66]. Hence, an optimum thickness of the MPL or GDL should be used.

The catalyst layer forms the most active part of a fuel cell. It should have maximum active surface area with minimum catalyst loading. Platinum is the most commonly and efficiently used catalyst material in a fuel cell; however, its high cost and instability at alkaline pH limits its extensive use. Other than high reactive surface area, catalyst layer should have proper water management to prevent blockage or dryness of the membrane.

Hydrophobic polytetrafluoroethylene (PTFE) was first developed by Union Carbide in the year 1960, which was used as binder and GDL in membrane electrode assembly of a fuel cell [26, 61, 67, 68]. Later on, PTFE was substituted by hydrophilic perfluorosulphonic acid. PTFE or perfluorosulphonic acid is used to enhance contact between electrolyte and catalyst layer and optimize water retention inside the MEA. A membrane electrode assembly should have efficient proton transfer ability, transporting protons from anode through the membrane to the cathode for oxygen reduction. This property can be achieved by using Nafion ionomer or other proton exchanging compounds in the catalyst layers.

Use of proton conducting compounds improves charge transfer efficiency thereby reducing the need of higher catalyst loading; hence, reducing cost of fuel cells [13, 20, 26, 61, 69–72].

Recent research has mainly focused on developing cheaper and effective catalysts. Binary or tertiary alloys or metal oxides are majorly used recently as a substitute for platinum [26, 62, 64, 70, 73–83]. Oxygen reduction reaction can be conducted via two different pathways (a) two electron pathway forming water, (b) four electron pathway producing hydrogen peroxide. The two electron pathway is more desirable because prolonged exposure to hydrogen peroxide reduces durability and structural integrity of the membrane [83, 84]. Hence, while choosing a catalyst, it is important to know the mechanism of the reaction it supports.

Solid polymer electrolyte or proton exchange membrane (PEM) is important in fuel cell as it allows only the proton to be transported through it and not the electrons [15, 38, 85–89]. Nafion 117 is the most commonly used proton exchange membrane, which is a polyfluorinated polymer with sulfonic groups as proton donors/acceptors. High chemical and thermal stability (can be used for short time at 140 °C), along with good ionic conductivity makes Nafion such a popular choice for PEM. Presence of sulfonic group that has an ionic bond between oxygen and proton in the Nafion is responsible for its ionic conductivity. Water can penetrate into the space between the polymeric chains thus giving the protons a high mobility. N117 membrane is also able to conduct ions in alkaline media by exchanging the protons present in the sulfonic group with sodium ions which are mobile in the alkaline fuel cells. Main characteristics of N117 membrane are presented in Table 1 [90].

DuPont is the major producer of these Nafion membranes, which should be given a pretreatment as mentioned below before use in a PEMFC:

1. treatment with 3% H_2O_2 for at least 30 min in order to eliminate all possible organic residues from inside of the membrane;
2. cleaning of the treated membrane with distilled water;
3. treatment with 1 M H_2SO_4 solution for at least 1 h, in order to obtain the membrane in "H" form, when protons are linked onto the SO_3^- groups;
4. cleaning with distilled water.

To use the membrane in alkaline media, the protons in the sulphonic group in the membrane should be replaced by sodium ions prior to use by boiling the N117 membrane for at least 1 h in 0.5 M NaOH solution [91, 92]. While using in a fuel

Table 1 Characteristics of Nafion 117

Main polymeric chain	Perfluorosulfonic polymer (PFSA polymer acid)
Active group	H^+
Thickness (μm)	183
Conductivity (S cm^{-2})	0–10
Acidity (meq g^{-1})	>90
Maximum working temperature (°C)	140
Working pH	2–11

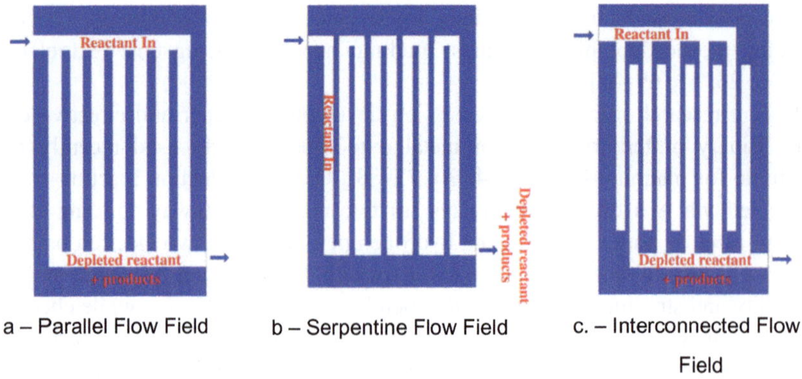

Fig. 4 Usual flow-fields used in PEMFC construction

cell, it is also possible to replace the protons with Na by recirculation of alkaline solution in the fuel cell for some time.

To achieve interconnected or stacked fuel cell, proper design and development of gas diffusion electrode is necessary. Reactants are supplied at the GDL from which it reaches the active catalyst sites by diffusion through GDL and MPL. Reactants can be supplied by using flow-fields technique, which also removes reaction products and ensures electrical contact, in order to collect and transport electrons into the external circuit [93, 94]. Design and construction of flow-fields have been modified from the day of inception when simple parallel channels drilled into graphite plates were used [64, 88, 94]. Later on the simple parallel channels were modified into serpentine and interconnected structures. Figure 4 shows a schematic representation of the different types of flow-fields.

When the reactants are transported from the inlet to the outlet of the flow-fields, there is a considerable pressure drop, hence the catalyst layer near the outlet is not optimally utilized, thus necessitating improved and modified design of flow-fields to reduce pressure drop. Rostami et al. [95] demonstrated that pressure drop observed along the flow-field channel is accompanied by the appearance of a reverse pressure drop responsible for more losses inside the channels [96–98]. Friction between reactant molecules and the walls of the flow-field, or with the internal channel of gas diffusive electrodes cause pressure drop [1, 4, 88, 99]. Inefficient transfer of water produced in the active catalyst sites causes blockage of channels, reactants are diverted to the neighboring channels, which also contribute to internal pressure loss.

When the cell is flooded, water blocks the passage of gases through the flow channels; thus causing mass transport limitation which in turn reduces fuel cell efficiency [1, 4, 100, 101]. Barbir et al. used pressure drop as a diagnostic tool for fuel cell flooding, co-relating it with presence of liquid water inside the cathode channels [99, 102]. With time, design of flow-fields are being further modified to more complex structures, reducing pressure drops to improve performance of fuel cells [88, 99, 103–105].

4 PEMFC Types, Electrode Reactions, and Cell Potential

4.1 H_2/O_2 PEMFC

This particular type of fuel cell was developed in 1963 by General Electric for the space mission Gemini. This cell was fueled with pure hydrogen and pure oxygen, and delivered a maximum power of 1 kW [34]. The following reactions take place at the electrodes:

$$\begin{array}{lll} \text{Anode}\,(-) & H_2 \rightarrow 2\,H^+ + 2e^- & E_a^0 = 0.\,\text{V} \\ \text{Cathode}\,(+) & 1/2\,O_2 + 2H^+ + 2e^- \rightarrow H_2O & E_c^0 = 1.229\,\text{V} \\ \text{Overall reaction:} & H_2 + 1/2\,O_2 \rightarrow H_2O & U^0 = 1.22\,\text{V} \end{array}$$

4.2 Direct Methanol Fuel Cells (DMFC)

Direct methanol fuel cells are a subset of polymer electrolyte fuel cells which are able to directly convert the chemical energy stored in methanol to electrical energy. Methanol can be easily transported using similar network like gasoline transport, thus making DMFC advantageous. As compared to PEMFC fueled with hydrogen, DMFC gives a lower power output and lower efficiency due to slow reaction kinetics of methanol oxidation and methanol crossover from the anodic chamber to cathode [106, 107]. Crossover of methanol from anode to cathode becomes the limiting factor when methanol concentration is above 2 M, and it is oxidized at the cathode leading to decreased performance of the cell [108]. Proper care should be taken while handling DMFC as methanol vapors are toxic [108, 109]. The following reactions take place at the anode and cathode of a DMFC:

$$\begin{array}{lll} \text{Anode}\,(-) & CH_3OH + H_2O \rightarrow CO_2 + 6H^+ + 6e^- & E_a^0 = 0.02\,\text{V} \\ \text{Cathode}\,(+) & 3/2\,O_2 + 6H^+ + 6e^- \rightarrow 3H_2O & E_c^0 = 1.23\,\text{V} \\ \text{Overall reaction:} & CH_3OH + 3/2O_2 \rightarrow CO_2 + 2H_2O & E_{cell} = 1.21\,\text{V} \end{array}$$

4.3 Direct Ethanol Fuel Cells (DEFC)

As DMFC, the direct ethanol fuel cells are also a subset of PEMFC which are able to directly convert ethanol into energy. Usage of ethanol as fuel is attractive because it is a nontoxic compound, which can be supplied by using the infrastructure used for petrol products distribution and can also be produced from food and nonedible crop residues in an environment friendly way [40, 71, 110, 111]. Low crossover of ethanol as compared to methanol, because of bigger molecule size (smaller permeability) is another advantage of DEFC [26, 110, 111]. Though ethanol oxidation is slow reaction, but reaction kinetics of both ethanol oxidation at anode and oxygen reduction at cathode can be improved by replacing acidic anodic media with an alkaline one [110]. This improved kinetics of oxygen reduction into the alkaline media was experimentally demonstrated by An et al. [112], Modestov et al. [113, 114], and An et al. [115] by comparative tests using classical acidic fuel cells and also alkaline one. An et al. [39, 116] suggested a direct ethanol fuel cell combining acidic and alkaline fuel cells, with alkaline anodic media and an acidic cathodic media.

In classical system, the following reactions take place in acidic media at the electrode in a DEFC:

$$\begin{array}{lll} \text{Anode } (-) & CH_3CH_2OH + 3H_2O \rightarrow 2CO_2 + 12H^+ + 12e & E_a^0 = 0.085\,V \\ \text{Cathode } (+) & 3O_2 + 12H^+ + 12e^- \rightarrow 6H_2O & E_c^0 = 1.229\,V \\ \text{Overall reaction:} & CH_3CH_2OH + 3O_2 \rightarrow 2CO_2 + 3H_2O & E_{cell} = 1.144\,V \end{array}$$

4.4 Direct Formic Acid Fuel Cells (DFAFC)

Direct formic acid fuel cells are becoming more and more attractive because of their high power output, which is a result of higher open circuit potential, correlated with faster kinetics of formic acid oxidation [117–119] and also with low crossover rate of formic acid [118, 120] than direct methanol fuel cells. Due to smaller number of electrons exchanged during formic acid oxidation, crossover of formic acid is only 50% as compared to methanol [108]. In case of usage of formic acid as fuel into a classical PEMFC, the semi-reactions are as follows [108]:

$$\begin{array}{lll} \text{Anode } (-) & CHOOH \rightarrow CO_2 + 2H^+ + 2e^- & E_a^0 = 0.25\,V \\ \text{Cathode } (+) & 1/2\,O_2 + 2\,H^+ + 2e^- \rightarrow H_2O & E_c^0 = 1.229\,V \\ \text{Overall reaction:} & CHOOH + 1/2O_2 \rightarrow CO_2 + H_2O & E_{cell} = 1.479\,V \end{array}$$

4.5 Direct Borohydride Fuel Cells (DBFCs)

DBFCs were developed in 1960 by Snyder and Irving [48]. Due to high energy density, sodium borohydride represents a potential fuel for direct electro-oxidation in DBFC [121]. Nontoxic by-products obtained from anodic oxidation of BH_4^- are major advantage of DBFC [45, 48]. Borohydride being a stable solid compound, it is easy to store and transport [79]. In addition, during electrochemical oxidation of borohydride CO_2 is not produced and also the catalytic activity is not affected because no CO intermediates are produced during the oxidation process [45, 79]. Borohydride, however, is not stable in acidic media and hence during operation of DBFC, a strong alkaline media is required to be used [45, 79, 121]. Among the different types of PEMFC, direct borohydride fuel cell generates maximum potential. When oxygen is used as oxidant, the half-cell reactions in a DBFC are as follows [121]:

Anode (−)	$BH_4^- + 8HO^- \rightarrow BO_2^- + 6H_2O + 8e^-$	$E_a^0 = -1.24\,V$
Cathode (+)	$2O_2 + 4H_2O + 8e^- \rightarrow H_2O$	$E_c^0 = 0.40\,V$
Overall reaction:	$BH_4^- + 2O_2 \rightarrow BO_2^- + H_2O$	$E_{cell} = 1.64\,V$

5 Fuel Cell Thermodynamics

Like any other energy converting device, fuel cell follows the second law of thermodynamics [1, 4, 6]. The electrode reactions are taking place at interface of the electrode–electrolyte, where three different phases are in contact: electrolyte, electrode (catalyst layer), and reactant (fuel or oxidant). The cathodic and anodic reactions take place simultaneously during power production [1, 2, 4–6, 9, 11, 38, 87]. To have a better understanding of fuel cell functioning, it is important to consider both thermodynamic and kinetic aspects [1, 4, 6, 12].

Let us consider that the fuel cell overall reaction is:

$$aA + bB \rightarrow cC + dD$$

For the above reaction, Gibbs free energy is calculated by using following Eq. 1.

$$\Delta G = c\mu_c + d\mu_d - a\mu_a - b\mu_b \tag{1}$$

where μ_c, μ_d, chemical potentials of reaction products, and μ_a, μ_b, chemical potentials of reactants. Chemical potential can be calculated as (Eq. 2):

$$\mu_i = \mu_i^0 + \mathrm{RT}\ \ln a_i \tag{2}$$

by replacing the expression of chemical potential into the expression of free Gibbs energy variation, we obtain Eq. 3:

$$\begin{aligned}\Delta G = {} & c(\mu_c^0 + \mathrm{RT}\ \ln a_C) + d(\mu_d^0 + \mathrm{RT}\ \ln a_d) - a(\mu_a^0 + \mathrm{RT}\ \ln a_a) \\ & - b(\mu_b^0 + \mathrm{RT}\ \ln a_b)\end{aligned} \tag{3}$$

or, it could be expressed as Eq. 4:

$$\Delta G = \Delta G^0 + \mathrm{RT}\ \ln \frac{a_c^c a_d^d}{a_a^a a_b^b} \tag{4}$$

But, the quantity of energy furnished by an electrochemical device into the external circuit in isotherm–isobar conditions is represented by Gibbs free energy variation of cell reaction [4, 6, 11, 15, 122, 123]:

$$W = \Delta G = -nFE \tag{5}$$

$$\rightarrow E = -\frac{1}{nF}\Delta G \tag{6}$$

where:

E electromotive force of the cell
n number of electrons involved into the cell reaction
F Faraday's Number
ΔG Gibbs free energy variation

From Eqs. 4 and 6, the following relation describing fuel cell electromotive force can be obtained:

$$E = -\frac{1}{nF}\Delta G^0 - \frac{\mathrm{RT}}{nF}\ln\frac{a_c^c a_d^d}{a_a^a a_b^b} \tag{7}$$

$$E = E^0 + \frac{\mathrm{RT}}{nF}\ln\frac{a_a^a a_b^b}{a_c^c a_d^d} \tag{8}$$

Let us evaluate the theoretical potential of a classical H_2/O_2 fuel cell. From Eq. 6, the maximum electrical output of a cell depends on the variation of Gibbs free energy, number of electrons involved in the reaction, and Faraday's Number [1, 4, 6, 8, 11, 15, 122]. In this particular case, considering the overall cell reaction as $H_2 + \frac{1}{2}O_2 \rightarrow H_2O$, the Gibbs free energy can be expressed as (Eq. 9):

$$\Delta G = \Delta H - T\Delta S \tag{9}$$

where ΔH-change in enthalpy during reaction, and ΔS-change in entropy during reaction. Change in enthalpy can be expressed as [124]:

$$\Delta H = \int_{H_i}^{H_f} \mathrm{d}H = H_F - H_i \tag{10}$$

Suppose reaction takes place at 25 °C, at this temperature the standard heat of formation of liquid water is −286.02 kJ mol^{-1}, and that of H_2 and O_2 is zero [4, 124].

$$\Delta H^0 = H^0_{H_2O} - H^0_{H_2} - H^0_{O_2} = -286.02 - 0 - 0 = -286\,\text{kJ mol}^{-1} \tag{11}$$

Similarly, change in entropy for water formation can be estimated from Eq. 12 [4, 124]:

$$\begin{aligned}\Delta S^0 &= S^0_{H_2O} - S^0_{H_2} - \frac{1}{2}S^0_{O_2} = 0.06996 - 0.13066 - \frac{0.20517}{2} \\ &= -0.163285\,\text{kJ mol}^{-1}\text{K}^{-1}\end{aligned} \tag{12}$$

From Eqs. 11 and 12, the Gibbs free energy for water formation can be calculated [4, 124]:

$$\Delta G^0 = -286 - 298.15 \times (-0.163285) \tag{13}$$

$$\rightarrow \Delta G = -237.32\,\text{kJ mol}^{-1} \tag{14}$$

From Eq. 6, the maximum electromotive force of H_2/O_2 fuel cell can be estimated:

$$E = -\frac{\Delta G^0}{nF} = \frac{237.32 \times 10^3}{2 \times 96485} = 1.2298\,\text{V} \tag{15}$$

When the reaction products are in gaseous state, the cell voltage would be different. In this case, at 25 °C, heat of formation of water is $H^0_{H_2O} = -241.818$ kJ mol^{-1}, and entropy is $S^0_{H_2O} = 0.18882$ kJ mol^{-1} K^{-1} [124]. By using these values, changes in enthalpy and entropy can be calculated as:

$$\Delta H = -241.818\,\text{kJ mol}^{-1} \tag{16}$$

and

$$\Delta S = 0.044425\,\text{kJ mol}^{-1}\text{K}^{-1} \tag{17}$$

That means the Gibbs free energy is more positive: $\Delta G = -228.57$ kJ mol^{-1}. By using the new value of ΔG, it is possible to determine the maximum cell voltage when the water is in vapor form:

$$E = -\frac{\Delta G}{nF} = \frac{228.57 \times 10^3}{2 \times 96485} = 1.1845\,\text{V} \tag{18}$$

When the reaction product is in vapor form, the maximum cell voltage that can be obtained is lower than that with liquid reaction product. This difference between the two maximum cell voltages is associated with the Gibbs free energy consumed for the vaporization of water molecules [4, 6, 122, 124].

Since the cell voltage depends on Gibbs free energy, hence, all the parameters namely temperature, pressure and reactant concentration; influencing Gibbs free energy, affects cell voltage.

5.1 Effect of Temperature

Gibbs free energy is given by the difference between ΔH and $T\Delta S$. Enthalpy of a substance increases with increasing temperature. Change in enthalpy with change in temperature at constant pressure can be calculated as [124, 125]:

$$\Delta H = \int_{T_i}^{T_f} C_p \mathrm{d}T \tag{19}$$

Similarly, change in entropy can be expressed as [124, 125]:

$$\Delta S = \int_{T_i}^{T_f} \frac{C_p}{T} \mathrm{d}T. \tag{20}$$

Substituting Eqs. 19 and 20 in Eq. 9 we get

$$\Delta G = \int_{T_i}^{T_f} C_p \mathrm{d}T - T \int_{T_i}^{T_f} \frac{C_p}{T} \mathrm{d}T \tag{21}$$

Heat capacity itself is not a constant term and depends on temperature: $C_p = f(T)$. An empirical expression for molar heat capacity is given by Eq. 22 [124, 125]:

$$C_p = a + bT + c\frac{1}{T^2} \tag{22}$$

where a, b, and c are empirical constants which are independent of temperature. By substituting Eq. 21 with Eqs. 19 and 20, we obtain the temperature dependence relationship among ΔG, ΔH, and ΔS.

Using the constants available in Atkins' Physical Chemistry [125], the values of ΔH, ΔS, and ΔG for different temperatures can be calculated. From the values of Gibbs free energy, enthalpy and entropy at different temperatures [124] the maximum theoretical fuel cell performance at a particular temperature can be estimated [1, 6, 124] as shown in Table 2.

Table 2 Estimated maximum fuel cell performance

Temperature (K)	E_{cell} (V)
298.15	1.229
333.15	1.200
353.15	1.184
373.15	1.167

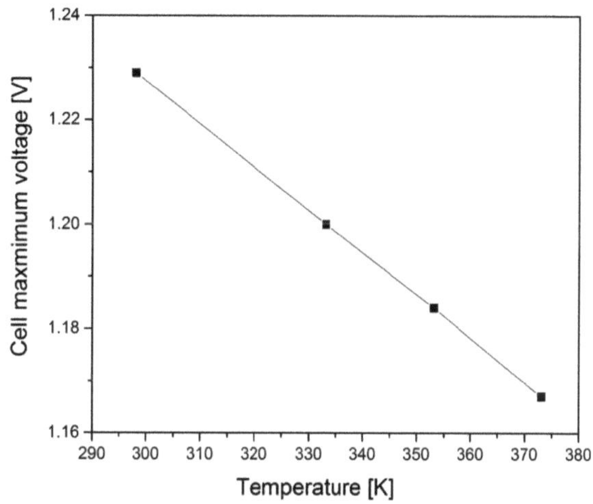

Fig. 5 Ideal potential of H_2/O_2 fuel cell as function of temperature

From Eq. 21 and Table 1, a strong dependence between the maximum voltage that can be obtained in a fuel cell and temperature of operation is evident. With increase of temperature, maximum cell voltage decreases as shown in Fig. 5.

5.2 *Effect of Pressure*

In order to evaluate the effect of pressure, the relationship of Gibbs free energy with pressure at constant temperature is important [4]. It is assumed that all gases involved are ideal gases. Applying the first and second laws of thermodynamics for a reversible reaction, the relationship between entropy and enthalpy can be obtained [124, 125]:

$$TdS = dH - VdP \tag{23}$$

Enthalpy is an extensive quantity which depends on temperature and pressure:

$$dH = \left(\frac{\partial H}{\partial T}\right)_P dT + \left(\frac{\partial H}{\partial P}\right)_T dP \tag{24}$$

By substituting Eq. 24 in Eq. 23:

$$TdS = \left(\frac{\partial H}{\partial T}\right)_P dT + \left(\frac{\partial H}{\partial P}\right)_T dP - VdP \tag{25}$$

Substituting $C_P = \left(\frac{\partial H}{\partial T}\right)_P$:

$$dS = \frac{1}{T} C_P dT + \frac{1}{T}\left[\left(\frac{\partial H}{\partial P}\right)_T - V\right] dP \tag{26}$$

By definition of prefect gases, enthalpy is only a function of temperature, that is $\left(\frac{\partial H}{\partial P}\right)_T = 0$, hence Eq. 26 becomes:

$$dS = \frac{1}{T} C_P dT - \frac{1}{T} VdP \tag{27}$$

using the perfect gase state:

$$dS = \frac{C_P}{T} dT - R\frac{dP}{P} \tag{28}$$

$$\Delta S = C_p \ln\frac{T}{T_0} - R \ln\frac{P}{P_0} \tag{29}$$

Similarly change in enthalpy can be obtained as:

$$\Delta H = C_P(T - T_0) \tag{30}$$

By substituting Eqs. 29 and 30 in Eq. 9:

$$\Delta G = \left(C_P - S^0\right)(T - T_0) - C_p T \ln\frac{T}{T_0} + \text{RT} \ln\frac{P}{P_0} \tag{31}$$

From the above equations, the variation of Gibbs free energy with pressure can be obtained as [4]:

$$\Delta G = \left[\sum_i n_i C_{p,i} - \sum_i n_i S_{i,0}\right](T - T_0) - \left(\sum_i n_i C_{p,i}\right) T \ln\frac{T}{T_0} + \text{RT} \sum_i{}' n_i \ln\frac{p_i}{p_{i,0}}, \tag{32}$$

where G—Gibbs free energy, C_p—molar heat at constant pressure, S—entropy, P—pressure, and T—temperature.

By substituting ΔG value in Eq. 6, Nernst Equation can be obtained [4, 6]:

$$E = E^0 + \frac{\mathrm{RT}}{nF} \ln \frac{P_{H_2} P_{O_2}^{1/2}}{P_{H_2O}} \tag{33}$$

When water is present in liquid form, it is necessary to replace the partial pressure of water with 1. In this case, we can conclude that the increase of reactant pressure leads to an increase of fuel cell potential.

5.3 *Effect of Concentration of Reactant*

Concentration of reactants also affects cell voltage. For example in a H_2/O_2 fuel cell, if air is used instead of pure oxygen, there will be a drop in performance. The partial pressure of a gas is proportional to its concentration in the gas mixture. Since, partial pressure of oxygen in air is only 0.21 of that of pure oxygen which results in a decreased cell voltage when air is used instead of pure oxygen.

6 Fuel Cell Efficiency

Thermal efficiency of any conversion device is defined as the ratio between the quantity of useful energy produced and the energy input. This energy input can be evaluated as the change in enthalpy between the products and reactants [1, 4, 6, 8, 122]. Classical conversion devices use thermo-mechanical conversion systems, in which the maximum efficiency is limited by the initial and final temperatures of the fluid [4, 6]. For electrochemical devices which convert chemical energy directly into electrical energy, the produced electrical energy can be calculated directly from the Gibbs free energy:

$$\eta_{\max}[\%] = \frac{\Delta G}{\Delta H} \times 100 \tag{33}$$

Under standard conditions (298 K temperature and 1 atmospheric pressure), thermal energy for overall reaction in a H_2/O_2 fuel cell is $\Delta H = -286\,\mathrm{kJ\ mol^{-1}}$, and the free Gibbs energy is $\Delta G = -237.32\,\mathrm{kJ\ mol^{-1}}$. By substituting these values in Eq. 33, the maximum cell efficiency can be obtained under standard conditions:

$$\eta_{\max}[\%] = \frac{-237.32\,\mathrm{kJ\ mol^{-1}}}{-286\,\mathrm{kJ\ mol^{-1}}} \times 100 = 82.98\% \tag{34}$$

6.1 Losses in Actual System

The actual cell potential obtained during operation of a fuel cell is less than the ideal theoretical potential at standard conditions. This is due to polarization taking place at the electrodes. These polarizations are known as overpotential or overvoltage [4, 6], and they are dependent on the fuel cell load, which are classified into three types:

- activation overpotential;
- ohmic polarization losses;
- mass transport overpotential.

6.2 Activation Overpotential

It appears at lower values of current density, as a consequence of low electrode kinetics. Such overpotential appears when the electrochemical reaction speed is controlled by charge transfer. Activation overpotential can also be influenced by the nature of the electrode, adsorption of the reactants on electrode surface, and the slow reaction kinetics between adsorbed intermediates. Activation overpotential can be calculated by Tafel equation, obtained as a particularization of Butler–Volmer Equation [4].

For the anodic reaction activation overpotential can be expressed as:

$$\eta_{\text{act}}^{a} = \frac{\text{RT}}{\alpha zF} \ln \frac{i}{i_0} \tag{35}$$

Similarly for the cathodic reaction activation overpotential can be expressed as:

$$\eta_{\text{act}}^{c} = \frac{\text{RT}}{(1-\alpha)zF} \ln \frac{i}{i_0} \tag{36}$$

where α—anodic electron transfer coefficient, $1-\alpha$—cathodic electron transfer coefficient, i_0—exchange current density.

6.3 Ohmic Polarization Losses

These losses occur due to the resistance encountered by the ionic flow while passing through the electrolyte and the resistance encountered by the electrons when they pass through cell electrodes. The ohmic losses are directly related to the internal resistance of the electrolyte, and can be reduced by increasing the ionic conduction of the polymeric membrane, as well as also by reducing the distance between anode and cathode (reducing the thickness of the membrane). Ohmic losses also depend on the internal resistance of the electrodes. Experimental data confirm that the electrolyte and the electrodes obey Ohm's Law (Eq. 37):

$$\eta_{\text{ohm}} = IR \tag{37}$$

6.4 *Mass Transport Overpotential*

It appears because the reaction actually takes place at the electrode–electrolyte interface. Consumption of the reactants reduces the availability of reacting species at the electrode surface. At the same time, presence of reaction products at the interface reduces the concentration of active reaction species. Hence, a concentration gradient is formed between the electrode surface and bulk liquid which causes mass transport overpotential.

Mass transport overpotential is caused because of:

- slow diffusion of gaseous phase into the active reaction sites;
- slow reactants and products diffusion through electrolyte to and from the place where the electrochemical reaction occurs;
- slow dissolution of reactants and reaction products into the electrolyte.

As a consequence of the concentration gradient, a sharp variation of the cell potential appears under practical conditions especially at high current densities, low fuel concentration, and low oxidant concentration, representing a significant loss of the cell potential. At practical current densities observed in fuel cells, the reactants are supplied at electrodes by diffusion; hence, the rate of mass transport can be described by Fick's Law of diffusion (Eq. 38):

$$\eta_{\text{diff}} = \frac{\text{RT}}{zF}\ln\left(1 - \frac{j}{j_{\text{lim}}}\right) \tag{38}$$

where j_{lim}—diffusion limiting current.

Activation and mass transport overpotentials can appear at both anode and cathode:

$$\eta_{\text{anode}} = \eta_{\text{act,anod}} + \eta_{\text{conc,anod}} \tag{39}$$

$$\eta_{\text{cathode}} = \eta_{\text{act,cathode}} + \eta_{\text{conc,cathode}} \tag{40}$$

Hence, the anode potential becomes more positive and the cathode potential becomes more negative as compared to ideal conditions:

$$E_{\text{anod}} = E_{\text{anod}} + \eta_{\text{anod}} \tag{41}$$

$$E_{\text{cathode}} = E_{\text{cathode}} - |\eta_{\text{cathode}}| \tag{42}$$

From the above equations overall cell potential can be estimated as:

$$E_{\text{cel}} = E_{\text{cathode}} - |\eta_{\text{cathode}}| - (E_{\text{anod}} + \eta_{\text{anod}}) - iR \tag{43}$$

$$E_{\text{cel},I \neq 0} = E_{\text{rev}} - \sum |\eta| - iR \tag{44}$$

When a cell is delivering energy to the external circuit; increase of current density leads to a decrease of cell voltage, due to activation, concentration and ohmic polarization losses. Future research must focus on minimization of all these internal losses in order to get a closer value of $E_{\text{cel},I \neq 0}$ to the cell reversible potential.

7 Conclusion

Fuel cells represent a viable option for clean energy production by direct conversion of chemical energy into electrical energy. Maximum theoretical cell voltage obtained for direct conversion of hydrogen is 1.2298 V with a cell efficiency of 82.98%. Due to several losses the real cell performance obtained during system operation is less than the ideal theoretical cell performance at standard conditions. The cell voltage is dependent on Gibbs free energy and in turn on: temperature, pressure, and partial pressure of reactants. Future research should deal with minimization of all internal losses in order to increase the real cell potential to a closer value to the theoretical cell potential.

References

1. Barbir F (2005) PEM fuel cells: theory and practice. Elsevier Academic Press
2. Energy USDo (2013) Fuel cell technologies overview (cited 10 Mar 2016)
3. History NMoA (cited 2016 15.04.2016); Available from: http://americanhistory.si.edu/fuelcells/origins/origins.htm
4. Oniciu L (1976) Fuel cells. Abacus Press
5. Scott K et al (2012) Biological and microbial fuel cells. In: Sayigh A (ed) Comprehensive renewable energy. Elsevier, Amsterdam, pp 257–280
6. EG&G Technical Services I (2004) Fuel cell handbook, 7th edn. U.S. Department of Energy, Morgantown, West Virginia 26507 - 0880
7. Mekhilef S, Saidur R, Safari A (2012) Comparative study of different fuel cell technologies. Renew Sustain Energy Rev 16(1):981–989
8. Kirubakaran A, Jain S, Nema RK (2009) A review on fuel cell technologies and power electronic interface. Renew Sustain Energy Rev 13(9):2430–2440

9. Merle G, Wessling M, Nijmeijer K (2011) Anion exchange membranes for alkaline fuel cells: a review. J Membr Sci 377(1–2):1–35
10. Liu Y et al (2016) A review of high-temperature polymer electrolyte membrane fuel-cell (HT-PEMFC)-based auxiliary power units for diesel-powered road vehicles. J Power Sources 311:91–102
11. Lepiller C (2016) Fuel cell basics. Available from: http://www.pragma-industries.com/wp-content/themes/default/images/fuel_cell_basics.pdf
12. Cao D, Sun Y, Wang G (2007) Direct carbon fuel cell: fundamentals and recent developments. J Power Sources 167(2):250–257
13. Duteanu N et al (2007) A parametric study of a platinum ruthenium anode in a direct borohydride fuel cell. J Appl Electrochem 37(9):1085–1091
14. Scott K et al (2008) Performance of a direct methanol alkaline membrane fuel cell. J Power Sources 175(1):452–457
15. Sharaf OZ, Orhan MF (2014) An overview of fuel cell technology: Fundamentals and applications. Renew Sustain Energy Rev 32:810–853
16. Srinivasan S et al (1991) Proceedings of the third space electrochemical research and technology conference. High energy efficiency and high power density proton exchange membrane fuel cells–electrode kinetics and mass transport. J Power Sources 36(3):299–320
17. Straßer K (1990) PEM-fuel cells: state of the art and development possibilities. Ber Bunsenges Phys Chem 94(9):1000–1005
18. Gurau V et al (2007) Characterization of transport properties in gas diffusion layers for proton exchange membrane fuel cells: 2. Absolute permeability. J Power Sources 165(2): 793–802
19. Prater KB (1992) Proceedings of the second Grove fuel cell symposium. Progress in fuel cell commercialisation. Solid polymer fuel cell developments at Ballard. J Power Sources 37(1): 181–188
20. Wilson MS, Gottesfeld S (1992) Thin-film catalyst layers for polymer electrolyte fuel cell electrodes. J Appl Electrochem 22(1):1–7
21. Wasmus S, Küver A (1999) Methanol oxidation and direct methanol fuel cells: a selective review. J Electroanal Chem 461(1–2):14–31
22. Biyikoglu A (2005) Review of proton exchange membrane fuel cell models. Int J Hydrogen Energy 30(11):1181–1212
23. Strasser K (1992) Proceedings of the second Grove fuel cell symposium. Progress in fuel cell commercialisation. Mobile fuel cell development at Siemens. J Power Sources 37(1):209–219
24. Wang ZH, Wang CY, Chen KS (2001) Two-phase flow and transport in the air cathode of proton exchange membrane fuel cells. J Power Sources 94(1):40–50
25. Gurau V et al (2006) Characterization of transport properties in gas diffusion layers for proton exchange membrane fuel cells: 1. Wettability (internal contact angle to water and surface energy of GDL fibers). J Power Sources 160(2):1156–1162
26. Zhou W et al (2003) Pt based anode catalysts for direct ethanol fuel cells. Appl Catal B 46 (2):273–285
27. Shukla S et al (2015) Analysis of low platinum loading thin polymer electrolyte fuel cell electrodes prepared by inkjet printing. Electrochim Acta 156:289–300
28. Oh H-S et al (2009) Corrosion resistance and sintering effect of carbon supports in polymer electrolyte membrane fuel cells. Electrochim Acta 54(26):6515–6521
29. Alcaide F et al (2009) Pt supported on carbon nanofibers as electrocatalyst for low temperature polymer electrolyte membrane fuel cells. Electrochem Commun 11(5): 1081–1084
30. Guha A et al (2007) Surface-modified carbons as platinum catalyst support for PEM fuel cells. Carbon 45(7):1506–1517
31. Subramanian NP et al (2009) Nitrogen-modified carbon-based catalysts for oxygen reduction reaction in polymer electrolyte membrane fuel cells. J Power Sources 188(1):38–44
32. Calvillo L et al (2009) Effect of the support properties on the preparation and performance of platinum catalysts supported on carbon nanofibers. J Power Sources 192(1):144–150

33. Calvillo L et al (2007) Platinum supported on functionalized ordered mesoporous carbon as electrocatalyst for direct methanol fuel cells. J Power Sources 169(1):59–64
34. Sebastian D et al (2009) Carbon nanofibers as electrocatalyst support for fuel cells: effect of hydrogen on their properties in CH4 decomposition. J Power Sources 192(1):51–56
35. Andersen SM et al (2013) Durability of carbon nanofiber (CNF) & carbon nanotube (CNT) as catalyst support for proton exchange membrane fuel cells, pp 94–101
36. Sebastian D et al (2010) Influence of carbon nanofiber properties as electrocatalyst support on the electrochemical performance for PEM fuel cells. Int J Hydrogen Energy 35(18): 9934–9942
37. Li YH et al (2016) Preparation of platinum catalysts supported on functionalized graphene and the electrocatalytic properties for ethanol oxidation in direct ethanol fuel cell. J Mater Sci Mater Electron 27(6):6208–6215
38. Peighambardoust SJ, Rowshanzamir S, Amjadi M (2010) Review of the proton exchange membranes for fuel cell applications. Int J Hydrogen Energy 35(17):9349–9384
39. An L, Chen R (2016) Direct formate fuel cells: a review. J Power Sources 320:127–139
40. Antolini E, Gonzalez ER (2010) Alkaline direct alcohol fuel cells. J Power Sources 195(11): 3431–3450
41. Rimbu GA, Jackson CL, Scott K (2006) Platinum/carbon/polyaniline based nanocomposites as catalysts for fuel cell technology. J Optoelectron Adv Mater 8(2):611–616
42. Zhao TS, Li YS, Shen SY (2010) Anion-exchange membrane direct ethanol fuel cells: Status and perspective. Front Energy Power Eng Chin 4(4):443–458
43. Zheng Y et al (2016) Platinum nanoparticles on carbon-nanotube support prepared by room-temperature reduction with H_2 in ethylene glycol/water mixed solvent as catalysts for polymer electrolyte membrane fuel cells, 448–453
44. Antolini E (2015) Composite materials for polymer electrolyte membrane microbial fuel cells. Biosens Bioelectron 69:54–70
45. Bayatsarmadi B, Peters A, Talemi P (2016) Catalytic polymeric electrodes for direct borohydride fuel cells. J Power Sources 322:26–30
46. Cheddie D, Munroe N (2005) Review and comparison of approaches to proton exchange membrane fuel cell modeling. J Power Sources 147(1–2):72–84
47. Giddey S et al (2012) A comprehensive review of direct carbon fuel cell technology. Prog Energy Combust Sci 38(3):360–399
48. Indig ME, Snyder RN (1962) Sodium borohydride, an interesting anodic fuel (1). J Electrochem Soc 109(11):1104–1106
49. Iwan A, Malinowski M, Pasciak G (2015) Polymer fuel cell components modified by graphene: Electrodes, electrolytes and bipolar plates. Renew Sustain Energy Rev 49: 954–967
50. Carrette L, Friedrich KA, Stimming U (2001) Fuel cells—fundamentals and applications. Fuel Cells 1(1):5–39
51. Divisek J et al (1998) Components for PEM fuel cell systems using hydrogen and CO containing fuels. Electrochim Acta 43(24):3811–3815
52. Litster S, McLean G (2004) PEM fuel cell electrodes. J Power Sources 130(1–2):61–76
53. Cindrella L et al (2009) Gas diffusion layer for proton exchange membrane fuel cells—a review. J Power Sources 194(1):146–160
54. Oedegaard A et al (2004) Influence of diffusion layer properties on low temperature DMFC. J Power Sources 127(1–2):187–196
55. Neergat M, Shukla AK (2002) Effect of diffusion-layer morphology on the performance of solid-polymer-electrolyte direct methanol fuel cells. J Power Sources 104(2):289–294
56. Feser JP, Prasad AK, Advani SG (2006) Experimental characterization of in-plane permeability of gas diffusion layers. J Power Sources 162(2):1226–1231
57. Song M, Kim HY, Kim K (2014) Effects of hydrophilic/hydrophobic properties of gas flow channels on liquid water transport in a serpentine polymer electrolyte membrane fuel cell. 19714–19721
58. Zhang J et al (2006) High temperature PEM fuel cells. J Power Sources 160(2):872–891

59. Pan YH (2006) Advanced air-breathing direct methanol fuel cells for portable applications. J Power Sources 161(1):282–289
60. Escribano S, Aldebert P (1995) Electrodes for hydrogen/oxygen polymer electrolyte membrane fuel cells. Solid State Ionics 77:318–323
61. Fischer A, Jindra J, Wendt H (1998) Porosity and catalyst utilization of thin layer cathodes in air operated PEM-fuel cells. J Appl Electrochem 28(3):277–282
62. Chandan A et al (2013) High temperature (HT) polymer electrolyte membrane fuel cells (PEMFC)—a review. J Power Sources 231:264–278
63. Escribano S, Aldebert P (1995) Solid state protonic conductors vii electrodes for hydrogen/oxygen polymer electrolyte membrane fuel cells. Solid State Ionics 77:318–323
64. Shao Y et al (2007) Proton exchange membrane fuel cell from low temperature to high temperature: material challenges. J Power Sources 167(2):235–242
65. Liang Y et al (2006) Preparation and characterization of carbon-supported PtRuIr catalyst with excellent CO-tolerant performance for proton-exchange membrane fuel cells. J Catal 238(2):468–476
66. George MG et al (2016) Composition analysis of a polymer electrolyte membrane fuel cell microporous layer using scanning transmission X-ray microscopy and near edge X-ray absorption fine structure analysis. J Power Sources 309:254–259
67. Lobato J et al (2008) Influence of the teflon loading in the gas diffusion layer of PBI-based PEM fuel cells. J Appl Electrochem 38(6):793–802
68. Guo Z, Faghri A (2006) Development of planar air breathing direct methanol fuel cell stacks. J Power Sources 160(2):1183–1194
69. Bose S et al (2011) Polymer membranes for high temperature proton exchange membrane fuel cell: Recent advances and challenges. Prog Polym Sci 36(6):813–843
70. Fathirad F, Afzali D, Mostafavi A (2016) Pd-Zn nanoalloys supported on Vulcan XC-72R carbon as anode catalysts for oxidation process in formic acid fuel cell
71. Chu YH, Shul YG (2010) Combinatorial investigation of Pt-Ru-Sn alloys as an anode electrocatalysts for direct alcohol fuel cells. Int J Hydrogen Energy 35(20):11261–11270
72. Kim HS et al (2016) Platinum catalysts protected by N-doped carbon for highly efficient and durable polymer-electrolyte membrane fuel cells. Electrochim Acta (in press)
73. Jongsomjit S, Prapainainar P, Sombatmankhong K (2016) Synthesis and characterisation of Pd-Ni-Sn electrocatalyst for use in direct ethanol fuel cells. Solid State Ionics 288:147–153
74. Arashi T et al (2014) Nb-doped TiO_2 cathode catalysts for oxygen reduction reaction of polymer electrolyte fuel cells. Catal Today Catal Mater Catal Low Carbon Technol 233:181–186
75. Han S, Chae GS, Lee JS (2016) Enhanced activity of carbon-supported PdCo electrocatalysts toward electrooxidation of ethanol in alkaline electrolytes. Korean J Chem Eng 33 (6):1799–1804
76. Kumar A, Ramani VK (2013) RuO_2-SiO_2 mixed oxides as corrosion-resistant catalyst supports for polymer electrolyte fuel cells. Appl Catal B 138–139:43–50
77. Patru A et al (2016) Pt/IrO_2–TiO_2 cathode catalyst for low temperature polymer electrolyte fuel cell—application in MEAs, performance and stability issues. Catal Today 262:161–169
78. Uehara N et al (2015) Tantalum oxide-based electrocatalysts made from oxy-tantalum phthalocyanines as non-platinum cathodes for polymer electrolyte fuel cells. Ubiquitus Electrochem 146–153
79. Yi L et al (2015) Enhanced activity of Au-Fe/C anodic electrocatalyst for direct borohydride-hydrogen peroxide fuel cell. J Power Sources 285:325–333
80. Kil KC et al (2014) The use of MWCNT to enhance oxygen reduction reaction and adhesion strength between catalyst layer and gas diffusion layer in polymer electrolyte membrane fuel cell. Int J Hydrogen Energy 39:17481–17486
81. Chen Z et al (2011) A review on non-precious metal electrocatalysts for PEM fuel cells. Energy Environ Sci 4(9):3167–3192
82. Reshetenko TV, Kim H-T, Kweon H-J (2007) Cathode structure optimization for air-breathing DMFC by application of pore-forming agents. J Power Sources 171(2):433–440

83. Wang B (2005) Recent development of non-platinum catalysts for oxygen reduction reaction. J Power Sources 152:1–15
84. Kinumoto T et al (2006) Durability of perfluorinated ionomer membrane against hydrogen peroxide. J Power Sources 158(2):1222–1228
85. Smitha B, Sridhar S, Khan AA (2005) Solid polymer electrolyte membranes for fuel cell applications—a review. J Membr Sci 259(1–2):10–26
86. Prater KB (1994) Polymer electrolyte fuel cells: a review of recent developments. J Power Sources 51(1):129–144
87. Sousa R Jr, Gonzalez ER (2005) Mathematical modeling of polymer electrolyte fuel cells. J Power Sources 147(1–2):32–45
88. Owejan JP et al (2007) Effects of flow field and diffusion layer properties on water accumulation in a PEM fuel cell. Int J Hydrogen Energy 32(17):4489–4502
89. Neburchilov V et al (2007) A review of polymer electrolyte membranes for direct methanol fuel cells. J Power Sources 169(2):221–238
90. DuPont (2016) Nafion—product bulletin P-12 (cited 10 May 2016). Available from: https://www.chemours.com/Nafion/en_US/assets/downloads/nafion-extrusion-cast-membranes-product-information.pdf
91. Yu EH, Scott K (2004) Development of direct methanol alkaline fuel cells using anion exchange membranes. J Power Sources 137(2):248–256
92. Yu EH, Scott K (2005) Direct methanol alkaline fuel cells with catalysed anion exchange membrane electrodes. J Appl Electrochem 35(1):91–96
93. Lim BH et al (2016) Effects of flow field design on water management and reactant distribution in PEMFC: a review. Ionics 22(3):301–316
94. Nguyen TV (1996) A gas distributor design for proton—exchange—membrane fuel cells. J Electrochem Soc 143(5):L103–L105
95. Rostami L, Mohamad Gholy Nejad P, Vatani A (2016) A numerical investigation of serpentine flow channel with different bend sizes in polymer electrolyte membrane fuel cells. Energy 97:400–410
96. Baek SM et al (2012) Pressure drop and flow distribution characteristics of single and parallel serpentine flow fields for polymer electrolyte membrane fuel cells. J Mech Sci Technol 26(9):2995–3006
97. Hsieh S-S, Her B-S, Huang Y-J (2011) Effect of pressure drop in different flow fields on water accumulation and current distribution for a micro PEM fuel cell. Energy Convers Manag 52(2):975–982
98. Yang H, Zhao TS, Ye Q (2005) Pressure drop behavior in the anode flow field of liquid feed direct methanol fuel cells. J Power Sources 142(1–2):117–124
99. Cho K-S (2015) The flow-field pattern optimization of the bipolar plate for PEMFC considering the nonlinear material. Int J Electrochem Sci 10:2564–2579
100. Beale SB (2015) Mass transfer formulation for polymer electrolyte membrane fuel cell cathode. Int J Hydrogen Energy 40:11641–11650
101. Diedrichs A et al (2013) Effect of compression on the performance of a HT-PEM fuel cell. J Appl Electrochem 43(11):1079–1099
102. Choi K-S, Kim H-M, Moon S-M (2011) Numerical studies on the geometrical characterization of serpentine flow-field for efficient PEMFC. Int J Hydrogen Energy 36(2):1613–1627
103. Arvay A et al (2013) Nature inspired flow field designs for proton exchange membrane fuel cell. Int J Hydrogen Energy 38(9):3717–3726
104. Li X, Sabir I (2005) Review of bipolar plates in PEM fuel cells: Flow-field designs. Int J Hydrogen Energy 30(4):359–371
105. Wang J, Wang H (2012) Flow-field designs of bipolar plates in pem fuel cells: theory and applications. Fuel Cells 12(6):989–1003
106. Aricò AS, Baglio V, Antonucci V (2009) Direct methanol fuel cells: history, status and perspectives. In: Electrocatalysis of direct methanol fuel cells. Wiley-VCH Verlag GmbH & Co. KGaA, pp 1–78

107. Aricò AS, Srinivasan S, Antonucci V (2001) DMFCs: from fundamental aspects to technology development. Fuel Cells 1(2):133–161
108. Yu X, Pickup PG (2008) Recent advances in direct formic acid fuel cells (DFAFC). J Power Sources 182(1):124–132
109. Demirci UB (2007) Direct liquid-feed fuel cells: thermodynamic and environmental concerns. J Power Sources 169(2):239–246
110. Kamarudin MZF et al (2013) Review: direct ethanol fuel cells. Int J Hydrogen Energy 38 (22):9438–9453
111. Lamy C, Coutanceau C, Leger JM (2009) The direct ethanol fuel cell: a challenge to convert bioethanol cleanly into electric energy. In: Catalysis for sustainable energy production. Wiley-VCH Verlag GmbH & Co. KGaA, pp 1–46
112. An L et al (2010) Performance of a direct ethylene glycol fuel cell with an anion-exchange membrane. Int J Hydrogen Energy 35(9):4329–4335
113. Modestov AD et al (2009) MEA for alkaline direct ethanol fuel cell with alkali doped PBI membrane and non-platinum electrodes. J Power Sources 188(2):502–506
114. Fujiwara N et al (2008) Direct ethanol fuel cells using an anion exchange membrane. J Power Sources 185(2):621–626
115. An L et al (2011) Alkaline direct oxidation fuel cell with non-platinum catalysts capable of converting glucose to electricity at high power output. J Power Sources 196(1):186–190
116. An L, Zhao TS, Xu JB (2011) A bi-functional cathode structure for alkaline-acid direct ethanol fuel cells. Int J Hydrogen Energy 36(20):13089–13095
117. Ha S, Dunbar Z, Masel RI (2006) Characterization of a high performing passive direct formic acid fuel cell. J Power Sources 158(1):129–136
118. Jeong K-J et al (2007) Fuel crossover in direct formic acid fuel cells. J Power Sources 168(1): 119–125
119. Miesse CM et al (2006) Direct formic acid fuel cell portable power system for the operation of a laptop computer. J Power Sources 162(1):532–540
120. Rice C et al (2002) Direct formic acid fuel cells. J Power Sources 111(1):83–89
121. Boyaci San FG et al (2014) Evaluation of operating conditions on DBFC (direct borohydride fuel cell) performance with PtRu anode catalyst by response surface method. Energy 71:160–169
122. Lucia U (2014) Overview on fuel cells. Renew Sustain Energy Rev 30:164–169
123. Mahapatra MK, Singh P (2014) Fuel cells: energy conversion technology A2. In: Letcher TM (ed) Future energy, 2nd edn (Chap. 24). Elsevier, Boston, pp 511–547
124. Davidescu CM (2002) Introducere in termodinamica chimica. Editura Politehnica, Timisoara
125. Atkins P, de Paula J (2005) Atkins' physical chemistry. Oxford University Press

Chapter 9
Organic/Inorganic and Sulfated Zirconia Nanocomposite Membranes for Proton-Exchange Membrane Fuel Cells

Sharf Ilahi Siddiqui and Saif Ali Chaudhry

Abstract Proton-exchange membrane fuel cells (PEMFCs) are one of the most promising commercial technologies used to produce clean energy with high efficiency, energy density, and low emission of harmful gases. Proton-exchange membrane (PEM) is a major part of fuel cell that plays an important role. However, these membranes are very expensive, thermally degradable at high temperature, and conduct protons only in aqueous condition that limits the performance of PEMFC. However, at lower operating temperature, the performance of fuel cell is affected, due to slow electrode kinetics. Recently, great attention has been paid to develop high temperature-tolerant PEM with high proton conductivity. The organic–inorganic composite PEM that can work at high temperature with high proton conductivity is being developed. These organic–inorganic membranes are created by incorporation of metal oxide nanoparticles in the polymer host such as Nafion with strong acid site. These membranes provide high proton conductivity, and chemical and thermal stability to PEMs. Sulfated zirconia (S–ZrO_2), a strongest super acid possessing protogenic groups, is being used as an inorganic filler for composite membranes which showed improved operation at elevated temperature. This chapter presents an overview of the commonly used polymer hosts and inorganic additives. The available literature on S–ZrO_2 nanohybrid membrane technology has been discussed in view of catalyzing the future research to develop more suitable PEM for fuel cell.

Keywords Fuel cells · Proton-exchange membrane · Perfluorinated sulfonic acid membrane · Organic/inorganic hybrid nanocomposite · Sulfated metal oxide · Sulfated metal oxide-doped membrane · Sulfated zirconia/polymer hybrid membrane

S.I. Siddiqui · S.A. Chaudhry (✉)
Department of Chemistry, Faculty of Natural Science, Jamia Millia Islamia,
New Delhi 110025, India
e-mail: saifchaudhry09@gmail.com

S.I. Siddiqui
e-mail: sharf_9793@rediffmail.com

Inamuddin et al. (eds.), *Organic–Inorganic Composite Polymer Electrolyte Membranes*, DOI 10.1007/978-3-319-52739-0_9

Abbreviation

C-PAMPS	Cross-linked poly 2-acrylamido-2-methylpropane-sulfonic acid
NMPA	Nitrilotrimethyl triphosphonic acid
PBI	Polybenzimidazol
PDMS	Polydimethylsiloxane
PVA	Polyvinyl alcohol
PVPA	Polyvinylphosphonic acid
PWA	Phosphotungstic acid
S-CNT	Sulfated carbon nanotubes
SHNTs	Sulfonated halloysite nanotubes
SPAES	Sulfonated polyarylene ether sulfone
SPEEK	Sulfonated polyether ether ketone
SPES	Sulfonated polyether sulfone
SPSU	Sulfonated polysulfone
SSA	Sulfosuccinic acid
ZrP	Zirconium phosphate

1 Introduction

Proton-exchange membrane (PEM) is a major part of fuel cell system that plays an important role in the production of clean and efficient electricity [1, 2]. Fuel cells are electrochemical system in which, through electrochemical reaction with an oxidant, the chemical energy of fuel is converted into electrical energy without combustion (Fig. 1). Therefore, in the fuel cell system, production of electricity occurs [3]. Fuel cells are assembly of different parts; in which central part, where electrochemical processes take place, is most important and it contains the two sub parts namely electro catalyst and membrane. The central part is also known as separation part which acts as sandwich in between two electrodes, i.e., anode and cathode [4]. A schematic diagram of fuel cell is shown in Fig. 2. During electrochemical process, ions generated at anode move towards the cathode through the electrolyte membrane and produce water and heat with free electrons. Varieties of

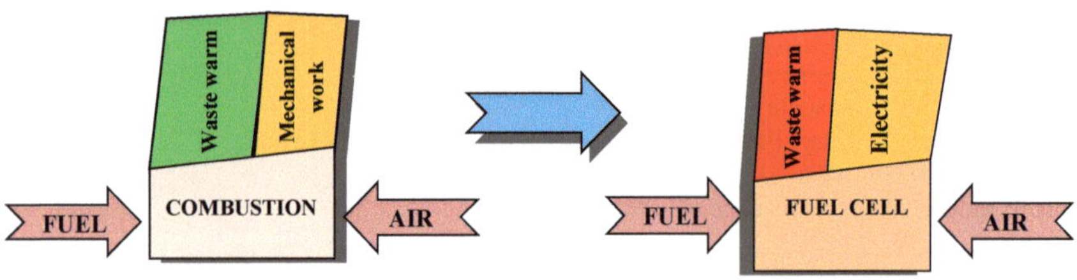

Fig. 1 Advantages of fuel cell reaction over combustion reaction

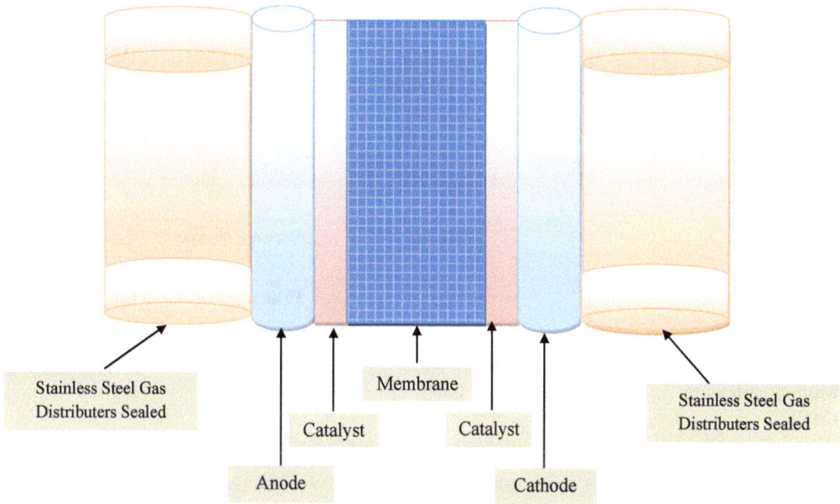

Fig. 2 Schematic representation of PEMFC

fuel cells are available in market, on the basis of different variables like type of fuel, exchange ion, oxidants and electrolytes, operating temperature, and pressure. However, the most important factor for various fuel cells is electrolyte used, which is responsible for the electrochemical process and separation between two electrodes in cell. On the basis of electrolytes, fuel cells are categorized as alkaline fuel cell, direct methanol fuel cell, phosphoric acid fuel cell, proton-exchange membrane fuel cell, solid oxide fuel cell, and molten carbonate fuel cell, in respect of electrolytes such as alkaline electrolyte solution, methanol electrolyte solution, phosphoric acid electrolyte solution, polymer electrolyte, solid oxide electrolyte, and molten carbonate electrolyte, respectively [5]. The short summary of most common types of electrolyte-based fuel cells is presented in Table 1. Among these, the proton-exchange membrane fuel cells (PEMFCs) are most useful for the production of clean electricity [1].

1.1 *Proton-Exchange Membranes (PEMs)*

Generally, hydrogen is used as fuel in proton-exchange membrane electrochemical cell (PEMFCs) [6]. PEMFCs contain both anode and cathode electrodes that are separated by proton-exchange membrane and on electrochemical reaction, with ejection of electron, H_2 changes into H^+ ion at anode and ejected electrons migrate towards the cathode through circuit to produce the electrical current. At the same time oxygen from air, passing through the cathode, combines with free electrons and hydrogen ions to produce water [3]. A schematic diagram of production of electricity in PEMFC is shown in Fig. 3.

Table 1 Characterization of different fuel cells

Fuel cell (FC)	Mobile ion	Operating temperature (°C)	Start up time	Power density (Kw/m^2)
Alkaline fuel cell (AFC)	OH^-	~50–200	Minute	0.7–1.8
Proton-exchange membrane (PEMFC)	H^+	~50–120	Second/minute	3.8–2.6
Direct methanol (DMFC)	H^+	~30–80	Second/minute	2.5–1.5
Phosphoric acid (PAFC)	H^+	~220	Hours	0.8–1.9
Solid oxide (SOFC)	O^{2-}	~600–100	Hours	1.5–2.6
Molten carbonate (MCFC)	CO_3^{2-}	~650	Hours	0.1–1.5

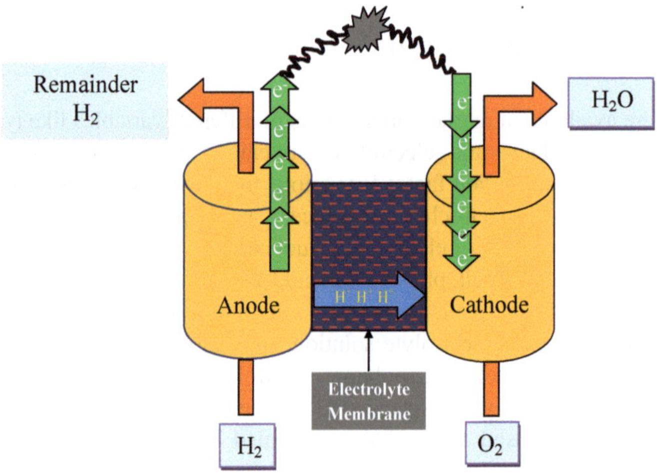

Fig. 3 A schematic diagram of production of electricity in PEMFC

$$\begin{aligned} &\textbf{Anode reaction}: && 2H_2 \rightarrow 4H^+ + 4e^- \\ &\textbf{Cathode reaction}: && O_2 + 4H^+ + 4e^- \rightarrow 2H_2O \\ &\textbf{Overall cell reaction}: && 2H_2 + O_2 \rightarrow 2H_2O + \text{electrical energy} + \text{heat} \end{aligned}$$

Therefore, the fuel cell in which fuel is hydrogen, charge carrier is hydrogen ion (proton), and hydrogen ions permeate across the electrolyte towards cathode to produce heat, water, and electricity, is known as proton-exchange membrane fuel cell. Thus, PEMFCs are operated by proton-exchange membrane [7]. This proton-exchange membrane helps in carrying only the proton from anode to cathode and does not permit electron and thus separates electrons from hydrogen. Another feature of this membrane is that it is impermeable to the gases and serves as solid electrolytes for variety of electrochemical applications. Basically, solid polymeric electrolytes are used in PEMFC as PEMs, which have ability to transfer proton in moist condition instead of dry condition [8]. These solid polymeric electrolytes have gel-like structure and specific for cations or anions. Therefore, these polymeric electrolytes can be classified as cationic electrolytes, containing the negatively charged group, $-SO_3^-$, $-COO^-$, $-PO_3^{2-}$, $-PO_3H^-$, and $-C_6H_4O^-$, attached to the membrane backbone and anionic electrolytes containing the positively charged group, $-NH_3^+$, $-NRH_2^+$, NR_2H^+, NR_3^+, SR_2^+, and PR_3^+ [9]. These membranes have been proved as key feature for variety of fuel cell applications like automobile, back-up power, and portable power. Before polymeric electrolyte membrane or proton-exchange membrane electrolysis, alkaline electrolysis technology was being used for the same purpose. However, there were some drawbacks of alkaline electrolysis technology such as partial load, and low current density. To overcome these drawbacks polymeric electrolyte membranes or proton-exchange membranes (PEMs) were developed by General Electric (GE) company in 1960s. At that time, the performance of PEMFC obtained was 1.88 V at 1.0 A/cm^2, which was efficient as compared to alkaline electrolysis technology [10]. However, till the end of 1970s, the performance of alkaline electrolysis technology was improved and reported as 2.06 V at 0.215 A/cm^2 [11]. To operate PEMFC, first chemically stable polymeric membrane named as “Nafion” which is based on sulfonated polytetrafluoroethylene was developed by DuPont company in 1970s [12].

Most of the PEMs were developed as pure polymer membranes or polymer composite membranes [13, 14]. However, fluoropolymer Nafion-based PEMs were most common and commercially available [15]. Higher ion-exchange capacity (IEC), proton conductivity (σ), gas permeability, durability, water uptake, and thermal stability were the major requirements for significant performance of PEMFCs [16–20]. The advantages of Nafion-based PEMFCs were good thermal, chemical and mechanical stability, relatively high proton conductivity, high efficiency, and clean production of electricity without emission of polluting gases like CO_2, NO_2, SO_2, and CO. But the drawbacks of Nafion-based polymeric membranes were high production cost, and low operational temperature, which hindered the wide spread commercialization of fuel cell [3, 21, 22]. However, one more drawback found in currently used PEMFC technology, based on expensive perfluorinated membrane, is that it could operate effectively only under fully hydrated environment. For researchers, the problems to resolve include the lowering of cost of polymer membrane, increasing the operational temperature of PEMFC, operating the polymer membrane fuel cell in warm condition, and opening the new windows for PEMs. To solve all these problems, researchers are finding new technology for mass production of fuel cell with low cost. They are focusing on developing PEMs with high proton conductivity, low electronic conductivity, low permeability of fuel, low electro-osmotic drag coefficient, good chemical/thermal stability, good mechanical properties, and low green house gases emission. Performance, durability, and cost are major issues for the production of proton-exchange membranes. To overcome these issues, great efforts have been devoted for the development of new materials or modified polymeric membranes. The modified polymer

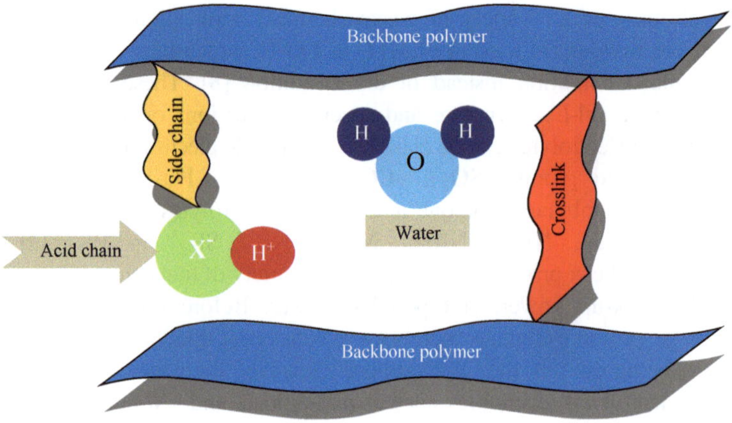

Fig. 4 Major components of polymer membrane

membranes as an alternative of PEMs are valid for performance of PEMFCs at low or medium temperature [21]. Therefore, it is necessary to develop high temperature membrane which can help to perform PEMFCs at high temperature. This so-called "high temperature membranes" could be developed either by the modification of polymer by acids, hygroscopic oxides (SiO_2 and TiO_2), and inorganic solid acid (ZrO_2 or TiO_2/SO_4^{2-}) or by developing completely new system of membrane [23, 24]. However, the lifetime stability of high temperature membrane is still major problem. Recently, attempts to resolve this problem, by right combination of elements of major components of PEMs, are underway. Polymer backbone, chemical cross-links, side chains, and pendant acid groups are four major components of membrane [25]. Figure 4 shows these components of membranes. The fluorine-containing polymer, hydrocarbon-based polymer, inorganic polymer, or a partially fluorinated polymer containing both C–H and C–F bonds in the polymer chain comprising a sulfonic, a carboxyl, a phosphoric acid, or a phosphonate groups may be the backbone. The carboxylic acid (–COOH), phosphonic acid ($–PO_3H_2$), sulfonic acid ($–SO_3H$), and sulfonylimide ($–SO_3NHSO_2CF_3$) are acidic groups bounded covalently to the fluorinated hydrocarbon such as polytetrafluoroethylene (PTFE). PTFE (Teflon) backbone provides stability to membranes in both oxidative and reductive environments. The thermomechanical properties, and inertness, of the membrane could be resolved by cross-linking of polymer backbone, and the number and strength of acid groups attached to backbone determine the electrolytic properties of membrane [26]. Researchers are working for the development of better alternative of PEMFC. Perfluorosulfonic acid membrane such as Nafion, sulfonated aromatic polymer membrane such as sulfonated poly(ether ether ketone), SPEEK, and sulfonated poly(ether ketone), SPEK, are commonly used modified membranes. These membranes contain the extremely hydrophobic backbone and hydrophilic functional groups. Extremely hydrophobic backbone provides the morphological stability to the membrane and hydrophilic functional groups act as

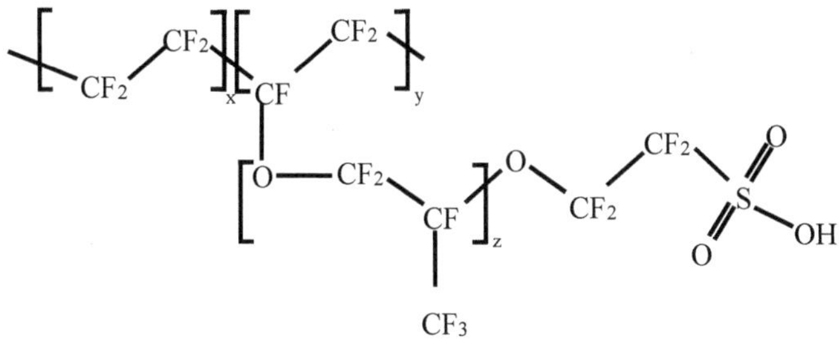

Fig. 5 Chemical structure of Nafion membrane

water reservoirs on aggregation. The perflourinated PEMs have been proved most chemically inert membranes [27, 28]. Hence, perfluorosulfonic acid membranes (Nafion PEMs) are the most commonly and significantly used PEMs that demonstrated fuel cell lifetimes over 2500 days below 100 °C [27]. Structure of Nafion membrane is shown in Fig. 5. The replacement of sulfonic acid group ($–SO_3H$) of Nafion with sulfonyl imide group (SO_2 $NHSO_2$ CF_3) increased the water uptake [29]. Similarly, Nafion with phosphonic acid group has improved thermal stability [30]. These results revealed that with the modification of acid groups, desire PEMs could be developed [31]. Higher proton conductivity, electrochemical stability, and mechanical strength of perflourinated PEMs (Nafion) could be achieved by polytetrafluoroethylene (PTFE) backbones [32]. The development of hydrocarbon backbone-based polymer demonstrated the high temperature performance of fuel cell at low relative humidity [33]. However, due to short lifetime, hydrocarbon backbone-based polymer membranes are not used these days. To overcome this disadvantage, polymer membranes such as polyether sulfones (PES), polyether ether ketone (PEEK), sulfonated polyether ether ketone (SPEEK), polyphosphazene (PP), sulfonated poly (phenylene sulfide) (SPPS), poly (4-benzoyl-1, 4-phenylene) (PPBP), polyphenylene oxide (PPO), poly (phenyl quinoaniline) (PPQ), polyimide, alkylsulfonated polybenzimidazol (AS-PBI), sulfoarylated polybenzimidazol (SA-PBI), and sulfonated poly (oxy-1, 4-phenyleneoxy-1, 4-phenylenecarbonyl-1, 4-phenylene) have been designed. Stable and cost-effective sulfonated trifluorostyrene, α, β, β-trifluorostyrene monomer substituted and a series of substituted-α, β, β-trifluorostyrene co-monomers membranes have also been developed [34, 35]. The solvent in PEMs fuel cell could be responsible for deciding the operating temperature. The operating temperature of fuel cell could be increased by replacing solvent like water (Bronsted base) with lower volatile solvent like H_3PO_4. Furthermore, H_3PO_4 and ionic liquid such as molten salt 1-butyl, 3-methyl imidazolium triflate (BMITF)-doped Nafion provided good conductivity at elevated temperature under relatively low humidity [36, 37]. However, there are some shortcomings such as H_3PO_4 being used is corrosive and ionic liquid is partly mobilized. To fulfill the requirement of PEMs, an alternate membrane was developed by Malhotra and Datta [38]. In order to improve water retention and simultaneously increasing acidic nature, they incorporated the inorganic acidic materials within the polymer (Nafion). The resultant membrane worked above 120 °C under low relative humidity. This alternate membrane was known as "high temperature membrane" and became very attractive approach for research. "High temperature membranes" are generally organic/inorganic hybrid nanocomposite membranes. The detailed study of polymer nanocomposite fuel cells with respect to the requirements of polymer electrolyte membranes, criteria for the selection of inorganic nanomaterials, preparation techniques of polymer nanocomposites, the requirements of electrolyte membrane applicable for fuel cell, and types of polymer membranes have been reviewed by Lade et al. [39].

2 Organic/Inorganic Hybrid Membranes

One of the unique functional properties of nature is to create the sophisticated materials that could carry out multiple functions within small volume. Organic/inorganic materials are one of these. On behalf of the nature, researchers are also producing the multifunctional synthetic organic/inorganic materials. This could be achieved by different methods; however, the sol–gel method has been used frequently. These synthetic materials, known as hybrid nanocomposites, are prepared by the nanoscale mixing of two phases, organic part such as polymer that provides flexibility and inorganic part such as inorganic acids and metal oxides that provides thermal and mechanical stability [40]. The features of organic and inorganic parts of a composite are illustrated in Fig. 6. Incorporation of inorganic phase into the organic phase brings changes in transparency, refractive index, adhesion, viscosity, flame resistance, solvent resistance, and decorative properties of the polymer [41–47]. Due to these advantages, of introducing inorganic material into the organic materials, hybrid nanocomposites are being used in different fields such as solid electrolytes, modified electrodes, electrochromic devices, functional biomolecules, bioreactor, biosensor, diagnostic devices, contact lenses, in bone

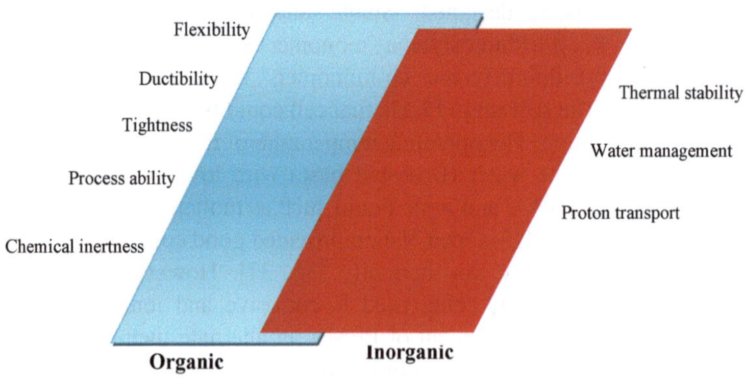

Fig. 6 Features of organic and inorganic parts in hybrid composite

replacement application, filling for dental restoration, and in the polymer electrolyte membrane and proton-exchange membrane fuel cells. The use of hybrid materials in the production of PEMs, is the advance technique and, fulfills the requirement of PEMs. PEMs with hybrid material have shown improved proton conductivity, high durability, better chemical and thermal stability, good mechanical properties, reduced electronic conductivity, decreased permeability of fuel, low production cost, and optimum performance at high temperature. Organic and inorganic hybrid membranes could be prepared by the incorporation of many types of inorganic fillers including inorganic acid and nano-sized metal oxides into the polymer matrix [24]. Liquid super acids may enhance the proton conductivity of membrane but they could not be applicable for the fuel cells because of their mobilization characteristics within PEMs [25]. Thus the solid acids became first choice as fillers. First hybrid membrane was developed by doping of inorganic solid acid (heteropoly acid), phosphotungstic acid (PTA), in ion-exchange polymer membrane (Nafion), which showed high proton conductivity and water retention at high temperature (120 °C) and low relative humidity (RH) [38]. However, PTA showed high solubility in water due to which it leached out from PEMs [48]. Recently, Nafion–PTA membranes were stabilized by heat and leaching out of PTA was reduced [49]. Because of high acidic and hydrophilic nature, different heteropoly acids (HPA), and heteropoly acid-derived salt-doped polymer membranes have been used in fuel cell. These membranes performed well at low humidity and high operating temperature with high water retention [50, 51]. However, the biggest drawback was the exploitation of acidity and hydrophilicity properties of HPA within polymer matrix. To overcome this drawback and to improve the performance of PEMs fuel cell at higher operating temperature and lower relative humidity; various nano-sized metal oxides have been added to the polymer matrix. Among these, MO_2 (M = Ti, Si or Zr) have frequently been used as inorganic filler [52]. The morphology of metal oxides plays an important role in the performance of polymer–metal oxide nanohybrid membrane at high temperature. However, the agglomeration of filler particles restricts the uniform dispersion of filler on the entire membrane, and thus resultant proton cannot transport effectively through the entire membrane. Therefore, uniform dispersion of inorganic fillers in polymer matrix for effective performance of PEMs is necessary. Different methods have been adopted for the fabrication of hybrid nanocomposite membranes with several advantages such as lower processing temperature, high homogeneity, favourable thermal or mechanical, and chemical stabilities, high durability, and reasonable purity of resulting material. As compared to conventional pathways, sol–gel method is more suitable for preventing the agglomeration of inorganic fillers which tend to reduce the size of inorganic parts and thereby the organic–inorganic phase separation. Sol–gel methods involve the hydrolysis of metal alkoxide [M(OR)n, M = Al, Si, Ti, Zr] in organic solvent followed by polycondensation. Hybrid nanocomposite membranes could be achieved by one of the approaches given below [53]:

1. The simultaneous hydrolysis and condensation of inorganic filler in the presence of desired polymer.
2. Performing the sol–gel process in the presence of desired polymer with functionalities.
3. The simultaneous polymerization of organic component with the formation of inorganic phase.

To achieve temperature-tolerant membrane, different nano-sized oxides such as SiO_2, TiO_2, Pt, Pt–SiO_2, and Pt–TiO_2 were doped in Nafion polymer [54–56]. These modified membranes showed much higher water uptake than Nafion. Under relatively low humidity, at 80 °C, these modified membranes showed lower resistance than Nafion. This was due to the suppression of H_2 crossover by Pt and subsequent sorption of water produced on the doped oxide. Jalani et al. [52] synthesized metal oxide (Zr, Ti, and Si)-doped hybrid membrane by sol–gel method. These synthesized nanohybrid membrane was completely transparent, homogeneous, and showed comparatively higher water retention than virgin Nafion at higher temperature (120 °C) and lower relative humidity (40%). However, only Nafion–ZrO_2 composites exhibited enhanced conductivity than virgin Nafion at 90 and 120 °C under low relative humidity [52]. This could be possible by enhanced acidity of Zr-based membrane, which was attributed to the decrease of equivalent weight in comparison to Ti- and Si-based nanocomposites. Miyake et al. [57] studied water content and conductivity of the Nafion–silica hybrid membrane and observed that modified membrane has higher water content than unmodified Nafion. The increase in water uptake at higher temperature was due to reorganization of the structure of the Nafion around the particles. The water uptake of Nafion–silica nanocomposite membrane changed with temperature, particle distribution in the membrane, porosity, and size of silica particle. The proton conductivity of Nafion and Nafion–silica nanocomposite membranes decreased with decreasing relative humidity, but was independent of silica particle size [58]. The improved thermal stability was observed for hollow silica sphere (HSS)-modified Nafion [59]. This composite showed higher water uptake and lower swelling degree within temperature range 40–100 °C. Higher water uptake efficiency was observed with smaller diameter of hallow silica sphere in Nafion. The improved proton conductivity was observed at 60–100 °C with 3–5% wt of HSS in composite membrane. Nano-sized SiO_2-embedded Nafion membrane fuel cell could operate at 130 °C, whereas, unmodified PEMs showed thermal degradation [60]. This was due to the high rigidity of modified Nafion. Similarly, inorganic additives such as TiO_2- and ZrO_2-filled PEMs performed well at higher temperature [61–66]. Presently, TiO_2 is considered most promising material because of high thermal, mechanical, photostability, powerful oxidation strength, and non-toxicity. Polymer/TiO_2 exhibits high conductivity, and the morphology of TiO_2 filler plays an important role in fuel cell performance at higher temperature. Recently, SnO_2 nanopowder-embedded polymer membrane fuel cell has been found suitable to operate at 120 °C under low humidity [67–69]. However, continuous ionic channels are another very important issue for water transport and high proton conductivity but still most of the membranes have short channel. Short channels restrict the content of absorbed water that creates discontinuation of hopping routs of proton. However, in the organic/inorganic hybrid membranes, protogenic groups of both inorganic fillers and polymer matrix were distributed on the interfaces and produced long-range ionic channels which were responsible for high proton conductivity of PEMs. One more important thing to consider for good PEMs is that the inorganic content in polymer should be balanced, as increase of inorganic filler in the polymer matrix results in decrease of proton conductivity (due to low proton conductivity of inorganic content) which may block the diffusion of the H^+ ions. It can be resolved by the incorporation of the acid functionalized inorganic fillers into polymer that can improve or preserve the proton conductivity of PEMs. Nawn et al. [70] developed polybenzimidazole–ZrO_2 nanocomposite membrane with varying oxide content (0–22 wt%) and studied the structure–property relationship of acid-modified and acid-unmodified membranes. Improved acid uptake and thermal and mechanical stabilities were observed from 0 to 8 wt%, while reverse effect was observed from 8 to 22 wt%. Acid-doped ZrO_2-modified polybenzimidazole (PBI) membrane fuel cell worked at high temperature, and low relative humidity, and showed ionic conductivity (0.104 S/cm) at very high temperature (185 °C). Similarly, phosphate–ZrO_2-doped Nafion membrane was developed which showed good results [71]. Nafion/ZrO_2–phosphate composite membranes showed enhanced thermal stability at 130 °C [72] and doping of 10% wt of zirconia–phosphate in polymer improved the thermal stability up to 140 °C at 90% relative humidity [73]. To improve the membrane proton conductivity at different temperatures and relative humilities, phosphated zirconia-sulfonated polyether sulfone nanohybrid membrane (ZrP–SPES) was synthesized by embedding zirconia in sulfonated polyether sulfone followed by treatment with H_3PO_4 [74]. Under 50% relative humidity, ZrP–SPES showed 0.01 S/cm proton conductivity at 90 °C, and at 90% relative humidity conductivity increased to 0.19 S/cm [74]. Recently, Ozdemir et al. [75] compared the proton conductivity of acid-modified polybenzimidazol (PBI) nanocomposite with the same percentage content of TiO_2, SiO_2, and ZrP inorganic fillers. To develop high temperature membranes, these inorganic fillers were dispersed in PBI and doped with H_3PO_4. Resultant, hybrid nanocomposites were thermally stable and showed improved acid retention capability. However, only SiO_2- and ZrP-dispersed nanocomposites showed improved proton conductivity. The highest proton conductivity (0.200 S/cm) was observed with PBI/ZrP having highest doped level of H_3PO_4. The composite membranes with TiO_2 showed low proton conductivity due to nonuniform membrane structure. Therefore, the various types of acid-modified inorganic filler-doped hybrid membranes are being produced to improve the membrane quality. Out of these, sulfated inorganic filler-embedded membranes showed excellent results. The comparative study of proton conductivity of different nanofiller-doped membranes is summarized in Table 2.

Table 2 Comparative proton conductivity of different hybrid nanocomposites

Polymer	Inorganic nanofiller	Ionic conductivity	Ref.
Nafion 115	–	41–61 mS/cm	[99]
	ZrP	24–60 mS/cm at 25 °C, 100% RH	
Nafion	S-CNT	0.01 S/cm	[100]
Nafion	S–SiO_2	59 mS/cm at 140 °C	[76]
SPEEK	–	0.0152 S/cm	[81]
	SHNTs	0.0245 S/cm	
SPEEK	Silane/silica	5 mS/cm at 90 °C, 100% RH	[101]
SPEEK	SiO_2/ZrP	0.09 S/cm	[102]
SPEEK	S–ZrO_2	3.88 mS/cm	[96]
PBI	S–ZrO_2	0.104 S/cm at 185 °C	[70]
PBI	ZrP + H_3PO_4	9×10^{-2} S/cm at 200 °C	[75]
		0.200 S/cm	
SPES	ZrP + H_3PO_4	10^{-2} S/cm at 50% RH, 90 °C	[74]
		0.19 S/cm at 90% RH, 90 °C	
PVPA	S–TiO_2	0.03 S/cm at 150 °C	[77]
C-PAMPS	S–ZrO_2	2.4×10^{-1} S/cm at 80% RH, 100 °C	[90]
PDMS	ZrO_2 + PWA	8×10^{-2} S/cm at 150 °C	[103]
SPSU	S–TiO_2 + NMPA	0.002 S cm^{-1} at 150 °C	[104]
PVA/SSA	SiO_2	10^{-3}–10^{-2} S/cm	[105]
PVA/glycerin	ZrP	10^{-4}–$^{-3}$ S/cm	[106]

3 Organic-Sulfated Metal Oxide Hybrid Membrane

Many types of acid-modified oxides such as sulfated silica (S–SiO_2) [76], titania (S–TiO_2) [77], and zirconia (S–ZrO_2) [78] are frequently used as super acid fillers to incorporate in polymers, and the sulfated nanoparticle-incorporated polymer membranes have shown higher proton conductivity. The acid-modified nanoparticles provide continuous ionic channel within polymer matrix and the resultant proton pathways allow the migration of protons with low resistance that improve proton conductivity. These acid-modified metal oxide-doped polymer membranes are thermally, and mechanically, stable than the other solid super acids. Sulfated silica (S–SiO_2)-modified nanocomposite membrane have shown much higher proton conductivity than the virgin polymer at various temperatures (40–140 °C) [76]. Higher proton conductivity for S–SiO_2-modified nanocomposite membrane was 0.059 S/cm at 140 °C [76]. Recently, S–SiO_2-modified Nafion nanohybrid membrane showed good proton conductivity at high temperature and low relative humidity. The silica and sulfonic groups responded for improvement in membrane hydration and proton conductivity, respectively. Moreover, proton conductivity was much affected by silica particle size as 20–30% improved conductivity was observed with less than 50 nm particle size under 100% humidity at 80 °C, while 22–42% improvement was observed with small size particles at 120 °C under same humidity.

However, at 120 °C under 50% humidity, the sulfonated particles were less efficient [79]. These S–SiO_2-modified nanocomposite membranes were fabricated by modification of Stöber silica particles with mercapto propyltriethoxysilane dispersed in pre-concentrated solution of Nafion followed by oxidation of mercapto groups to sulfonic acids by heating the membrane in 10 wt% hydrogen peroxide for 1 h. Several sulfated titania (S–TiO_2)-modified polymer membrane-based fuel cells have been developed for use at higher temperature. Poly(vinylphosphonic acid)-sulfated titania (PVPATS)-based nanocomposite membrane has shown more thermal and chemical stabilities at much higher temperature [77]. TGA and DSC characterization techniques confirmed the thermal stability of nanocomposite membranes up to 200 °C. The formation of phosphonic acid anhydride was verified by TGA study and on increasing the sulfated nanoparticles content in polymer, the shift in *Tg* of membrane towards higher temperature was observed by the DSC study. The proton conductivity of PVPATS was 0.03 S/cm at 150 °C. Recently, to improve thermal stability of polymer membrane, Aslan and Bozkurt mixed poly (vinylphosphonic acid)-sulfated titania (PVPATS) with poly (1-vinyl 1, 2, 4-triazole) (P (VTri))/sulfated nano-titania (TS) (PVTriTS) [80]. They observed that the temperature-tolerant power of newly modified binary membrane increased up to ~300 °C and *Tg* of binary membrane decreased with the increase of PVPATS content in PVTriTS–PVPATS binary membrane. However, at 150 °C, the proton conductivity was 0.003 S/cm. Moreover, ~60% enhanced proton conductivity was observed with thermally and mechanically stable sulfonated halloysite nanotubes (SHNTs) incorporated in sulfonated poly(ether ether ketone) (SPEEK) membrane [81]. Among all the solid acids, sulfated zirconia (SO_4^{2-}/ZrO_2) is considered strongest super acid, which exhibited highest acidity and proton conductivity [82, 83]. However, little information on sulfated zirconia (S–ZrO_2) and its composite materials is available. S–ZrO_2/Nafion and S–ZrO_2/SPEEK are examples of the modified membranes. Nafion/S–ZrO_2 membranes, prepared by several methods, having a structure with homogeneous surface were stable up to 120 °C. There is good ion-exchange membrane formation between S–ZrO_2 and Nafion as indicated by FTIR and SEM analyses. Its screening has demonstrated the highest conductivity and water uptake at high temperature.

4 Sulfated Zirconia Nanocomposite Membranes

Sulfated zirconia (S–ZrO_2) is a strongest super acid as compared to ZrO_2 due to the protogenic groups available on S–ZrO_2 [84, 85]. Basically, S–ZrO_2 has Hammett acid strength of −16.03 which is much higher than H_2SO_4 (−11.9). Super acids are those which have stronger acidity than 100% perchloric or sulfuric acid [86]. Higher acidic strength is responsible for super acidity of S–ZrO_2 which is claimed by changing the color of Hammett indicator at $pKa < -12$ [87]. Moreover, it has been reported that strong acid site is responsible for the isomerization and long-range ionic channel [88]. Because of high acidity and water affinity, sulfated zirconia nanoparticles have been used as an alternative of Nafion [78]. Thus, S–ZrO_2-based cell performed as good ionic conductor but the maximum power density of sulfated zirconia-based cell was about 1/3 as compared to Nafion-based cell [78]. To improve the cell performance, researchers attempted for S–ZrO_2/Nafion hybrid membrane and observed that improved results could be achieved by two different approaches either by treating zirconia particles in sulfuric acid or by the thermal decomposition of $ZrOCl_2$ and $(NH_4)_2SO_4$. However, it is very difficult to dope sulfate groups on ZrO_2 surface. To attach significant sulfate groups onto the ZrO_2 surface, multistep approach has been attempted [89]. Yao et al. [90] prepared S–ZrO_2 nanoparticles by electrospinning of PVP/Zr(OPr)$_4$ composite nanofiber mats [91] followed by the compression [92], heat treatment, immersion of mats in H_2SO_4, and calcination. The successful formation and structure of S–ZrO_2 was confirmed by the XRD and FTIR studies. Because of need of high temperature-tolerant membrane, Yang et al. prepared [71] S–ZrO_2/Nafion PEMs and used as electrolyte in fuel cell. First, they prepared fine powder of S–ZrO_2 by the reaction of dry $Zr(OH)_2$ with boiled 0.5 M H_2SO_4 followed by filtration and calcination at 110 °C and 700 °C, respectively, and then crushed and added this prepared fine powder to Nafion solution to recast into a membrane (Fig. 7). The prepared S–ZrO_2/Nafion membrane showed good conductivity at high temperature. Zhai et al. [93] developed S–ZrO_2/Nafion hybrid membrane by the recasting of different S–ZrO_2 contents, and Nafion, and then used as electrolyte in the cell. The membrane physiochemical properties were examined by the ion-exchange capacity

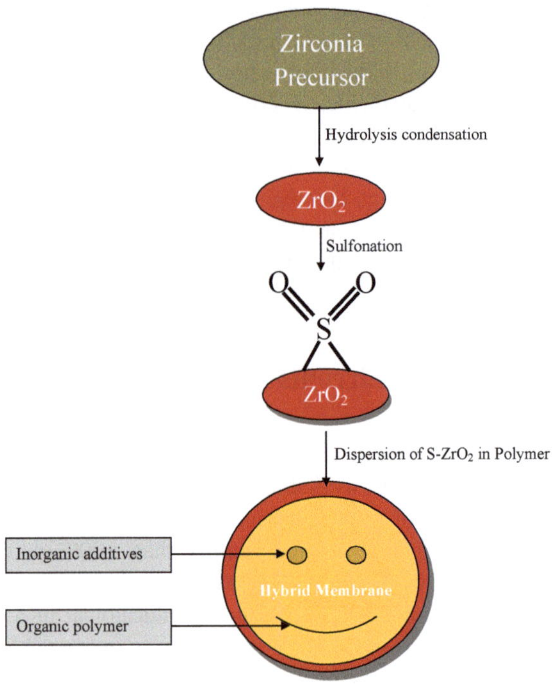

Fig. 7 Preparation steps of thermally stable organic–inorganic hybrid membrane

(IEC), and water uptake measurements. SEM, XRD, TGA, and FTIR studies revealed that S–ZrO_2 well dispersed in Nafion matrix and increased zirconia content in the matrix improved crystallinity, IEC, and initial degradation temperature of the hybrid membrane. The best performance of cell was observed with 15 wt% S-zirconia that showed lower resistance than Nafion membrane at high temperature and low relative humidity. At fewer than 2 atm pressures, IECs based on H_2/O_2 were 1.35 and 0.99 W/cm^2 at 80 °C and 120 °C, respectively. Navarra et al. [94] demonstrated the effect of super acid zirconia content in Nafion matrix that significantly enhanced the hydration level, and acidity, along with improvement in thermal, and transport properties, of composite membranes. High conductivity was realized with 5 wt% of super acid content in S-zirconia/Nafion membranes [94]. Proton conductivity of S-zirconia/Nafion composite membranes was attributed to the surface properties of S-zirconia. Hydrophilic nature of filler surface was confirmed by the higher water uptake capacity for composite membrane. The presence of SO_4^{2-} group on S-zirconia-embedded Nafion showed the highest water diffusion coefficient. Lin et al. [95] reviewed that on doping electrospun super acidic fiber (S–ZrO_2) in Nafion a large amount of protogenic groups aggregated in the interfacial region between acidic fiber and the Nafion matrix that formed regular channel for facile proton transport. Nafion is not only the polymer used in the fabrication of high temperature composite membrane but other polymers have also been used. However, S–ZrO_2-modified sulfonated poly(ether ether ketone) (SPEEK) has exhibited better performance. Results revealed that S–ZrO_2 content in SPEEK is responsible for the enhanced proton conductivity and oxidative stability of membrane. Moreover, SPEEK sulfonation time also showed attractive effect on the proton conductivity and oxidative stability of membrane. Recently, Mossayebi et al. [96] optimized the effects of SPEEK sulfonation time and S–ZrO_2 content. Optimum proton conductivity and oxidative stability were found to be 0.00388 S/cm and 102 min, respectively, after 6.9 h of sulfonation time and 5.94 wt% content of sulfated zirconia. Water uptake and acidity of S–ZrO_2/SPEEK-based membrane could be improved by doping Pt catalyst in S–ZrO_2 [97]. This modification in S–ZrO_2/SPEEK membrane provided the self-humidifying nature in membrane that showed highest proton conductivity. In comparison to virgin SPEEK membrane, the SPEEK/Pt–S–ZrO_2 showed improved power density from 0.54 to 0.95 Wcm^{-2} under the dry condition. S–ZrO_2 fiber-filled cross-linked poly 2-acrylamido-2-methylpropane-sulfonic acid (C-PAMPS) hybrid PEM was fabricated from poly 2-acrylamido-2-methylpropane-sulfonic acid (PAMPS) polymer matrix for high water uptake capacity [90]. C-PAMPS alone showed high proton conductivity [98] and on the incorporation of S–ZrO_2 in C-PAMPS, S–ZrO_2/C-PAMPS showed improved proton conductivity. An S–ZrO_2 fiber/C-PAMPS hybrid membrane was developed by immersing S–ZrO_2 fiber in 2-acrylamido-2-methylpropane-sulfonic acid (AMPS) monomer, followed by the addition of azobisisobutyronitrile (ABIN), where ABIN was used as initiator and ethylene glycol diacrylate (EGD) as the cross-linker. Polymerization provided the long-range channels to the membrane due to significant distribution of protogenic groups of both S–ZrO_2 and C-PAMPS at the inorganic/organic interfaces.

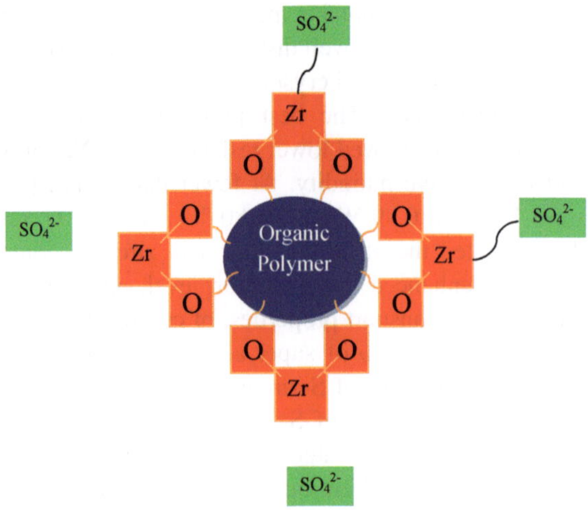

Fig. 8 $S–ZrO_2$/polymer hybrid nanocomposite membrane

These long-range channels helped to attract water molecules and provided the hopping pathway to protons which was responsible for the high proton conductivity. As compared to the C-PAMPS conductivity (0.087 S/cm), $S–ZrO_2$/C-PAMPS nanohybrid showed much higher conductivity (0.24 S/cm) with introduction of 20% $S–ZrO_2$ at 100 °C and under 80% of humidity and 20% of cross-linking degree. However, the existence of $S–ZrO_2$ and cross-linking degree of C-PAMPS decreased the ion-exchange capacity of membrane while, the fiber-induced interconnected ionic channels kept the proton conductivity and water content at a high level over the entire temperature range (Fig. 8).

The diameter of the fiber affected the ionic channel, with the decrease in diameter of fiber, and the number of ionic channel was increased that enhanced hopping pathway of proton which led to the proton conductivity. The highest proton conductivity (0.34 S/cm) was observed with thinner fiber (85 nm). The investigations suggested that the great improvements in PEMFC using sulfated zirconia polymer nanohybrid membranes are possible.

5 Conclusion and Future Prospects

Doping of sulfated metal oxides, especially sulfated zirconia ($S–ZrO_2$), into polymer matrix of Nafion or other polymers enhances the chemical and mechanical stability of polymer membranes and also increases the proton conductivity. These are required conditions for the working of fuel cell. The sulfated zirconia ($S–ZrO_2$)-modified membranes provide continuous ionic channels and thus enhance proton pathways to allow more protons to transport with low resistance that lead to improve proton conductivity. Moreover, polyaromatic structure membrane such as SPEEK with acidic metal oxides has been considered as preferred choice for high temperature membranes. The modification of S–ZrO_2/SPEEK with self-humidifying additive caused improvement in the self-humidifying nature of membranes which improved water uptake and acidity of S–ZrO_2/SPEEK. The chapter concludes that PEM with S–ZrO_2 provides high-performance fuel cell as compared to membranes with other additives.

However, increase in content of fillers in polymer or imbalanced dispersion of fillers causes lowering of performance of PEM fuel cell. Therefore, a future prospective based on this study is to develop more useful PEMs with high proton conductivity, and high thermal stability, and need to design polymer membrane with controlled structure and morphology.

Acknowledgements We gratefully acknowledge the financial support from Jamia Millia Islamia University, New Delhi,110025, India.

References

1. Babir F, Gomez T (1996) Efficiency and economics of proton exchange membrane (PEM) fuel cells. Int J Hyd Energy 21:891–901
2. Bakangura E, Wu L, Ge L, Yang Z, Xu T (2016) Mixed matrix proton exchange membranes for fuel cells: state of the art and perspectives. Prog Polym Sci 57:103–152
3. Peighambardoust SJ, Rowshanzamir S, Amjadi M (2010) Review of the proton exchange membranes for fuel cell applications. Int J Hyd Energy 35:9349–9384
4. Ahmad MI, Zaidi SMJ, Rahman SU (2006) Proton conductivity and characterization of novel composite membranes for medium-temperature fuel cells. Desalination 193:387–397
5. Shaari N, Kamarudin SK (2015) Chitosan and alginate types of bio-membrane in fuel cell application: an overview. J Power Sources 289:71–80
6. Kreuer KD (2001) On the development of proton conducting polymer membranes for hydrogen and methanol fuel cells. J Membr Sci 185:29–39
7. Tripathi BP, Shahi VK (2011) Organic–inorganic nanocomposite polymer electrolyte membranes for fuel cell applications. Prog Polym Sci 36:945–979
8. Sharma S, Pollet BG (2012) Support materials for PEMFC and DMFC electrocatalysts—a review. J Power Sources 208:96–119
9. Strathmann H (1995) Chapter 6 Electrodialysis and related processes. Membr Sci Technol 2:213–281
10. Russell JH, Nuttall LJ, Fiket AP (1973) ACS division of fuel chemistry preprints 18:24–40
11. Leroy RL, Janjua MB, Renaud R, Leuenberger U (1979) Analysis of time-variation effects in water electrolyzers. J Electro Chem Soc 126:1674–1682
12. Grot WG (1973) US Patent 3,770,567, 1973
13. Bose S, Kuila T, Nguyen TXH, Kim NH, Lau KT, Lee JH (2011) Polymer membranes for high temperature proton exchange membrane fuel cell: recent advances and challenges. Prog Polym Sci 36:813–843
14. Krishnan P, Park JS, Kim CS (2006) Preparation of proton-conducting sulfonated poly(ether ether ketone)/boron phosphate composite membranes by an in situ sol–gel process. J Membr Sci 279:220–229
15. Mauritz KA, Moore RB (2004) State of understanding of nafion. Chem Rev 104:4535–4586

16. Savadogo O (2004) Emerging membranes for electrochemical systems: Part II. High temperature composite membranes for polymer electrolyte fuel cell (PEFC) applications. J Power Sources 127:135–161
17. Wang JT, Savinell RF, Wainright J et al (1996) A H_2/O_2 fuel cell using acid doped polybenzimidazole as polymer electrolyte. Electrochim Acta 41:193–197
18. Mecerreyes D, Grande H, Miguel O, Marcilla R, Cantero I (2004) Porous Polybenzimidazole. Membranes doped with phosphoric acid: highly proton-conducting solid electrolytes. Chem Mater 16:604–607
19. Silva VS, Schirmer J, Reissner R, Ruffmann B, Silva H, Mendes A (2005) Proton electrolyte membrane properties and direct methanol fuel cell performance. II. Fuel cell performance and membrane properties effect. J Power Sources 140:41–49
20. Xue S, Yin G (2006) Methanol permeability in sulfonated poly(etheretherketone) membranes: a comparison with Nafion membranes. Eur Polym J 42:776–785
21. Gubler L, Prost N, Gursel SA, Scherer GG (2005) Proton exchange membranes prepared by radiation grafting of styrene/divinylbenzene onto poly(ethylene-alttetrafluoroethylene) for low temperature fuel cells. Solid State Ionics 176:2849–2860
22. Xu T (2005) Ion exchange membranes: state of their development and perspective. J Membr Sci 263:1–29
23. Vona MLD, Sgreccia E, Donnadio A, Casciola M, Chailan JF, Auer G, Knauth P (2010) Composite polymer electrolytes of sulfonated poly-ether-ether ketone (SPEEK) with organically functionalized TiO_2. J Membr Sci 369:536–544
24. Kontou E, Niaounakis M (2006) Thermo-mechanical properties of LLDPE/SiO_2 nanocomposites. Polymer 47:1267–1280
25. Thampan T, Jalani NH, Choi P, Datta R (2005) Systematic approach to design higher temperature composite PEMs. J Electrochem Soc 152:316–325
26. Guo Q, Pintauro PN, Tang H, Connor S (1999) Sulfonated and crosslinked polyphosphazene-based proton-exchange membranes. J Membr Sci 154:175–181
27. Steck A (1995) Proceedings of the first international symposium on new materials for fuel cell systems. In: Savagodo O, Roberge PR, Veziroglu TN (eds), Montreal, Canada, pp 74–94
28. Wang Z, Tang H, Zhang H, Lei M, Chen R, Xiao P, Pan M (2012) Synthesis of Nafion/CeO_2 hybrid for chemically durable proton exchange membrane of fuel cell. J Membr Sci 421–422:201–210
29. Des Marteau DD (1995) Novel perfluorinated ionomers and ionenes. J Fluorine Chem 72:203–208
30. Kotov SV, Pedersen SD, Qiu W, Qiu Z-M, Burton DJ (1997) Preparation of perfluorocarbon polymers containing phosphonic acid groups. J Fluorine Chem 82:13–19
31. Savett SC, Atkins JR, Sides CR, Harris JL, Thomas BH, Creager SE, Pennington WT, Des Marteau DD (2002) A Comparison of Bis[(perfluoroalkyl) sulfonyl] imide ionomers and perfluorosulfonic acid ionomers for applications in PEM fuel-cell technology. J Electrochem Soc 149:1527–1532
32. Liu W, Ruth K, Rusch G (2001) Membrane durability in PEM fuel cells. J New Mater Electrochem Syst 4:227–232
33. Savadogo O (1998) Emerging membranes for electrochemical systems: (I) solid polymer electrolyte membranes for fuel cell systems. J New Mater Electrochem Syst 1:47–66
34. Hogdon RB (1968) Polyelectrolytes prepared from perfluoroalkylaryl macromolecules. J Polymer Sci Part A Polym Chem 6:171–191
35. Che Q, Chen N, Yu J, Cheng S (2016) Sulfonated poly(ether ether) ketone/polyurethane composites doped with phosphoric acids for proton exchange membranes. Solid State Ionics 289:199–206
36. Savinell R, Yeager E, Tryk D, Landau U, Wainright J, Weng D, Lux K, Litt M, Rogers C (1994) A polymer electrolyte for operation at temperature up to 200 °C. J Electrochem Soc 141:46–48

37. Doyle M, Choi SK, Proulex G (2000) High-temperature proton conducting membranes based on perfluorinated ionomer membrane-ionic liquid composites. J Electrochem Soc 147:34–37
38. Malhotra S, Datta R (1997) Membrane supported non-volatile acid electrolyte allow High-T operations for PEMFC. J Electrochem Soc 144:23–26
39. Lade H, Kumar V, Arthanareeswaran G, Ismail AF (2016) Sulfonated poly(arylene ether sulfone) nanocomposite electrolyte membrane for fuel cell applications: a review. Int J Hydrogen Energy. doi:10.1016/j.ijhydene.2016.10.038
40. Siuzdak DA, Mauritz KA (1999) Surlyn®/[silicon oxide] hybrid materials. 2. Physical properties characterization. J Polym Sci Part B: Polym Phys 37:143–154
41. Hsiue GH, Liu YL, Liao HH (2001) Flame-retardant epoxy resins: an approach from organic–inorganic hybrid nanocomposites. J Polym Sci Part A: Polym Chem 39(7):986–996
42. Khrenov V, Klapper M, Koch M, Muellen M (2005) Surface functionalized ZnO particles designed for the use in transparent nanocomposites. Macromol Chem Phys 206:95–101
43. Wang B, Wilkes GL, Hedrick JC, Liptak SC, McGrath JE (1991) New high-refractive-index organic/inorganic hybrid materials from sol–gel processing. Macromolecules 24:3449–3450
44. Ochi M, Zakahashi R, Terauchi A (2001) Phase structure and mechanical and adhesion properties of epoxy/silica hybrids. Polymer 42:5151–5158
45. Mackay ME, Dao TT, Tuteja A, Ho DL, Van Horn B, Kim HC, Hawker CJ (2003) Nanoscale effects leading to non-Einstein-like decrease in viscosity. Nat Mater 2:762–766
46. Ogoshi T, Itoh H, Kim KM, Chujo Y (2002) Synthesis of organic–inorganic polymer hybrids having interpenetrating polymer network structure by formation of ruthenium-bipyridyl complex. Macromolecules 35:334–338
47. Schottner G (2001) Hybrid sol–gel-derived polymers: applications of multifunctional materials. Chem Mater 13:3422–3435
48. Chandra S (1989) Super ionic solids and solid electrolytes recent trends. In: Laskar AL, Chandra S (eds), Material Science Series, Academic Press Inc, p 190
49. Ramani V, Kunz HR, Fenton JM (2004) Investigation of Nafion®/HPA composite membranes for high temperature/low relative humidity PEMFC operation. J Membr Sci 232:31–44
50. Zaidi SMJ, Mikhailenko SD, Robertson GP, Guiver MD, Kaliaguine S (2000) Proton conducting composite membranes from polyether ether ketone and heteropolyacids for fuel cell applications. J Membr Sci 173:17–34
51. Amirinejad M, Madaeni SS, Navarra MA, Rafiee E, Scrosati B (2011) Preparation and characterization of phosphotungstic acid-derived salt/Nafion nanocomposite membranes for proton exchange membrane fuel cells. J Power Sources 196:988–998
52. Jalani NH, Dunn K, Datta R (2005) Synthesis and characterization of Nafion®-MO 2 (M = Zr, Si, Ti) nanocomposite membranes for higher temperature PEM fuel cells. Electrochim Acta 51:553–560
53. Novak BM (1993) Hybrid nanocomposite materials—between inorganic glasses and organic polymers. Adv Mater 5:422–433
54. Watanabe M, Uchida H, Seki Y, Emori M, Stonehart P (1996) Self-humidifying polymer electrolyte membranes for fuel cells. J Electrochem Soc 143:3847–3852
55. Watanabe M, Uchida H, Seki Y, Emori M (1998) Polymer electrolyte membranes incorporated with nanometer-size particles of pt and/or metal-oxides: experimental analysis of the self-humidification and suppression of gas-crossover in fuel cells. J Phys Chem B 102:3129–3137
56. Uchida H, Ueno Y, Hagihara H, Watanabe M (2003) Self-humidifying electrolyte membranes for fuel cells—preparation of highly dispersed TiO_2 particles in Nafion 112. J Electrochem Soc 150:A57–A62
57. Miyake N, Wainright JS, Savinell RF (2001) Evaluation of a sol–gel derived Nafion/silica hybrid membrane for proton electrolyte membrane fuel cell applications. J Electrochem Soc 148:898–904

58. Muriithi B, Loy DA (2014) Proton conductivity of nafion/ex-situ sulfonic acid-modified stöber silica nanocomposite membranes as a function of temperature, silica particles size and surface modification. J Mater Sci 49:1566–1573
59. Yuan J, Zhou G, Pu H (2008) Preparation and properties of Nafion®/hollow silica spheres composite membranes. J Membr Sci 325:742–748
60. Adjemian KT, Lee SJ, Srinivasan S, Benziger J, Bocarsly AB (2002) Silicon oxide nafion composite membranes for proton-exchange membrane fuel cell operation at 80–140 °C. J Electrochem Soc 149:256–261
61. Abbaraju RR, Dasgupta N, Virkar AV (2008) Composite nafion membranes containing nanosize TiO_2/ SnO_2 for proton exchange membrane fuel cells. J Electrochem Soc 155:1307–1313
62. Mecheri B, D'Epifanio A, Traversa E, Licoccia S (2008) Sulfonated polyether ether ketone and hydrated tin oxide proton conducting composites for direct methanol fuel cell applications. J Power Sources 178:554–560
63. Mecheri B, D'Epifanio A, Pisani L, Chen F, Traversa E, Weise FC, Greenbaum S, Licoccia S (2009) Effect of a proton conducting filler on the physico-chemical properties of SPEEK based membranes. Fuel Cells 4:372–380
64. Liu ZL, Guo B, Huang J, Hong L, Han M, Gan LM (2006) Nano-Ti_{O2}-coated polymer electrolyte membranes for direct methanol fuel cells. J Power Sources 157:207–211
65. Chalkova E, Pague MB, Fedkin MV, Wesolowski DJ, Lvov SN (2005) Nafion®/TiO_2 proton composite membranes for PEMFCs operating at elevated temperature and reduced relative humidity. J Electrochem Soc 152:1035–1040
66. Baglio V, Aric'o AS, Blasi AD, Antonucci V, Antonucci PL, Licoccia S (2005) Nafion–TiO_2 composite DMFC membranes: physico-chemical properties of the filler versus electrochemical performance. Electrochim Acta 50:1241–1246
67. Sacca A, Carbone A, Passalacqua E, D'Epifanio A, Licoccia S, Traversa E (2005) Nafion-TiO_2 hybrid membranes for medium temperature polymer electrolyte fuel cells (PEFCs). J Power Sources 152:16–21
68. Trakanprapai C, Esposito V, Licoccia S, Traversa E (2005) Alternative chemical route to mesoporous titania from a titanatrane complex. J Mater Res 20:128–134
69. Silva V, Ruffmann B, Silva H, Gallego Y, Mendes A, Madeira L (2005) Proton electrolyte membrane properties and direct methanol fuel cell performance: I. Characterization of hybrid sulfonated poly(ether ether ketone)/zirconium oxide membranes. J Power Sources 140:34–40
70. Nawn G, Pace G, Lavina S, Vezzu K, Negro E, Bertasi F, Polizzi S, Di Noto V (2015) Nanocomposite membranes based on polybenzimidazole and ZrO_2 for high-temperature proton exchange membrane fuel cells. Chem Sus Chem 8:1381–1393
71. Yang P, Costamagna S, Srinivasan J, Benziger AB, Bocarsly AB (2001) Approaches and technical challenges to high temperature operation of proton exchange membrane fuel cells. J Power Sources 103:1–9
72. Costamagna P, Yang C, Bocarsly AB, Srinivasa S (2002) Nafion® 115/zirconium phosphate composite membranes for operation of PEMFCs above 100 °C. Electrochim Acta 47:1023–1033
73. Alberti G, Casciola M, Capitani D, Donnadio A, Narducci R, Pica M, Sganappa M (2007) Novel Nafion–zirconium phosphate nanocomposite membranes with enhanced stability of proton conductivity at medium temperature and high relative humidity. Electrochim Acta 52:8125–8132
74. Anilkumar GM, Nakazawa S, Okubo T, Yamaguchi T (2006) Proton conducting phosphated zirconia-sulfonated polyether sulfone nanohybrid electrolyte for low humidity, wide-temperature PEMFC operation. Electrochem Commun 8:133–136
75. Ozdemir Y, Uregen N, Devrim Y (2016) Polybenzimidazole based nanocomposite membranes with enhanced proton conductivity for high temperature PEM fuel cells. J Hydrogen Energy. doi:10.1016/j.ijhydene.2016.04.132

76. Gomes D, Buder I, Nunes SP (2006) Sulfonated silica-based electrolyte nanocomposite membranes. J Polym Sci Part B: Polym Phys 44:2278–2298
77. Aslan A, Bozkurt A (2012) Nanocomposite polymer electrolyte membranes based on poly (vinylphosphonic acid)/sulfated nano-titania. J Power Sources 217:158–163
78. Tominaka S, Akiyama N, Croce F, Momma T, Scrosati B, Osaka T (2008) Sulfated zirconia nanoparticles as a proton conductor for fuel cell electrodes. J Power Sources 185:656–663
79. Muriithi B, Loy DA (2016) Proton conductivity of Nafion/ex-situ sulfonic acid-modified stöber silica nanocomposite membranes as a function of temperature, silica particles size and surface modification. Membranes 6(12):1–14
80. Aslan A, Bozkurt A (2013) An investigation of proton conductivity of nanocomposite membranes based on sulfated nano-titania and polymer. Solid State Ionics 239:21–27
81. Zhang H, Ma C, Wang J, Wang X, Bai H, Liu J (2014) Enhancement of proton conductivity of polymer electrolyte membrane enabled by sulfonated nanotubes. Int J Hyd Energy 39:974–986
82. Grecea ML, Dimian AC, Tanase S, Subbiah V, Rothenberg G (2012) Sulfated zirconia as a robust superacid catalystfor multiproduct fatty acid esterification. Catal Sci Technol 2:1500–1506
83. Wang Y, Ma J, Liang D, Zhou M, Li F, Li R (2009) Lewis and Brønsted acids in super-acid catalyst $SO_4^{2-}/ZrO_2–SiO_2$. J Mater Sci 44:6736–6740
84. Arata K (1990) Solid Super acids. Adv Catal 37:165–211
85. Nagarale RK, Shin W, Singh PK (2010) Progress in ionic organic–inorganic composite membranes for fuel cell application. Polym Chem 1:388–408
86. Conant JB, Hall NF (1927) A study of super acid solutions, II. J Am Chem Soc 49:3062–3070
87. Hammett LP, Deyrup AJ (1932) A series of simple basic indicators. I. The acidity functions of mixtures of sulphuric acid and perchloric acids with water. J Am Chem Soc 54:2721–2739
88. Shimizu K, Kounami N, Wada H, Shishido T, Hattori H (1998) Isomerization of n-butane by sulfated zirconia: the effect of calcination temperature and characterization of its surface acidity. Catal Lett 54:153–158
89. Yadav GD, Nair JJ (1999) Sulfated zirconia and its modified versions as promising catalysts for industrial processes. Microporous Mesoporous Mater 33:1–48
90. YaoY Guo B, Ji L, Jung KH, Lin Z, Alcoutlabi M, Hamouda H, Zhang X (2011) Highly proton conductive electrolyte membranes: fiber-induced long-range ionic channels. Electrochem Commun 13:1005–1008
91. Ostermann R, Li D, Yin Y, McCann JT, Xia Y (2006) V_2O_5 nanorods on TiO_2 nanofibers: a new class of hierarchical nanostructures enabled by electrospinning and calcinations. Nano Lett 6:1297–1302
92. Choi J, Lee KM, Wycisk R, Pintauro PN, Mather PT (2008) Nanofiber network ion-exchange membranes. Macromolecules 41:4569–4572
93. Zhai Y, Zhang H, Hu J, Yi B (2006) Preparation and characterization of sulfated zirconia (SO42-/ZrO_2)/Nafion composite membranes for PEMFC operation at high temperature/low humidity. J Membr Sci 280:148–155
94. Navarra MA, Croce F, Scrosati B (2007) New, high temperature super acid zirconia-doped Nafion composite membranes. J Mater Chem 17:3210–3215
95. Lin Z, Yao Y, Zhang X (2014) Reviews the recent advances in development of functional electrospun nano fibers. In: Ding B, Yu J (eds) Springer. doi:10.1007/978-3-642-54160-52
96. Mossayebi Z, Saririchi T, Rowshanzamir S, Parnian MJ (2016) Int J Hyd Energy (in press)
97. Zhang Y, Zhang H, Zhai Y, Zhu X, Bi C (2007) Investigation of self-humidifying membranes based on sulfonated poly (ether ether ketone) hybrid with sulfated zirconia supported Pt catalyst for fuel cell applications. J Power Sources 168:323–329
98. Diao H, Yan F, Qiu L, Lu J, Lu X, Lin B, Li Q, Shang S, Liu W, Liu J (2010) High performance cross-linked Poly(2-acrylamido-2-methylpropanesulfonic acid)-based proton exchange membranes for fuel cells. Macromolecules 43:6398–6405

99. Yang C, Srinivasan S, Aric AO, Cret P, Baglio V, Antonucci V (2001) Composite Nafion/zirconium phosphate membranes for direct methanol fuel cell operation at high temperature batteries and energy conversion. Electrochem Solid-State Lett 4:31–34
100. Kannan R, Kakade B, Pillai V (2008) Polymer electrolyte fuel cells using Nafion-based composite membranes with functionalized carbon nanotubes. Angew Chem Int 120: 2693–2696
101. Ponce M, Prado L, Silva V, Nunes S (2004) Membranes for direct methanol fuel cell based on modified heteropolyacids. Desalination 162(383–391):0011–9164
102. Bonnet B, Jones D, Rozière J, Tchicaya L, Alberti G, Casciol M, Massinelli L, Bauer B, Peraio A, Ramunni E (2000) Hybrid organic–inorganic membranes for a medium temperature fuel cell. J New Mater Electrochem Syst 3:87–92
103. Kim JD, Honma I (2004) Proton conducting polydimethylsiloxane/metal oxide hybrid membranes added with phosphotungstic acid(II). Electrochim Acta 49:3429–3433
104. Aslan A, Bozkurt A (2014) Nanocomposite membranes based on sulfonated polysulfone and sulfated nano-titania/NMPA for proton exchange membrane fuel cells. Solid State Ionics 255:89–95
105. Kim D, Park H, Rhim J, Lee Y (2004) Preparation and characterization of crosslinked PVA/SiO_2 hybrid membranes containing sulfonic acid groups for direct methanol fuel cell applications. J Membr Sci 240:37–48
106. Vaivars G, Azens A, Grangvist CG (1999) Proton conducting polymer composites for electrochromic devices. Solid State Ionics 119:269–273

Chapter 10
Electrochemical Promotional Role of Under-Rib Convection-Based Flow-Field in Polymer Electrolyte Membrane Fuel Cells

Hyung-Man Kim and Vinh Duy Nguyen

Abstract Literature data on the promotional role of under-rib convection for polymer electrolyte membrane fuel cells (PEMFCs) fueled by hydrogen and methanol are structured and analyzed, with the aim of providing a guide to improve fuel cell performance through the optimization of flow-field interaction. Data are presented for both physical and electrochemical performance showing reactant mass transport, electrochemical reaction, water behavior, and power density enhanced by under-rib convection. Performance improvement studies ranging from single cell to stack are presented for measuring the performance of real operating conditions and large-scale setups. The flow-field optimization techniques by under-rib convection are derived from the collected data over a wide range of experiments and modeling studies with a variety of components including both single cell and stack arrangements. Numerical models for PEMFCs are presented with an emphasis on mass transfer and electrochemical reaction inside the fuel cell. The models are primarily used here as a tool in the parametric analysis of significant design features and to permit the design of the experiment. Enhanced flow-field design that utilizes the promotional role of under-rib convection can contribute to commercializing PEMFCs.

Keywords Electrochemical reaction · Electrochemistry · Exchange interactions · Flow-field design · Fuel cell performance · Mass transfer · Polymer electrolyte membrane fuel cell · Under-rib convection · Water management

H.-M. Kim (✉) · V.D. Nguyen
Department of Mechanical Engineering and High Safety Vehicle Core Technology Research Center, INJE University, 607 Eobang-Dong, Gimhae-Si, Gyongsangnam-Do 621-749, Republic of Korea
e-mail: mechkhm@inje.ac.kr

Inamuddin et al. (eds.), *Organic–Inorganic Composite Polymer Electrolyte Membranes*, DOI 10.1007/978-3-319-52739-0_10

Nomenclatures

Δp_{tot}	Total pressure drop
z_{max}	Total length of the path
K_{ch}	Channel permeability derived from the Hagen–Poiseuille equation
Δc_{tot}	Total concentration loss
q_{rib}	Under-rib convection flowrate
K_{gdl}	Flow permeability of the GDL
$\bar{\dot{v}}_{spec}$	Mean specific volume flow
$\dot{V}$	Inlet volume flow
$\dot{V}_{local}^{meander}$	Integral of the local volume flow (l/h)
$\bar{\dot{v}}_{spec}^{meander}$	Specific volume flow in the meander channel
$\bar{\dot{v}}_{spec}^{diff}$	Total specific flow through the diffusion layer
P_{cell}	Cell output power density
W_P	Pressure drop loss
AFC	Alkaline fuel cell
BOP	Balance of plant
CFD	Computational fluid dynamic
CL	Catalyst layer
DMFC	Direct methanol fuel cell
EIS	Electrochemical impedance spectroscopy
GDL	Gas diffusion layer
GFF	Grid flow-field
MCFC	Molten carbonate fuel cell
MEA	Membrane electrolyte assembly
MFF	Original design mixed parallel and serpentine
MPL	Microporous gas diffusion layer
MSFF	Multi-serpentine
OCV	Open-circuit voltages, E_0
ORR	Oxygen reduction reaction
PAFC	Phosphoric acid fuel cell
PEMFC	Polymer electrolyte membrane fuel cell
PFF	Parallel flow-field
PTFE	Polytetrafluoroethylene
SFF	Single serpentine
SSFF	Single serpentine flow-field
SOFC	Solid oxide fuel cell
VOF	Volume of fluid
q_{tot}	Total flow rate

1 Introduction

Fuel cells generate fewer harmful emissions and are more efficient compared with the Carnot efficiency of heat engines, because of converting the chemical energy of fuels directly into electricity without combustion [1]. Polymer electrolyte membrane fuel cells (PEMFCs) were first employed in the Gemini space program in the early 1960s. Fuel cells are expected to play a significant role in the strategy to produce positive global change, increase fuel efficiency, and decrease dependency on traditional fossil fuels. Fuel cells and direct electrochemical fuels, particularly hydrogen, provide the promise of being one of the several possible long-term solutions to the improvement of energy efficiency, energy sustainability, energy security, and the reduction of greenhouse gases. Significant environmental benefits are expected for fuel cells, particularly in the area of energy conversion for electric power generation and transportation. There are five types of fuel cells, which are differentiated on the basis of their electrolytes: (i) the polymer electrolyte membrane fuel cell (PEMFC), (ii) the phosphoric acid fuel cell (PAFC), (iii) the alkaline fuel cell (AFC), (iv) the molten carbonate fuel cell (MCFC), and (v) the solid oxide fuel cell (SOFC). While all five fuel cell types are based on the same underlying electrochemical principles, they operate at different temperature regimens, incorporate different materials, and often differ in their fuel tolerance and performance characteristics [2].

The hydrogen-fed proton exchange membrane fuel cells and liquid methanol-fed direct methanol fuel cells (DMFCs) using Nafion®-based polymer electrolyte membranes operate at low temperature (typically less than 100 °C). These low-temperature operating fuel cells are well suited for transportation, portable, and micro-fuel cell applications because of the requirement of fast start-up and dynamic operation in those applications. Some of the sulfonated hydrocarbon polymer membranes show high proton conductivity for potential operation at 100–120 °C [3]. The cathodes in DMFCs fueled by methanol, are similar to the cathodes of PEMFCs fueled by hydrogen; However, anode catalysts in DMFCs are typically high-loading Pt/Ru on high-surface blacks and are used at higher electrode loadings than those of PEMFCs [4]. The basic processes of both PEMFCs and DMFCs are shown in Fig. 1. In addition to the effects of electrochemical oxidation and reduction processes, the performance of the DMFC is affected by methanol permeation from the anode to the cathode, where the methanol is chemically oxidized. Although PEMFCs and DMFCs have similar theoretical open-circuit voltages, E_0, of 1.229 and 1.214 V, respectively, the methanol permeation that occurs during the process lowers the DMFC cell voltage by several hundreds of mV [5].

Technical progress in the design and manufacture of PEMFCs has been dramatic in recent years. In the 1970s, a chemically stable cation-exchange membrane, Nafion®, based on sulfonated polytetrafluoroethylene, was developed by Dupont which led to its large-scale use in the chlor-alkali production industry, energy storage, and fuel cells. During the past two decades, research on the development of PEMFCs with a Nafion® membrane as the electrolyte have received much attention.

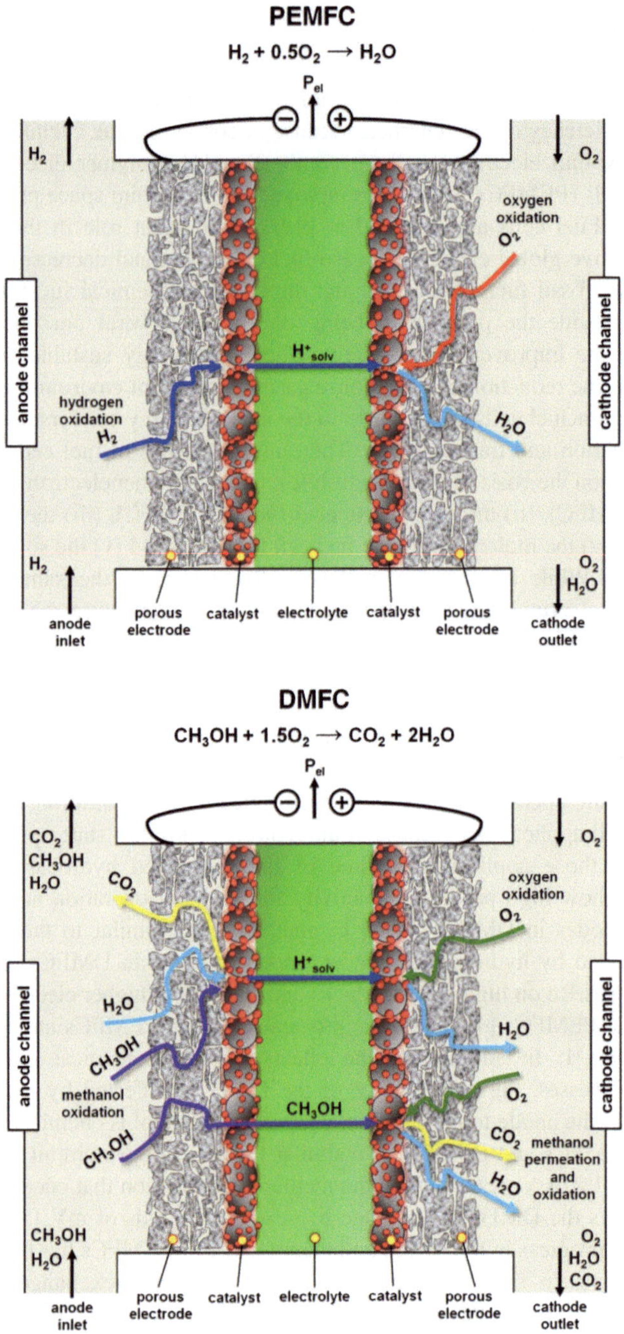

Fig. 1 Principles of PEMFC and DMFC with proton exchange membranes. *Source* Ref. [5]

However, PEMFCs will need to be competitive with the established and highly developed internal combustion engine and other forms of power generation on an economic and consumer basis. Although much progress has been made in the development of the PEMFC, significant technical challenges still remain in a number of areas including reliability, durability, cost, operational flexibility, simplification, and integration of the underlying technology, fundamental understanding, and life cycle impact. To close technological gaps fundamental understanding, new advanced materials and associated engineering design and modeling are required. Many researchers have therefore recently focused on single cells of PEMFC and their components that involve novel membrane electrolytes, catalysts, and structure, electrochemical reaction mechanisms and kinetics, as well as electrode materials and preparation [6].

For a given membrane electrode assembly (MEA), the power density of a fuel cell stack can be significantly increased by reducing the profile of the bipolar plates. Bipolar plate design as a whole and flow channel layout configuration in particular are potential areas of research to make this alternative clean power source compatible to its counterparts. As one step forward, the interdigitated flow-field, in which baffles are added at the ends of some channels, was proposed. The baffle design forces the reactants to flow through the gas diffusion layer (GDL), and the actual force of this reactant flow helps blow out the liquid water trapped in the inner layers of the electrodes. With the interdigitated flow-field design, the mass transport rates of the reactants from the flow channel to the inner catalyst layer were improved, and the water flooding problem at the cathode was significantly reduced [7]. An increase of as much as 50% in the output power density was achieved by appropriate distribution of the gas flow-field alone [8]. Other more complex flow-field patterns and designs that combine more than one of the common patterns have also been reported, particularly for larger MEAs.

In early studies on flow-field development, cross convection has been widely ignored. In two of the first few studies to address this effect, two borderline cases for a serpentine design were distinguished: one in which all the fluid follows through the channel and one in which all the fluid passes through the GDL. The latter reduces the pressure drop over the flow-field. The authors concluded that, for GDLs with high permeability, it is important to keep the pressure drop along the channel low to reduce the risk of unequal reactant distribution due to dominant cross convection [5]. Recently, a number of researchers [9–42] have experimentally and numerically shown that convection in the porous diffusion layer affects transportation of mass and heat, liquid water removal and pressure drop, in PEMFCs. The convection transport mechanism has been referred to as gas–liquid two-phase flow [9], convection through the GDL [14], channel-to-channel gas crossover or cross convection [15–18], cross-leakage flow [21], and sub- or under-rib convection [27–38]. In this paper, we hereafter refer to this particular transport mechanism as 'under-rib convection.' Under-rib convection has recently been recognized as a non-negligible transport process that influences the performance of PEMFCs and DMFCs as a result of the higher GDL permeability that it produces.

The aim of this study is to provide a guide to the promotional role of under-rib convection in PEMFCs that covers a wide spectrum of the relevant scientific, engineering, and technical aspects of this phenomenon within the highly interdisciplinary nature of the fuel cell field. To achieve this, a large body of under-rib convection related data were screened and structured. Our motivations for undertaking the literature review of a particular aspect of PEMFC flow-field design are manifold. First, an analysis of the literature shows that under-rib convection between neighboring channels feeds reactants through GDL to the catalyst layer for mass transport and electrochemical reaction and enables more effective utilization of electrocatalysts by increasing reactant concentrations. Second, a number of particular aspects of PEMFC performance, such as liquid water removal, uniformity of concentration, pressure, temperature and current density, and output power, are promoted by under-rib convection, and these parameters have been screened and structured empirically through experimental and numerical studies. Finally, this work is intended to provide a basis for the optimization of flow-fields in which higher power density is achieved using the promotional role of under-rib convection; this optimization includes state-of-the-art designs that are likely to change as this technology continues to develop.

2 General Description of Performance Improvements in PEMFCs

Before presenting a review of the promotional role of under-rib convection, a brief analysis of recent progress in PEMFC design is presented that includes information on new advanced materials and their associated engineering design and modeling. Extensive surveys of the reliability, durability, cost, operational flexibility, technology simplification and integration, fundamental understanding, and life cycle impact of PEMFCs are already available [43–64]. As discussed later in this study, the analysis undertaken here nevertheless provides a better understanding of the PEMFC based on the collected literature data. The challenges that must be faced in application and commercialization of PEMFCs are emphasized. These challenges include continuous massive advancements in fundamental science and engineering research and in the technological development of PEMFCs.

The main components of a PEMFC power source are illustrated in Fig. 2. The PEMFC consists of (i) a single cell containing porous gas diffusion electrodes, a proton exchange membrane, catalyst layers, and current collectors with the reactant flow-fields (ii) a stack of cells in series with the current collectors, also serving as the bipolar plates, (iii) cell stacks connected in series or in parallel depending on the voltage and current requirements for specific applications, and (iv) the necessary auxiliaries for thermal and water management as well as for the compression of the gases. The unique feature of the PEMFC compared with the other types of fuel cells is that it has a solid proton-conducting electrolyte.

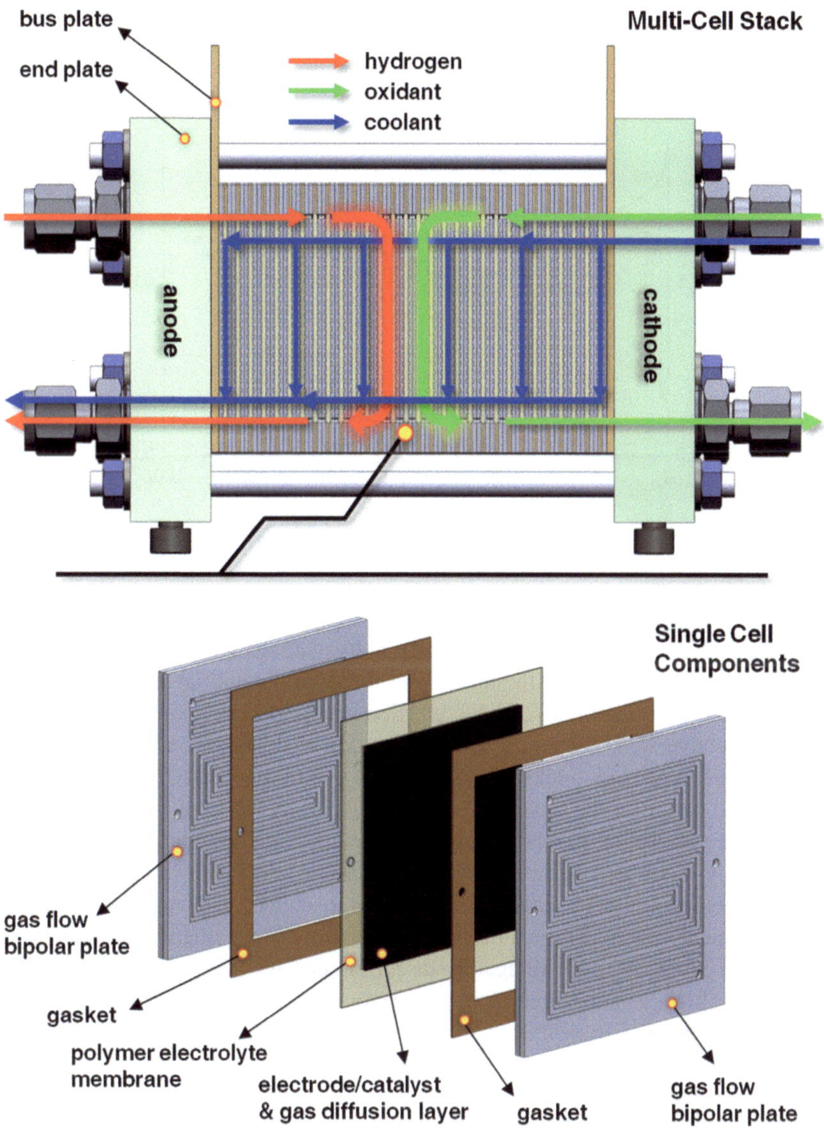

Fig. 2 Schematic of PEMFC single cell components and multi-cell stack

PEMFCs generate a specific power (W/kg) and power density (W/cm^2) higher than any other type of fuel cell [43].

The development of new components with improved characteristics for fuel cell applications requires quantitative determination of their electrochemical performance under relevant fuel cell conditions. The most straightforward approach to this is to construct an MEA and measure the cell parameters in a single cell

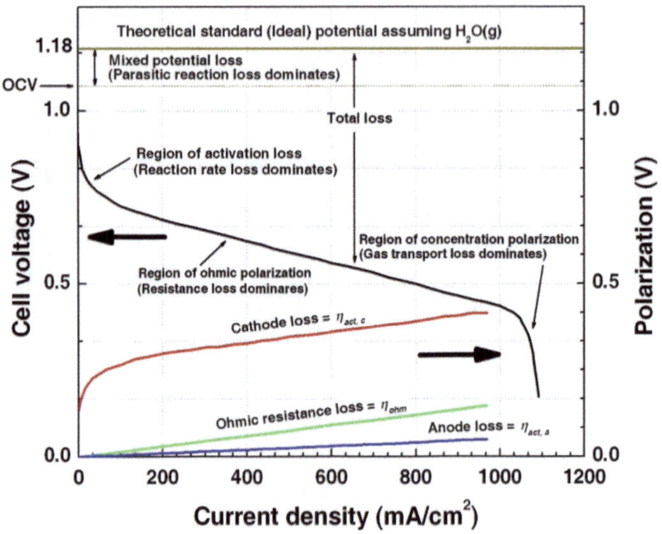

Fig. 3 A typical curve of PEM fuel cell performance. *Source* Ref. [18]

configuration. A membrane electrode assembly includes an anode, a cathode, a membrane disposed between the anode and the cathode, and an extended catalyst layer between the membrane and the electrodes. Apart from the operation conditions, the conversion efficiency of a given MEA depends on several factors, including the type and thickness of both the membrane and the gas diffusion material, the nature of the binder used in the electrodes, and the binder-to-catalyst ratio. Single cell testing is relatively straightforward, and operation conditions can be accurately monitored because it allows specific control over humidity, reactant flow, and temperature. In addition, the ability to collect data from an operating electrochemical system can be alluring. Cell performance is often described by the polarization curve, i.e., cell voltage versus current density. A typical curve is shown in Fig. 3 [41]. In general, three main polarization losses can be identified: (i) activation losses, arising from charge transfer and other reaction kinetics; (ii) ohmic resistances, arising from the electrical resistances of the cell materials and interfaces, and (iii) mass transport limitations, arising from the limitations of mass transport. At low current densities, the shape of the curve is primarily determined by activation polarization; which gives the curve its characteristic logarithmic shape. Activation polarization plays an important role in cell performance because of the reaction rate on the electrode surface is restricted by sluggish electrode kinetics. Like a chemical reaction, the electrochemical reaction must overcome an activation barrier. This barrier usually depends on the electrode material. When pure hydrogen is used as fuel, the activation losses at the anode are negligible because the rate of the hydrogen oxidation reaction is much higher than the rate of the cathode reaction. Hence, the main source of activation overpotential is the cathode, at which oxygen reduction occurs. When current density increases, the shape of the curve becomes approximately linear, reflecting the effect of ohmic losses. This is caused by both the resistance due to the migration of ions within the electrolyte and by the resistance due to the flow of electrons. It can be expressed by the product of cell current and the overall cell resistance. When current density is increased further, the curve begins to bend down due to mass transport overpotentials, which result from limitations in the availability of reactants at the catalyst surfaces. The main source of losses is again the cathode side because the diffusivity of oxygen is significantly lower than that of hydrogen due to the larger molecular size of oxygen [60].

Fuel cell efficiency, on the other hand, is directly proportional to the power density, which can be linked directly to the chemistry of the polymer membrane. Higher achievable power density directly translates to smaller, thus less expensive, fuel cells. A swift comparison of the obtained data those obtained with unmodified membranes is expected to provide useful information about the influence of the inorganic phase on the nanocomposite efficiency. Their effectiveness as a catalyst binder may be evident from an investigation of the interfacial effects of membranes on electrodes and catalysts. In the case of Class I membranes intended for high temperature operations, the methanol crossover flux versus methanol feed concentration can be collected, and the suitability of the membrane for DMFC applications may be determined from these data. The time stability of the membranes under different operating conditions may also be studied. Single cell testing will be of great advantage in fine-tuning hybrid membrane properties in order to give them commercial viability [64]. Henceforth, a major technical challenge is up-scaling single cell performance to stacks.

2.1 Proton Exchange Membrane

The proton exchange membrane (PEM) is the vital component of a PEMFC that encourages fuel cell to attain high power densities. The role of the membrane between the electrodes is to conduct protons produced by the electrochemical reaction from the anode to the cathode. In the 1970s, DuPont developed a perfluorosulfonic acid membrane, Nafion®, which not only showed a twofold increase in specific conductivity but also extended the lifetime of the cell by four orders of magnitude [43]. Perfluorosulfonic acid membranes soon became a standard for PEMFC that remain continued till today. The Nafion® membrane consists of a copolymer of 3, 6-dioxo-4, 6-octane sulfonic acid with polytetrafluorethylene; the Teflon backbone of this structure creates the hydrophobic nature of the membrane, and hydrophilic sulfonic acid groups $\left(HSO_3^-\right)$ have been grafted chemically onto the backbone structure. The hydrophilic sulfonic acid groups cause the absorption of a large amount of water by the polymer, leading to hydration of the polymer [60]. The level of hydration and membrane thickness are two important factors that affect the performance of proton exchange membranes; both play important roles in the suitability of membranes for application in fuel cells.

PEMFC performance, which depends on the proton conductivity, also depends on the degree of humidity of the membrane. Higher proton conductivity is achieved at higher membrane humidity. One way to avoid water drag or water crossover is to reduce the membrane thickness, thereby enabling an improvement in fuel cell performance. Other advantages of reduced thickness include lower membrane resistance, cost effectiveness, and rapid hydration. However, because of the difficulties with durability and fuel bypass, there is a limit to the extent to which membrane thickness can be reduced. To achieve high efficiency in fuel cell applications, the polymer electrolyte membrane must possess the following desirable properties: high proton conductivity to support high currents with minimal resistive losses, zero electronic conductivity, adequate mechanical strength and stability, chemical and electrochemical stability under operating conditions, control of moisture in the stack, extremely low fuel or oxygen bypass to maximize columbic efficiency and production costs compatible with the intended application [53]. The Dow Chemical and Asahi Chemical Companies have synthesized advanced perfluorosulfonic acid membranes with shorter side chains and a higher ratio of SO_3H to CF_2 groups. Compared to Nafion®, the lower equivalent weights of these membranes account for their higher specific conductivities and result in significant improvements in the PEMFC performance, i.e., approximately 50–100 mV increase in cell potential at 1 A/cm^2 over the Nafion® 115, which has approximately the same thickness ($\sim$100 μm) [66].

Proton exchange membrane is the key component of fuel cell systems that limit the lifetime of PEMFCs. Thus, the enhancement of the durability of PEMs is critical to the commercial viability of PEMFCs. In the past decade, membrane degradation mechanism studies have become a focus of attention. In recently published reports, membrane degradation is primarily classified as either chemical/electrochemical or physical. With respect to the former, hydrogen peroxide generated during fuel cell operation and its decomposition intermediate products, both of which have strong oxidative characteristics, have been considered one of the important factors resulting in membrane degradation. The formation of H_2O_2 was confirmed using a microelectrode in an operating fuel cell [67]. It was detected in the outlet stream of a cell with a Nafion® membrane [68]. Membrane durability was subsequently evaluated by both an ex situ Fenton test [69] and an in situ OCV accelerated test [70].

One of the major issues to be addressed in the development of proton-conducting nanocomposite and hybrid membranes for fuel cell applications is their high temperature stability since low temperature mitigates degradation [71]. Although fuel cell performance degradation was considerable under conditions in which the MEA was allowed to lose much water, no lowering of the open-circuit potential was observed, suggesting that no increase in hydrogen crossover occurred. On the other hand, a combination of high temperature and reduced humidity promotes the degradation rate [72]. Studies of membrane stability at even more elevated temperatures (e.g., 120 °C) are often carried out at reduced humidity (<50%). It is expected that chemical degradation will be faster under these conditions than under ideal conditions. Hence, Asahi Glass has reported operation of a new membrane for 400 h without membrane failure or even significant fluoride release [73].

Desulfonation is generally studied by means of thermogravimetric analysis (TGA), differential thermal analysis (DTA), Fourier transform infrared spectroscopy (FTIR), and TGA-mass spectrometry (MS). In Nafion®-based composite, the decomposition behavior is attributed to the loss of sulfonic acid groups present in the unmodified Nafion® membrane [74]. The primary mechanism of degradation of these membranes is the degradation of the polymeric backbone; a secondary mechanism is the degradation of the pendant groups or inorganic compounds inside the membranes, which occurs at higher temperatures than primary degradation. Modification of polyaromatic membranes with acidic oxides results in an increase in membrane thermal stability at elevated temperatures. In efforts to improve membrane performance for high-temperature fuel cell applications, various polymers have been synthesized and tested for their proton conductivity, mechanical stability, electrode–membrane interface, and connectivity. These efforts, aimed at the commercialization of such membranes and the reduction of the cost of using PEM at elevated temperatures, seem to continue with insightful vision [3]. The future design concept of high-temperature PEMFCs will open new promising avenues for further research and development.

2.2 *Electrode and Catalyst*

In PEMFCs, as in the case of other low- or intermediate-temperature fuel cells such as the phosphoric acid fuel cell (PAFC) and the alkaline fuel cell (AFC), Pt and Pt alloys are the best electrocatalysts discovered to date for both hydrogen oxidation and oxygen reduction. In the types of fuel cells mentioned, the overpotential for the former reaction is considerably lower than that for the latter reaction. For example, in a PEMFC operating at a current density of 1 A/cm^2, the overpotential at the hydrogen electrode is about 20 mV while the overpotential at the oxygen electrode is about 400 mV. About one-half of the overpotential at the oxygen electrode is due to its loss at open circuit. The departure of the potential of the PEMFC from the reversible value is due to the extremely low exchange current density for oxygen reduction (about 10^{-9} A/cm^2) on smooth platinum electrodes. Due to this very low exchange current density value, competing anodic reactions are responsible for setting up a mixed potential of about 1.0 V for the oxygen electrode at open circuit. Oxygen reduction is considerably more complex than hydrogen oxidation due to (i) the strong O–O bond and the formation of highly stable Pt–O or Pt–OH species, (ii) the fact that it is a four-electron transfer reaction, and (iii) the possible formation of a partially oxidized species [43].

One of the major problems encountered with Pt electrocatalysis for hydrogen electrodes is its low tolerance to CO in H_2 from reformed fuels. Furthermore, according to the US Department of Energy, an increase in the cell potential to approximately 0.75–0.8 V is necessary for PEMFCs to compete with compression-injection direct ignition engines in order to meet the goal of 45% efficiency in fuel consumption. The improvement can only be achieved by reduction of the oxygen overpotential by 50–100 mV. Recent studies have demonstrated that such an improvement is possible by using intermetallic electrocatalysts of platinum with a transition metal [75, 76], as used in state-of-the-art PAFCs.

The issue of carbon corrosion seems to be of even greater concern. Graphitic carbon is more stable than the conventionally used carbon black but has a lower surface area. This limits the minimum metal particle size, which in turn reduces the activity. In combination with the requirement for further reduction of the Pt loading, this is not a promising situation. Carbon corrosion may also be mitigated by promoting competing reactions. Development of catalysts for oxygen evolution has been suggested, but this concept is still unproven and may introduce new durability issues. The same holds for alternative supports. The electrode concept is based upon nonconductive organic whiskers coated by Pt [9]. In this concept, the support is not exposed to the electrolyte and Pt forms a continuous structure that is much less susceptible to dissolution and shows a high specific activity toward the oxygen reduction reaction (ORR). In spite of the low active surface area that is obtained and anticipated mass transport problems [77], may be a superior concept from durability point of view. Thus, in spite of recent research that has led to many new insights, testing protocols and characterization methods, improvement of fuel cell electrodes to meet existing durability targets still appears to be a formidable task.

2.3 Gas Diffusion Layer

The gas diffusion layer (GDL) is responsible for the transportation of heat and gaseous phase, electronic contact, and water removal in a fuel cell. GDL consisting of carbon fibers, is a macroporous layer which has to some extent made hydrophobic by a Teflon coating. The microporous gas diffusion layer (MPL), which is positioned between the GDL and the catalyst layer, consists of carbon black particles and Teflon as a binder. Unlike the carbon black in the catalyst layer, the carbon black in the MPL is not susceptible to electrochemical corrosion and does not contain Pt to catalyze oxidation reactions but its chemical surface oxidation by water or even loss of carbon due to oxidation to CO or CO_2 cannot be excluded [78, 79]. These processes are responsible to increase hydrophilicity of the MPL. The carbon fibers of the GDL may be more stable but are otherwise susceptible to the same reactions. Decomposition of polytetrafluoroethylene (PTFE) used as a binder or hydrophobic coating has also been suggested. This idea was set forth on the basis of XPS data, but a mechanism has not been proposed [80].

As a result of chemical surface oxidation degradation, both the GDL and the MPL lose their hydrophobic character [80, 81] and the pore structure of the materials also changes. Both these phenomena have substantial effect on the water content of the GDL and MPL and therefore on their mass transport properties. Increased liquid water content of the GDL and MPL will impede gas phase mass transport because pores initially used for gas phase mass transfer will be increasingly blocked by water. The relation between microstructure and surface properties on the one hand and mass transport properties on the other has been the subject of several recent experimental [13, 82] and modeling [83–86] studies. The results of these studies indicate that mass transport can indeed be seriously affected by the hydrophobicity of the GDL and MPL as well as by the pore size.

The properties of the GDL can also be changed by mechanical degradation arising from the compression forces in a fuel cell. From ex situ tests in which the material was aged under a compression force, it was concluded that compressive strain increased with applied pressure but even more strongly with temperature [87]. From this, it was concluded that GDL strain was influenced by PTFE stability. Properties such as in-plane electrical resistivity, surface contact angle, bending stiffness, and porosity were not affected. However, it was found that convective air flow through the GDL can lead to loss of material [58].

2.4 Membrane Electrode Assembly

The membrane electrode assembly (MEA) is the 'heart' of the PEMFC and thus its structure and composition are of vital importance for the following reasons: (i) minimizing all forms of overpotential and maximizing the power density, (ii) minimizing the noble metal loading in the gas diffusion electrodes by high utilization of the surface areas of nanosized particles of the electrocatalyst (iii) for effective thermal and water management, and (iv) attaining the lifetimes of PEMFCs necessary for power generation, transportation, and portable power applications. Major progress in the designing of the MEA was made in the late 1980s and the early 1990s.

A breakthrough in achieving a tenfold reduction in platinum loading (from about 4 mg/cm^2 as used in the Gemini space flights to 0.4 mg/cm^2 or less in the PEMFC) in the 1980s and 1990s arose out of an invention [88] by workers at Los Alamos National Laboratory (LANL). The breakthrough was made possible by using platinum supported on high-surface-area carbon (e.g., Vulcan XC72R) as electrocatalyst rather than pure Pt black crystallites as in the Gemini fuel cells and by impregnation of a proton conductor (e.g., Nafion®) into the active layer of the porous gas diffusion electrode. The main factors that make it possible to reduce the platinum loading from more than 4 to 0.4 mg/cm^2 include (i) the considerably higher BET surface area of the carbon-supported electrocatalyst (particle size about 30 Å) than that of the unsupported previously developed PEMFC electrocatalyst (particle size about 100–200 Å) and (ii) extension of the three-dimensional zone in the electrode by impregnation of the proton conductor so that the utilization of the electrocatalyst might be similar to that in a fuel cell with a liquid electrolyte (e.g., phosphoric acid or potassium hydroxide).

In the late 1990s, significant increase in power densities with even further reduction in platinum loading to a level of about 0.05 mg/cm^2 for the hydrogen electrode and 0.1 mg/cm^2 for the oxygen electrode were achieved by deposition of thin active layers of the supported electrocatalyst and proton conductor on an uncatalyzed electrode [89] or on the proton-conducting membrane [90]. These active layers are only about 10–20 μm in thickness and, unlike conventional electrodes, contain no Teflon. Because the active layers are considerably thinner than those of conventional electrodes (10 vs. 50 μm), the ohmic and mass transport overpotentials of the electrodes, which are generally predominant at intermediate and high current densities, are greatly minimized. An equally important advantage of such types of electrodes is the increase in platinum utilization from about 20–25% to 50–60%. The high utilization of platinum is essential from the point of view of reducing the platinum loading and hence the cost of the platinum in the electrode.

2.5 Bipolar Plate

Bipolar plates constitute the backbone of a fuel cell power stack, conduct current between cells, facilitate water and thermal management through the cell, and provide conduits for reactant gases, namely hydrogen and oxygen. In a PEMFC stack, the bipolar plates are key elements because they account for large fractions of the total weight, volume, and cost of the stack. Bipolar plates may represent up to 80% of the total weight and 45% of the total cost of a PEMFC stack [91]. Furthermore, these components perform vital functions in the stack such as carrying electric current away from each cell, distributing fuel and oxidant homogeneously within individual cells, separating individual cells and facilitating water management within the cell [92]. Because the plates perform such a number of functions, a variety of materials have been proposed for use in the manufacture of bipolar plates.

The gold-coated titanium and niobium used by General Electric for manufacturing bipolar plates in the 1960s, were substituted by graphite in the early 1970s because of its high corrosion resistance and low cost [44]. Graphite bipolar plates were manufactured starting from high-surface-area graphitic carbon powder mixed with ligand resins; after molding at high temperature and pressure, the gas distribution channels were introduced into the graphite blocks. However, due to the lack of graphite durability under mechanical shocks and vibration combined with cost effectiveness concerns regarding its high volume manufacturability, considerable research work is currently underway to develop metallic bipolar plates with high corrosion resistance, low surface contact resistance, and inexpensive mass production. Alternatives to pure graphite plates are composite bipolar plates based on the mixture of polymers and graphite particles. This class of materials allows mass production at a reasonable cost using manufacturing processes such as injection molding for thermoplastics [93]. There are several examples of graphite-based composite bipolar plates that use polypropylene (PP), polyphenylene sulfide (PPS), phenolic, or vinyl ester resins as matrices [94]. The polymer matrix gives flexibility to the bipolar plate, thus improving its mechanical strength. The chemical stability is also not seriously affected by the incorporation of polymer in graphite. On the other hand, electrical conductivity is proportionally diminished because polymers are insulating materials. Thus, it is mandatory to formulate a composite bipolar plate with care so as to attain mechanical performance without sacrificing electrical conductivity.

In spite of the advantages of graphite-based composite bipolar plates associated with their low weight, high production, and chemical stability, comparison of the overall performance of graphite-based plates with that of metal bipolar plates reveals two major drawbacks, lower mechanical resistance and lower electrical conductivity. However, a handicap that may significantly decrease the performance of metal bipolar plates is their susceptibility to corrosion in the acidic and humid environment of PEM fuel cells. Metals operating in a fuel cell within a pH range of 2–4 and temperatures around 80 °C may suffer dissolution. The ions leached may poison the MEA, decreasing the power output of the fuel cell [95]. Furthermore, passive layers formed during operation increase the electrical resistivity of metal bipolar plates. Consequently, the fuel cell efficiency is negatively affected due to increasing interfacial contact resistance as the oxide layer grows. These effects offset the advantage of high electrical conductivity [96]. The problems outlined above may be minimized by protecting metal bipolar plates from corrosive fuel cell operating conditions with coatings [61]. A wide variety of alternatives have been proposed in research toward this objective.

2.6 Single Cell and Stack

The wide range in power output of fuel cells implies significant variation in fuel cell operating conditions, active area, gas diffusion media, bipolar plate, and MEA properties. The performance of a PEMFC stack varies significantly with respect to the number of cells, operating conditions, material properties, and flow-field characteristics. Understanding how the PEMFC stack performs, including the transport behavior of each particular design with respect to overall performance analysis and the local examination of electrochemical variables, temperature and water distribution, will lead developers toward improved designs and enhanced durability. This understanding requires both experimental study and numerical analysis [97]. The 55-partner-strong FCTESTNET thematic network was established to define harmonized test procedures applicable to the component level (single cells, multi-cell stacks, Balance of Plant or BoP), subsystems, and entire fuel cell systems. The purpose of this test module was to characterize the performance of a PEMFC stack under constant current conditions. The module is used for measuring the voltage and power of a stack as a function of drawn current. If properly instrumented, cell voltages, different temperatures, reactant flows, relative humidity, stack fluid pressures, and pressure drops can be measured. These modules are accessible at the FCTES^{QA} website [98].

2.6.1 Water and Heat Management

Water plays an important role in fuel cells. Its functions in fuel cells include as a reactant at the anode in the creation of protons, hydrating the PEM membrane to promote proton transport toward the cathode, and representing a product of the consumption of those protons at the cathode. The flux of water toward the anode under fuel operating conditions can lead to so-called 'water flooding,' requiring a balance between membrane hydration and flooding avoidance [99]. The maximum degree of hydration of the membrane electrolyte is vital for the PEMFC to attain its highest performance. If sufficient hydration is not achieved, the ohmic overpotential in the membrane could be a major source of loss of efficiency in the PEMFC. The flux of water is measured by the electro-osmotic coefficient, which equals the ratio of the number of transported water molecules to transported protons. With improvements due to the invention of Nafion®, the electro-osmotic drag coefficient of water in a fuel cell is about 2.5. Thus, because the oxidation of 1 mol of methanol in the presence of water at the anode generates 6 protons, 2.5×6 mol of water will be dragged through a Nafion® 117 membrane toward the cathode. This value will be higher if the membrane is equilibrated in a water-methanol mixture [100], in which the loss of 16 water molecules from the anode occurs for every mole of methanol oxidized.

The potential applications of PEMFCs in electric vehicles have stimulated the development of internal hydration techniques. In these techniques, the water produced by the electrochemical reaction is used for the hydration of the membrane, allowing the elimination of an external humidification subsystem. This method can also reduce the volume and weight, and thus the size, of the overall system. Several methods have been proposed which include (i) the use of porous carbon blocks for the bipolar plates (due to capillary condensation, such plates are able to retain the water produced by the electrochemical reaction and also assist water transport from the cathodic to the anodic side of the fuel cell) in this type of cell stack, the carbon blocks are not channeled, and the gases are humidified by forcing them to flow through the wet porosities [101]; (ii) impregnation of the membrane solution into the electrode, forming a thin recast film on the surface, followed by hot-pressing of two impregnated electrodes onto each other; with this procedure, very thin electrolyte films that show very small ohmic resistance and thus allow operation even under unfavorable conditions, such as low pressure and temperature and without external humidification, can be prepared; and (iii) impregnation of thin Nafion® recast membranes with a small amount (approximately 5–6 wt%) of nanosized Pt particles. In this case, Pt catalyzes the production of water from the crossover flux of H_2 and O_2 across the membrane, thereby ensuring a satisfactory hydration level [102].

Proper thermal management is also recognized as a critical issue in the commercialization of PEMFCs. The ambient temperature directly affects the heat exchanger fan power consumption and the maximum power, with a statistically significant effect on net efficiency [103]. PEMFCs are operated at temperatures in the range of 70–80 °C to prevent dehydration of the PEMs. A cell temperature below 60 °C leads to water condensation and flooding at the electrodes, accompanied by voltage loss. These stringent thermal requirements present significant heat transport problems. Heat generation in PEMFC arises from the entropic heat of reactions and the irreversibility of the electrochemical reactions and ohmic resistances, as well as from water condensation [104]. Thermal management in a DMFC is intimately tied to the water and methanol-transport processes. The heat generation in DMFCs is comparatively higher than in PEMFCs due to lower energy efficiency (20–25% at 0.3 and 0.4 V). High-cell temperatures promote methanol oxidation and increase the methanol crossover rate, which reduces fuel cell efficiency and energy density.

2.6.2 Fuel Crossover, Oxidation, and CO Poisoning

There are two technical challenges for PEMFC technologies: (i) high methanol crossover (10^{-6} mol/cm^2 s) and its further reaction with the Pt catalyst sites on the cathode, which reduces the fuel cell efficiency (50–100 mA/cm^2) [105]; (ii) insufficient activity of the anode catalyst, which results in high overpotential loss (about 350 mV) for DMFC compared with that for PEMFC (60 mV) [106]. Slow anode kinetics due to methanol crossover reduces the power density of DMFC approximately 3–4-fold in comparison with a hydrogen fuel cell [104]. Pt–Ru and several other anode catalysts have been developed [107, 108], and their effects on electrochemical anode reaction and cell performance have been experimentally studied [109].

The presence of CO in the fuel gas (sometimes termed CO poisoning) degrades PEMFC performance by its preferential adsorption on the platinum surface, resulting in blockage of active sites. The following three methods were reported to mitigate CO poisoning effects: (i) the use of a platinum alloy catalyst; (ii) higher fuel cell operating temperature; and (iii) introduction of oxygen in the fuel gas [110]. The poisoning effect is temperature dependent, being less pronounced at high temperature [111]. To avoid CO poisoning, PEMFCs require operation in pure hydrogen and DMFCs must be operated at high temperature with an efficient electrocatalyst.

In addition to the problems discussed above, various other factors also affect fuel cell performance; these include membrane dehydration, reduction in membrane conductivity, and mechanical stability [112]. Also, under oxidizing conditions, free radicals (oxygen, hydroxide, and peroxide) attack the alkyl chains of the membrane, resulting in loss of functionality and reduced overall membrane performance. Thus, for successful fuel cell operation, membranes that are highly thermally, mechanically, and oxidatively stable over a wide pH range are urgently required.

2.6.3 Scale-up and Long-Term Experiments

Unlike other types of fuel cells, PEMFCs show some loss of efficiency and power density with scale-up in the area of the electrodes and the increase in number of cells in a stack. The main reason for this is that removal of the product (i.e., liquid water) becomes more difficult in larger systems. Furthermore, a high water vapor pressure in the reactant flows causes an increase in overpotential, especially at the cathode [113]. An issue strongly related to water management is that of thermal management in stacks. On one hand, temperatures that are too low cause water condensation problems, as discussed above. On the other hand, an even more important factor is that high-cell temperature, even in confined areas, leads to membrane dehydration and consequent loss of performance. Although a PEMFC is a very efficient system, 40–50% of the energy produced is still dissipated as heat. This loss of electrochemical performance is due to the irreversibility of the cathodic reaction, to ohmic resistance and mass transport overpotentials. To prevent drying out of the membrane and the rise in the cell temperature, the generated heat must be removed from the cell. Because the temperature difference between the cell and the surroundings being approximately about 50 °C, it is impossible to rely on natural convection and air cooling for efficient heat removal [114].

Fuel cell and stack design engineering aimed to improve the performance of devices in terms of power density and specific power. Single cell design is realized on the basis of experimental purposes. In addition, fuel cell stacks are often designed for worldwide applications in order to reach enough power. However, stack design is more complicated because power and overall voltage target must be taken into account and MEA loss of performance must be minimized. Moreover, the stack final application, specific geometrical requests and the cooling system must also be considered [115]. Consequently, innovative PEMFC stack research involves all the components of the stack from the membrane to stack auxiliaries, with special attention to materials, hardware design (channels geometry, manifold, sealing, and the like) and fuel cell component coupling.

The polarization curves of an H_2/O_2 single cell with a composite membrane are shown in Fig. 4. It is obvious that the performance of the single cell reached its maximum at 60 h and then degraded significantly during subsequent testing.

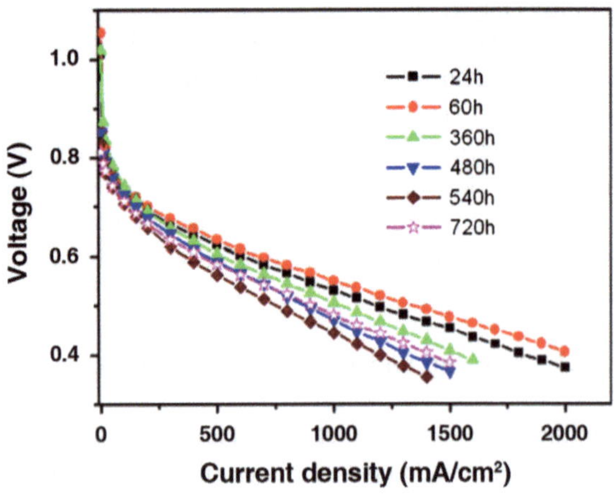

Fig. 4 Polarization curves of the H_3PO_4/Nafion®-PBI composite membrane single cell at different test times. *Source* Ref. [60]

For example, the single cell voltage at 1000 mA/cm^2 fell from 0.55 V at 60 h to 0.45 V at 540 h, although it increased slightly at 720 h. The degradation in the performance of the single cell during the test resulted from degradation of the electrocatalyst and the membrane [116]. The small increase at 720 h appears due to an increase in the membrane proton conductivity resulting from the increased permeability of the membrane [117]. The long-term stability of membranes under different operating conditions can also be examined. In short, MEA testing will be of great advantage in fine-tuning the properties of hybrid membranes to establish their commercial viability. However, a major technical challenge is the scaling up single cell performance to stacks [65].

Long-term experiments can be indicator of the severity of degradation of membranes and relative contribution to performance loss under various conditions. The compilations of PEMFC experiments longer than 1000 h are meant to give a qualitative understanding of the possible relationship between operating conditions and voltage decay rates; such tests are difficult to compare quantitatively because of differences in materials, flow-fields, start-up procedures, and other parameters. Note that the voltage decay rates comprise irreversible as well as reversible losses. In durability studies, it must always be remembered that part of the decay may be reversible. This is particularly true in the case of under- or over-saturation of the gases. Both drying out and flooding can have a detrimental effect. During a 26,000 h test conducted by Gore [118], a voltage loss of 110 mV was observed at a current density of 800 mA/cm^2. The sensitivity of a fuel cell system to dual-stack parallel and series array operation has been evaluated experimentally. The system net efficiency was lower for the parallel arrangement than that of the series arrangement because connecting the stacks in parallel equalized the stack voltages. The weaker stack depresses the polarization curve of the stronger stack, while the stronger stack boosts the polarization curve of the weaker stack [119].

3 Structured Techniques for Flow-Field Optimization

The performance of a PEMFC is primarily determined by the intrinsic electrochemical efficiency of the MEA. Nevertheless, other factors such as flow-field design, thermal and water management, and operational control are also important [51]. The flow plate is one of the key components of a PEMFC and serves as both the current collector and the reactant distributor. The reactants, as well as the products, are transported to and from the cell through the flow channels. The essential requirements for the flow-field are uniform distribution of reactants over the entire electrode surface and effective removal of products from the cell, to minimize the concentration polarization. Flow plates contain either a few very long channels or a large number of channels. These channels make up the flow-field through which the reactants are distributed to the entire surface of the MEA. An optimal flow-field design is critical for obtaining high power density in a fuel cell and thus is extremely important. Since the early development of PEMFCs, variety of flow-field designs has been introduced. Generally three major types of flow-fields are in use: parallel, serpentine, and interdigitated. The serpentine flow-field is the most commonly used [2].

Much efforts have been put forth to design an optimum flow-field for PEMFCs that can both efficiently distribute reactants to the reaction sites and remove products through the outlet. The presence of water in the products has been one of the main concerns. To operate a PEMFC within a narrow band of water content is a challenging task considering the vast range of power demand and environmental conditions that pertain to vehicular applications. This makes the design of the gas channels a very critical factor. Under different flow rates, relative humidities of air intake and different current demands, complex two-phase flow regimes have been observed in the gas channels. Therefore, it is essential to design blockage-resistant gas channels with an acceptably small power loss to pump fuel and oxidant.

Data from a set of investigations on flow-field optimization that include both experimental [39, 63, 98, 120–137] and modeling approaches [59, 68, 138–157] to achieving uniform distribution of reactants were analyzed with the validation of experimental and numerical results [97, 158–161]. The subject of these investigations ranged from single cells to stacks in PEMFCs. These studies, which are schematically shown in Fig. 5, are being described in this section in order to provide guidance to the optimization of flow-field design for efficient operational control in PEMFCs.

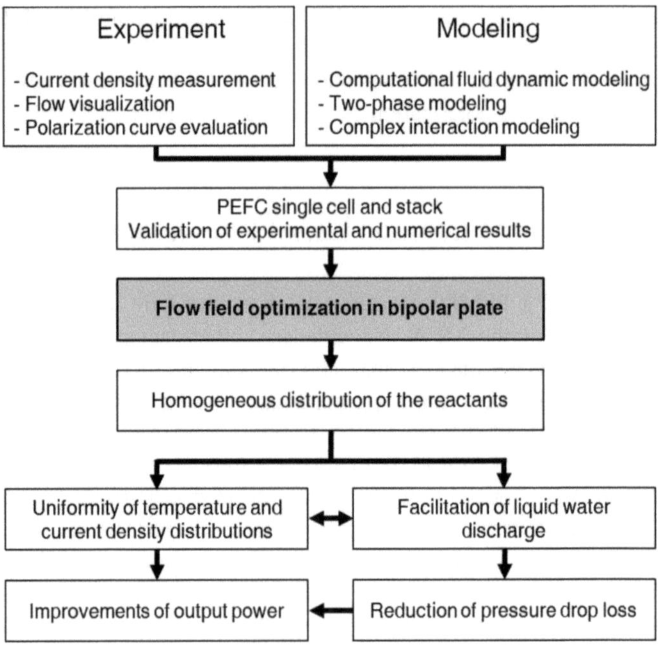

Fig. 5 Flow diagrams for the optimization of flow-field in PEMFCs

3.1 Experimental Approaches to Flow-Field Optimization

3.1.1 Current Density Measurement

It is desirable to operate a PEMFC at uniform current density over the MEA because nonuniform current distribution in a PEMFC could result in poor reactant and catalyst utilization, low energy efficiency, and possible corrosion inside the fuel cell. Local current distribution in a fuel cell can be strongly affected by the operating conditions as well as by the organization of the reactant flow arrangement between the anode and cathode streams, especially in practical PEMFCs of large cell size [120]. For instance, reactant depletion along the flow channel leads to current variation from the channel inlet to the exit and degrades the cell performance and flow arrangements between the anode and cathode streams, such as co-, counter-, and cross-flow, can exacerbate this effect considerably, resulting in complex current distribution patterns over MEA surfaces. During normal PEMFC operation, nonuniform current distribution cannot be measured directly because only integral values such as cell voltage, current density, and impedance can be measured. Therefore, considerable effort has been made to measure and understand the current density distribution in PEMFCs using a variety of approaches.

The segmenting flow-field is one of the most popular techniques for measuring the current distribution in PEMFCs. This technique measures the current distribution in a PEMFC using a segmented bipolar plate and printed circuit board. The anode GDL and catalyst layers were also segmented. Three other segmentation methods were also proposed; these included a partial MEA approach, current mapping, and the use of sub-cells to measure the current distribution [121]. Using the same principles, the current distribution was measured along the length of a single flow channel in a PEMFC with higher resolution of spatial and time. The time delay of local currents was observed after inlet reactant gas changes due to the mass transport of gas through GDL. Simultaneous measurements were made for species concentration and current distribution. Furthermore, a PEMFC permitting simultaneous evaluation of current, temperature, and water distribution in the cell under various operating conditions was designed [122, 124]. Consequently, the experimental setup allowed to perform a simultaneous evaluation of current, temperature, and water distribution in a polymer electrolyte fuel cell under operation. The test fuel cell has a segmented anode flow-field for current distribution measurement. The cathode end plate was made of an optical window, which was transparent for infrared as well as for visible wavelengths. This allowed infrared thermography and optical surveillance of water droplets and flow-field flooding.

The uniformity of current density distribution was evaluated experimentally using third single cells and the PEMFC test station developed and installed at Power System and Sustainable Energy Laboratory of INJE University, as shown in Fig. 6 for quantitative analysis in performance test of PEMFC. As shown in Fig. 6, reaction gases such as hydrogen and oxygen must be carefully moved from the container to the entrance of fuel cell without any changes of temperature, pressure,

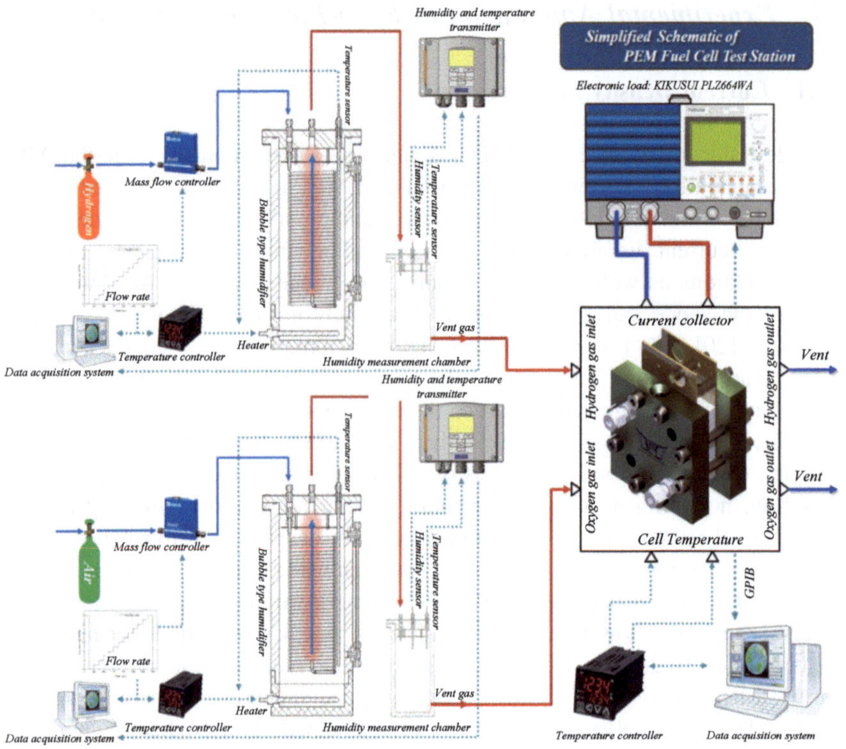

Fig. 6 Schematic diagram of the main experimental modules installed at Power System and Sustainable Energy Laboratory of INJE University

and flow rate. The performance of PEMFC is greatly affected by temperature, pressure, flow rate, and humidity. The segmented current collector was inserted between the back side of the anode flow-field plate and the current collector, as shown in Fig. 7. The graphite block is filled with an electrically insulating epoxy between the anode bipolar plate, the sensor plate, and the current collector to minimize electrical contact resistances [123].

Other techniques for measuring local current distribution in PEMFCs have been reported in the literature. The potential distribution at the interface between the GDL and the catalyst layer (CL) was measured to obtain sub-millimeter resolution of the current distribution measurement [125]. This research provided useful insight into mass transport issues in land and channel areas of the flow-field plate. A novel approach was developed to measure current densities under the channel as well as under the land separately in PEMFCs with a parallel flow-field. The cathode catalyst layer was designed to have either the area under the land or the area under the channel loaded with catalyst. This design yielded separate measurements of current densities under the land and the channel. An optimization study of current density

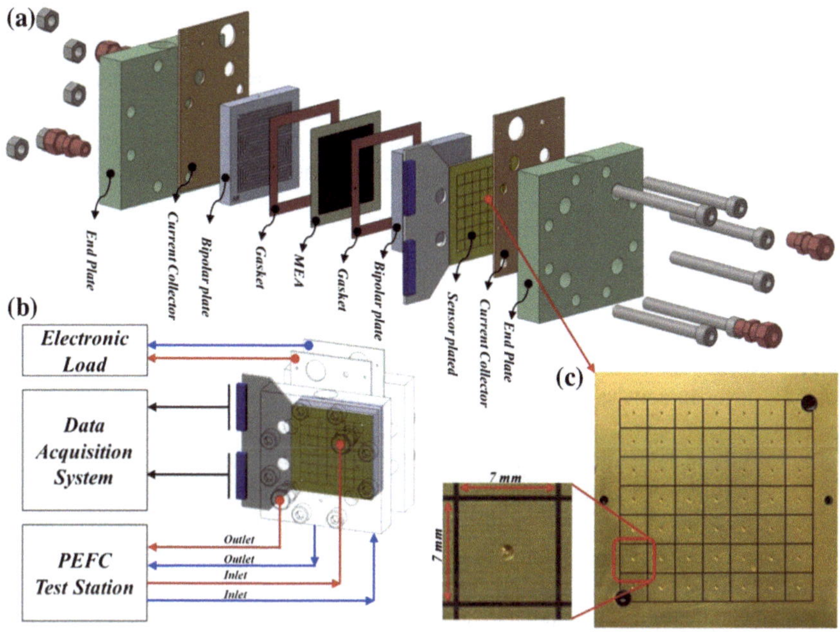

Fig. 7 Schematic of the experimental setup with the segmented Bipolar Plate installed at Power System and Sustainable Energy Laboratory of INJE University

distribution under the land and channel areas was conducted for a variety of serpentine flow-field geometries and operating conditions [126].

It is clear that considerable effort has been devoted to developing methods for current distribution measurements in PEMFCs for various reactant flow arrangements and under different operating conditions such as varying stoichiometry ratio, reactant pressure, cell temperature, and relative humidity. Currently, the local current distribution in a PEMFC with different flow arrangements and operating conditions is most often measured using the segmented bipolar plate or printed circuit board technique [127]. The design of the flow-field greatly affects the flow distribution and the final performance of the PEMFC system. An optimization model that takes the effect of the GDL deformation into consideration was proposed to obtain more uniform flow distribution for the flow-field configuration design [128]. It was found that the GDL deformation must be taken into consideration in the optimization of the flow-field due to its great influence on the flow distribution. In addition, both the uniformity of the flow-field and its sensitivity to the GDL intrusion should be considered in order to obtain a robust design. From the optimization results, it can be seen that in order to achieve more uniform flow distribution and high performance, the slots in the central channels should be shallower and wider than those in the side channels.

3.1.2 Flow Visualization

In PEMFCs, insufficiencies in the proton network and membrane degradation can occur as a result of drying of the membrane. Excess water, on the other hand, blocks gas flow channels and hinders gas access to active catalyst areas, resulting in significant drops in performance. Water management is, therefore, one of the most important issues in the design of PEMFCs. Oxygen partial pressures and water blockages were visualized simultaneously in a triple serpentine PEMFC, and the relationship between the two was investigated [129]. Air can be supplied from the other channels through the GDL, and power generation continues at the catalyst layer, even under the blocked channel. Thus, water continues to be produced at the catalytic layer under the channel blocked by water. The water layer produced at the catalytic layer under the blocked channel might not be easily removed because of the limitation of gas flow due to the blockage. In this situation, even though the water droplet in the channel is ejected from the cell, the catalyst layer is still wet; thus, the oxygen consumption is low in the channel and thus the oxygen partial pressure remains high. Such a flooded area may expand with operation time, causing significant performance and efficiency losses in the fuel cell, as shown in Fig. 8.

Novel diagnostic techniques designed especially for the visualization of water have been intensively applied to PEMFCs to elucidate the nature of water transport inside the cell. There has been a rapid and continual growth in the number of publications on in situ and ex situ visualization techniques in PEMFCs. These publications cover topics such as magnetic resonance imaging, neutron radiography, X-ray imaging, fluorescence microscopy, infrared visualization, and direct optical visualization. These techniques play complementary roles in achieving an understanding of water transport in PEMFCs because each technique has an individual capability of detecting the presence of water in different materials at different spatial and temporal resolutions. The state of the art in relation to in situ

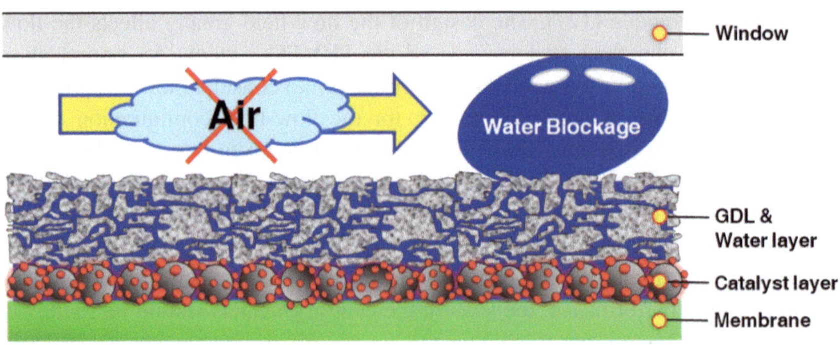

Fig. 8 Schematic drawing of the channel, gas diffusion layer, and catalyst layer with water blockage. *Source* Ref. [128]

visualization techniques for use as diagnostic tools in studies on water transport in PEMFCs has been comprehensively reviewed [63].

Because of the importance of water transport to the functioning and efficiency of fuel cells, many researchers focus on water transport mechanism and water equilibrium technique in PEMFCs. For liquid water transport in gas channels of PEMFCs, direct observation of transparent method has been conducted by many researchers. The influence of operating conditions on liquid water formation and the fuel cell performance is one point that is often emphasized. During the start-up and shut-down processes of a PEMFC, platinum particles are lost from the catalyst layer at the cathode due to corrosion of the carbon supports. During gas exchange, the distribution of oxygen partial pressures at the anode was visualized using our real-time/space visualization system, which clearly showed the location of H_2- and O_2-rich areas along the gas flow channel from the inlet to the outlet. The gas exchange rate was found to be much slower than that predicted from simple replacement. In addition, it was also related to the proton transfer derived from carbon corrosion of the cathode catalyst layer. From the visualization results, it was found that the shut-down process produces a more serious effect than the start-up process. The oxygen partial pressure at the cathode was visualized during cell operation after degradation. Because the MEA was degraded mainly near the inlet and outlet of the reactant gases in the cell, oxygen was consumed primarily in the middle of the MEA [130]. Transparent acrylic materials were used to make various fuel cell models for the experiments. The parameters considered in the experiments were the rate of water injection into the models, the velocity and temperature of the humidified gas in the cathode channels, the type of flow-field, and the temperature. It was found that the parallel and interdigitated flow channels were easily flooded under certain conditions. Fuel cells with two different types of flow channels and two different electrode sizes were made, and their performances were compared with some of the flooding results obtained from the transparent physical models [131].

The effects of four operating parameters, namely air stoichiometry, hydrogen stoichiometry, cell temperature, and electric load, on the formation and extraction of water from flow channels were investigated [132]. These results showed that hydrogen and air stoichiometry contributed almost equally to the process of water formation of water in the cathode channels. However, contrary to hydrogen, changing the air stoichiometry proved capable of extracting all the water from the cathode channels without dehydrating the membrane. Increasing the operating temperature of the cell was found to be very effective in the water extraction process; no water was present in the anode flow channels under any of the examined operating conditions. The liquid water flow patterns in the fuel cell gas channels and GDLs were also addressed by the researchers. Visualization of the two-phase flow occurring at the anode side was carried out using fluorescence microscopy. In correlating the observed flow patterns with the corresponding current density of the polarization curve, a strong influence of the two-phase flow on the performance of a fuel cell at high current densities became apparent. The functionality of a flexible direct methanol micro-fuel cell was also investigated under different bent conditions. The tests showed an insignificant drop in electrical performance under bending due to an inhomogeneous contact resistance. Characteristics of liquid water removal from GDL were investigated experimentally by measuring the nonsteady pressure drop in a cell in which the GDL was initially wet with liquid water [133]. The thickness of the GDL was carefully controlled by inserting metal shims of various thicknesses between the plates. It was found that severe compression of the GDL could result in an excessive pressure drop from channel inlet to channel outlet. Removing liquid water from GDLs with high compression levels and low inlet air flow rates by cross-flow is difficult. However, effective water removal can still be achieved at high GDL compression levels if the inlet air flow rate is high. Water removal characteristics due to the cross-flow through the GDL at different levels of compression were investigated for a transparent cell with a serpentine flow channel layout. Experimental measurements revealed that the total pressure drop from the channel inlet to the channel outlet is reduced substantially when compared with the case without the GDL [134]. These studies show that flow visualization under the land area confirms that the cross-flow has significant effects on water removal even at low air flow rates. Thus, the effects of cross-flow and the compression level of the GDL should be considered in the design and optimization of practical PEMFCs.

Visualization of liquid water transport has been conducted using transparent flow channels, and liquid water removal from GDL under the land was observed for all the tested inlet air flow rates, confirming that cross-flow is practically effective in removing the liquid water accumulated in GDL under the land area. To date, there have been few publications that focused on direct observation of liquid water flow state at the turns of gas channels of PEMFCs. In fact, literature concerning the relationship among the liquid water production rate, the fuel cell internal resistance and performance is hardly found. However, water transport in the channels of transparent PEMFCs has been studied under different operating conditions using a high-speed camera [135]. First, the spot distribution for liquid water droplet emergence in the channels of transparent PEMFCs with multi-straight channel flow-fields was observed and analyzed; second, the flow state of liquid water at the turns of a transparent PEMFC with serpentine channel flow-field was studied under different gas flow rates; finally, the water transport in a transparent PEMFC with a multi-straight channel flow-field was examined. The fuel cell performance and the internal resistance were measured under different operating conditions, the liquid water accumulated in the gas channels was simultaneously recorded and estimated, and the relationship among the fuel cell performance, the internal resistance and the liquid water mass accumulated in the channels was analyzed. This work led to some beneficial conclusions.

3.1.3 Polarization Curve Evaluation

The performance of a fuel cell is mainly characterized by its polarization curve (a plot of cell potential versus current density). In polarization curves, three different regions are observed that correspond to different phenomena (electrochemical kinetics, ohmic resistance, and reactant mass transport) limiting the cell potential, which depends on the current density being drawn (Fig. 3). The test is used to determine the performance of a PEMFC stack and the stability of test input and output parameters as a function of test duration, current density, and stack power [98]. At minimum, stability criteria for the stack current, the stack temperature and pressure, and the stack voltage should be defined and stated in the test report. The stability check of the test outputs should follow the stability check of the test inputs. If all stability criteria are fulfilled, the verification may also be carried out offline during data post-processing.

The polarization curve test method was used to investigate a non-isothermal tapered flow channel installed with a baffle plate for enhancing cell performance in the cathodic side of a PEMFC. The effect of parameters including tapered ratio (0.25–1.0) and gap ratio (0.005–0.2) on cell performance were explored in detail [39]. The results indicate that the stronger composite effect of tapered flow channels and baffle blockage provides better convection heat transfer performance and higher fuel flow velocity and thus enhances cell performance. The effects of three different anode and cathode flow-field designs (single serpentine (SFF), multi-serpentine (MSFF), and an original mixed parallel and serpentine design (MFF)) on the performance of a DMFC was investigated experimentally. The studies were conducted on a DMFC with 25 cm^2 of active membrane area, working near ambient pressure and using two values of methanol concentration, fuel cell temperature, methanol, and air flow rate. With respect to the anode flow-field design, it was found that the use of MFF has a positive effect on cell voltage and power at the two values of methanol flow rate tested, at the lower value of fuel cell temperature and at the lower value of methanol concentration [136]. However, in case of the cathode flow-field design, MSFF leads to a better. For the higher value of methanol concentration tested, a very important condition for portable applications, the use of MSFF or MFF as anode flow-field design and MSFF or SFF as cathode flow-field design produces enhanced fuel cell performance. Most of the reported experiments were conducted close to room temperature, thus providing information and results that can be used for designing a portable DMFC to be used under less severe operating conditions.

AC impedance spectroscopy is helpful for measuring different fuel cell parameters and for understanding the limiting processes affecting overall fuel cell performance, as well as for cell state-of-health determination. Experimental performance results, including polarization curves and Electrochemical Impedance Spectroscopy (EIS) analysis of the fuel cell, were obtained for a 50-cm^2 PEMFC [137]. A cell with a serpentine flow-field was found to perform better than a cell with a parallel flow-field. In the former, the polarization curve was higher and the membrane and contact resistances were lower. Fuel cell operation with pure oxygen improved the performance parameters as well. The resistance of the oxygen reduction reaction at the cathode was found to be significantly lower when operating with pure oxygen. Reactant gases humidification also improved cell operation, resulting in higher polarization curves and membrane conductivities and preventing the MEA from drying out. A special tubular cathode has been prepared using graphite-doped mesocarbon microbeads (MCMB/G), dip-coating the gas diffusion and catalyst layers with subsequently wrapping the sintering tube with the Nafion membrane. The sol-gel flux phase was prepared using sol-gel technology. The impedance of the specially shaped DMFC with a sol-gel flux phase was investigated and the effects of tube wall thickness, gas diffusion layer loading, working temperature, and sol viscosity on the cell polarization performance were examined [138]. The results showed that the impedance of the specially shaped cell was much higher than that of the traditional flat electrode; however, an obvious decrease in impedance was observed after activation. The cell performance improved with the decrease in sol viscosity and with the increase in working temperature.

3.2 Modeling Approaches to Flow Optimization

3.2.1 Computational Fluid Dynamic Modeling

Because of the many difficulties involved in observing and measuring species movement and distribution inside a PEMFC, numerical modeling has become an indispensable research tool in the design and assessment of PEMFCs. The first application of computational fluid dynamic (CFD) methods to PEMFCs focused on a one-dimensional model that included several important principles related to the electrochemical reaction process that occurs in fuel cells [68]. Two-dimensional models that emphasize the effects of two-phase water and heat transport have also been developed by some researchers. Although two-dimensional models represent significant advances in fuel cell modeling, the transport processes that occur in a PEM fuel cell are intrinsically three dimensional. Therefore, it is not possible to accurately predict cell performance unless a realistic, three-dimensional description of the cell geometry is considered. Two-dimensional models are inadequate because they only consider areas where cell efficiency is the highest thus; these models strongly overestimate the limiting current unless heavy model tuning is performed. Obviously, this impedes their use as design tools. Recently, two-phase numerical models that incorporate the coupled effects of electrochemical reaction and mass transport in three-dimensional domains have been developed by many researchers [59]. These models are extremely complicated because they take into account the biphasic phenomenon which occurs at high current density (>1 A/cm^2) and offers a small effect on cell performance. This high current interval is characterized by power drop and is therefore less used in industrial applications.

Common flow-field patterns include serpentine, parallel, interdigitated, and straight patterns (see Fig. 9) as well as their combinations. A three-dimensional single-phase isothermal model was developed to describe the steady-state and transient response of four PEMFCs with different flow-field designs including serpentine, parallel, multi-parallel, and interdigitated types [139]. The modeling results showed that the transient responses of serpentine and parallel designs were faster but the performances were lower than those of the other two designs.

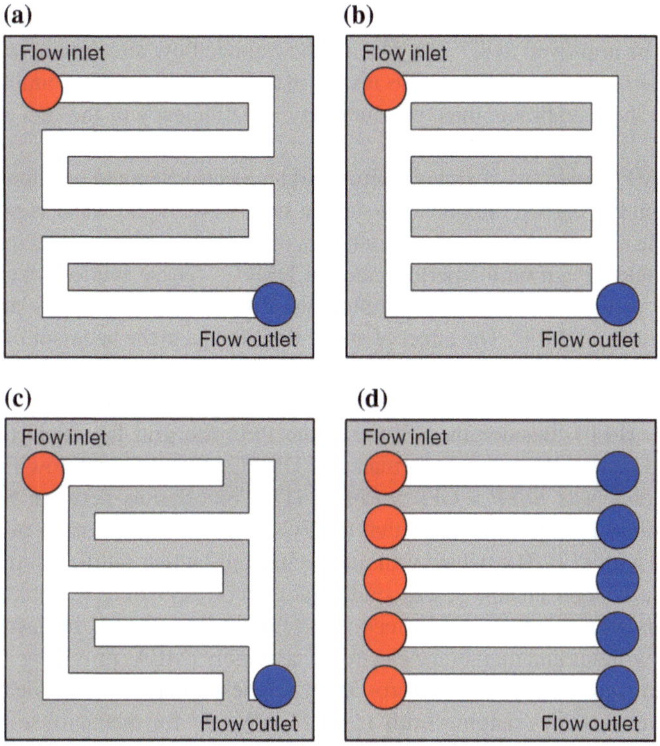

Fig. 9 Major flow-field patterns of PEMFCs: **a** serpentine; **b** parallel; **c** interdigitated; **d** straight. *Source* Ref. [4]

However, this study did not consider water accumulation in the GDLs. Due to the heavy computational method employed; the model was not able to simulate PEMFCs with different anode and cathode flow-field designs. Lumped models that assume uniform reactions within the entire area of the fuel cell do not consider the spatial distribution of reactants. In practice, fuel cell performance and water distribution depend on flow-field design. Moreover, water generation and accumulation inside the fuel cells are distributed unevenly along the flow channels. A model that can describe the distributed properties of a single cell, especially that of a cell with a complex flow pattern, is needed. In order to solve this problem, a segmented model was developed to predict the distribution of liquid water accumulation, current density, membrane water content, and relative humidity in the flow channels [140]. In the segmented model, three single cells with different flow-field patterns are designed and fabricated. These three flow-field designs are simulated using the model, and the predicted results are validated by comparison with experimental data. The segmented model can be used to predict the effect of flow-field patterns on water and current distributions before the cells are machined. Under low-humidity working conditions, the current density of the first segment, which is near the inlet, takes approximately 3.9% of the current load in traditional flow-field designs. The value can be improved to 4.7% by placing the cathode flow direction opposite to the anode. The counterflow design also results in a more uniform distribution of water content in the membrane; thus, the durability and efficiency of the fuel cell can be improved.

In a DMFC operated at ambient temperature, electrochemical reactions generate large amount of carbon dioxide at the anode side, while liquid water is produced at the cathode side. Studies using different flow-fields have compared the influence of the flow-field design on the performance of DMFCs. These studies concluded that flow-field patterns can significantly affect the distribution of gas flow and current density within a DMFC. The effect of anode flow-fields on the behavior of methanol and carbon dioxide in DMFCs has been extensively studied. DMFCs with single serpentine flow-fields (SSFF) showed better performance than those with parallel flow-fields (PFF). Besides the SSFF and the PFF, the grid flow-field (GFF) was introduced into DMFCs; the performance of DMFCs with different flow-fields, from best to worst, was SSFF > GFF > PFF [141]. Comparisons between SSFFs and PFFs in micro-DMFCs also show that DMFCs with SSFF have better performance than those with PFF. The behavior of micro-channels, which is different from that of conventional larger channels, was studied at the corresponding Reynolds number. Flow-fields with micro-channels need to be further investigated in micro-DMFCs [142]. Flow-fields and membrane electrolyte assembly (MEA) providing a 2.25-cm^2 active area were assembled in micro-DMFCs. These micro-DMFCs yielded maximum power densities ranging from 11 to 23 mW/cm^2 for methanol solution concentrations of 1, 2, 3, 4, and 5 M at a temperature of 20 ± 1 °C. The maximum power densities observed imply that under the ambient temperature and low flow rate of methanol solutions, the performance of micro-DMFCs with different flow-fields ranked as double-channel serpentine > single-channel serpentine > mixed multichannel serpentine with wide channels > mixed multichannel serpentine with narrow channels [143].

A complete three-dimensional simulation based on a single domain approach for species transport in PEMFC with straight flow channels was examined [144]. Numerical simulations show that water vapor is mainly concentrated in the gas diffusion layer under the current collecting land due to the deceleration created by collector contraction effects, but the water vapor is still far from the catalyst layer. This result draws more attention to the role played by the porous media of GDL, which prevents severe liquid flooding on the cathode side. Oxygen concentration in the reaction sites is significantly affected by an increase in pressure, which increases the power of fuel cells but also requires more auxiliary energy consumption, which must be reduced from the power of the cells' output. One of the critical issues in developing efficient PEMFCs is to achieve better water management within the MEA. Although electrochemical reactions are only feasible in moist environments, excessive liquid water blocks the passage of air into the active sites of the catalyst layer. Because explicit water front tracking is unnecessary with single domain formulations, fewer complexities exist in mathematical modeling as well as fewer difficulties in computations, especially for high-dimensional problems.

A new method of single domain formulation, termed the pseudo-phase-equilibrium approach, has been recently proposed [145]. The validity of the pseudo-phase-equilibrium approach was evaluated over an extensive polarization range under specified operating temperature, pressure, and inlet humidity conditions. The solutions obtained using the pseudo-phase-equilibrium approach and the exact phase-equilibrium equations were compared over a wide range of parameter values. In addition, the influences of important transport parameters such as water transport coefficient, gas diffuser porosity, and absolute liquid permeability have been evaluated.

The fundamental algorithms and methods described here have been used for decades, have proven to be very efficient, and are being used by many modeling groups along with user-defined coding. One cannot say which strategy will be the best for efficiently simulating PEMFC and stack models. From a mathematical/numerical point of view, novel mathematical methods and innovative numerical schemes will continuously be ameliorated and implemented in CFD software, and add-on modules are often used for PEMFC modeling. Enhanced discretization tools will lead to more efficient meshes, producing a better quality result while lowering the number of cells (e.g., polyhedral mesh).

3.2.2 Two-Phase Modeling for Water Management

The two-phase flow in PEMFCs is a unique multiphase flow that occurs due to several factors such as large gas to liquid ratios, the water produced by electrochemical reaction, and water condensing in the flow channels from humidified reactants and other operating conditions. Another important distinction is that water is introduced into the air flow-field channels from a porous GDL instead of each phase being introduced together via a common inlet. Furthermore, the coupled gas and liquid flow rates (via Faraday's law) and the contact angles of each wall (flow-field walls and GDL in the same channel) make two-phase flow studies in PEMFCs a challenge. Liquid water in flow channels can cause channel blockage, which can increase the pressure drop in the channel and result in the formation of a liquid film on the GDL surface, which blocks the reactant gas from reaching active catalyst sites. Experiments have documented the following flow patterns in parallel cathode flow-field channels: slug, film, corner, and mist flow. Film flow is considered a desirable flow pattern for water removal in fuel cells due to water traveling on the sidewalls instead of on the GDL surface. However, a specific combination of surface properties and superficial gas and liquid velocities must be met in order to ensure the desired pattern. Parallel channels have shown to have the potential for high performance when no flooding occurs, but this flow-field configuration is flooding prone, and more research is needed for improved water management.

CFD has been used extensively to model multiphase flow in the gas channels of PEMFCs. Multiphase flow through the cathode side of an interdigitated flow-field was studied using a multi-fluid model [146]. The model employs the so-called multi-fluid approach, which solves one complete set of transport equations for each phase. The physics of phase change has now been implemented, and the model also accounts for the Kelvin effect. In the interdigitated design, more water was found to be carried out of the cell in the vapor phase compared to the straight channel design, indicating that liquid water management might be less problematic. This effect also leads to the finding that, in the interdigitated design, more waste heat is carried out of the cell in the form of latent heat, which reduces the load on the coolant. Simulation of a fuel cell stack is performed by applying a general numerical model with the Volume of Fluid (VOF) method that was successfully applied to the single PEMFC model to investigate the fluid dynamics, mass transport, flooding phenomenon, and the effects of liquid water on the stack performance [147]. The performance of three single cells in series connection in the fuel cell stack was examined according to the presence of liquid water in different single cells. The distributions of fluid flow, species concentration and the current density were presented to illustrate the effects of liquid water on the performance of each single cell. The numerical results show that the low distribution of species in the flooding cell certainly degrades the performance of this cell. Using the multiphase free-energy lattice Boltzmann method (LBM), the formation of a water droplet was simulated emerging through a micro-pore on the hydrophobic GDL surface in a PEMFC and its subsequent movement on the GDL surface under gas shear stresses [148]. The results showed that water droplet removal was facilitated by high gas flow velocity on a more hydrophobic GDL surface. A highly hydrophobic surface was shown found to be capable of lifting the water droplet from the GDL surface, resulting in the availability of more GDL surface for the transport of gas reactant. A Multiphase-Mixture (M2)-based model was developed for two-phase flow computations in the cathode channels of a PEMFC [149]. Because in a real PEMFC stack, geometry is very difficult to achieve and hence the model has been extended to non-orthogonal hexahedral and tetrahedral meshes, which can be used to mesh any three-dimensional geometry. Additionally, in order to reduce the meshing effort, an immersed body approach has been tested successfully on this model. The resulting two-phase flow model that is valid for arbitrary flow-field geometries is fast and accurate.

The pressure drop hysteresis is a recently studied two-phase flow phenomenon. This behavior occurs when the gas and liquid flow rates (determined by a given current density) are increased along a set path and then decreased along the same path with differing pressure drops. Two-phase flow pressure drop hysteresis was studied in a nonoperational PEMFC to understand the effect of stoichiometry, GDL characteristics, operating range and initial conditions (dry vs. flooded) for flow conditions typical of an operating fuel cell [150]. This hysteresis is noted when the air and water flow rates are increased and then decreased along the same path, exhibiting different pressure drops. The hysteresis was found to be greater when water breakthrough occurs at higher simulated current densities. The operating range had to reach a critical simulated current density between the ascending and descending approach in order to create a pressure drop hysteresis zone. Two-phase flow pressure drop hysteresis was studied in an operating PEMFC. The variables studied include air stoichiometry (1.5, 2, 3 and 4), temperature (50, 75, and 90 °C), and the inclusion of a microporous layer [151]. The cathode channel pressure drops can differ in PEMFCs when the current density is increased along a path and then decreased along the same path (pressure drop hysteresis). The results showed that the hysteresis occurs with or without the inclusion of a microporous layer.

In terms of fuel and oxidant delivery schemes, DMFCs can be categorized into active-feed DMFCs and passive-feed DMFCs. In passive DMFCs, fuel and oxidant are passively delivered without any moving parts such as pumps, fans, or blowers. A two-dimensional, two-phase, non-isothermal model was developed to investigate the water transport characteristics in a passive liquid-feed DMFC. The two-phase mass transports of liquid–gas in the porous anode and cathode were formulated based on a multi-fluid model in porous media. Moreover, water and methanol crossover through the membrane were considered with the effects of diffusion, electro-osmotic drag, and convection [152]. The results showed that for the free-breathing cathode, gas species concentration and temperature showed evident differences between the cell and the ambient air. The use of a hydrophobic air filter layer at the cathode helped to achieve water recovery from the cathode to the anode, although the oxygen transport resistance was increased to some extent. It was further revealed that the water transport can be influenced by the ambient relative humidity. A one-dimensional, steady-state, two-phase DMFC model was developed to precisely investigate the complex physiochemical phenomena inside DMFCs [153]. The developed model was validated against readily available experimental data in the literature. Next, a parametric study was carried out to investigate the effects of the operating temperature; methanol feed concentration, and properties of the backing layer. The results of the numerical simulation clarify the detailed influence of these key designs and operating parameters on the methanol crossover rate as well as on cell performance and efficiency. The results emphasize that the material properties and design of the anode backing layer play a critical role in the use of highly concentrated methanol fuel in DMFCs.

3.2.3 Complex Flow-field Interaction Modeling

In the literature, many models have been developed to take into consideration multi-physical phenomena in PEMFCs. A sequential approach was developed to study the pressure effects by combining the mechanical and electrochemical phenomena in fuel cells [154]. In this research, the pressure causes the deformation of the GDL and impacts the mass transfer and the electrical contact resistance. The GDL porosity in the PEMFC was also affected by the clamping pressure; therefore, a global value of GDL porosity was taken into account (not a local porosity distribution. There is an optimal width of the rib of the bipolar plate that gives a small amount of contact resistance and a good porosity of the GDL. At the same time, the effect of the clamping force on the performance of PEMFCs was investigated with interdigitated gas distributors considering the interfacial contact resistance, the nonuniform porosity distribution of the GDL and the GDL deformation [155]. The clamping force between the GDL and the bipolar plate is one of the important factors that affect the performance and efficiency of the fuel cell system. It directly affects the structure deformation of the GDL and the interfacial contact electrical resistance, and in turn influences the reactant transport and the Ohmic overpotential in the GDL. The finite element method and the finite volume method were used, respectively, to study the elastic deformation of the GDL and the mass transport of the reactants and products. An optimal clamping force was found to exist for obtaining the highest power density for the PEMFC with the interdigitated gas distributors. The effects of mechanical solicitations were studied on the performance of a PEMFC, as was the coupling between the physicochemical phenomena and the mechanical behavior. A finite element method was first developed to analyze the local porosity distribution and the local permeability distribution inside the gas diffusion layer induced by different pressures applied on deformable graphite or steel bipolar plates [156]. Taking into account the chemical phenomena and the effects of the mechanical compression of the fuel cell to examine a multi-physical approach was carried out to more precisely the deformation of the gas diffusion layer, the changes in the physical properties and the mass transfer in the gas diffusion layer. The effects of varying porosity and permeability fields on the polarization as well as on the power density curves were reported, and the local current density was also investigated. Unlike other studies, this model accounts for a porosity field that varies locally in order to correctly simulate the effect of inhomogeneous compression in the cell.

On the other hand, a computational fuel cell dynamics model was presented in order to elucidate the three-dimensional interactions between mass transport and electrochemical kinetics in PEMFCs with straight and interdigitated flow-fields, respectively [157]. This model features a detailed MEA submodel in which water transport through the membrane with spatially variable transport properties along with spatial variations of the reaction rate and the ionic resistance through the catalyst layer are accounted for. Emphasis was on understanding the way three-dimensional flow and transport phenomena in the air cathode impact the electrochemical process in both types of flow-fields. Fully three-dimensional results of the flow structure, species profiles, and current distribution were presented for PEMFCs with an interdigitated cathode flow-field. The model results indicated that forced convection induced by the interdigitated flow-field substantially improves mass transport of oxygen to, and water removal from, the catalyst layer, thus leading to a higher mass transport limiting current density as compared to that of the straight flow-field. The effects of nonuniform compression of the GDL and GDL intrusion into a channel due to the channel/rib structure of the flow-field plate were investigated numerically [158]. The focus was placed on accurately predicting two-phase transport between the compressed GDL near the ribs and the uncompressed GDL near the channels and on understanding its associated effects on cell performance. A GDL compression model was newly developed and incorporated into the comprehensive, three-dimensional, two-phase PEMFC model developed earlier. To assess the effects of GDL compression and intrusion in isolation, the new fuel cell model was applied to simple single straight channel fuel cell geometry. Numerical simulations with different levels of GDL compression and intrusion were carried out, and simulation results revealed that GDL compression and intrusion considerably increase the nonuniformity—particularly the in-plane gradient in liquid saturation, oxygen concentration, membrane water content, and current density profiles—which in turn resulted in significant ohmic and concentration polarizations. The three-dimensional GDL compression model yields realistic species profiles and cell performance that help to identify the optimal MEA, gasket, and flow channel designs in PEMFCs.

3.3 Validation of Experimental and Numerical Results

The wide range in power output implies a variation of fuel cell operating conditions, active area, gas diffusion media, bipolar plate, and MEA properties. Therefore, the performance of a PEMFC stack varies significantly with respect to the number of cells, operating conditions, material properties, and flow-field characteristics. The design principles of PEMFC stacks were established for portable applications [97]. A combination of experiments and numerical simulations was used to enhance our understanding of the behavior of this portable PEMFC stack. A three-dimensional CFD-based methodology was utilized to make predictions about the current and temperature distributions of this portable PEMFC stack. The results explained how the baseline operation and original design of this stack impact the local temperature, water content, water transport, and kinetic variables inside the individual cells.

In developing the flow-field design, CFD simulation, experimental test and in situ multichannel impedance analysis were performed in order to optimize the serpentine channel design for a 300-cm^2 active area PEMFC stack [159]. The CFD simulation results successfully predicted the channel variation effect on the performance, showing the same trend as that of the experimental test. In situ multichannel impedance analysis not only successfully predicted the experimental result but also explained the performance of each channel within the heterogeneous stack. The experiments carried out in this study also proved that the in situ multichannel impedance analysis with a heterogeneous channel stack is very effective in predicting the channel performance and analyzing the reasons behind the difference in the performance between channels. Finally, it was shown that in situ multichannel impedance has the advantages of reducing the analysis time and gathering synchronous data from several cells. The operating characteristics of 5 W class DMFC stacks, using the flow-field patterns of serpentine, parallel, and square spot, were investigated to compare how capable these characteristics are with mass transport and water removal in the cathode [160]. The stability of the stack was predicted through the simulation results of the flow-field patterns on the pressure drop and the water mass fraction in the cathode of the stack. It was then estimated through the performance and the voltage distribution of the stack. According to the simulation results, although the square spot pattern shows the lowest pressure drop, the square spot pattern had a much higher water mass fraction in the central region of the channel compared to the other flow-field patterns.

The flow-field structures can exert a significant influence on both flow velocity and temperature distributions of the DMFCs. Thus, proper flow-field constructions are very important for the improvement of the DMFC's performance. Anodic flow velocity and temperature distributions based on four different designs, including double serpentine, parallel, helix, and single serpentine have been simulated in three-dimensional models [161]. Computational fluid dynamics was used to investigate the effects of flow-field structures on the DMFC's performance. Simulated results indicated that the double-serpentine flow-field showed better flow velocity distribution and more uniform temperature distribution, which might be helpful for better performance of the DMFC. Further experimental investigation of four types of flow-fields also confirmed that the DMFC with a double-serpentine flow-field structure exhibited a maximal power density at a variety of inlet velocities, which was in good agreement with the simulated results. The CFD simulations showed that the initial solution of the fuel cell model developed for a 50-cm^2 PEMFC with parallel and serpentine flow-fields was not fully converged in the mass transport polarization region for a particular case and therefore the flow-field did not reach a steady-state solution [162]. This point is especially critical for the multiphase flow variables, as these are generally difficult to converge. An increase in the under-relaxation factors resulted in a better convergence with a fully developed flow-field, and also worked best for the multiphase-related variables. The polarization curve predicted by the CFD model showed better accuracy compared with the experimental data once the fully converged solution was achieved. Enhanced convergence controls for PEMFC CFD simulations are still needed, such as the monitoring of the variation of water content and water saturation in both CL and GDL. These enhanced controls will ensure that further modifications in the solver parameters do not modify the flow-field, especially for the multiphase-related variables.

4 New Flow-field Optimization Approaches Utilizing Under-Rib Convection

In principle, the channel structure offers a homogeneous flow distribution with low-pressure drops. But, on the other hand, the formation of liquid water droplets at the cathode can flood one or more channels with the consequence of stopping the gas flow there. In order to effectively utilize the area, flow distribution should be as homogeneous as possible. In addition, pressure losses should be minimized with regard to the power demand of auxiliary components such as pumps and compressors. In PEMFCs and DMFCs, the flow-field is in direct contact with the diffusion layer. The main task of the diffusion layer is to distribute the reactants from the flow-field toward the catalyst layer. The diffusion layer is general highly porous and provides high reactant fluxes to prevent diffusion overvoltage. Consequently, the flow distribution in the flow-field can be superpositioned by a flow in the diffusion layer. The interaction between the diffusion layer and the flow-field has been experimentally examined for DMFC [5]. To avoid depletion of reactants in specific regions of the cell, the geometry of the meander should be chosen with regard to the permeability of the diffusion layers. Unsuitable combinations of meander geometry and diffusion layer properties lead to even lower flow homogeneities compared to the flow passing only in the diffusion layer. High permeabilities require meander structures with low-pressure losses; otherwise, the flow homogeneity decreases. Additional simulation studies will help in understanding these processes and to determine suitable combinations of flow-fields and diffusion layers.

The presence of a convective flow in the under-rib regions enables more effective utilization of electrocatalysts by increasing reactant concentrations and facilitating liquid water removal in those regions [32]. The mechanism of under-rib convection was explained with three neighboring channel regions A, B, and C in the exemplary serpentine flow-field shown in Fig. 10a. The path-lengths Z_A, Z_B, and Z_C are defined as the distances traveled by the reactants from an inlet to the three regions along the channel as illustrated by the dotted lines in Fig. 10a. The path-length, z, can be viewed as a coordinate whose upper limit is the same as the total length of the path, z_{max}. The pressure and the reactant concentration in PEMFCs are expected to change linearly along the flow channel, i.e., with respect to z. The total flow rate, q_{tot}, depends on the total pressure drop, Δp_{tot}, across the flow-field expressed as follows:

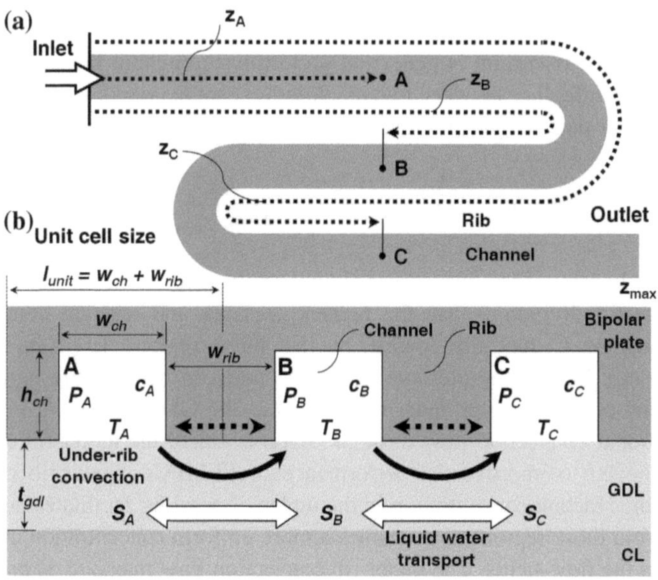

Fig. 10 Under-rib convection and transport in serpentine flow-field: **a** definition of path-length, z; **b** under-rib convection and transport mechanisms. *Source* Ref. [32]

$$q_{tot} = \frac{K_{ch}}{\mu} A_{ch} \frac{\Delta p_{tot}}{Z_{max}} \tag{1}$$

where μ is the viscosity, A_{ch} is the cross-sectional area of the channel (height h_{ch} × width w_{ch}), and K_{ch} is the channel permeability derived from the Hagen–Poiseuille equation. Note that Eq. (1) assumes that the flow rate is constant along the channel with a low under-rib convection flow rate ($q_{rib} \approx 0$) and that the secondary (nonlinear) pressure loss in 90° or 180° corners is negligible. The pressure difference between two neighboring channel regions is expressed in terms of the path-length difference as:

$$|p_A - p_B| = \frac{\Delta p_{tot}}{Z_{max}} |z_A - z_B| \tag{2}$$

Let us assume that the flow-field shown in Fig. 10a is for the cathode of a PEMFC. The reactant concentration, c (mol/m^3), then denotes the molar concentration of oxygen, and the total concentration loss across the flow-field, $\Delta c_{tot} = c_{in} - c_{out}$, is calculated from the mass balance of oxygen according to:

$$(c_{in} - c_{out}) q_{tot} = \Delta c_{tot} q_{tot} = \frac{I_{tot}}{4F} \tag{3}$$

Here, I_{tot} (A) is the current generated over the entire cell area and F is the Faraday constant (96,487 Cmol^{-1}). Note that Eq. (3) ignores the change in the total flow rate, q_{tot}, due to oxygen consumption. In fact, the reactant flow rate in the cathode vicinity is relatively constant because the humidified air is supplied at high excess air ratios. If current is generated uniformly throughout the active area, the concentration difference between two neighboring channel regions is also expressed in terms of the path-length difference, i.e.:

$$|c_A - c_B| = \frac{\Delta c_{tot}}{z_{max}} |z_A - z_B| = \frac{I_{tot}}{4F q_{tot} z_{max}} |z_A - z_B| \tag{4}$$

Figure 10b shows the cross-sectional view of the three adjacent channel regions (A, B, and C). Region A has the highest pressure and reactant concentration, whereas region C has the lowest of the three regions ($p_A > p_B > p_C$ and $c_A > c_B > c_C$). Due to the pressure difference between neighboring regions, convective flow occurs through under-rib regions in the GDL, as shown in Fig. 10b. This additional convective flow through GDLs is called the under-rib convection and is believed to improve the performance of PEMFCs. Under-rib convection increases the reactant concentration in the under-rib regions, facilitates liquid water removal from those regions, and enables a more uniform concentration distribution throughout the flow-fields. The under-rib convection flow rate, q_{rib}, is proportional to the pressure difference between two neighboring regions. For example, q_{AB} rib between regions A and B in Fig. 15b can be expressed as:

$$q_{\text{rib}}^{\text{AB}} = \frac{K_{\text{gdl}}}{\mu} A_{\text{rib}}^{\text{AB}} \frac{|p_{\text{A}} - p_{\text{B}}|}{w_{\text{rib}}} = q_{\text{tot}} \frac{K_{\text{gdl}} A_{\text{rib}}}{K_{\text{ch}} A_{\text{ch}}} \frac{|z_{\text{A}} - z_{\text{B}}|}{w_{\text{rib}}} \tag{5}$$

where K_{gdl} is the flow permeability of the GDL, w_{rib} is the rib width, and $A_{\text{rib}}^{\text{AB}}$ is the flow area for the under-rib convection. Note that $A_{\text{rib}}^{\text{AB}}$ is determined by multiplying the GDL thickness, t_{gdl}, together with some length, l_{ch}, of the regions along the channel (the under-rib convection through the catalyst layer is ignored).

Figure 10b shows that other transport mechanisms also become operative when channel regions that have different path-lengths are in close contact. The local temperature and liquid water saturation in GDLs depend on the electrochemical reaction rate, and this rate is generally governed by the reactant concentration, which varies along the channel (with respect to z). The temperature gradient due to closely placed channels of different path-lengths facilitates conductive heat transfer by the under-rib or cross-rib paths, and the water saturation gradient facilitates the capillary transport of water by the under-rib path. These transport mechanisms may be termed as under-rib transports to distinguish them from under-rib convection. Whereas under-rib convection occurs in only one direction (from a high-pressure region to a low-pressure region), under-rib transport may occur in both directions, as shown in Fig. 10b. In fact, the successful dry operation of a PEMFC was attributed to the under-rib transport of water in the opposite direction of the under-rib convection [163].

In the following, new flow-field approaches by the above-stated under-rib convection aim to enhance the uniformity of operative conditions in the flow-fields, including the reactant and product concentrations, temperature and liquid water saturation. The flow-field pattern that activates the internal mass transfer mechanism of PEMFCs is able to improve its performance from a mechanical engineering perspective. In other words, this mechanical method can improve the performance of the fuel cells without changing the electrochemical materials. The promotional role of under-rib convection is (a) to provide a homogeneous distribution of the reactants, (b) ensure the uniformity of temperature and current distributions, (c) facilitate liquid water discharge, (d) reduce the pressure drop and parasitic losses, and (e) improve cell output power. These studies are listed schematically in Table 1 and are described in this section in order to provide guidance to the optimization of flow-field design by under-rib convection.

4.1 Homogeneous Distribution of the Reactants

In a PEMFC, anode and cathode gases usually flow through each channel, and the reactant gas diffuses to the interface of MEA through GDL. GDL is porous, and it plays a role in helping hydrogen and oxygen to move to the electrode catalyst, in order to support MEA and to pass the electron. However, in the case of a large-scale cell, the differential pressure between adjoining channels is thought to increase,

Table 1 Experimental and numerical studies on under-rib convection used in this article

Flow-field details	PEFC types	Experiment/Simulation	Features	Refs.
100 cm^2 meander structure	PEMFC/DMFC	Experiment/Simulation	Flow interaction between the diffusion layer and the flow-field, validation of experiments using DMFC cells	[5]
Parallel mini-channels	PEMFC	Experiment	Flow maldistribution and flow hysteresis of gas and liquid two-phase flow, presence of gas shorting between channels	[9]
Square serpentine channels	PEMFC	Simulation	Convective transport as a function of GDL permeability	[10]
25 cm^2 serpentine, parallel, semi-serpentine separator	PEMFC	Simulation	Effect of gas channel depth on current density distribution using oxygen transfer and thermal flow analysis models	[11]
20 straight channels connected in a serpentine fashion	PEMFC	Simulation	Three-dimensional flow and mass transfer model, water transport by both electro-osmosis and diffusion processes	[12]
Parallel flow-field	PEMFC/DMFC	Experiment/simulation	Inefficient usage of the electrochemically active area due to poor water management	[13]
Single serpentine flow-field	PEMFC	Experiment	Convection contributes to the limiting current.	[14]
Serpentine flow channels with a square cross-section	PEMFC	Simulation	The effect of channel-to-channel gas crossover on the pressure and temperature distribution	[15]
Serpentine channel system	PEMFC	Simulation	Pressure distribution and flow crossover through the GDL	[16]
Serpentine/parallel flow-fields	PEMFC	Simulation	Significant cross-flow can lead to more effective evacuation of the water vapor	[17]
Serpentine flow-fields	PEMFC/DMFC	Experiment/simulation	An increase in power density by optimal use of the cross convection effect	[18]
Flow-field parameters	PEMFC	Simulation	The extent of reactant bypass through the GDL from one channel to the other enhances reaction rates	[19]
Double-path serpentine with 36 channels	PEMFC	Simulation	Provides a unique opportunity to explore the rich physics behind the complex flow and transport phenomena	[20]
Serpentine flow channels	PEMFC	Simulation	Cross-leakage flow can help in the reduction of overpotential and the effective removal of liquid water	[21]

(continued)

Table 1 (continued)

Flow-field details	PEFC types	Experiment/Simulation	Features	Refs.
Single serpentine flow-field	PEMFC	Simulation	Convective bypass can be increased by simply increasing the length of the flow channel	[22]
Flow-field parameters	DMFC	Experiment	The reduced cell performance, as a result of thinning the backing layer, may be attributed mainly to the increased under-rib mass transport polarization	[23]
Serpentine flow channel	PEMFC	Experiment/simulation	Cross-flow through the gas diffusion layer between the adjacent flow channels helps drive the liquid water	[24]
50, 100, 200, 300, and 441 cm^2 serpentine flow channels	PEMFC	Experiment/simulation	The effectiveness in water removal is confirmed by two different experimental observations	[25]
Simple straight channel	PEMFC	Simulation	Thoughts on which combination of geometric design parameters would be optimal to get the best performance	[26]
Single serpentine flow-field	DMFC	Experiment/simulation	Under-rib convection plays an important role in making the reactant evenly distributed over the entire catalyst layer	[23]
Serpentine/parallel flow-fields	DMFC	Experiment/simulation	The overall methanol-transfer coefficient can be significantly increased due to the enhanced under-rib convection	[27]
Convection-enhanced serpentine flow-field	PEMFC/DMFC	Experiment/simulation	Multiple serpentines in parallel do not lose under-rib convection and result in a smaller pressure drop	[28]
Novel serpentine-baffle flow-field	PEMFC	Simulation	The baffled design increases the limiting current density and improves the cell performance relative to conventional design	[29]
Single serpentine flow-field	PEMFC	Simulation	Reduced channel heights enhance sub-rib convection to transport oxygen to and liquid water out of the diffusion layer	[30]
Single serpentine flow-field	DMFC	Experiment/simulation	A comparison of the influence of two-phase flow and under-rib mass transfer on DMFC performance	[31]
Multi-pass serpentine flow-fields	PEMFC	Simulation	Multi-pass serpentine flow-fields lead to significantly higher under-rib convection intensities	[32]

(continued)

Table 1 (continued)

Flow-field details	PEFC types	Experiment/Simulation	Features	Refs.
Single and triple serpentine flow-fields	PEMFC	Simulation	The sub-rib convection is much higher for single serpentine flow than the triple serpentine flow cells	[33]
Parallel/interdigitated/serpentine flow-fields	PEMFC	Simulation	Flow design that enhances cell performance also increases pressure drop over a working PEMFC	[34]
Single serpentine flow-field	PEMFC	Simulation	The role of sub-rib convection on cell performance and optimization of the cathode flow-field design	[35]
New tapered single serpentine flow-field	DMFC	Simulation	Tapered flow-field enhances the pressure difference between the adjacent flow channels and improves mass transport efficiency	[36]
1/3/5 parallel path serpentine flow-fields	PEMFC	Simulation	Enhanced under-rib convection intensity was the key reason for the performance enhancement	[37]
Parallel/interdigitated flow-fields	PEMFC	Simulation	The effect of the channel to rib width ratio on the transient characteristics of the parallel but less on the interdigitated fields	[38]
Tapered flow channel installed with a baffle plate	PEMFC	Simulation	The tapered flow channel with a baffle plate enhances the overall cell performance	[39]
Split serpentine flow-field	PEMFC/DMFC	Experiment/simulation	The splitting of the channel with enhanced cross-flow in selected regions more prone to localized flooding	[40]
Serpentine flow channel	PEMFC	Simulation	The findings make it possible to optimize the design of gas flow channels for a more efficient PEMFC	[41]
Serpentine flow-field with sub-channels/by-passes	PEMFC	Experiment/simulation	The flow-field pattern that activates the internal mass transfer mechanism can improve its power density from a mechanical engineering perspective	[42]

allowing the supplied gas to flow through GDL. As a result, the cell endurance is assumed to decrease because the temperature and the humidity conditions are not uniform. However, the oxygen transfer rate to the electrode and the output density can be expected to increase via such gas flow in the GDL. Therefore, it is important to examine the influence of this gas flow on cell performance and to optimize operating conditions and cell shape [11]. Cell areas supplied with low amounts of reactants are affected by diffusion overvoltages in the electrodes, which lower the current density. Below, the parameters characterizing the flow homogeneity are defined.

In a system consisting of a meander structure and a diffusion layer, there are two borderline cases. The first assumes low permeabilities of the diffusion layer. In this case, the overall flow is guided through the meander only. The second borderline case describes a high permeability of the diffusion layer compared to that of the meander. This case results in the total flow passing through the diffusion layer. The mean specific volume flow $\bar{\dot{v}}_{\mathrm{spec}}$ is equivalent to the inlet volume flow $\dot{V}$ divided by the cell area A_{cell}. This parameter will be used for the calculation of the flow homogeneity by comparison with the local flow rates. By integrating the local volume flow $\dot{V}_{\mathrm{local}}^{\mathrm{meander}}$ (l/h) over the entire length of the meander channel L_{M}, the specific volume flow in the meander channel $\bar{\dot{v}}_{\mathrm{spec}}^{\mathrm{meander}}$ is obtained as

$$\bar{\dot{v}}_{\mathrm{spec}}^{\mathrm{meander}} = \frac{1}{A_{\mathrm{cell}}}\frac{1}{L_{\mathrm{M}}}\int\limits_{0}^{L_{\mathrm{M}}} \dot{V}_{\mathrm{local}}^{\mathrm{meander}}\,\mathrm{d}L_{\mathrm{M}} \tag{6}$$

The flow cross-section in the inlet or the outlet is substantially smaller than that in the middle of the cell. The total specific flow through the diffusion layer $\bar{\dot{v}}_{\mathrm{spec}}^{\mathrm{diff}}$ is obtained by mass conservation:

$$\bar{\dot{v}}_{\mathrm{spec}}^{\mathrm{diff}} = \bar{\dot{v}}_{\mathrm{spec}} - \bar{\dot{v}}_{\mathrm{spec}}^{\mathrm{meander}} \tag{7}$$

The overall flow homogeneity ψ results in

$$\psi = \left(\frac{\bar{\dot{v}}_{\mathrm{spec}}^{\mathrm{meander}}}{\bar{\dot{v}}_{\mathrm{spec}}}\right)^2 - \frac{1}{2}\left(\frac{\bar{\dot{v}}_{\mathrm{spec}}^{\mathrm{diff}}}{\bar{\dot{v}}_{\mathrm{spec}}}\right)^2 \tag{8}$$

This equation is a semiempirical expression valid for a meander flow-field combined with a porous diffusion layer [5]. Of course, if the bipolar plate itself is porous, such as porous carbon materials with a meander structure manufactured inside, then the resulting flow can be described equivalently. The influence of the permeability of the diffusion layer for four different cell geometries using the simulation is summarized in Fig. 11. The flow homogeneity ψ was calculated according to Eq. (8). The thickness of the diffusion layer was assumed to be

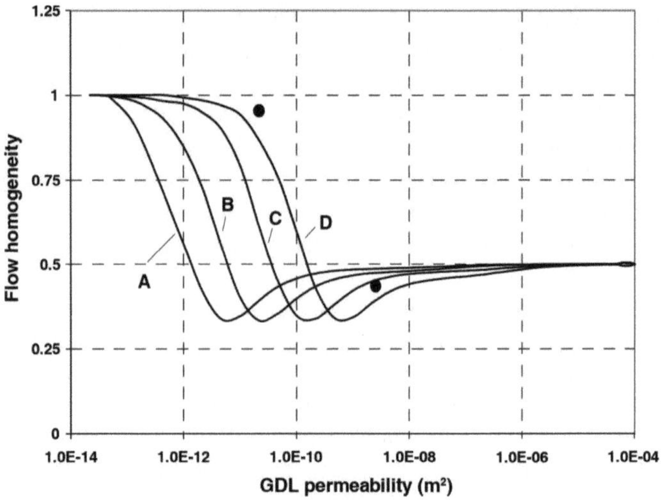

Fig. 11 Influence of the permeability of the diffusion layer on the flow homogeneity for different cell geometries. Results based on simulation. The *dots* represent the correlation with the experimentally tested flow-field geometries. *A* Cell area 200 mm × 200 mm, meander cross-section 1 mm × 1 mm. *B* Cell area 100 mm × 100 mm, meander cross-section 1 mm × 1 mm. *C* Cell area 200 mm × 200 mm, meander cross-section 2 mm × 2 mm. *D* Cell area 100 mm × 100 mm, meander cross-section 2 mm × 2 mm. *Source* Ref. [5]

0.4 mm in all cases. Flow homogeneity helps to evaluate different flow-fields, but it is not the sole criterion for the evaluation of flow-fields or for the decision regarding whether or not a flow-field is suitable. Another important parameter, especially for the cathode of fuel cells, is the pressure drop that requires compressor work and thus reduces the total system power. In general, a flow-field is a compromise between flow homogeneity and a low-pressure drop. In our simulation studies, the pressure drop will also be taken into account.

As shown in Fig. 12, two 25 cm^2 serpentine flow-fields of 5-passes and 4-turns were considered in this study. Through the previous geometrical characterization of serpentine flow channels with various heights and widths, a conventional-advanced serpentine flow-field (CASFF) without sub-channel was selected as a design standard [41]. The presence of under-rib convection enables more effective utilization of electrocatalysts by increasing reactant concentration and facilitating liquid water removal. As shown in Figs. 13b, c and 14b, c, in case of anode side of CASFF, generally uniform velocity vectors can be observed at main channel and rib; however, it shows minor change from the inlet to the outlet direction. In addition, the velocity vectors are significantly increased at the turn rib area due to the large pressure difference between adjacent main channels. Also, in case of cathode side, under-rib convection is generated from the inlet to the outlet direction due to high stoichiometry ratio. And in case of serpentine flow-field with sub-channels and by-passes (SFFSB), rib width is decreased therefore improving gas permeability and overall gas diffusion force as we add sub-channel to change

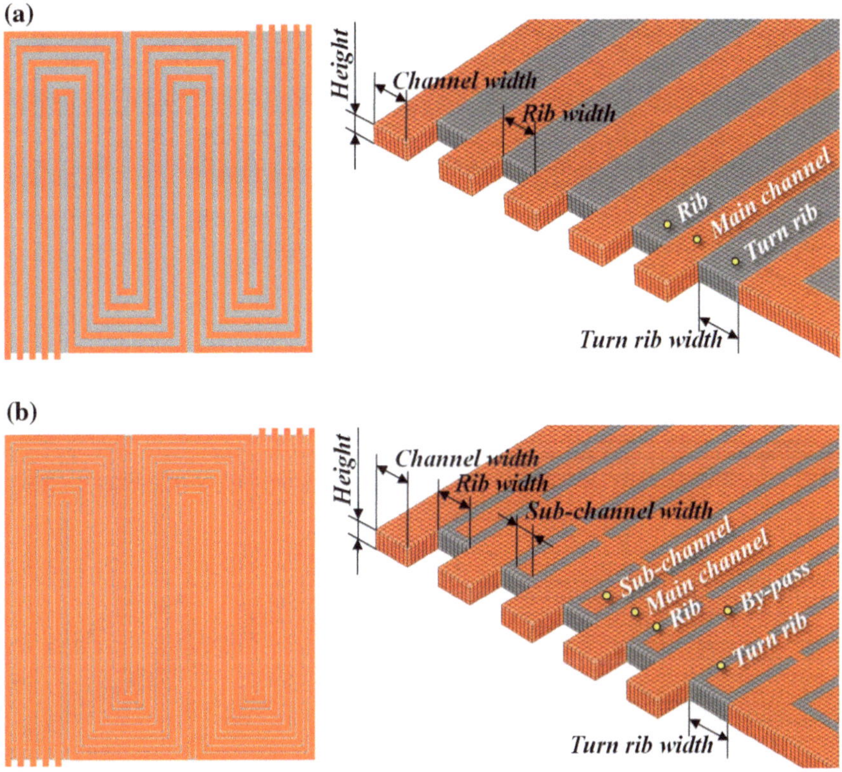

Fig. 12 Two 25 cm^2 serpentine flow-field patterns of 5-passes and 4-turns; **a** conventional-advanced serpentine flow-field (CASFF), **b** serpentine flow-field with sub-channels and by-passes (SFFSB)

the flow direction of under-rib convection. Therefore, in case of anode, the velocity vectors are large at the main channel inlet and under-rib convection flow direction converges to sub-channel through convection because the pressure of main channel is higher than sub-channel. The velocity vectors are uniform the main channel outlet due to the pressure decrease. In case of cathode side, the under-rib convection which has different size but similar tendency with anode side is generated, and the flow direction of the under-rib convection is changed from sub-channel to main channel at the channel outlet due to high stoichiometry ratio. This is because the internal pressure of sub-channel is increased as the reactant which traveled through sub-channel migrates to the outlet, so the flow direction is changed toward the main channel for smooth discharge. Therefore, we can illustrate the representative under-rib convection flow direction and liquid water transport mechanisms of CASFF and SFFSB as shown in Figs. 13a and 14a. From the velocity vectors shown in Figs. 13b, c and 14b, c, numerical analysis result of hydrogen and oxygen mass fractions as well as under-rib convection flow direction can be estimated.

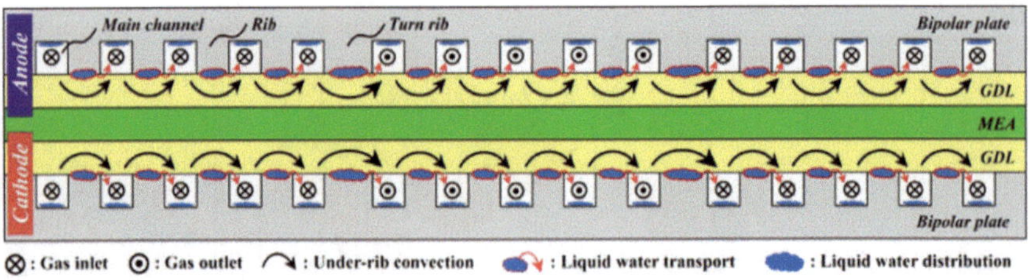

(a) Water behavior with CASFF

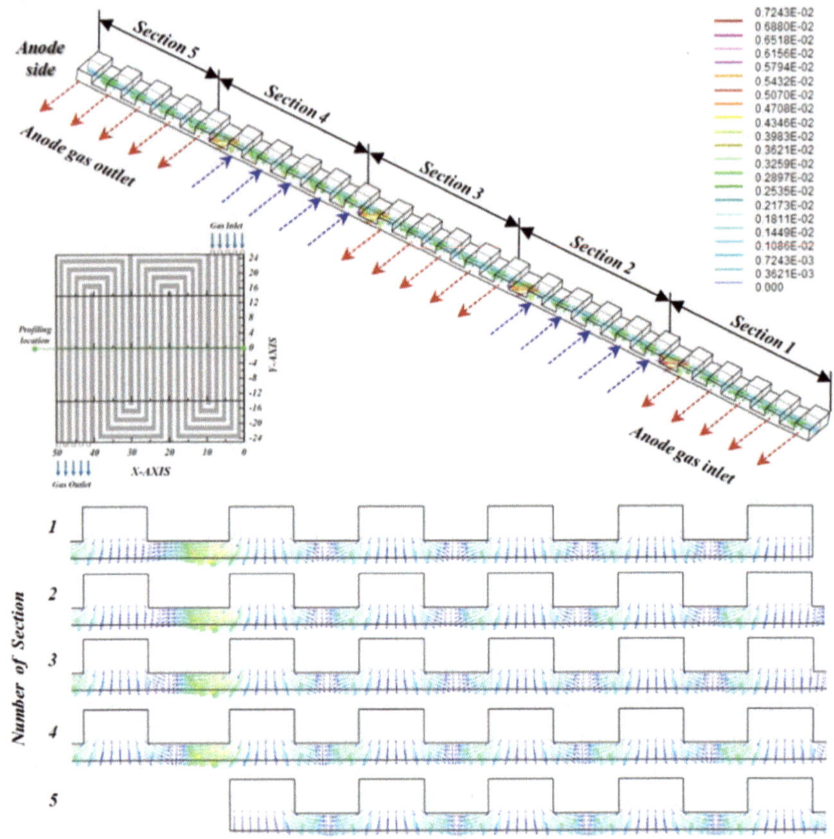

(b) Velocity vectors of anode side with CASFF

Fig. 13 Cross-sectional views illustrating the under-rib convection and liquid water transport mechanisms and velocity vectors of reactants in the conventional-advanced serpentine flow-field (CASFF) without sub-channel: **a** water behavior; **b** velocity vectors of anode side; **c** velocity vectors of cathode side

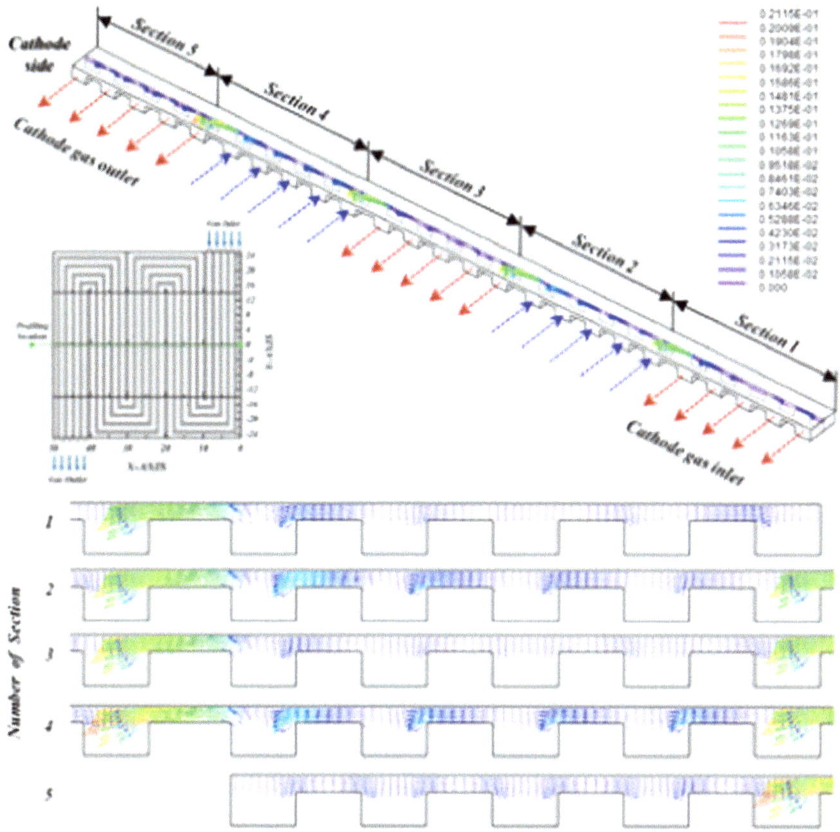

(c) Velocity vectors of anode side with CASFF

Fig. 13 (continued)

A new flow-field design based on the splitting of the channel with enhanced cross-flow in selected regions has been proposed. Its principal hydrodynamic features were demonstrated using CFD analysis [164]. It could be observed from the vectors heading from the straight section of the serpentine channel towards the U-bends that significant under-rib convection existed from the straight feeder channel into each of the U-bend portions through the GDL. Similar under-rib convection could be observed in all the U-bend regions and in the corresponding straight channels in the three serpentine channels. This cross-flow could also be seen to be higher towards the end channels. This effect on the flow rate through the individual serpentine channels is shown in Fig. 15, where the predicted velocity at mid-channel height and length across various serpentine portions of the three split serpentine channels is shown as a function of distance in the width direction of the flow-field. This contributes in reducing the overall pressure drop and increasing the stoichiometric ratio for the end channels.

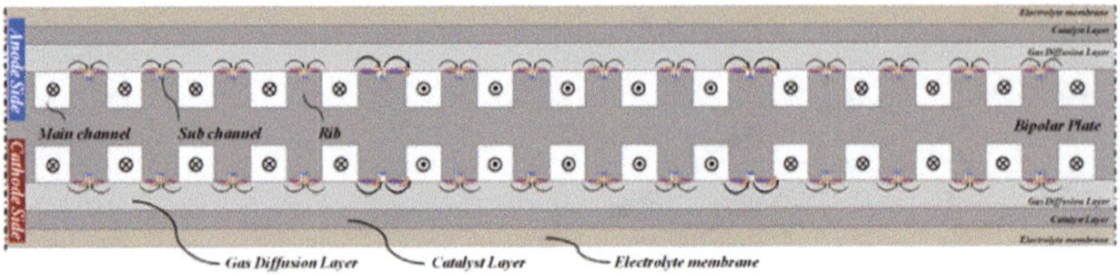

(a) Water behavior with SFFSB

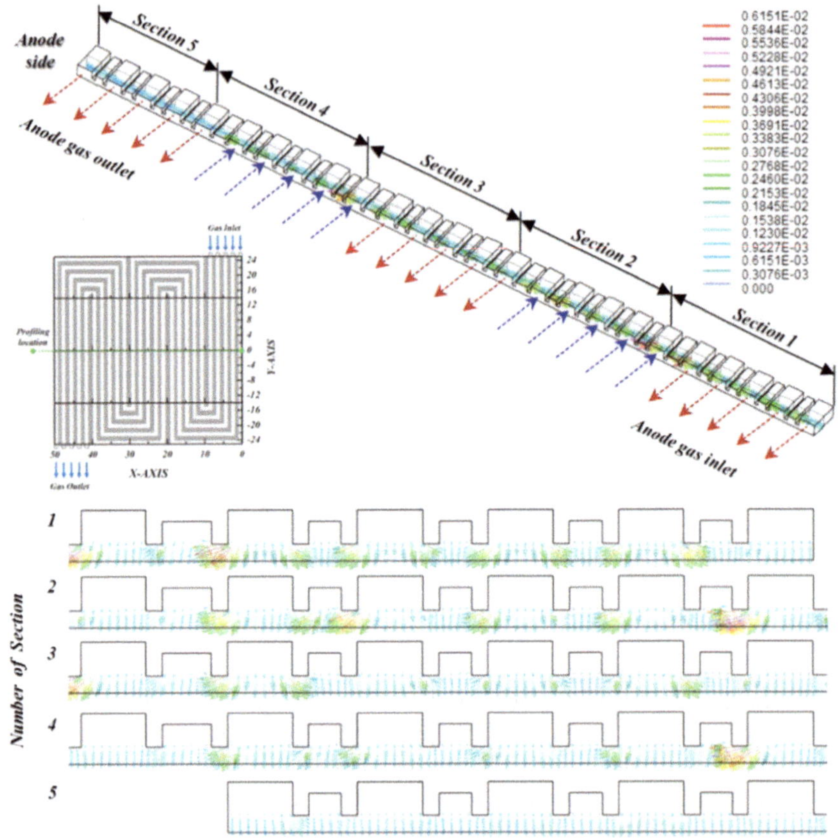

(b) Velocity vectors of anode side with SFFSB

Fig. 14 Cross-sectional views illustrating the under-rib convection and liquid water transport mechanisms and velocity vectors of reactants in the serpentine flow-field with sub-channels and by-passes (SFFSB) with sub-channel: **a** water behavior; **b** velocity vectors of anode side; **c** velocity vectors of cathode side

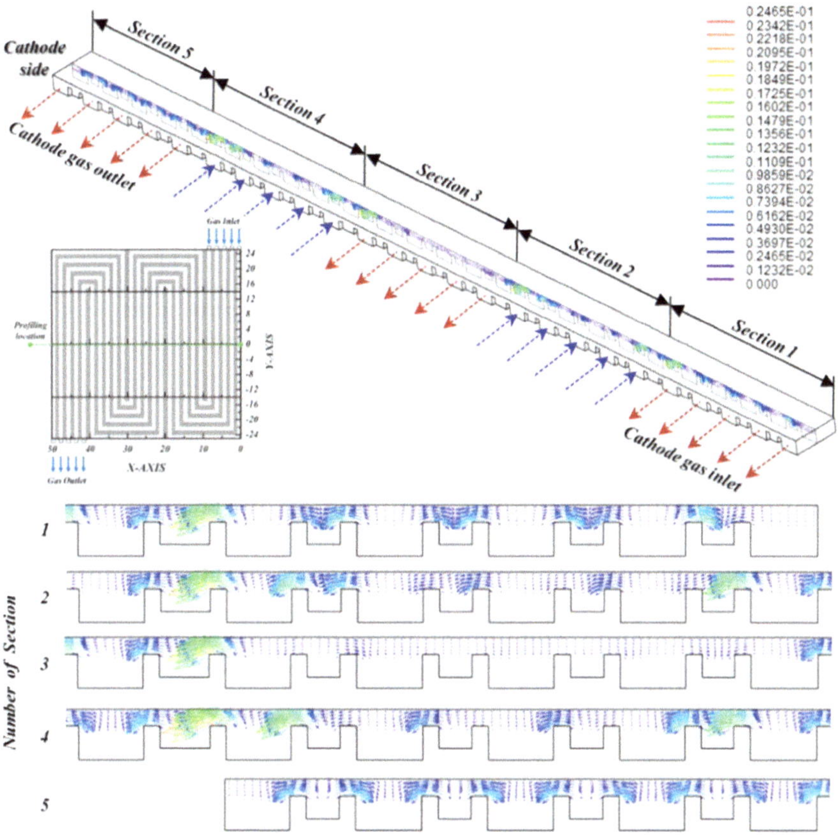

(c) Velocity vectors of anode side with SFFSB

Fig. 14 (continued)

Methanol crossover is driven by diffusion, electro-osmosis, and convection. These three factors contribute differently under different operating conditions. Nevertheless, all three depend on the local methanol concentration in the PEM. Hence, the maximum crossover flux is expected at the open-circuit voltage where the highest mass transfer of un-reacted methanol occurs. When the current density increases, the excess methanol at the interface between the anode catalyst layer and the PEM decreases. Accordingly, methanol crossover is reduced to almost zero after the current density approaches to its limiting point where nearly all methanol molecules are consumed during electrochemical reaction. This result indicates that diffusion dominates in methanol crossover, except in the region near the limiting current density. In this region, the methanol concentration at the interface of the catalyst layer and the PEM decrease due to the mass transport limitation and high current output.

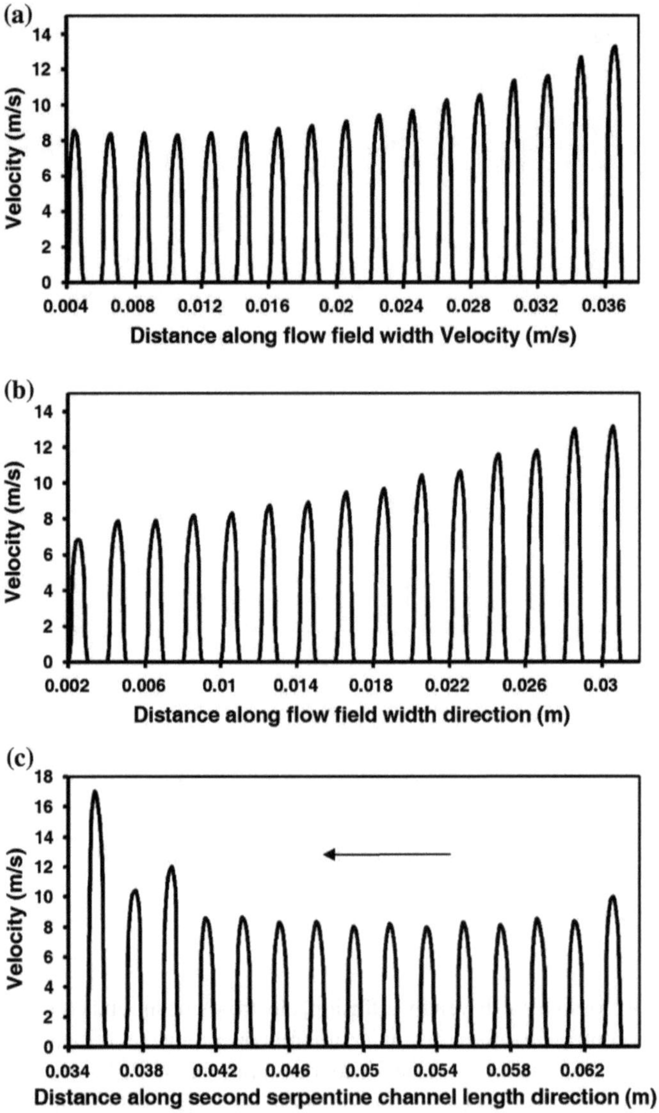

Fig. 15 Predicted velocity variations at mid-channel height showing the channel-to-channel variation of velocity in the mid-length of **a** first serpentine channel, **b** third serpentine channel and **c** second serpentine channel. *Source* Ref. [40]

Therefore, the diffusion of methanol flux governed by the concentration gradient in the PEM is markedly suppressed. The electro-osmosis drag flux, on the other hand, does not decrease significantly because it is proportional not only to the decrease in methanol concentration but also to the increase in cell current density. Once the current density reaches to its limiting point, the diffuse flux becomes zero because on the absence of concentration gradient. The electro-osmosis drag flux of methanol also reduces to zero, because there is no methanol content in the electro-osmosis water flux transferring through the membrane.

4.2 Uniformity of Temperature and Current Density Distributions

An ideal flow-field design would enable uniform gas mole fractions over the entire surface area of the cell. This ideal design would lead to a uniform distribution of current density and thus produce uniform distributions of temperature and water production. These uniform distributions cause less mechanical stresses on the MEA and lead to longer cell lifetimes. In the real fuel cells, mass transport losses are minimized by employing the intricate flow structures containing many small flow channels. Compared to a single-chamber design, a design that employs many small flow channels keeps the reactants constantly flowing across the fuel cell, facilitating uniform convection and homogeneous distribution. Small flow channel designs also provide more contact points across the surface of the electrode from which the fuel cell electrical current can be harvested. An intricate design promotes forced convection of the reactants through the GDL. This design has been the subject of recent attention because it provides far better water management, leading to an improved mass transport. The problem with the forced convection through the GDL is that it leads to significant pressure drop losses. However, this major disadvantage can be overcome by employing extremely small rib spacing [41].

With an appropriate design of the flow channel of the bipolar plate, designers can enhance the cell performance of a PEMFC system effectively. A projection finite element analysis with an element-by-element preconditioned conjugated-gradient method was used to investigate the non-isothermal tapered flow channel installed with a baffle plate for enhancing cell performance in the cathodic side of a PEMFC [39]. The stronger composite effect of tapered flow and blockage was seen to result in a lower and more uniform temperature distribution in the fuel flow channel. In general, the improvement seen in the single-phase convection heat transfer performance in the fuel flow channel may be caused by an increasing flow interruption, a reduction in the thermal boundary layer, or an increased velocity gradient near the GDL boundary. In order to improve the reduction of the cell performance in the downstream region, a non-isothermal tapered flow-field design was considered to reduce the reflection in the downstream region. The stronger composite effect of the tapered flow channel and baffle blockage results in better performance of convection heat transfer and a higher fuel flow velocity, which in turn improves the local current density and polarization characteristics with a penalty of high-pressure loss.

The PEMFC system consists of many parts, and the bipolar plate is one of the key components. Channels at the bipolar plate distribute air on the cathode side and hydrogen on the anode side, which are essential for the electric-chemical reaction to produce electricity. Bipolar plates may also be used as a path for the cooling of water to control the temperature of the cell. A parametric study was conducted in order to find the optimal design of the bipolar plate in the PEMFC system for automobiles [26]. First, the design parameters of a bipolar plate are defined. Then, an analysis is conducted on how the combination of those parameters would affect the performance of the fuel cell. Finally, the computation is done for the optimum combination of the parameters. The local current density, concentration and pressure of the reacting gases, cell temperature, and water distribution were obtained by CFD calculation using electrochemical variables fitted by a *V–I* curve from the real system. There is an optimal combination of the design parameters through a trade-off between the performance and the pressure drop in the system design [41].

In order to apply a numerical analysis to design an actual scale separator in PEMFC, the calculation size must be enlarged. Mass transfer and flow in GDL were calculated, and a GDL mass transfer approximate model based on the theoretical model was developed. Next, with this model, a PEMFC reaction and thermal flow analysis model was created. Furthermore, the effects of the separator channel depth on output performance and current density distribution were examined using this numerical analysis [11]. As a result, in the case of shallow channels, the oxygen transfer rate to the electrode increased because of gas flow in GDL. However, current density distribution and pressure drop also increased.

4.3 Facilitation of Liquid Water Discharge

Dynamic water balance and management have imposed one of the dormant technical challenges on PEMFC design and operation, and directly influence the performance and lifetime of PEMFC systems. The importance of the water balance and management arises from the fact that the PEM used in PEMFC requires full hydration in order to maintain good performance and lifetime. The guarantee of membrane hydration under full operating conditions has been achieved, with a good degree of success, by supplying fully humidified reactant gas streams for both the anode and cathode. However, water is produced at the cathode as a result of an electrochemical reaction, and water is also transported from the anode to the cathode through the membrane electrolyte via the electro-osmotic drag effect. Water transport may also occur due to the local gradients of the pressure and/or concentration. Consequently, liquid water flooding occurs if water is not removed adequately from the cell (the cathode in particular). Water flooding will make the PEMFC performance unpredictable, and unrepeatable under nominally identical operating conditions. As a result, elaborate experimental diagnostics and schemes need to be implemented in order to achieve consistent results with much lower performance and limiting current densities [25]. Therefore, water management has become a delicate task; both too much and too little water can adversely impact the performance and lifetime of PEMFCs. Maintaining a perfect water balance during the dynamic operation process has posed a significant challenge for PEMFC design and operation, and hence water management is a critical issue for PEMFCs.

The design of flow-fields grooved in the bipolar plates of polymer electrolyte membrane fuel cells becomes important at high current densities. Many flow-field designs are being studied to enhance the current density and homogeneity of the reactants in order to overcome the challenges of pressure losses and removal of water from the cathode [22]. One of the features of a serpentine flow-field is the high-pressure drop in the flow direction. This drop creates a significant pressure difference between two neighboring channels that may lead to the under-rib convection. Because the pressure difference across neighboring channels varies along the length, the strength of the cross-flow also varies significantly and is lower in the U-bend region. This variation may cause insufficient evacuation of water vapor in the region and lead to build-up and eventual local flooding. In the present paper, this scenario is verified by means of CFD simulations. This result would indicate that the measures that improve local under-rib convection may lead to a better design of the flow-field. If the electrode permeability or the pressure drop is high, then significant under-rib convection can be established which leads to more effective evacuation of the water vapor. The superior performance of the convection-enhanced serpentine flow-field may also be attributed to this process. Hence, under-rib convection can be used as an additional parameter in arriving at the optimum design of a flow-field [17].

The cathode flow-field design of PEMFC determines its reactant transport rates to the catalyst layer and the removal rates of liquid water from the cell. The cathode flow-field has been optimized for a single serpentine PEM fuel cell with 5 channels using the heights of channels 2–5 as search parameters. An optimization approach that integrates the simplified conjugated-gradient scheme and a three-dimensional, two-phase, non-isothermal fuel cell model has also been described [35]. Because most oxygen at the cathode side is consumed by the upstream regime and water accumulates at the downstream regime (channels 4–5 in Fig. 16) in the cell, there is a limited chemical reaction rate along channels 4–5. Additionally, the local water concentrations in regions A–D gradually decrease as the optimization process proceeds. At an optimal design, the local current density and liquid water are generally uniformly distributed over the entire membrane. The amount of water at the outlet in the optimal design is significantly less than that in the conventional design (step 1). The under-rib convection velocities are comparable to channel flow velocities; both increase with reduced channel heights. The strong convection flows enhance oxygen transport and water removal, leading to nearly uniform distributions of local current density in the membrane throughout the entire cell. Conversely, the final channel should be a diverging channel because strong under-rib convection occurring at the rib next to the outlet can result in considerable oxygen leakage and poor water removal for the final channel. The combined optimization has been proven effective in optimizing the flow-field design for a single serpentine PEMFC. The role of under-rib convection should be considered in cathode flow-field design.

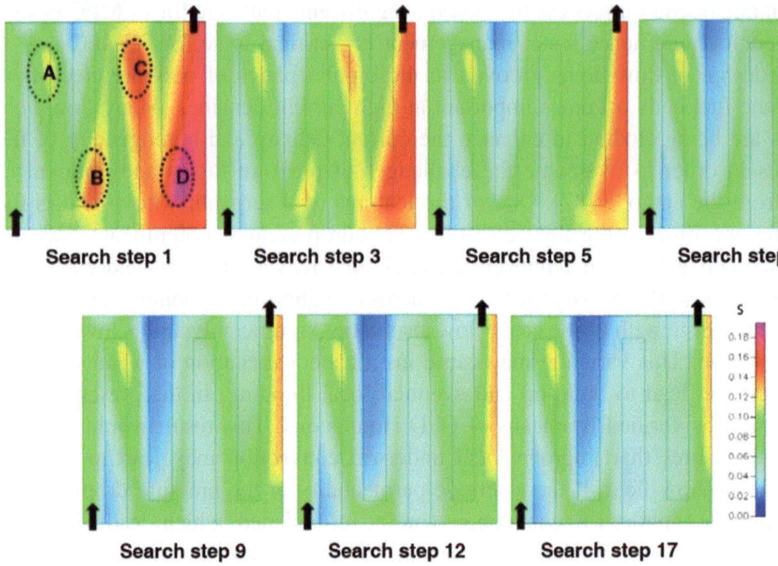

Fig. 16 Liquid water distributions on the interface between the cathode gas diffusion layer and the catalyst layer at various search steps. *Numbers in the figure* denote each search step. *Source* Ref. [35]

4.4 Reduction in Pressure Drop

The fuel flow channel has a GDL as a side-wall, and hence the GDL morphology is expected to influence the reactant gas transport from the channel to the catalyst surface and ultimately the cell performance. The design of this flow-field has a distinct influence on the oxygen transport from the GDL to the cathode catalyst layer. Although different designs have been examined, the choice of an optimal flow-field design still remains controversial. The parallel design avoids water accumulation in the corners of the flow-field and consequently avoids local starvation of the oxidant due to minimal pressure drop by equally distributing the flow rate into many straight parallel paths. Serpentine channel designs have been the most widely used because their increased channel length offers higher gas velocity in the channel with a moderate increase in parasitic loss by air pumping. This design ensures effective liquid water removal, which prevents channel clogging by condensate.

High power density requires high current density operation. Under these conditions, mass transport losses can be significant contributor to the overall loss mechanisms in the cell. When using air as the oxidant, the driving force for oxygen transport from channel to catalyst is low due to the low partial pressure. Oxygen needs to be transported from the feeding channel to the catalyst layer through a porous material. Besides, water that forms at the cathode catalyst can condense in the pores. A parameter study for was performed limiting current conditions;

this study emphasized the transport in the GDL between channels and did not require the complex description of the electrochemical kinetics [18]. The channel-rib geometry, channel-to-channel pressure difference and GDL characteristics were investigated in order to determine GDL design. To achieve a well-defined compression, the in-plane and the through-plane apparatuses were realized based on 30 mm thick stainless steel blocks. The distance was set by placing steel spacer plates between the blocks. A water column gauge with 1 Pa resolution was used to determine the pressure drop. To study the effects of cross convection in a serpentine channel, a 2D slice model of the GDL under the channel and rib was linked to a 3D model of a U-bend channel (simulating the serpentine flow-field). This 2D + 3D serpentine model was used to determine the local cross convection flow rates and the pressure distribution. To avoid excessive cross convection at the channel inlet where pressure gradients are high, a well thought out dimensioning of the channel-rib geometry and size of the additional GDL were imperative. The increased thickness under the rib also had the drawback of slightly increasing the electric resistance. In the light of the significant increase in the rib current and the relatively low Ohmic resistance of the GDL, the introduction of additional GDL material is expected to pay off in terms of overall performance.

The theory of the entrance region pressure drop was first introduced and validated in four-tube parallel channel geometry [13]. An experimental test section was designed to validate/modify the theoretical equations using water as the fluid. The equations derived for the developing laminar flow in circular ducts were used to describe the entrance region pressure drop within the tubes. The pressure drop in the tube entrance region was the summation of two components: a frictional pressure drop, which accounts for the greatest portion of the total pressure drop, and the minor pressure drop, caused by entrance region effects. For micro- and mini-channel applications, such as PEMFCs, laminar flow generally existed due to the relatively low flow rates and the small hydraulic diameters. A new technique was proposed to measure the instantaneous flow rate through individual channels in a parallel channel array. The experimental setup was specifically designed to allow calibration of the individual channel flow rates. The ability of this technique to accurately predict flow maldistribution in parallel channels was experimentally validated. The method was implemented in ex situ and in situ setups designed to study two-phase flow in gas diffusion channels of a PEMFC to develop effective water management strategies for PEMFCs. Based on the improvement of the local cross-flow conditions in a split serpentine flow-field [40], the overall flow rates in the three serpentine channels were within 5% of each other, showing that the flow distribution among the three serpentine channels was relatively uniform. The predicted pressure drop for the flow-field is only 2861 kPa, much less than the pressure drop of 25,665 kPa obtained with a single serpentine flow channel for the same air flow rate. Compared to the single serpentine, the flow rate through each of the three parallel paths was about 1/3. The path-length in each was also only one-third of the total serpentine path-length. Thus, the overall pressure drop, which was roughly equal to the pressure drop in one of the parallel paths, was 1/9 of the single serpentine.

To understand how a tapered flow channel and baffle plate affect the pressure drop, the dimensionless pressure drop between the gas fuel inlet and outlet across the channel was calculated [39]. Though, this design improved the fuel transport rate and enhanced the reaction at the catalyst layer, but the channel arrangement produced a larger pressure loss and required more pumping power for the delivery of the fuel. Therefore, with an appropriate gap and tapered ratios, the pressure drop as well as the pumping power can be considerably reduced. The convection-enhanced serpentine flow-field realized during this work showed superior performance compared to conventional serpentine flow-field due to its enhanced mass transport capability as a result of the enhanced under-rib convection associated with the increased pressure difference between adjacent flow channels [28]. Another striking feature of this new flow-field is that the pressure drop between the inlet and the outlet remains almost the same as that in the conventional design, because the channel dimensions, including width, depth, and total length, are the same for both flow-fields. The reduced gas flow rate will decrease the parasitic power in the fuel cell system, thus leading to higher system efficiency. It is worth mentioning that the convection-enhanced serpentine flow-field can also be readily extended to the formation of a flow-field consisting of multiple serpentines in parallel without losing the feature of enhanced under-rib convection and while resulting in a smaller pressure drop. The effects of the design parameters on the PEMFC performance have been examined using CFD calculation [26]. The numerical results based on a Taguchi L18 test matrix showed the output power and the pressure drop on the cathode and the anode sides for the current density of 1.2 A/cm^2, as shown in Fig. 17. It was evident that the power and the pressure drop results are affected by geometric design parameters. However, the appropriate pressure drop should be set because a high-pressure drop at the channel inevitably causes a high load on the Balance of Plants (BOP) system. Because more BOP resources result in lower net

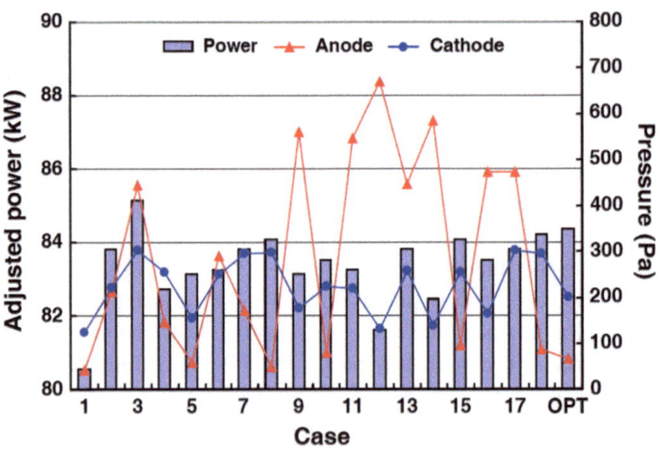

Fig. 17 Power and pressure drop curve; at 1.2 A/cm^2 condition. *Source* Ref. [26]

power output from the system, it was necessary to design the bipolar plate with the maximum power and the minimum pressure drop. Based on the calculation of results, the power reaches its maximum in case 3, but the pressure drop is not at its minimum in this case. The set of optimum parameters obtained from the Taguchi method showed high power output with an appropriate pressure drop. There is an optimal combination of the design parameters through a trade-off between the performance and the pressure drop in the system design. The design parameters should be selected considering their effects on both the performance and the pressure drop in the channel in order to meet the goals of the PEMFC system.

4.5 Improvement in Output Power

The cell output power increases with increasing active area, but not in proportional manner. This is attributed to the fact that long flow channels are commonly fabricated with cells of large active areas in order to provide high power which renders water removal difficult in maintaining a sufficient oxygen transfer rate at the cathode. In other words, the cell size effect is noticeable in PEMFC applications. However, to the best of our knowledge, at present the size effect has not yet been satisfactorily analyzed. Performance curves of PEMFCs were investigated for parallel flow-fields with various active areas [34]. The cathode and anode flow channels formed a rectangular shape with 1×1 mm^2 ribs. The active area of cell 1 was 11×11 mm^2 with 6 channels and 5 ribs; the active area of cell 2 was 21×21 mm^2 with 11 channels and 10 ribs; the active area of cell 3 was 31×31 mm^2 with 16 channels and 15 ribs; and the active of cell 4 was 41×41 mm^2 with 21 channels and 20 ribs. Moreover, because the reactant inlet flow rates significantly affect the cell performance, for convenience of comparison, specified inlet flow rates were fed to the unit active area for all cells in the present work with 0.4 cm^3/min humidified hydrogen to 1 mm^2 active area on the anode side, and 1.2 cm^3/min humidified air to 1 mm^2 active area on the cathode side. For operating voltages >0.7 V, the polarization curves and power density curves for four active areas almost coincided, indicating negligible cell size effects. However, at operating voltages <0.7 V, the size effect was profound. The average current density and output power density all decreased with an increase in the active area. As the active area was increased from 11×11 mm^2 to 41×41 mm^2 (by 13.9 times), the average current density decreased by 10.6% at 0.4 V and 11.6% at 0.3 V. Three different channel heights and widths were compared with the base flow-field design of the serpentine channel that was 1 mm wide and 0.34 mm high, each through a detailed numerical study of the distribution of temperature, pressure, water content, and local current density [41]. The output power density is lowered due to the decreased cell voltage and larger volume of PEMFC stack with the channel height. The cell voltage was decreased with the increase of channel width, and its extent of the reduction was larger than that with the increased channel height.

The optimal cathode flow-field design of a single serpentine proton exchange membrane fuel cell was obtained by adopting a combined optimization procedure including a simplified conjugate-gradient method and a completely three-dimensional, two-phase, non-isothermal fuel cell model, to look for optimal flow-field design for a single serpentine fuel cell 9×9 mm^2 in size with five channels [30]. The initial guess assumed all channel heights and widths of 1 mm, identical to the basic design with straight channels. The cell output power density P_{cell} was maximized and subjected to an optimal set of channel heights, H_1–H_5, and channel widths, W_2–W_5. The basic case with all channel heights and widths set at 1 mm yielded a P_{cell} = 7260 W/m^2. The optimal design displays a tapered characteristic for channels 1, 3, and 4, and a diverging characteristic in height for channels 2 and 5, producing a P_{cell} = 8894 W/m^2, about a 22.5% increment. The reduced channel heights of channels 2–4 significantly increased the under-rib convection and widths for effectively removing liquid water and oxygen transport in the gas diffusion layer. The final diverging channel minimized the leakage of fuel to outlet via under-rib convection from channels 4 to 5. The use of a straight, final channel of 0.1 mm height led to a 7.37% power loss, whereas designing all channel widths to be 1 mm with optimal channel heights obtained the above yields with only a 1.68% loss in current density.

The serpentine flow channel layout is not the ideal flow-field configuration, and this layout has a number of problems. It typically results in a relatively long reactant flow path, hence a substantial pressure drop, resulting in significant parasitic power loss associated with the cathode air supply (as much as 35% or more of the stack output power). A novel serpentine-baffle flow-field design was proposed to improve the cell performance compared to that of a conventional serpentine flow-field using a three-dimensional numerical model [29]. The flow-field design influenced not only the cell performance but also the pressure drop in the fuel cell. Larger pressure drops in the fuel cell meant that more power was needed to pump the reactants. Thus, the pressure drop is a significant issue to be considered in choosing the flow-field designs in addition to the I–V_{cell} curve. The pressure drops for baffled designs were higher than those for the conventional design, and then the cathode pressure drop loss, W_P, was calculated. Although the pressure drops for the baffled designs are larger than those for the conventional design, the pressure drop losses, W_P, were far lower than the cell output power, P_{cell}, for the miniature fuel cells analyzed.

A concise mathematical model was developed for quick numerical simulation to allow performance prediction and analysis of DMFC [31]. Two-phase flow at the anode and pressure-driven convection were both included in the proposed model and thereby enabled accurate and realistic performance prediction and analysis. To investigate the two-phase mass transport effect on cell performance, the I–V curve for 1 M input methanol at 40 °C was simulated. The liquid saturation S was assumed to be unity and thereby represented the transport of only liquid phase at the anode. The limiting current density 147.0 mA/cm^2 was much smaller than that assigned for two-phase mass transport, viz., 200.6 mA/cm^2 by under-rib mass transport effect. The results thus revealed that methanol transport in the gas phase is important for sufficient fuel supply to the reaction sites. Moreover, the cell voltage

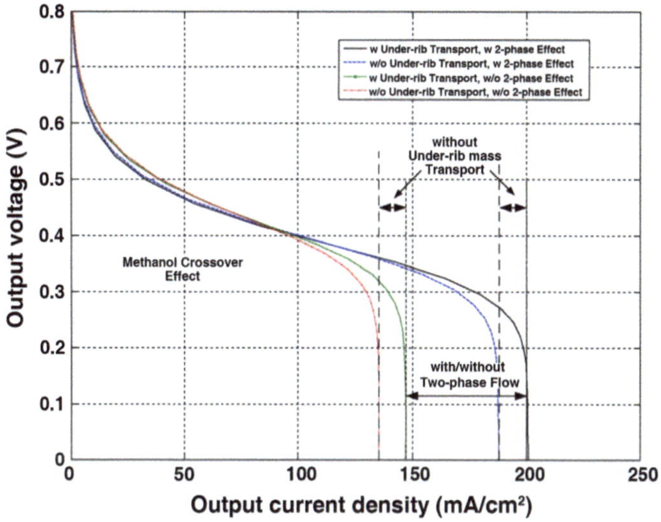

Fig. 18 Influence of two-phase flow phenomenon on DMFC performance, with and without under-rib mass transport consideration effects; 1 M input methanol solution with 2 ml/min flow rate at 40 °C operating temperature. *Source* Ref. [31]

reduction induced by methanol crossover was noted at a low current density where the liquid phase transport situation gave a higher cell output voltage compared with that for two-phase transport. Consequently, the limiting current density had a relatively small value of 135.5 mA/cm^2. That is reasonably close to the modeling result of 138.4 mA/cm^2. The difference between including or not including the under-rib mass transport effect implied that the facilitated methanol transport towards the catalyst layer was influenced by under-rib mass transport. The significance of under-rib mass transport on performance was, however, less than that caused by the two-phase effect. Performance data with the under-rib mass transport effect subtracted from the two-phase mass transport are presented in Fig. 18.

As shown in Fig. 19, numerical simulations were performed to compare the four configurations: configuration I in which CASFFs were used both at the anode and the cathode; configuration II in which a SFFSB and a CASFF were used at the anode and the cathode, respectively, configuration III in which SFFSBs were used at both the anode and the cathode, and configuration IV in which a CASFF and a SFFSB were used at the anode and the cathode, respectively. To verify the maximization of power density among the performance-related parameters, the polarization and power density curves for different flow-field configurations I–IV obtained by numerical simulation are compared, and the results are shown in Fig. 20a. The results reveal that the power densities of configurations III and IV are higher in comparison with the flow-field configurations I and II. Maximum power densities of the four flow-field configurations I–IV are 0.5199, 0.5278, 0.6122 and 0.6175 W/cm^2, respectively. The adoption of SFFSB at the cathode bipolar plate

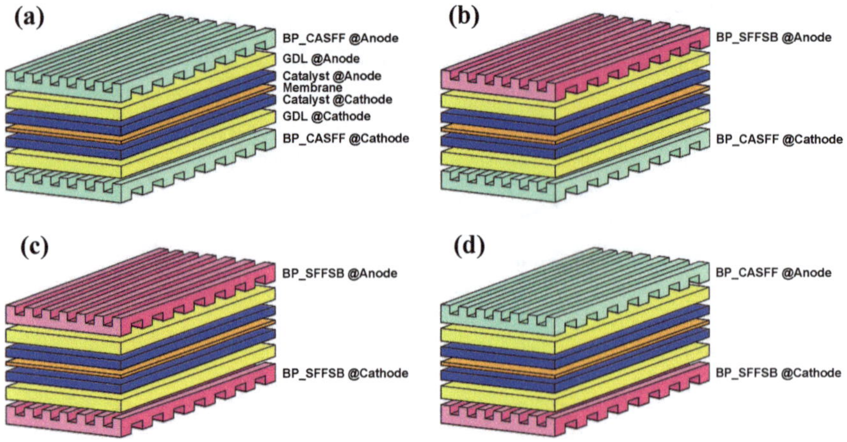

Fig. 19 Four flow-field configurations of CASFF and SFFSB applied on the anode and cathode bipolar plates: **a** configuration I in which CASFFs were used at both the anode and the cathode; **b** configuration II in which a SFFSB and a CASFF were used at the anode and the cathode, respectively; **c** configuration III in which SFFSBs were used at both the anode and the cathode; **d** configuration IV in which a CASFF and a SFFSB were used at the anode and the cathode, respectively

increases the output power density because the under-rib convection enables a more effective utilization of the electrocatalysts by increasing the mass transport rates of the reactants from the flow channel to the inner catalyst layer and by significantly reducing the water flooding at the cathode.

Figure 20b shows that the experimental results for the polarization and power density curves obtained with the CASFF and SFFSB have the same trend as the simulation results. The PEMFC performances with 5 passes and 4 turns on an active area of 25 cm^2 were experimentally evaluated for a new flow-field design, which stimulates under-rib convection by adding sub-channels and by-pass areas to the CASFF. Although the activation loss region and the ohmic resistance loss region exhibit almost the same characteristics, a small performance variation occurs in the mass transport loss region. This variation is caused by worsening water management conditions for moving toward the higher current density zone because the height of the channel is shorter due to the flow-field patterns, even though the cell went through an adequate activation process. The fuel cell performance is affected by the internal behavior, and it is especially sensitive to the behavior of the liquid water in the electrochemical reaction. The behavior of the liquid water inside the flow-field increases the water content of the MEA at the beginning of fuel cell operation, and it usually moves from the anode to the cathode. However, back diffusion sometimes occurs due to the liquid water mass fraction and the pressure drop of the cathode. Furthermore, the liquid water cools the heat caused by the electrochemical reaction. The behavior of the liquid water becomes one of the determining factors for cell performance. By adding the sub-channels and the

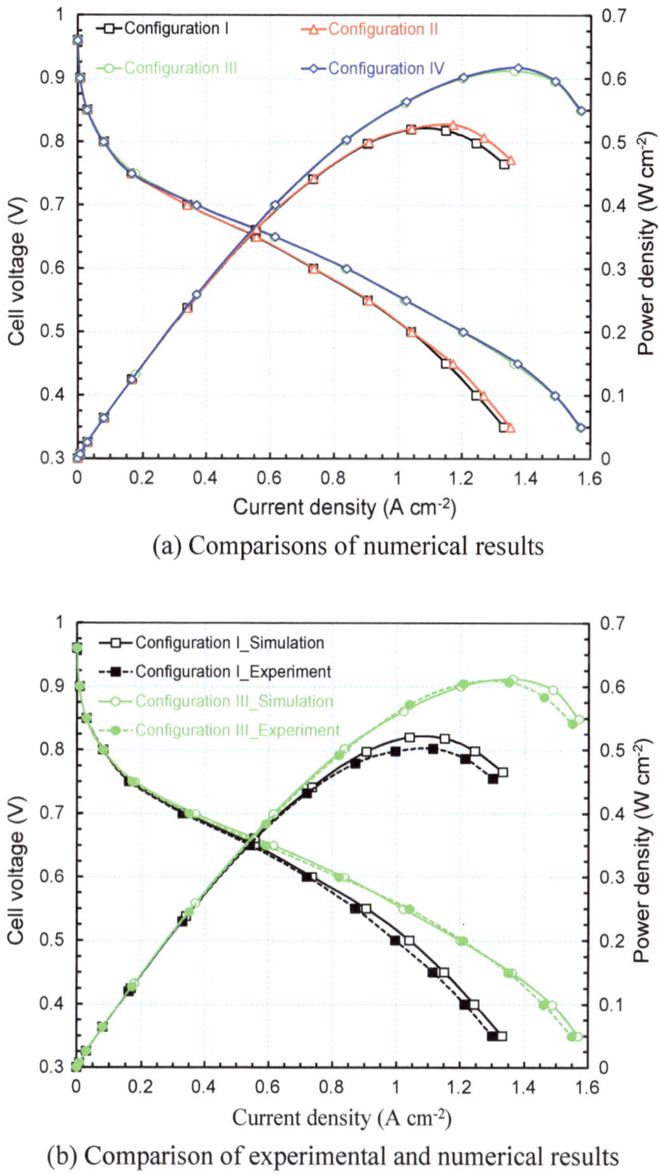

Fig. 20 Comparisons of the polarization and power density curves: **a** between the four flow-field configurations I–IV obtained by numerical simulation; between the experimental and simulation results of the flow-field configurations I and III

bypasses to the CASFF, the SFFSB improves the flow consistency of the inner reaction gas and the liquid water behavior, and it enhances the current density in both ohmic resistance loss region and the mass transport loss region [165].

5 Summary

In this article, the research and development in PEMFCs, ranging from their components (PEM, electrodes and catalysts, GDL, MEA and bipolar plates) and cells (single cell and multi-cell stack), as well as the experimental and modeling approaches for the optimization of flow-field, followed by the new flow-field optimization using under-rib convection have been discussed. Although the performance of PEMFCs is primarily determined by the intrinsic electrochemical efficiency of the MEA, the flow plate is one of the key components of a PEMFC, serving as both the current collector and the reactant distributor. The uniform distribution of reactants over the entire surface of the fuel cell catalyst layer also depends on the design of the porous diffusion layer, including pore size and the thickness of the porous diffusion layer. Recently, the effects of under-rib convection on mass and heat transport, liquid water removal as well as pressure drop have related to PEMFCs been actively investigated.

The under-rib convection enables more effective utilization of electrocatalysts by increasing reactant concentration and promotes the uniformity of temperature and current density distributions. If facilitates liquid water discharge, reduces the pressure drop, and improves output power. Although much progress has been made with respect to flow-field optimization, discussions of optimal flow-field designs still remained controversial. Development of new flow-fields via the promotional role of under-rib convection has been suggested; however, some challenges remain regarding the new flow plates issues such as the volume, weight, and cost of PEMFC stacks.

First, the electrochemical and physical properties, which are verified not by invalidated measurements but by standard measurements in an operating fuel cell, are essential to bridging the gap between modeling and experiments. The validation of experimental results by modeling provides qualitative comparison, and hence numerical models have been quantitatively verified. Recently, segment-wise pressure distribution data have revealed how to optimize the clamping pressure using the developed setup for a given GDL for uniform reactant distributions and pressure drop behaviors [166]. Simulation results have been compared quantitatively to the bubble velocities and geometric dimensions observed in the experimental data of both coated and uncoated channels [167].

Second, there is a need to apply the developed flow plate to a practical PEMFC stack and validate the effects of under-rib convection on water and heat management, pressure drops, and output power. Under-rib convection is a complex phenomenon in which the mass transport of reactants in flow plates as well as GDL and catalysts influence liquid water, pressure, temperature and current density.

Therefore, the advanced flow plates, which were developed to promote under-rib convection, should be confirmed as playing a promotional role in the flow plates at the same time in order to meet the goal of the PEMFC system. For example, this situation can be seen in comparing the output power and the pressure drop loaded on the BOP [26].

This review has shown that the new flow-field optimization technique by under-rib convection has great potential for improving cell performances inside operating PEMFCs. The flow-field that activates the internal mass transfer mechanism of PEMFCs is able to improve its power density from a mechanical engineering perspective. In other words, this mechanical method can improve the performance of the fuel cells without changing the electrochemical materials. An approach combining fundamental study and engineering development of under-rib convection would help us take the necessary steps to advance the performance and commercialize PEMFCs.

References

1. Barbir F (2005) PEM fuel cells: theory and practice. Elsevier, London
2. O'Hayre RP, Cha SW, Colelle W, Prinz FB (2006) Fuel cell fundamentals. Wiley, New York
3. Bose S, Kuila T et al (2011) Polymer membranes for high temperature proton exchange membrane fuel cell: recent advances and challenges. Prog Polym Sci 36(6):813–843
4. Gasteiger HA, Panels JE, Yan SG (2004) Dependence of PEM fuel cell performance on catalyst loading. J Power Sources 127(1):162–171
5. Dohle H, Jung R, Kimiaie N, Mergel J, Müller M (2003) Interaction between the diffusion layer and the flow field of polymer electrolyte fuel cells: experiments and simulation studies. J Power Sources 124(2):371–384
6. Wilkinson DP, Zhang J, Hui R, Fergus J, Li X (2010) Proton exchange membrane fuel cells: materials properties and performance. In: Lee S (ed) Green chemistry and chemistry engineering. CRC Press, New York
7. Nguyen TV (1996) A gas distributor design for proton-exchange-membrane fuel cells. J Electrochem Soc 143(5):103–105
8. Li X, Sabir I (2005) Review of bipolar plates in PEM fuel cells: flow-field designs. Int J Hydrogen Energy 30(4):359–371
9. Zhang L, Bi HT, Wilkinson DP, Stumper J, Wang H (2008) Gas–liquid two-phase flow patterns in parallel channels for fuel cells. J Power Sources 183(2):643–650
10. Pharoah JG (2005) On the permeability of gas diffusion media used in PEM fuel cells. J Power Sources 144(1):77–82
11. Inoue G, Matsukuma Y, Minemoto M (2006) Effect of gas channel depth on current density distribution of polymer electrolyte fuel cell by numerical analysis including gas flow through gas diffusion layer. J Power Sources 157(1):136–152
12. Dutta S, Shimpalee S, Van Zee JW (2001) Numerical prediction of mass-exchange between cathode and anode channels in a PEM fuel cell. Int J Heat Mass Transfer 44(11):2029–2042
13. Kandlikar SG, Lu Z, Domigan WE, White AD, Benedict MW (2009) Measurement of flow maldistribution in parallel channels and its application to ex-situ and in-situ experiments in PEMFC water management studies. Int J Heat Mass Transfer 52(7):1741–1752

14. Williams MV, Kunz HR, Fenton JM (2004) Influence of convection through gas-diffusion layers on limiting current in PEMFCs using a serpentine flow field. J Electrochem Soc 151(10):1617–1627
15. Oosthuizen PH, Sun L, McAuley KB (2005) The effect of channel-to-channel gas crossover on the pressure and temperature distribution in PEM fuel cell flow plates. Appl Therm Eng 25(7):1083–1096
16. Sun L, Oosthuizen PH, McAuley KB (2006) A numerical study of channel-to-channel flow cross-over through the gas diffusion layer in a PEM-fuel-cell flow system using a serpentine channel with a trapezoidal cross-sectional shape. Int J Therm Sci 45:1021–1026
17. Prasad KBS, Jayanti S (2008) Effect of channel-to-channel cross-flow on local flooding in serpentine flow-fields. J Power Sources 180(1):227–231
18. Tehlar D, Flückiger R, Wokaun A, Büchi FN (2010) Investigation of channel-to-channel cross convection in serpentine flow fields. Fuel Cells 10(6):1040–1049
19. Sun W, Peppley BA, Karan K (2005) Modeling the influence of GDL and flow-field plate parameters on the reaction distribution in the PEMFC cathode catalyst layer. J Power Sources 144(1):42–53
20. Wang Y, Wang CY (2005) Simulation of flow and transport phenomena in a polymer electrolyte fuel cell under low-humidity operation. J Power Sources 147(1):148–161
21. Kanezaki T, Li X, Baschuk JJ (2006) Cross-leakage flow between adjacent flow channels in PEM fuel cells. J Power Sources 162(1):415–425
22. Feser JP, Prasad AK, Advani SG (2006) On the relative influence of convection in serpentine flow fields of PEM fuel cells. J Power Sources 161(1):404–412
23. Ye Q, Zhao TS, Xu C (2006) The role of under-rib convection in mass transport of methanol through the serpentine flow field and its neighboring porous layer in a DMFC. Electrochim Acta 51(25):5420–5429
24. Park J, Li X (2007) An experimental and numerical investigation on the cross flow through gas diffusion layer in a PEM fuel cell with a serpentine flow channel. J Power Sources 163(2):853–863
25. Li X, Sabir I, Park J (2007) Review of bipolar plates in PEM fuel cells: flow-field designs. J Power Sources 30(4):359–371
26. Lee S, Jeong H, Ahn B, Lim T, Son Y (2008) Parametric study of the channel design at the bipolar plate in PEMFC performances. Int J Hydrogen Energy 30(20):5691–5696
27. Xu C, He YI, Zhao TS, Chen R, Ye Q (2006) Analysis of mass transport of methanol at the anode of a direct methanol fuel cell. J Electrochem Soc 153(7):1358–1364
28. Xu C, Zhao TS (2007) A new flow field design for polymer electrolyte-based fuel cells. Electrochem Commun 9(3):497–503
29. Wang XD, Duan YY, Yan WM (2007) Novel serpentine-baffle flow field design for proton exchange membrane fuel cells. J Power Sources 173(1):210–221
30. Wang WD, Huang YX, Cheng CH, Jang JY, Lee DJ, Yan WM, Su S (2009) Flow field optimization for proton exchange membrane fuel cells with varying channel heights and widths. Electrochim Acta 54(23):5522–5530
31. Yang Y, Liang YC (2009) Modelling and analysis of a direct methanol fuel cell with under-rib mass transport and two-phase flow at the anode. J Power Sources 194(2):712–729
32. Nam JH, Lee KJ, Sohn S, Kim CJ (2009) Multi-pass serpentine flow-fields to enhance under-rib convection in polymer electrolyte membrane fuel cells: design and geometrical characterization. J Power Sources 188(1):14–23
33. Wang XD, Duan YY, Yan WM, Lee DJ, Su A, Chi PH (2009) Channel aspect ratio effect for serpentine proton exchange membrane fuel cell: role of sub-rib convection. J Power Sources 193(2):684–690
34. Wang XD, Zhang XX, Yan WM, Lee DJ, Su A (2009) Determination of the optimal active area for proton exchange membrane fuel cells with parallel, interdigitated or serpentine designs. Int J Hydrogen Energy 34(9):3823–3832

35. Wang XD, Huang YX, Cheng CH, Jang JY, Lee DJ, Yan WM, Su A (2010) An inverse geometry design problem for optimization of single serpentine flow field of PEM fuel cell. Int J Hydrogen Energy 35(9):4247–4257
36. Zhang Y, Zhang P, Zhenyu Y, He H, Zhao Y, Liu X (2011) A tapered serpentine flow field for the anode of micro direct methanol fuel cells. J Power Sources 196(6):3255–3259
37. Baek SM, Koh SG, Kim KN, Kang JH, Nam JH, Kim CH (2011) A numerical study on the performance of polymer electrolyte membrane fuel cells due to the variation in gas diffusion layer permeabilities. J Mech Sci Technol 25(2):457–467
38. Wang XD, Xu JL, Yan WM, Lee DJ, Su A (2011) Transient response of PEM fuel cells with parallel and interdigitated flow field designs. Int J Heat Mass Transfer 54(11):2375–2386
39. Perng SW, Wu HW (2011) Non-isothermal transport phenomenon and cell performance of a cathodic PEM fuel cell with a baffle plate in a tapered channel. Appl Energy 88(1):52–67
40. Suresh PV, Jayanti S, Deshpande AP, Haridoss P (2011) An improved serpentine flow field with enhanced cross-flow for fuel cell applications. Int J Hydrogen Energy 36(10):6067–6072
41. Choi KS, Kim HM, Moon SM (2011) Numerical studies on the geometrical characterization of serpentine flow-field for efficient PEMFC. Int J Hydrogen Energy 36(2):1613–1627
42. Choi KS, Kim HM, Moon SM (2011) An experimental study on the enhancement of the water balance, electrochemical reaction and power density of the polymer electrolyte fuel cell by under-rib convection. Electrochem Commun 13(12):1387–1390
43. Costamagna P, Srinvasan S (2001) Quantum jumps in the PEMFC science and technology from the 1960s to the year 2000: Part I. Fundamental scientific aspects. J Power Sources 102(1):242–252
44. Costamagna P, Srinvasan S (2001) Quantum jumps in the PEMFC science and technology from the 1960s to the year 2000: Part II. Engineering, technology development and application aspects. J Power Sources 102(1):253–269
45. Gamburzev S, Appleby AJ (2002) Recent progress in performance improvement of the proton exchange membrane fuel cell (PEMFC). J Power Sources 107(1):5–12
46. Mehta V, Cooper JS (2003) Review and analysis of PEM fuel cell design and manufacturing. J Power Sources 114(1):32–53
47. Haraldsson K, Wipke K (2004) Evaluating PEM fuel cell system models. J Power Sources 126(1):88–97
48. Yao KZ, Karan K, McAuley KB, Oosthuizen P, Peppley B, Xie T (2004) Review of mathematical models for hydrogen and direct methanol polymer electrolyte membrane fuel cells. Fuel Cells 4(1):3–29
49. Brykoğlu A (2005) Review of proton exchange membrane fuel cell models. Int J Hydrogen Energy 30(11):1181–1212
50. Cheddie D, Munroe N (2005) Review and comparison of approaches to proton exchange membrane fuel cell modeling. J Power Sources 147(1):72–84
51. Faghri A, Guo Z (2005) Challenges and opportunities of thermal management issues related to fuel cell technology and modeling. Int J Heat Mass Transfer 48(19):3891–3920
52. Hermann A, Chaudhuri T, Spagnol P (2005) Bipolar plates for PEM fuel cells: a review. Int J Hydrogen Energy 30(12):1297–1302
53. Smitha B, Sridhar S, Khan AA (2005) Solid polymer electrolyte membranes for fuel cell applications: a review. J Membr Sci 259(1):10–26
54. Pettersson J, Ramsey B, Harrison D (2006) A review of the latest developments in electrodes for unitised regenerative polymer electrolyte fuel cells. J Power Sources 157(1):28–34
55. Tawfik H, Hung Y, Mahajan D (2007) Metal bipolar plates for PEM fuel cell: a review. J Power Sources 163(2):755–767
56. Neburchilov V, Martin J, Wang H, Zhang J (2007) A review of polymer electrolyte membranes for direct methanol fuel cells. J Power Sources 169(2):221–238
57. Djilali N (2007) Energy computational modelling of polymer electrolyte membrane (PEM) fuel cells: challenges and opportunities. Energy 32(4):269–280

58. De Bruijn FA, Dam VAT, Janssen GJM (2008) Review: durability and degradation issues of PEM fuel cell components. Fuel Cells 8(1):3–22
59. Siegel C (2008) Review of computational heat and mass transfer modeling in polymer-electrolyte-membrane (PEM) fuel cells. Energy 33(9):1331–1352
60. Peighambardoust SJ, Rowshanzamir S, Amjadi M (2010) Review of the proton exchange membranes for fuel cell applications. Int J Hydrogen Energy 35(17):9349–9384
61. Antunes RA, Oliveira MCL, Ett G, Ett V (2010) Corrosion of metal bipolar plates for PEM fuel cells: a review. Int J Hydrogen Energy 35(8):3632–3647
62. Krivobokov IM, Gribov EN, Qkunev AG (2011) Proton conducting hydrocarbon membranes: Performance evaluation for room temperature direct methanol fuel cells. Electrochim Acta 56(5):2420–2427
63. Tsushima S, Hirai S (2011) In situ diagnostics for water transport in proton exchange membrane fuel cells. Prog Energy Combust Sci 37(2):204–220
64. Sengupta R, Bhattacharya M, Bandyopadhyay S, Bhowmick A (2011) A review on the mechanical and electrical properties of graphite and modified graphite reinforced polymer composites. Prog Polym Sci 36(5):638–670
65. Chang JH, Park JH, Park GG, Kim CS, Park OO (2003) Proton-conducting composite membranes derived from sulfonated hydrocarbon and inorganic materials. J Power Sources 127(1):18–25
66. Wakizoe M, Velev OA, Srinivasan S (1995) Analysis of proton-exchange membrane fuel-cell performance with alternate membranes. Electrochim Acta 40(3):335–344
67. Liu W, Zuckerboard D (2005) In situ detection of hydrogen peroxide in PEM fuel cells. J Electrochem Soc 152(6):1165–1170
68. Scherer GG, Bunsen-Ges B (1990) Polymer membranes for fuel cells. Phys Chem 94(9): 1008–1014
69. Qiao J, Saito M, Hayamizu K, Tokadaz T (2006) Degradation of perfluorinated ionomer membranes for PEM fuel cells during processing with H_2O_2. J Electrochem Soc 153(6):967–974
70. Teranishi K, Kawata K, Tsushima S, Hirai S (2006) Degradation mechanism of PEMFC under open circuit operation. Electrochem Solid-State Lett 9(10):475–477
71. Trogadas P, Parrondo J, Ramani V (2008) Degradation mitigation in polymer electrolyte membranes using cerium oxide as a regenerative free-radical scavenger. Electrochem Solid-State Lett 11(7):113–116
72. Healy J, Hayden C et al (2005) Aspects of the chemical degradation of PFSA ionomers used in PEM fuel cells. Fuel Cells 5(2):302–308
73. Liu W, Ruth K, Rusch G (2001) Membrane durability in PEM fuel cells. J New Mat Electr sys 4(4):227–232
74. Kyu T, Hashiyama M, Eisenberg A (1983) Dynamic mechanical studies of partially ionized and neutralized Nafion polymers. Can J Chem 61(4):680–687
75. Mukerjee S, Srinivasan S (1993) Enhanced electrocatalysis of oxygen reduction on platinum alloys in proton exchange membrane fuel cells. J Electroanal Chem 357(1):201–224
76. Swette LL, LaConti AB, McCatty SA (1994) Proton-exchange membrane regenerative fuel cells. J Power Sources 47(3):343–351
77. Borup R, Meyers J et al (2007) Scientific aspects of polymer electrolyte fuel cell durability and degradation. Chem Rev 107(10):3904–3951
78. Stevens DA, Dahn JR (2005) Thermal degradation of the support in carbon-supported platinum electrocatalysts for PEM fuel cells. Carbon 43(1):179–188
79. Cai M, Ruthkosky MS et al (2006) Investigation of thermal and electrochemical degradation of fuel cell catalysts. J Power Sources 160(2):977–986
80. Schulze M, Wagner N, Kaz T, Friedrich KA (2007) Combined electrochemical and surface analysis investigation of degradation processes in polymer electrolyte membrane fuel cells. Electrochim Acta 52(6):2328–2336
81. St-Pierre J, Wilkins DP, Knights S, Bos ML (2000) Relationships between water management, contamination and lifetime degradation in PEFC. J New Mat Electr Syst 3:99–106

82. Jordan LR, Shukla AK, Behrsing T, Avery NR, Muddle BC, Forsyth M (2000) Diffusion layer parameters influencing optimal fuel cell performance. J Power Sources 86(1):250–254
83. Pasaogullari U, Wang CY (2004) Two-phase transport and the role of micro-porous layer in polymer electrolyte fuel cells. Electrochim Acta 49(25):4359–4369
84. Pasaogullari U, Wang CY, Chen KS (2005) Two-phase transport in polymer electrolyte fuel cells with bilayer cathode gas diffusion media. J Electrochem Soc 152(8):1574–1582
85. Weber AZ, Darling RM, Newman J (2004) Modeling two-phase behavior in PEFCs. J Electrochem Soc 151(10):1715–1727
86. Weber AZ, Newman J (2005) Effects of microporous layers in polymer electrolyte fuel cells. J Electrochem Soc 152(4):677–688
87. Lee C, Mérida W (2007) Gas diffusion layer durability under steady-state and freezing conditions. J Power Sources 164(1):141–153
88. Raistrick ID (1990) Impedance studies of porous electrodes. Electrochim Acta 35(10): 1579–1586
89. Dhar HP (1997) Study of combined electroreflectance and double layer effects at lead electrodes. Surf Sci 66(2):449–462
90. Gottesfeld S, Zawodzinski TA (1997) In polymer electrolyte fuel cells. Wiley, Weinheim
91. Tsuchiya H, Kobayashi O (2004) Mass production cost of PEM fuel cell by learning curve. Int J Hydrogen Energy 29(10):985–990
92. Cooper JS (2004) Design analysis of PEMFC bipolar plates considering stack manufacturing and environment impact. J Power Sources 129(2):152–169
93. Müller A, Kauranen P, Von Ganski A, Hell B (2006) Injection moulding of graphite composite bipolar plates. J Power Sources 154(2):467–471
94. Maheshwari PH, Mathur RB, Dhami TL (2007) Fabrication of high strength and a low weight composite bipolar plate for fuel cell applications. J Power Sources 173(1):394–403
95. Pozio A, Silva RF, Fancesco DM, Giorgi L (2003) Nafion degradation in PEFCs from end plate iron contamination. Electrochim Acta 48(11):1543–1549
96. Wind J, Späh R, Kaiser W, Böhm G (2002) Metallic bipolar plates for PEM fuel cells. J Power Sources 105(2):256–260
97. Shimpalee S, Ohashi M et al (2009) Experimental and numerical studies of portable PEMFC stack. Electrochim Acta 54(10):2899–2911
98. Malkow T, Thalau O et al (2010) PEFC power stack performance testing procedure, test module PEFC ST 5-6. Publications Office of the European Union, Luxembourg
99. Tripathi BP, Shahi VK (2011) Organic–inorganic nanocomposite polymer electrolyte membranes for fuel cell applications. Prog Polym Sci 36(7):945–979
100. Rhee CH, Kim HK, Chang H, Lee JS (2005) Nafion/sulfonated montmorillonite composite: a new concept electrolyte membrane for direct methanol fuel cells. Chem Mater 17(7): 1691–1697
101. Watanabe M, Satoh Y, Shimura C (1993) Management of the water vontent in polymer electrolyte membranes with porous fiber wicks. J Electrochem Soc 140(11):3190–3193
102. Miachon S, Aldebert P (1995) Internal hydration H2/O2 100 cm^2 polymer electrolyte membrane fuel cell. J Power Sources 56(1):31–36
103. Choi KS, Kim HM, Yoon YC, Forrest ME, Erickson PA (2008) Effects of ambient temperature and relative humidity on the performance of Nexa fuel cell. Energ Convers Manage 49(12):3505–3511
104. Kandlikar SG, Lu Z (2009) Thermal management issues in a PEMFC stack—a brief review of current status. Appl Therm Eng 29(7):1276–1280
105. Heinzel A, Barragán VM (1999) A review of the state-of-the-art of the methanol crossover in direct methanol fuel cells. J Power Sources 84(1):70–74
106. Wasmus S, Küver A (1999) Methanol oxidation and direct methanol fuel cells: a selective review. J Electroanal Chem 461(1):14–31
107. Liu L, Pu C, Viswanathan R, Fan Q, Liu L, Smotkin ES (1998) Carbon supported and unsupported Pt–Ru anodes for liquid feed direct methanol fuel cells. Electrochim Acta 43(24):3657–3663

108. Page T, Johnson R, Hormes J, Nording S, Rambabu B (2000) A study of methanol electro-oxidation reactions in carbon membrane electrodes and structural properties of Pt alloy electro-catalysts by EXAFS. J Electroanal Chem 485(1):34–41
109. Ren X, Springer TE, Gottesfeld S (2000) Water and methanol uptakes in Nafion membranes and membrane effects on direct methanol cell performance. J Electrochem Soc 147(1):92–98
110. Baschuk JJ, Li X (2001) Carbon monoxide poisoning of proton exchange membrane fuel cells. Int J Energy Res 25(8):695–713
111. Gang X, Qingfeng L, Hjuler HA, Bjerrum NJ (1995) Hydrogen oxidation on gas diffusion electrodes for phosphoric acid fuel cells in the presence of carbon monoxide and oxygen. J Electrochem Soc 142(9):2890–2893
112. Doyle M, Rajendran G (2003) Handbook of fuel cells. Wiley, England
113. Wilkinson DP, Voss HH, Prater K (1994) Water management and stack design for solid polymer fuel cells. J Power Sources 49(1):117–127
114. Mosdale R, Srinivasan S (1995) Analysis of performance and of water and thermal management in proton exchange membrane fuel cells. Electrochim Acta 40(4):413–421
115. Squadrito G, Barbera O, Giacoppo G, Urbani F, Passalacqua E (2008) Polymer electrolyte fuel cell stack research and development. Int J Hydrogen Energy 33(7):1941–1946
116. Zhai Y, Zhang H, Zhang Y, Xing D (2007) A novel H_3PO_4/Nafion–PBI composite membrane for enhanced durability of high temperature PEM fuel cells. J Power Sources 169(2):259–264
117. Zhai Y, Zhang H, Liu G, Hu J, Yi B (2007) Degradation study on MEA in H_3PO_4/PBI High-temperature PEMFC life test. J Electrochem Soc 154(1):72–76
118. Cleghorn SJC, Mayfield DK et al (2006) A polymer electrolyte fuel cell life test: 3 years of continuous operation. J Power Sources 158(1):446–454
119. Choi KS, Jang SH, Shin GS, Kim HM, Yoon HC, Forrest ME, Erickson PA (2010) Effects of stack array orientation on fuel cell efficiency for auxiliary power unit applications. Int J Automot Techn 11(3):429–434
120. Ghosh PC, Wuster T, Dohle H, Kimiaie N, Mergel J, Stolten D (2006) In situ approach for current distribution measurement in fuel cells. J Power Sources 154(1):184–191
121. Stumper J, Campbell SA, Wilkinson DP, Johnson MC, Davis M (1998) In-situ methods for the determination of current distributions in PEM fuel cells. Electrochim Acta 43(24):3773–3783
122. Hakenjos A, Muenter H, Wittstadt U, Hebling C (2004) A PEM fuel cell for combined measurement of current and temperature distribution, and flow field flooding. J Power Sources 131(1):213–216
123. Park K, Kim HK, Choi KS (2013) Numerical and experimental verification of the polymer electrolyte fuel cell performances enhanced by under-rib convection. Fuel Cells 13(5):927–934
124. Hogarth WHJ, Steiner J, Benziger JB, Hakenjos A (2007) Spatially-resolved current and impedance analysis of a stirred tank reactor and serpentine fuel cell flow-field at low relative humidity. J Power Sources 164(2):464–471
125. Freunberger SA, Reum M, Wokaun A, Büchi FN (2006) Expanding current distribution measurement in PEFCs to sub-millimeter resolution. Electrochem Commun 8(9):1435–1438
126. Higier A, Liu H (2010) Optimization of PEM fuel cell flow field via local current density measurement. Int J Hydrogen Energy 35(5):2144–2150
127. Alaefour I, Karimi G, Jiao K, Li X (2011) Measurement of current distribution in a proton exchange membrane fuel cell with various flow arrangements—a parametric study. Appl Energy 93:80–89
128. Peng L, Mai J, Hu P, Lai X, Lin Z (2011) Optimum design of the slotted-interdigitated channels flow field for proton exchange membrane fuel cells with consideration of the gas diffusion layer intrusion. Renew Energ 36(5):1413–1420
129. Takada K, Ishigami Y et al (2011) Simultaneous visualization of oxygen distribution and water blockages in an operating triple-serpentine polymer electrolyte fuel cell. J Power Sources 196(5):2635–2639

130. Ishigami Y, Takada K et al (2011) Corrosion of carbon supports at cathode during hydrogen/air replacement at anode studied by visualization of oxygen partial pressures in a PEFC-start-up/shut-down simulation. J Power Sources 196(6):3003–3008
131. Su Weng FB, Hsu CY, Chen YM (2006) Studies on flooding in PEM fuel cell cathode channels. Int J Hydrogen Energy 31(8):1031–1039
132. Ous T, Arcoumanis C (2009) Visualisation of water accumulation in the flow channels of PEMFC under various operating conditions. J Power Sources 187(1):182–189
133. Weinmueller C, Tautschnig G, Hotz N, Poulikakos D (2010) A flexible direct methanol micro-fuel cell based on a metalized, photosensitive polymer film. J Power Sources 195 (12):3849–3857
134. Jiao K, Park JW, Li X (2010) Experimental investigations on liquid water removal from the gas diffusion layer by reactant flow in a PEM fuel cell. Appl Energy 87(9):2770–2777
135. Zhan Z, Wang C, Fu W, Pan M (2012) Visualization of water transport in a transparent PEMFC. Int J Hydrogen Energy 37(1):1094–1105
136. Oliveira VB, Rangel CM, Pinto AMFR (2010) Effect of anode and cathode flow field design on the performance of a direct methanol fuel cell. Chem Eng J 157(1):174–180
137. Iranzo A, Muñoz M, Lòpez E, Pino J, Rosa F (2010) Experimental fuel cell performance analysis under different operating conditions and bipolar plate designs. Int J Hydrogen Energy 35(20):11437–11447
138. Ni HJ, Zhang CJ, Wang XX, Ma SY, Liao P (2010) Performance of special-shaped direct methanol fuel cell with sol–gel flux phase. J Fuel Chem Techno 38(5):604–609
139. Kumar A, Reddy RG (2006) Effect of gas flow-field design in the bipolar/end plates on the steady and transient state performance of polymer electrolyte membrane fuel cells. J Power Sources 155(2):264–271
140. Chen YS, Peng H (2011) Predicting current density distribution of proton exchange membrane fuel cells with different flow field designs. J Power Sources 196(4):1992–2004
141. Yang H, Zhao TS (2005) Effect of anode flow field design on the performance of liquid feed direct methanol fuel cells. Electrochim Acta 50(16):3243–3252
142. Hsieh SS, Her BS (2007) Heat transfer and pressure drop in serpentine μDMFC flow channels. Int J Heat Mass Transfer 50(25):5323–5327
143. Lu Y, Reddy RG (2011) Effect of flow fields on the performance of micro-direct methanol fuel cells. Int J Hydrogen Energy 36(1):822–829
144. Dokkar B, Settou NE, Imine O, Saifi N, Negrou B, Nemouchi Z (2011) Simulation of species transport and water management in PEM fuel cells. Int J Hydrogen Energy 36(6): 4220–4227
145. Hsuen HK, Yin KM (2011) A pseudo-phase-equilibrium approach for the calculation of liquid water saturation in the cathode gas diffuser of proton-exchange-membrane fuel cells. Int J Hydrogen Energy 36(9):5487–5499
146. Berning T, Odgaard M, Kær SK (2010) A study of multi-phase flow through the cathode side of an interdigitated flow field using a multi-fluid model. J Power Sources 195(15):4842–4852
147. Le D, Zhou B (2010) A numerical investigation on multi-phase transport phenomena in a proton exchange membrane fuel cell stack. J Power Sources 195(1):5278–5291
148. Hao L, Cheng P (2009) Lattice Boltzmann simulations of liquid droplet dynamic behavior on a hydrophobic surface of a gas flow channel. J Power Sources 190(2):435–446
149. Basu S (2011) A generalized multiphase mixture (M2) model for PEFC flow field design. Int J Hydrogen Energy 36(16):9855–9863
150. Anderson R, Wilkinson DP, Bi X, Zhang L (2010) Two-phase flow pressure drop hysteresis in parallel channels of a proton exchange membrane fuel cell. J Power Sources 195(13): 4168–4176
151. Anderson R, Wilkinson DP, Bi X, Zhang L (2011) Two-phase flow pressure drop hysteresis in an operating proton exchange membrane fuel cell. J Power Sources 196(19):8031–8040
152. Xu C, Faghri A (2010) Water transport characteristics in a passive liquid-feed DMFC. Int J Heat Mass Transfer 53(9):1951–1966

153. Ko J, Chippar P, Ju H (2010) A one-dimensional, two-phase model for direct methanol fuel cells—Part I: model development and parametric study. Energy 35(5):2149–2159
154. Zhou Y, Lin G, Shin AJ, Hu SJ (2009) Multiphysics modeling of assembly pressure effects on proton exchange membrane fuel cell performance. J Fuel Cell Sci Tech 6(1–7)
155. Zhou P, Wu CW, Ma GJ (2007) Influence of clamping force on the performance of PEMFCs. J Power Sources 163(2):874–881
156. Akiki T, Charon W, Iltchev MC, Accary G, Kouta R (2010) Influence of local porosity and local permeability on the performances of a polymer electrolyte membrane fuel cell. J Power Sources 195(16):5258–5268
157. Um S, Wang CY (2004) Three-dimensional analysis of transport and electrochemical reactions in polymer electrolyte fuel cells. J Power Sources 125(1):40–51
158. Chippar P, Oh K, Kim WG, Ju HC (2014) Numerical analysis of effects of gas crossover through membrane pinholes in high-temperature proton exchange membrane fuel cells. Int J Hydrogen Energy 39(6):2863–2871
159. Kwon J, Kang MS et al (2011) Development of flow field design of polymer electrolyte membrane fuel cell using in-situ impedance spectroscopy. Int J Hydrogen Energy 36(16): 9799–9804
160. Park YC, Peck DH et al (2011) Operating characteristics and performance stability of 5 W class direct methanol fuel cell stacks with different cathode flow patterns. Int J Hydrogen Energy 36(2):1853–1861
161. Wang SJ, Huo WW, Zou ZQ, Qia YJ, Yang H (2011) Computational simulation and experimental evaluation on anodic flow field structures of micro direct methanol fuel cells. Appl Therm Eng 31(14):2877–2884
162. Iranzo A, Muñoz M, Pino J, Rosa F (2011) Update on numerical model for the performance prediction of a PEM Fuel Cell. Int J Hydrogen Energy 36(15):9123–9127
163. Xu Z, Qi Z, He C, Kaufman A (2006) Combined activation methods for proton-exchange membrane fuel cells. J Power Sources 156(2):315–320
164. Choi KS, Kim BG, Park K, Kim HM (2012) Flow control of under-rib convection enhancing the performance of proton exchange membrane fuel cell. Comput Fluids 69:81–92
165. Vinh ND, Kim HM et al (2015) Dynamic simulations of under-rib convection-driven flow-field configurations and comparison with experiment in polymer electrolyte membrane fuel cells. J Power Sources 293:447–457
166. Radhakrishnan V, Haridoss P (2011) Effect of GDL compression on pressure drop and pressure distribution in PEMFC flow field. Int J Hydrogen Energy 36(12):14823–14828
167. Hutzenlaub T, Paust N, Zengerle R, Ziegler C (2011) The effect of wetting properties on bubble dynamics and fuel distribution in the flow field of direct methanol fuel cells. J Power Sources 196(19):8048–8056

Chapter 11
Methods for the Preparation of Organic–Inorganic Nanocomposite Polymer Electrolyte Membranes for Fuel Cells

Mashallah Rezakazemi, Mohtada Sadrzadeh, Toraj Mohammadi and Takeshi Matsuura

Abstract In the last decade, the organic–inorganic nanocomposite polymer electrolyte membranes (PEM) have gained high technical relevance in a wide range of fuel cells applications. The significance of nanocomposite membranes fabrication is particularly highlighted by the fact that one of the major challenges of this century is to provide well-performing and cost-effective membrane materials for fuel cells applications. Many efforts have been made in the development of advanced membranes with the aim to outperform the most commonly used polymer membranes. With the advances in nanomaterials and polymer chemistry, the innovative nanocomposite membranes with superior properties can be designed by various techniques including blending of nanoparticles in a polymer matrix, doping, or infiltration and precipitation of nanoparticles and precursors, self-assembly of nanoparticles, layer-by-layer fabrication method, and nonequilibrium impregnation

M. Rezakazemi (✉)
Department of Chemical Engineering, Shahrood University of Technology, Shahrood, Iran
e-mail: mashalah.rezakazemi@gmail.com

M. Sadrzadeh
Department of Mechanical Engineering, University of Alberta, Edmonton, AB T6G 2G8, Canada
e-mail: sadrzade@ualberta.ca

T. Mohammadi
Faculty of Chemical Engineering, Research and Technology Centre for Membrane Separation Processes, Iran University of Science and Technology (IUST), Narmak, Tehran, Iran
e-mail: torajmohammadi@iust.ac.ir

T. Matsuura
Advanced Membrane Technology Research Centre (AMTEC), Universiti Teknologi Malaysia, 81310 Skudai, Johor, Malaysia
e-mail: matsuura@uottawa.ca

Inamuddin et al. (eds.), *Organic–Inorganic Composite Polymer Electrolyte Membranes*, DOI 10.1007/978-3-319-52739-0_11

reduction. This study presents a brief overview of these techniques and discusses the encountered challenges, the problems to be overcome, the major findings and guidance for future developments.

Keywords Nanocomposite · Synthesis · Membrane · Preparation · Fuel cell

1 Introduction

A significant amount of research and development is currently underway throughout the world to protect environment via development of green power technology such as fuel cells (FCs). A fuel cell is an electrochemical device that converts chemical energy into an electrical energy by using various types of fuels such as hydrogen, methanol, ethanol, methylene blue, glucose, natural gas, etc., through a reaction with an oxidant (mainly oxygen). There are five different types of FCs that are primarily classified based on the nature of the polymer electrolyte in the FC: (1) alkaline fuel cells (AFCs) (alkaline solution), (2) phosphoric acid fuel cells (PAFCs) (acidic solution), (3) solid proton exchange membrane fuel cells (PEMFCs) (polymer electrolyte), (4) molten carbonate fuel cells (MCFCs) (molten carbonate salt electrolyte), and (5) solid oxide fuel cells (SOFCs) (ceramic ion conducting electrolyte) [1]. Among them, PEMFC is the most promising one since it can be applied to small-scale products such as automotive, portable power generation, and stationary equipment [1, 2].

The key component of the PEMFCs is the polymer electrolyte membrane (PEM), which facilitates transport of protons from the anode to the cathode and prevents the passage of the electrons and fuel. Ideal membranes for the PEMFCs should have high proton conductivity and oxidative stability, low fuel permeability, reasonable mechanical stability, adequate water transport (diffusion and electro-osmotic) properties, and low cost of fabrication [1, 3]. The widely applied Nafion membranes possess high thermal and chemical stability as well as high ionic conductivity. However, a decrease in ionic conductivity and relative humidity by increasing temperature limits the commercialization of FCs [1]. Hence, numerous research works have been conducted to synthesize high-performance PEMs.

Membrane for FC applications can be prepared from either neat polymer membranes or from nanocomposite membranes where other materials such as inorganic nanoparticles including silica, titania, tin phosphate, cerium phosphate, titanium phosphate, zirconia, sulfated zirconia, palladium, montmorillonite, zeolite, etc., are incorporated in a polymer matrix. Application of high-performance polymers, other than poly (perfluorosulfonic acid) (Nafion), for the synthesis of PEMs, have been widely investigated [1]. These polymers including polystyrenes, polyimides, polyphosphazenes, polybenzimidazoles, poly(arylene ether), polysulfones, and their derivatives are typically sulfonated to achieve conductivities higher than Nafion. High degree of sulfonation causes excessive membrane swelling and subsequently decreases mechanical stability and durability during operation.

In order to overcome this drawback, nanocomposite membranes are developed to provide a unique combination of polymers (high processability and electrical conductivity) and inorganic materials (high thermal and chemical stability, high mechanical strength, and low fuel permeability through the membrane) [1].

During the past few years, several advances on the fabrication of nanocomposite membranes have been made [4]. The general idea is to induce the thermal, electrical, mechanical properties of these nanomaterials to the base membrane. However, the incorporation of nanomaterials into a polymer, with the aim of fabricating a robust hybrid membrane, may not induce the desired functionality to the polymer and even deteriorate its properties. The major challenges to this are the severe aggregation of the nanomaterials in monomer containing solutions and the weak compatibility of encapsulated nanomaterials with polymers. The nonuniform dispersion of nanomaterials within the host polymer forms nonselective voids at the interface of the polymer and nanomaterials, reducing the permeation properties significantly. Furthermore, the desired thermal stability is not achieved, since the predicted reduction in chain mobility by the addition of nanomaterials is mitigated by the formation of these voids. This article aims to (i) present various techniques that have been used for fabrication of nanocomposite membranes for FC application, (ii) discuss the challenges encountered for each technique, (iii) provide guidelines for overcoming the challenges, and (iv) assess the future trend in the fabrication of high-performance nanocomposite membranes.

2 Methods for Preparation of Nanocomposite Polymer Electrolyte Membranes

Remarkable advancements have been recently performed in the preparation of organic–inorganic nanocomposite PEM materials, development of basic principles for the synthesis processes, and application of PEMs in various industries. The nanocomposite membranes can be synthesized by various techniques through physical blending (mixing) of inorganic nanoparticle or precursor (oligomer and monomer) in the three-dimensional (3D) network or linear polymer structure (Fig. 1) [5]. In the case of incorporating pre-synthesized nanoparticles into the polymer matrices, the main challenge is proper dispersion and distribution of nanoparticles in the membrane. Hence, strategies must be developed for surface functionalization of nanoparticles to improve their dispersion in both aqueous and organic solvents. In the case of using inorganic precursors and monomers, crosslinking and polymerization of inorganic or organic monomers is needed. Inspired by the mechanism of mineralization, NPs can be self-assembled via a hydrolysis and condensation reaction of inorganic precursors within the polymer matrix during membrane formation [6]. In this technique, the formation of membrane and nanoparticles could occur at the same time when the membrane is synthesized by the in situ polymerization reaction of two reactive monomers. Upon

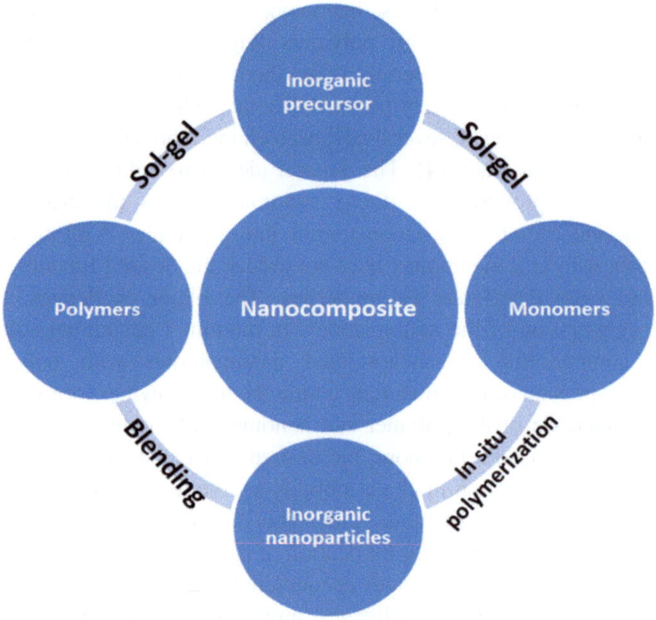

Fig. 1 Representative of various synthesis routes of nanocomposite membranes

contact of solutions containing inorganic precursors and monomers, the monomers will react to form the polymer, and simultaneously, the nanoparticles will be synthesized and entrapped in the polymer matrix. The main challenge is the timing of the polymerization reaction and hydrolysis/condensation reaction to ensure the incorporation of nanoparticles inside the polymer matrix. The techniques used to prepare organic–inorganic membranes are explained in the following.

2.1 Blending of Nanoparticles in Polymer Matrix

Physical blending of nanoparticles into the polymer structure is the simplest way to synthesize nanocomposite PEMs. Physical blending can be accomplished by either melt mixing or dissolution in a solvent, where inorganic nanoparticles are uniformly dispersed in the molten or dissolved polymer. The melt mixing is more favorable owing its environmentally friendly nature and efficiency. One major challenge is the uniform dispersion of nanoparticles in the polymer since the nanoparticles tend to minimize their surface energy by aggregation. Tuning the surface chemistry of nanoparticles by modifying their surface with proper functional groups that have a better interaction with the polymer yields defect-free nanocomposite membranes. Surface treatment such as "*Whisker*" surface modification, irradiation, Grignard reagents treatment, grafting, ion-exchange treatment, etc., have been widely used to change the surface chemistry of nanoparticles and break down its agglomerates in preparing hybrid inorganic–organic nanocomposite electrolyte membrane [7].

Physical blending at liquid state is extensively used to synthesize nanocomposite membranes due to its relatively simple synthesis route accomplished mainly through well-known phase inversion and solution casting methods [8, 9]. Such nanocomposite films have been widely applied for ultrafiltration [2, 3], pervaporation [10–12], gas separation [13, 14], and PEMs for fuel cell applications [6–9].

The blending of nanoparticles into the polymer matrix technique is typically conducted by phase inversion and hot press techniques, which are discussed in the following sections [15–17].

2.1.1 Phase Inversion Method for Preparation of PEMs

During the phase inversion process, a thermodynamically stable polymer solution is subjected to a liquid–liquid demixing during which the cast polymer film separates into a polymer-rich (membrane matrix) and a polymer lean phase (membrane pores). Polymeric membranes can be prepared by means of phase inversion from virtually all polymers that are soluble at a certain temperature in an appropriate solvent. Based on the way the polymer solution is solidified, phase separation technique can be distinguished into four techniques [18, 19]:

1. **Thermally induced phase separation (TIPS)**. In this method, a low molecular weight liquid that acts as solvent at high temperature and as nonsolvent at low temperature is required. A system of polymer and solvent is used which has a high critical solution temperature. The solution is cast at high temperature and cooling leads to demixing/precipitation [18]. Evaporation of the solvent often leads to the formation a skinned microfiltration membrane [20].
2. **Nonsolvent-induced phase separation (NIPS)**. This method is also called immersion precipitation or wet method. Most commercially available membranes are prepared by this technique. A polymer solution is cast on a suitable support and is subsequently immersed in a coagulation bath containing a non-solvent (typically water). Demixing and precipitation occur due to the exchange of solvent and non-solvent, that is, the solvent and non-solvent must be miscible [18, 20, 21].
3. **Vapor-induced phase separation (VIPS)**. This technique is based on the fact that absorption of nonsolvent from vapor can also cause phase separation of a polymer solution. The polymer solution is exposed to a vapor atmosphere containing a nonsolvent (typically water), saturated with the same solvent. The high solvent concentration prevents the evaporation of solvent from the cast film. Diffusion of nonsolvent into the cast film causes demixing/precipitation [18, 20].

4. **Evaporation-induced phase separation (EIPS)**. The polymer in this method is dissolved in a volatile solvent and nonsolvent mixture. The evaporation of the solvent causes a higher concentration of nonsolvent in the casting solution that eventually results in precipitation of polymer and formation a skinned membrane [18, 20].

The TIPS and the NIPS process are most widely used to prepare all membrane geometries because of the compatibility with very short residence times between formation of the liquid polymer films and the onset of phase separation. The slower VIPS and EIPS processes are much less frequently used, but they can also be used in combination with the NIPS process [18].

A few studies have been conducted on the synthesis of PEMFC using phase inversion technique [10, 11]. The main challenge that limits the application of phase inversion membranes for proton exchanging in FCs is the high porosity of the synthesized membranes. Phase inversion process typically results in porous membranes that possess poor separation properties. However, dense structures can also be obtained by smart control of the kinetics of the solvent–nonsolvent demixing process [12]. Hence, in order to yield suitable phase inversion membranes for FC application, synthesis parameters must be adjusted to levels that slow down the precipitation of polymer. In addition, selection of slower phase inversion techniques (e.g., EIPS) is critical to obtain dense membranes.

Synthesis of mixed matrix phase inversion membranes is also more complicated than it sounds. Due to the good interaction of most nanoparticles with the nonsolvent, a considerable amount of them are leached out of the polymer matrix to the nonsolvent during phase separation process, thereby reducing the real loading of nanoparticles in the membrane. Hence, the surface of nanoparticles must be modified with proper functional groups that have good interaction with the polymer to reduce their tendency toward nonsolvent.

2.1.2 Solution Casting Method

The solution casting method is the simplest technique of synthesizing inorganic–organic hybrid PEMs. In this method, the following general stages are involved to fabricate reproducible and homogenous cast PEMs:

(a) Homogeneous dispersion of nanofillers in the solvent.
(b) Incorporation of a certain amount of polymer to the mixtures.
(c) Mixing the mixtures by ultrasonic equipment or stirrer.
(d) Casting the mixture uniformly on an appropriate substrate.
(e) Drying in an inert or in a vacuum environment.

Both phase separation and solution casting methods require a coating of a polymer film on a suitable substrate. The coating is conducted by film applicators, roll-coaters and spin-coaters. Applying manual film applicators, the homogenous polymer solutions are cast by adjusting the clearance gap at a specific thickness.

Casting speed can be adjusted by a motorized film applicator [12]. In roll-coating polymer solution is poured into a feed tank, which is placed under a roller, and then roll-coated on the surface of a suitable support (e.g., polyester fabric) [13]. In spin-coating, the polymer solution is poured drop by drop on an appropriate surface and housed in a spin coater, which is then rotated at a specified speed. The speed of the rotation and the concentration of the solution affect the membrane thickness. This technique is, however, only applicable for the low-to-medium viscosity of the polymer solution and would not be useful for a sol–gel like a mixture since the rotation of spin coater is not sufficient to distribute the mixture droplet radially on the substrate surface to make a thin membrane.

2.1.3 Hot Press

The hot press device for fabrication of PEMFC is shown in Fig. 2. A certain amount of polymer, nanofillers, and salt are poured into the mortar and blended for an hour. The sample is then pressed between two Mylar sheets and is placed into a

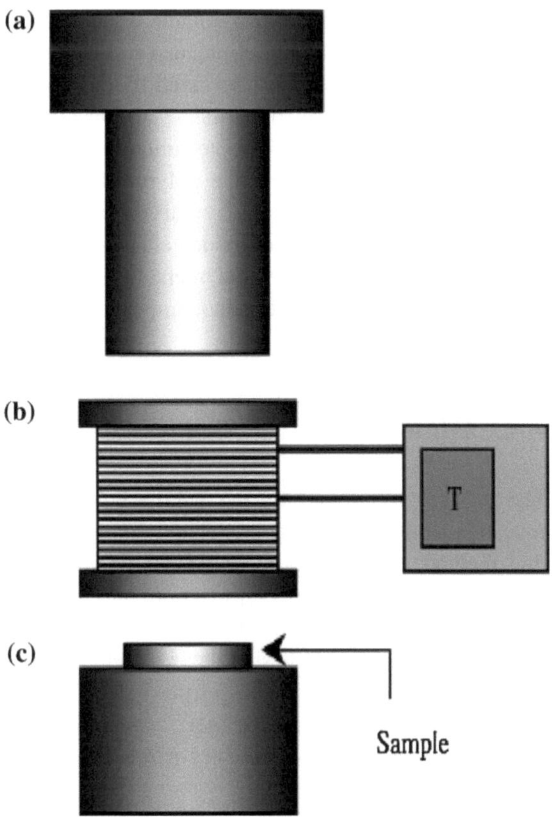

Fig. 2 Representation of hot press device: **a** weighing cylinder, **b** heater, **c** base, and *T* temperature controller [22]

box. Afterward, the box is heated at temperatures higher than the melting point of the polymer. Subsequently, the mixture is pressed for a certain time at an adjustable pressure, which is set by weighing cylinder. After heating and pressing processes, the film is naturally cooled down to ambient temperature. Finally, the film is detached from the Mylar sheets and kept in a glove box. This method is promising since it can provide robust membranes with the repeatable fabrication process. However, there are a few records in literature for the synthesis of nanocomposite membranes by hot press method. The main challenges are the preparation of monodispersed mixtures of polymer and nanoparticles at such a high viscosity and unknown/noncontrollable behavior of the nanoparticles at high temperatures [22].

2.2 Doping or Infiltration and Precipitation of Nanoparticles and Precursors

Infiltration or doping method and precipitation of nanoparticles and precursors have generally been used to prepare polymer electrolyte membranes to change the separation performance by incorporating precursors or nanofillers. In this method, the preformed film is swelled in a suitable organic solvent to enhance the free volume before infiltration or doping of precursor or nanofiller. After infiltration process, the hybrid nanocomposites must be cured by radiation, thermal annealing or chemical grafting of precursor/nanoparticles, to achieve covalent bonding within the nanocomposite networks. This technique suffers extremely from the leaching of precursor/nanoparticles from the polymer network. However, the hybrid polymer membranes can be achieved by mixing single or mixed metal alkoxides with polymers or by infiltration of oxide gels with polymerized monomers. Impregnation of porous oxide gels with polymers through the in situ irradiation or thermal polymerization forms stable nanocomposite membranes [5, 23].

2.3 Self-assembly of Nanoparticles

The sol–gel technique for the synthesis of hybrid materials has been widely investigated due to its environmental containment and low-temperature process. In this method, the functional groups of polymer matrix enhance bonding of two phases, followed by a sol–gel reaction, and form a ceramic like phase.

It is vital to gain fundamental sciences of the physical and chemical properties of the nanocomposites synthesized by sol–gel technique. Numerous efforts have been recently accomplished on sol–gel process to understand the chemistry and mechanism of sol–gel reaction. The "sol" in general is a dispersion of colloidal particles in a liquid. The "gel" can be categorized into four groups:

(i) The polymer matrix shaped by physical aggregation;
(ii) The covalent disordered polymer matrix;
(iii) The disordered buildings of organic and inorganic materials; and
(iv) The well-ordered lamellar structures of nanoparticles/precursor with polymers, e.g., organopolysilsesquioxanes.

The well-known chemistry of the sol–gel reactions contain two serial phases: (I) the hydrolysis of single or mixed metal alkoxides to form hydroxyl functional groups, (II) the polycondensation of the hydroxyl functional groups to produce a 3D structure (Fig. 3).

A typical sol–gel reactions commonly initiates in an organic solvent with low molecular weight in the presence of metal alkoxide precursors $M(OR)_n$, where M is the element in the structure which can be Zr, Al, Ti, B, Si, etc., and R can be water or an alkyl group. When the hydrolysis and polycondensation reactions are accomplished, some undesirable byproducts such as water or alcohol are produced that must be removed, otherwise cause shrinkage during the formation of the nanocomposites [5].

The hydrolysis and polycondensation reactions are classified as "nucleophilic substitution" methods having three stages: (1) nucleophilic incorporation, (2) transferring proton in the transition modes, and (3) exclusion of the protonated

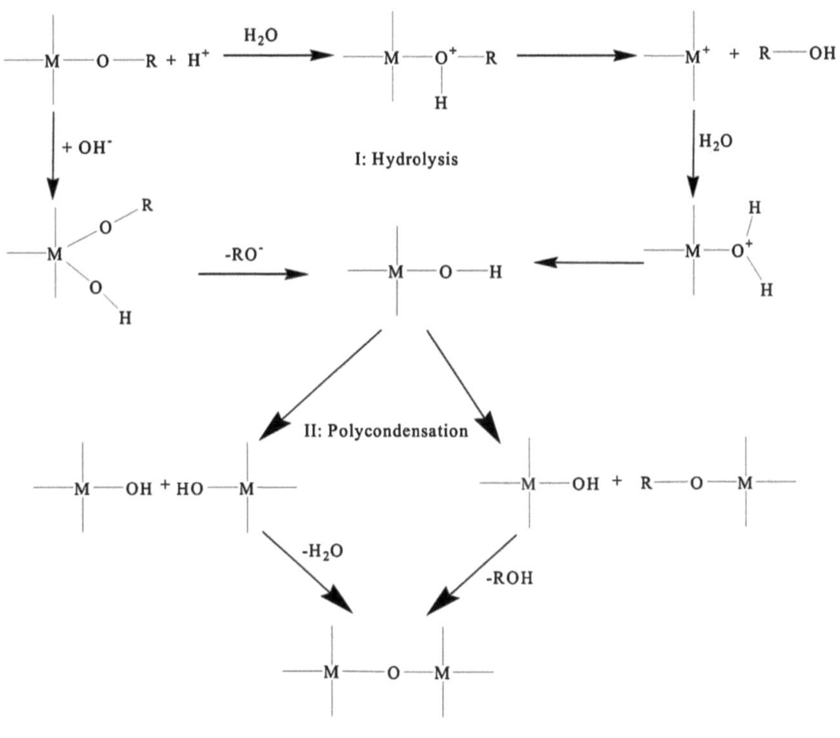

Fig. 3 Schematic representation of the sol–gel mechanism for metal alkoxide [24]

components. The hydrolysis and polycondensation reactions can be accomplished without the addition of a catalyst particularly for non-silicate-based metal alkoxides, whereas for silicon metal alkoxides base or acidic catalysts are needed. When a catalyst is used, the morphology and structure changes severely which is related to the nature of the catalyst [5].

Several parameters affect the kinetics and mechanism of the hydrolysis and the polycondensation reactions such as the solvent, ratio of silane/water, temperature, and the catalyst nature. The preparation approaches used in the sol–gel method to synthesize nanosilica based nanocomposite membranes take account of the in situ establishment of a nanoparticle/precursor in a polymer, and the simultaneous creation of MO_2 nanoparticles and polymer, results in the new generation of membranes with interpenetrating networks (IPNs). These nanocomposite membranes have attracted great attentions in different applications that are not well served by conventional membranes. So far, the commercial use of sol–gel based nanocomposite membranes is not practiced yet, but there is a high potential for novel scientific advancement.

2.4 Non-hydrolytic Sol–Gel (NHSG) Method

The non-hydrolytic sol–gel (NHSG) method is primarily used to fabricate the organic–inorganic nanocomposite PEMFC [25]. This method provides a water-free way to synthesize nanocomposite membranes by the interaction of an oxygen donor (e.g., ether, alkoxide, alcohol, etc.) with an appropriate metalloid or metal precursor (e.g., metal halide). Using a metal halide as a precursor, an alkyl halide, and inorganic oxide by-products are generated from this nonaqueous process. The electron transfer mechanism which controls the NHSG reactions is different from the hydrolytic sol–gel reactions. In the NHSG method, the reaction progresses through the coordination between the metal halide and the oxygen donor, and consequently, splitting the C–O bond, instead of metal–O bond splitting that takes place through the NHSG reactions. The key factor in the NHSG method is the change in the reactivity of metals, like transition metals and silicon.

2.5 Layer-by-Layer Fabrication Method

The layer-by-layer self-assembly of nanosized fillers with polymer forms one type of organic–inorganic nanocomposite membranes which are structured spontaneously with non-covalent interactions. These hybrid membranes show superior properties due to the high interfacial area, nanosized constituent phases, and synergetic interactions [5]. This technique is established according to the electrostatic association between deposited, oppositely charged species.

2.6 Nonequilibrium Impregnation Reduction

The nonequilibrium impregnation-reduction (NEIR) method is another encouraging technique to prepare multifunctional nanocomposite PEMs. In this technique, one side of the membrane is faced to an impregnation solution containing proper nanofillers followed by a reduction process. Finally, the organic–inorganic nanocomposite PEM is soaked in an acid solution and is washed with the deionized water. In this method, the loading of nanofillers in the polymer can be controlled by the impregnation time [26].

2.7 Surface Patterning Method

The surface patterning method is a facile and low-cost technique to prepare large-scale platforms with well-ordered patterns of a membrane by means of an elastomer mold at room temperature without hot pressing. The property at the interface of membrane and cathode is controlled by a surface patterned structure of the PEM, which affects the electrochemically active surface area (ECSA). This indicates the electrochemical properties develop the membrane electrode assembly performance [27].

Koh et al. [27] used surface patterning method to prepare patterned Nafion membranes by using polydimethylsiloxane (PDMS) mold as shown in Fig. 4. In this method, the patterned PDMS was used as a mold (not as a stamp) to make a pattern on Nafion membranes. A 0.5 mm thick patterned silicon master was used to fabricate the patterned PDMS mold. The prepared Nafion solution was dropped onto the patterned PDMS mold surfaces. For this purpose, the 5 wt% Nafion solution in a mixture of water and lower aliphatic alcohols was used. This low concentration is due to the small size of the pattern on the PDMS mold, and easy penetration of the solution into the micropatterns. Afterward, the Nafion solution can be casted on the surface of the patterned PDMS to create the same thickness of the Nafion on the patterned surface. Then, the sample should be dried overnight at

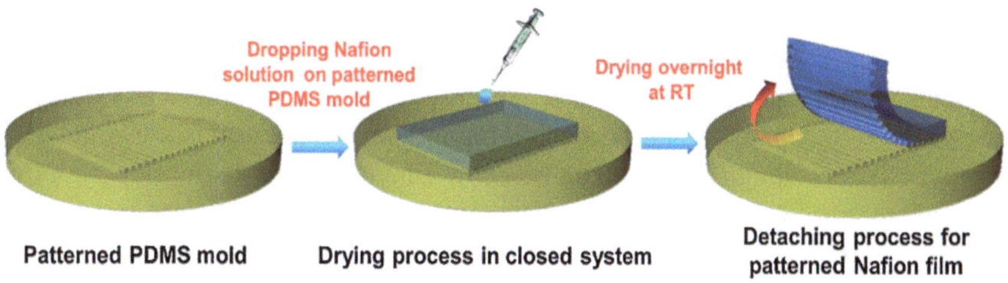

Fig. 4 A representation of the preparation process of membranes with surface patterned method [27]

room temperature in a closed system. Subsequently, the patterned membrane must be cautiously detached from the PDMS.

There are numerous advantages when the surface patterned method is used: (i) the thickness of the film can be easily controlled; (ii) it is possible to make a large surface patterned area; (iii) many patterned membranes can be fabricated in a short time, repeatedly; (iv) it is probable to fabricate a desired number of films by using only one silicon master, representing excellent processability and reproducibility without any defection in the mold and the membranes; (v) the surface patterned mold can be reused without any treatments or washing, which is very promising for industrial applications.

3 Future Directions and Conclusion

The full potential of the synthesis of organic–inorganic nanocomposite PEMs has yet to be assessed. With the developments in nanoscaled materials and polymer chemistry, the advanced nanocomposite membranes with promising performance can be prepared by different methods. Some of the important methods are incorporation of nanoparticles in polymer, doping or infiltration and precipitation of nanoparticles and precursors, self-assembly of nanoparticles, layer-by-layer fabrication method and nonequilibrium impregnation reduction. The main challenge in making hybrid nanocomposite membranes is to enable efficient dispersion and distribution of nanoparticles in the polymer matrix. For this purpose, effective strategies for surface functionalization of nanoparticles should be developed for all above-mentioned methods except for self assembly of nanoparticles. These strategies rely on the balance of intermolecular forces between surface functionalized nanoparticles including attractive forces (e.g., covalent and hydrogen bonding, electrostatic attraction between oppositely charged ligands, and dipole–dipole interactions) and repulsive forces (e.g., steric forces and electrostatic repulsion between ligands of similar charge). Hence, nanoparticles surface modification has become an active research area for the development of nanocomposite membranes. In the self-assembly technique, controlling of the hydrolysis and condensation reactions yields uniform distribution of small size nanoparticles in the polymer. However, the synthesized nanoparticles are not crystallized since the hydrolysis and condensation reactions are typically conducted at room temperature. Although the synthesized membranes by this method cannot be used in some particular membrane processes that require crystallized (e.g., antibacterial properties of TiO_2 nanoparticles in the nanofiltration and reverse osmosis membranes), however, the resulting IPNs membranes show a high capacity to develop high flux membranes. Furthermore, the surface patterning method is a simple and fast method to produce a large surface area with desirable thickness in commercial scale within a short time. Other techniques presented in this study are at the early stage of development and their main challenge is finding the pathway for larger scale fabrication and commercialization.

As far as the future directions for fabrication of nanocomposite membranes in fuel cell applications are concerned, the type and characteristics of the utilized nanoparticles are the most important parameters need to be considered. In fact, the incorporation of nanomaterials to a polymer matrix, with the aim of fabricating a robust hybrid membrane, may not induce the desired functionality to the polymer and even deteriorate its permeation properties. The most important factors to select an appropriate functional additive are (1) the hydrophilicity of the inorganic additive; (2) the acid site density on the inorganic nanoparticle surface; (3) the size or the specific surface area of the nanoparticles; and (4) the nanoparticles loading. Hence, a compelling path for future research is to properly utilize advanced nanomaterials, which can primarily enhance the water uptake, proton conductivity, and thermal stability.

Despite the extensive research conducted on nanocomposite membranes, several disadvantages such as removal of inorganic material from the polymer matrix after operation, inefficient dispersion and distribution of inorganic fillers throughout the membrane, and the miscibility with the catalyst layer are the main obstacles in the path of progress toward commercialization. Therefore, the main technical areas where future directions are predicted are as follows:

- Improving the water retention and proton conductivity of the synthesized nanocomposite membranes at higher temperatures.
- Improving the mechanical strength and dimensional stability of hybrid membranes in the hydrated state and at high and medium temperatures (~80–150 °C).
- Selecting and utilizing suitable inorganic nanofillers and efficient dispersion in the membrane.
- Developing methods for surface modification of nanofillers to obtain the ideal property of these materials.
- Improving the chemical interaction between organic and inorganic phase to enhance the compatibility of these two phases.
- Testing the long term stability of the synthesized nanocomposite membranes by conducting accelerated aging tests or long term exposure in a fuel cell application.

References

1. Kim DJ, Jo MJ, Nam SY (2015) A review of polymer-nanocomposite electrolyte membranes for fuel cell application. J Ind Eng Chem 21:36–52
2. Hashemi F, Rowshanzamir S, Rezakazemi M (2012) CFD simulation of PEM fuel cell performance: effect of straight and serpentine flow fields. Math Comput Model 55:1540–1557
3. Kraytsberg A, Ein-Eli Y (2014) Review of advanced materials for proton exchange membrane fuel cells. Energy Fuels 28:7303–7330
4. Ng LY, Mohammad AW, Leo CP, Hilal N (2013) Polymeric membranes incorporated with metal/metal oxide nanoparticles: a comprehensive review. Desalination 308:15–33

5. Tripathi BP, Shahi VK (2011) Organic–inorganic nanocomposite polymer electrolyte membranes for fuel cell applications. Prog Polym Sci 36:945–979
6. Zhang H, Mao H, Wang J, Ding R, Du Z, Liu J, Cao S (2014) Mineralization-inspired preparation of composite membranes with polyethyleneimine–nanoparticle hybrid active layer for solvent resistant nanofiltration. J Membr Sci 470:70–79
7. Rezakazemi M, Ebadi Amooghin A, Montazer-Rahmati MM, Ismail AF, Matsuura T (2014) State-of-the-art membrane based CO_2 separation using mixed matrix membranes (MMMs): an overview on current status and future directions. Prog Polym Sci 39:817–861
8. Rezakazemi M, Vatani A, Mohammadi T (2015) Synergistic interactions between POSS and fumed silica and their effect on the properties of crosslinked PDMS nanocomposite membranes. RSC Advances 5:82460–82470
9. Rezakazemi M, Vatani A, Mohammadi T. Synthesis and gas transport properties of crosslinked poly(dimethylsiloxane) nanocomposite membranes using octatrimethylsiloxy POSS nanoparticles. J Nat Gas Sci Eng
10. Rezakazemi M, Iravaninia M, Shirazian S, Mohammadi T (2013) Transient computational fluid dynamics (CFD) modeling of pervaporation separation of aromatic/aliphatic hydrocarbon mixtures using polymer composite membrane. Polym Eng Sci 53:1494–1501
11. Rezakazemi M, Shahverdi M, Shirazian S, Mohammadi T, Pak A (2011) CFD simulation of water removal from water/ethylene glycol mixtures by pervaporation. Chem Eng J 168:60–67
12. Rezakazemi M, Razavi S, Mohammadi T, Nazari AG (2011) Simulation and determination of optimum conditions of pervaporative dehydration of isopropanol process using synthesized PVA–APTEOS/TEOS nanocomposite membranes by means of expert systems. J Membr Sci 379:224–232
13. Rostamizadeh M, Rezakazemi M, Shahidi K, Mohammadi T (2013) Gas permeation through H_2-selective mixed matrix membranes: Experimental and neural network modeling. Int J Hydrogen Energy 38:1128–1135
14. Rezakazemi M, Mohammadi T (2013) Gas sorption in H_2-selective mixed matrix membranes: Experimental and neural network modeling. Int J Hydrogen Energy 38:14035–14041
15. Rezakazemi M, Shahidi K, Mohammadi T (2012) Hydrogen separation and purification using crosslinkable PDMS/zeolite A nanoparticles mixed matrix membranes. Int J Hydrogen Energy 37:14576–14589
16. Shahverdi M, Baheri B, Rezakazemi M, Motaee E, Mohammadi T (2013) Pervaporation study of ethylene glycol dehydration through synthesized (PVA–4A)/polypropylene mixed matrix composite membranes. Polym Eng Sci 53:1487–1493
17. Rezakazemi M, Shahidi K, Mohammadi T (2012) Sorption properties of hydrogen-selective PDMS/zeolite 4A mixed matrix membrane. Int J Hydrogen Energy 37:17275–17284
18. Basile A, Gallucci F (2011) Membranes for membrane reactors: preparation, optimization and selection. Wiley, UK
19. Baheri B, Shahverdi M, Rezakazemi M, Motaee E, Mohammadi T (2015) Performance of PVA/NaA mixed matrix membrane for removal of water from ethylene glycol solutions by pervaporation. Chem Eng Commun 202:316–321
20. Mulder M (1996) Basic principles of membrane technology, 2nd edn. Kluwer Academic Publishers, The Netherlands
21. Khulbe KC, Feng C, Matsuura T (2008) Synthetic polymeric membranes: characterization by atomic force microscopy. Springer, Berlin
22. Sadhukhan S (2011) Preparation and characterization of polymer electrolyte. National Institute of Technology, Rourkela
23. Caseri W (2007) Nanocomposites of polymers and inorganic particles. In: Hybrid materials, Wiley-VCH Verlag GmbH & Co. KGaA, pp 49–86
24. Tripathi BP (2010) Organic–inorganic nanocomposite polymer electrolyte membranes for electrochemical processes. Chemistry Department, Bhavnagar University
25. Hay JN, Raval HM (2001) Synthesis of organic−inorganic hybrids via the non-hydrolytic sol −gel process. Chem Mater 13:3396–3403

26. Pethaiah SS, Ulaganathan M, Ramalinga Viswanathan M, Chan SH (2015) Fabrication and electrochemical characterization of Pt–Pd impregnated nanocomposite polymer electrolyte membranes for high concentration DMFCs. RSC Adv 5:981–987
27. Koh JK, Jeon Y, Cho YI, Kim JH, Shul Y-G (2014) A facile preparation method of surface patterned polymer electrolyte membranes for fuel cell applications. J Mater Chem A 2: 8652–8659

Chapter 12
An Overview of Chemical and Mechanical Stabilities of Polymer Electrolytes Membrane

Izazi Azzahidah Amin, Joon Ching Juan and Chin Wei Lai

Abstract Fuel cells provide high efficiency, clean energy and low/zero emission compared to fossil-based energy. The most important component of fuel cell systems is polymer electrolyte membrane (PEM), which acts as a charge carrier that transports proton (H^+) ion from anode to cathode, as well as a barrier for anode fuel and cathode oxidant gas. Therefore, the basic requirements in terms of PEM performance include (1) good mechanical strength and toughness, (2) high thermal and chemical stability, (3) a good barrier for anode H_2 and cathode O_2, (4) a good proton conductance and (5) low electron conductance. In this chapter, the important attributes of PEM, such as chemical and mechanical stabilities have been described and reviewed. Efforts have been made to highlight the responses of chemical and mechanical stabilities of membrane at different temperatures and relative humidities of the fuel cell operation that lead to cell failure. A literature review regarding the chemical and mechanical degradation of membrane as well as the mitigation for the membrane degradation has also presented.

1 Introduction

Nowadays, among the assorted type of fuel cells, PEMFC has received tremendous attention due to its applicability as a promising high fuel/energy efficiency, high energy/power density, zero/low emission devices, compact size and low operating

I.A. Amin · J.C. Juan (✉) · C.W. Lai
Nanotechnology & Catalysis Research Centre (NANOCAT), Institute of Graduate Studies, University of Malaya, 50603 Kuala Lumpur, Malaysia
e-mail: jcjuan@um.edu.my

I.A. Amin
e-mail: azzahidah_87@siswa.um.edu.my

C.W. Lai
e-mail: cwlai@um.edu.my

J.C. Juan
School of Science, Monash University, Sunway Campus, Jalan Lagoon Selatan, 47500 Bandar Sunway, Malaysia

Inamuddin et al. (eds.), *Organic–Inorganic Composite Polymer Electrolyte Membranes*, DOI 10.1007/978-3-319-52739-0_12

temperature. To sum up, they are made for mobile, portable, stationary, transportation and automobiles power applications [1–4]. Furthermore, PEMFC has been leading in providing clean and efficient energy conversion during the twenty-first century [5].

Therefore, because of the wide range of applications, PEM has to fulfil the following basic requirements in terms of their performance [6]:

1. Good mechanical strength and toughness;
2. High thermal and chemical stability;
3. Good barrier for anodic fuel H_2 and cathodic fuel O_2;
4. Good proton conductance between the electrodes;
5. Low electron conductance and good electrical insulator in order to prevent short-circuiting of both electrodes.

Most of the current studies are focused on the development of high ionic conductivity and the stability of PEM. PEM conducts specific ions during the electrochemical reactions that occur at anode and cathode. Due to that reason the mechanisms in the fuel cell are influenced by the properties of PEM [6, 7].

The performance of PEMFC is closely related to the performance of membrane electrode assembly (MEA). MEA is the most important central core in the polymer electrolyte membrane fuel cell (PEMFC), which comprises of a polymer electrolyte membrane (PEM) located at its centre with anode and the cathode catalyst layers (CLs) are located next on both sides of the PEM, then two gas diffusion layers (GDLs) are located on the outer surfaces of CLs. To conclude, based on the location of PEM in MEA, PEM plays several roles in the fuel cell viz. (1) barrier for separating chemical reactions occur at anode and cathode, (2) barrier for protecting reaction between fuel and oxidant gas, (3) charge carrier that transports the proton (H^+) ion from anode to cathode, (4) electron flow controller, (5) barrier for maintaining chemical and mechanical stabilities [6, 8].

PEM conducts specific ions during the electrochemical reactions that occur at anode and cathode, due to that, mechanisms in the fuel cell are influenced by the properties of PEM [7]. The membranes used for PEM are broadly classified on the basis of type of materials (perfluorinated ionomers, partially fluorinated polymers, non-fluorinated membranes with aromatic backbones, non-fluorinated hydrocarbons) and preparation methods (acid–base blends and others). Figure 1 provides the classification of the membranes [9].

PEMs have the ability to transfer the proton as well as carrying fixed positive and negative charges. These types of electrolytes have gel-like polymeric structure which gets swollen easily. PEMs have the ability to transfer proton under moist condition but do not have the ability under dry conditions. The ionic conductivity of PEM increases when the water molecule in the polymeric structure increases [10]. Since PEM conducts proton in hydrated condition, the controlling of polymer structure and morphology is required to maximise the conductivity of PEM even at minimum water content [11].

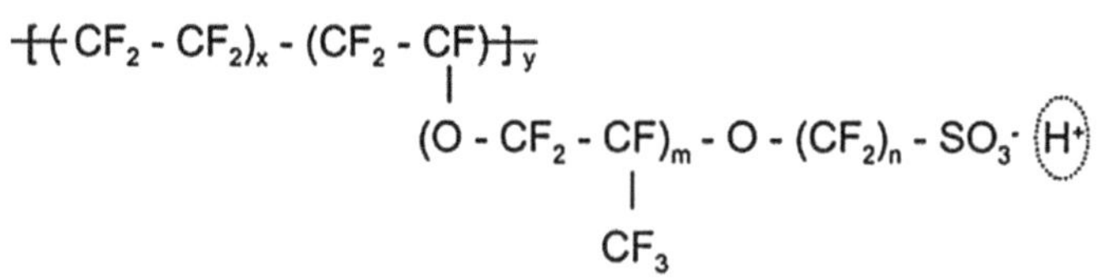

Fig. 1 Chemical structure of perfluorinated sulfonated polymer (Nafion®) (modified and reprinted with permission from [35]). Although PFSA-membranes have been a benchmark for PEM, but still facing the problem of chemical degradation during the operation of fuel cell. Thus, chemical stability of PFSA is one of the most serious obstacles in developing fuel cell application

The maximum performance of PEMFC is achieved when PEM retains an optimum quantity of proton carrier regardless of whether it is in operating or non-operating condition [6]. An increase in the quantity of proton carrier weakens the mechanical property of membrane, while a decrease in the quantity of proton carrier reduces the proton conductivity of membrane. Proton carrier is the second part that forms the chemical structure of PEM which includes water, H_3PO_4 and other ionic medium (e.g. ionic liquids).

High-temperature PEMFC refers to the fuel cell working at a temperature range from 100 to 200 °C and great effort is needed for its development. However, the elevated temperature has been a challenge on the coupled stability, that is, chemical and mechanical stabilities of PEM [12]. It is crucial to develop a PEM with high mechanical strength for high-temperature fuel cell, since the mechanical properties of PEM like tensile strength, tensile strain and toughness, have high tolerance to the impact force from the high-flow-rate fuel gas.

The development of PEM was first started by General Electric (GE) in 1959 with the usage of phenolic membranes, prepared by polymerisation of phenol–sulfonic acid with formaldehyde. This is the first application of fuel cell as an auxiliary power source for the NASA's Gemini space flights. However, the membrane encountered insufficient proton conductivity to reach a power density even as low as 100 mW cm^{-2} [13–15]. Other than that, their tendency to degrade in fuel cells operation has limited the shelf life of these membranes [16].

In the 1970s, DuPont Co. developed the first polymer electrolytes membrane based on the perfluorosulfonic acid (PFSA) called Nafion®. The Nafion® membrane has been the standard membrane since then, and became the state of the art in PEMFC. Currently, PFSA-membranes have been the benchmark for membrane performance, and are the leading and most widely used PEMs in fuel cells because of their excellent mechanical, thermal and chemical stabilities in the PEMFC operating systems as well as having good and outstanding proton conductivity [17]. It also showed a twofold increase in conductivity and possesses a lifetime extension of up to four order of magnitude (10^4–10^5 h) [9].

Nafion® is the well-known PEM that exhibits good proton conductivity at low temperatures ($T \sim 80$ °C) and high chemical/mechanical resistance among other types of membranes [18]. Other than that, Nafion® shows a high proton conductivity (~0.1 S/cm at operation temperature), a moderate water uptake (38 mass% water uptake and ~15% swelling ratio), a high strength (~25 MPa tensile strength) and flexibility (~180% elongation at break) and a good chemical stability. These excellent properties of Nafion® promise high operational performance and long-term durability of PEMFC [19].

However, the production of Nafion® involves high cost [20, 21]. In addition, Nafion® has several drawbacks such as a drop in membrane performance as well as a decrease in proton conductivity at elevated temperature above 80 °C and low relative humidity; Higher temperature caused the dehydration of membranes, which exacerbates and affects the proton conductivity and mechanical stability [22–24]. The afore mentioned drawbacks have become a barrier for the commercialisation of PEMFC.

Accordingly, it has been reported that many studies have been carried out in search of alternative materials for PEM. The studies are concerned with the membrane performance with the aim to improve the chemical and mechanical stabilities of PEM and to accelerate the commercialisation process of PEMFC. For example, the modification of Nafion® and the development of hydrocarbon-based polymers like sulfonated aromatic polymers based on poly(ether ketone) PEEK, poly(ether sulfone) (PES) and polybenzimidazole (PBI) are for use in PEMFC [25, 26].

The aim of this chapter is to provide an overview of the chemical and mechanical stabilities of polymer electrolyte membranes towards the performance and lifetime of fuel cell.

2 Durability of Polymer Electrolyte Membrane (PEM)

Researchers have given attention to the PEM structural response towards the performance and operating principle of fuel cells especially in terms of electrochemical efficiency. The major challenges in the development and commercialisation of PEMFC, are the insufficiency of durability related to the severe coupled mode of degradation as well as chemical and mechanical degradation. The performance of PEMFC gradually reduces with the failure of cell component, especially in MEA, as membrane degradation is one of the major factors that caused the cell failure. Membrane degradation including chemical and mechanical degradation led to the formation of pinhole, membrane thinning and fuel crossover. The formation of pinholes can lead to exothermic combustion and to a short circuit of cell through direct contact of oxygen and hydrogen [1, 27, 28].

In addition, chemical and mechanical degradation of PEM can cause parasitic degradation of other components, total degradation to fuel cell operation and a serious problem that restricted the commercialisation of the system application based on the fuel cells. The determination of the durability of PEMFC can be achieved by a deep analysis and understanding of the PEM degradation phenomena [29].

PEM is sandwiched between anode and cathode to transport the protons, thus the length of the proton carrier pathway and the resistance (*R*) depend on the thickness (*L*) of the PEM. Increasing the membrane thickness may lead to a longer proton carrier pathway, a larger *R* and thus reduces the fuel cell performance. Furthermore, reducing the membrane thickness could create the possibility of a crossover between fuel and oxidant gas through the membrane, and thus reducing the fuel cell output power. The mechanical properties of PEM like tensile strength, tensile strain and toughness, have a high tolerance to the impact force from the high-flow-rate of fuel gas. Thus, the development of PEM with high mechanical strength, high conductivity and optimum thickness are important for a maximum fuel cell performance [6]. The aforementioned, Nafion® (perfluorosulfonic acid (PFSA) membrane) is the most widely used PEMFC. The PFSA structure consists of hydrophobic polytetrafluoroethylene (PTFE) backbone that provides thermal and chemical stabilities, while the periodic perfluorinated side chain terminated with hydrophilic sulfonic acid (–SO_3H) provides the medium for proton conduction [30]. Table 1 lists the different types of membranes and their characteristics including durability and conductivity.

3 Proton Conductivity of PEM

As mentioned earlier, high proton conductivity is one of the important requirements of PEM in order to achieve a high efficiency of fuel cell. Proton conductivity depends on the concentration of the ion-conducting unit in PEM, known as ion exchange capacity (IEC) [25]. A well-connected hydrophilic ionic channel in PEM is shown by high IEC value. Other than that, high proton conductivity is achieved at a higher prevailing level of hydration. The characteristic of the commonly used PEM, perlfluorinated sulfonic acid type membranes (PFSA) [32], known by the trade names of Nafion®, Flemion® and Hyflon® are the requirements of water molecule. The presence of water as proton solvent is important for the dissociation and formation of mobile proton [11]. Mobile proton from the fuel anode will pass through the proton-conducting PEM heading to the cathode, reducing oxygen [33]. Therefore, a good water transport is important in PEM because proton conductance relied on the PEM hydration [34]. A reduction of PEM thickness can avoid water drag and crossover. Moreover, reducing the thickness will lower the PEM resistance, rapid in hydration, thus increasing the membrane conductivity [9]. Table 2 sums up properties of some commercial PEMs. To conclude, operating parameters such as water content, temperature and thickness can influence the proton conductance in Nafion®.

Table 1 Summary of the characteristics of some membranes (modified and reprinted with permission from [31])

Membrane type	Nafion (perfluorinated polymer)	Aciplex (perfluorinated polymer)	Flemion (perfluorinated polymer)	S-FPAE (perfluorinated polymer)	PBI (nonfluoriated polymer	SPEEK (aromatic polymer)
Conductivity (S/cm)	0.05–0.15 (80–100 °C; RH 50–100%)	0.11 (80 °C; RH 100%)	0.14 (80 °C; RH 100%)	0.012 (80 °C; RH 100%)	0.05 (180 °C RH 10%) 0.13 (160 °C) 0.1 (200 °C)	0.03 (100 °C 100% RH) 0.11 (150 °C 100% RH)
Water uptake (% water)	20	Up to 20	35	100	22	60–100
Permeability (mol/cm s Pa)	NAFION 117 $H_2$15 × 10^{-17}; O_2 3 × 10^{-17}	H_2 2 × 10^{-8} (cm^3 cm/cm^2 s cm Hg)	H_2 5 × 10^{-8}; O_2 1 × 10^{-8} (cm^3 cm/cm^2 s cm Hg)	n/a	H_2 1.6–4.3 × 10^{-17} e 1.2–4.0 × 10^{-15} at 120 and 180 °C respectively O_2 5–10 × 10^{-19} e 3–9.4 × 10^{-16} at 120 and 180 °C respectively	n/a
Electrochemical stability	Very good	Good	Good	Poor	Very good	Good
Durability	40,000–60,00 h	≤ Nafion	≤ Nafion	<<Nafion	20,000 h	3000–4000 h
Power density (mW/cm^2)	400–1200	750	450–700	430 at 80 °C	150–250, 300, 390	325–350 at 95 °C
Working temperature range (°C)	80–130 for P = 1–3 bar respectively	80	80	80–150	130–230	80–130

Table 2 Properties of commercial polymer electrolyte membranes (modified and reprinted with permission from [9])

Membrane	Membrane type	IEC (mequiv./g)	Thickness (mm)	Gel water (%)	Conductivity (S/cm) at 30 °C and 100% RH
Asahi Chemical Industry Company Ltd., Chiyoda-ku, Tokyo, Japan					
K 101	Sulfonated polyarylene	1.4	0.24	24	0.0114
Asahi Glass Company Ltd., Chiyoda-ku, Tokyo, Japan					
CMV	Sulfonated polyarylene	2.4	0.15	25	0.0051
DMV	Sulfonated polyarylene	–	0.15	–	0.0071
Flemion	Perfluorinated	–	0.15	–	–
Ionac Chemical Company, Sybron Corporation, USA					
MC 3470	–	1.5	0.6	35	0.0075
MC 3142	–	1.1	0.8	–	0.0114
Ionics Inc., Watertown, MA 02172, USA					
61AZL386	–	2.3	0.5	46	0.0081
61AZL389	–	2.6	1.2	48	–
61CZL386	–	2.7	0.6	40	0.0067
Du Pont Company, Wilmington, DE 19898, USA					
N 117	Perfluorinated	0.9	0.2	16	0.0133
N 901	Perfluorinated	1.1	0.4	5	0.01053
Pall RAI Inc., Hauppauge, NY 11788, USA					
R-1010	Perfluorinated	1.2	0.1	20	0.0333

4 Chemical Stabilities and Degradation of PEM

As mentioned earlier, the most suitable and commonly used PEM in fuel cell is Nafion® which has perfluorinated structures that are attached with sulfonic acid groups. The ionic properties are applied to Nafion® after addition of a pendant side chain with hydrophilic terminated sulfonic acid group into the polymer (tetra-fluoroethylene)-like hydrophobic main backbone. Figure 1 shows the chemical structure of perfluorinated sulfonated Nafion®, where the sulfonic acid group is shown in anhydrous form, SO_3H. The hydrolysed form ($SO_3^-H_3O^+$) will appear when Nafion® is exposed to water, thus enabling the proton conductance across membrane from anode to cathode [17, 33]. Sulfonic acid groups play a role as a proton conductance group in the Nafion® structure [34].

The chemical stability of PEM is one of the crucial prerequisites that need to be considered when determining the performance of PEMFC. The chemical stability of the membrane can offer significant effects to the performance of PEM [34]. The most common phenomenon that caused PEM to lose chemical stability is the membrane degradation. Membrane degradation is caused by the chemical degradation due to the reactive radical species attack [36].

PEM degradation resulting from chemical degradation mechanism can result in the formation of pinhole together with fuel crossover. Many studies have been conducted focusing on alternative PEM and improvement of chemical stability of the current PEM in order to meet the performance target required for application in fuel cell [14]. The studies are concerned with the mitigation of the membrane degradation [24]. The degree of membrane degradation is influenced by certain factors, such as temperature, humidity, current density, cell voltage, MEA structure and catalyst. However, the operation of PEMFC at a higher temperature and lower relative humidity leads to cell failure through faster degradation [29, 37]. Recent studies [38] have confirmed that the free radicals formed during fuel cell operation contribute to the chemical degradation of the ionomer molecular structure in the PEM.

At the interface of anode-membrane and cathode-membrane, the oxidation of hydrogen and reduction of oxygen PEM takes place. The fuel gas, H_2 also interacted with oxidant gas, O_2 on the surface of Pt catalyst, which generated hydroxyl HO· and peroxyl HOO· radical species. The radical species attack the polymeric membrane and bring it to chain scission, resulting to membrane degradation [11].

The reduction of oxidant gas, O_2 at cathode and chemical combination of H_2 and O_2 have generated hydroxyl radicals and hydrogen peroxide. Due to unpaired electron, the highly reactive and short-dated radicals attack the polymer backbone and weak carboxylic acid end of PFSA membrane [37, 39, 40]. Furthermore, the generation of radical species and chemical degradation are accelerated under low humidity conditions as well as under open circuit voltage (OCV) [29]. Other than that, hydroxyl radical can attack carboxylic and sulfonic acids easily and the generated –COO· and $–SO_3·$ radicals initiate the unzipping reaction of the PFSA membrane. The hydroxyl radicals can also attack the C–H and C–C of the hydrocarbon polymer membrane and lower chemical stability of PEM.

Poornesh and Cho [27], stated that the polymer degradation mechanism can be divided into cleavage and unzipping. Cleavage is the scissoring of the functional group of the polymer backbone, while unzipping is the releasing the weak side chain of the polymer backbone. Figure 2 illustrated the cleavage and unzipping mechanisms of Nafion®.

Thus, one of the mitigation for chemical degradation is by removing hydroxyl, OH· radical species in PEM. Kim et al. [22] incorporated mesoporous ceria–silica as a radical species scavenger inside the hydrocarbon sulfonated poly(arylene ether sulfone) (sPES)/mesoporous ceria–silica polymer composite membrane. The oxidation stability against hydroxyl, OH· radical species as well as the antioxidant effect of ceria was investigated using Fenton's test and the hydrogen peroxide exposure. The composite membrane exhibited higher oxidation stability compared to the pristine sPES membrane. The degradation percent ratio for pristine sPES was 20%, whereas for the composited membrane was in the range of 0.53–16.2%.

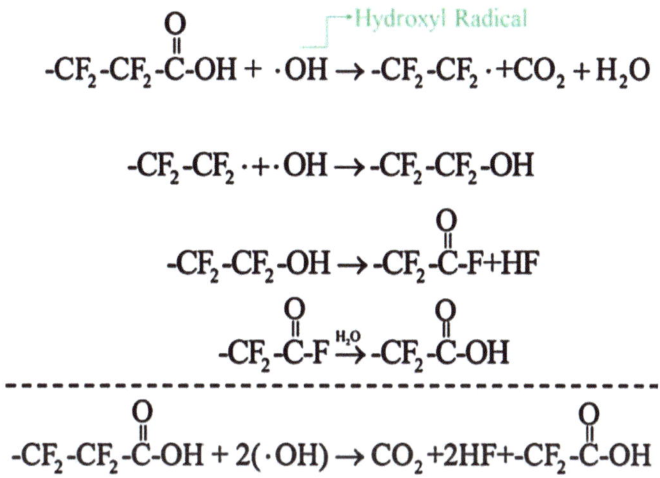

Fig. 2 Structure of Nafion® polymer chain showing the degradation site (reprinted with permission from [27])

Tanuma and Itoh [24] proposed the cerium and manganese to be incorporated in PFSA membrane. The cerium and manganese redox couples reduced the radical species as well as mitigate the membrane degradation. Effects of OCV on PEM without additives showed a faster deterioration rate of 0.46×10^{-6} (V/S), while the deterioration rates of PEM with additives ($Ce_2(CO_3)_3$) and $MnCO_3$ were around 0.03×10^{-6} (V/S). During OCV test on PEM, the crossover of hydrogen and oxygen occurred at anode and cathode forming hydrogen peroxide, then the cerium and manganese were scavenged which reduced the hydroperoxyl, OOH· radical species. These indicate that cerium and manganese are effective in preventing membrane degradation caused by a radical species attack.

5 Mechanical Stability and Degradation of PEM

The mechanical stability, such as swelling dimensional stability and mechanical strength give significant influence to the performance of fuel cell [25]. As PEM is a key component of fuel cell, it must show durability towards a wide range of operating conditions, such as humidity in the range of ambient to 100% relative humidity (RH), and also in a large range of temperatures from −40 to 120 °C [41]. Water transport, humidification and hydration in the membrane as well as clamping effect using bipolar plates have considerable stresses on the PEM, which may lead to the reduction of mechanical stability and crack issues [42]. Water uptake (WU) is important in PEM because it is needed for the mobile phase of proton conductance.

However, the mechanical stability of PEM will be affected by absorbing water and becoming plasticiser, lowering the *Tg* and modulus of the membrane. The management of water uptake is important to reduce the severe effects of swelling and mechanical degradations of PEM in humidity; it also reduces the inducing stress between PEM and electrode [43].

The mechanical degradations of PEM are caused by the swelling and shrinking of membrane. Furthermore, changes in the membrane humidity could lead to membrane fatigue and creep while the mechanical degradation exacerbates the effect of membrane chemical degradation [44]. Therefore, the mechanical stability of membrane becomes an important issue in determining the performance of PEM especially at elevated temperature operation [45].

According to Tripathi and Shahi [46] under low humidification, Nafion® showed poor performance as well as drying at elevated temperature. Thus, this trend did not fulfil the demanding requirement of the high performance of PEM with good mechanical stability in both dry and hydrated conditions. Nonetheless, in recent years, substantial efforts have been made to overcome the mechanical degradation of PEM in commercialisation of the PEMFC especially under low humidity and at elevated temperatures [47]. The strategies used to overcome the poor performance of PEM include the modification of Nafion® and the development of novel PEM as an alternative to Nafion® [32]. The methods used are cross-linking and physically as well as by mechanical reinforcement with porous polymer matrix, fiber or inorganic reinforcement. Reinforced membrane exhibits a longer lifetime than non-reinforced membrane even at the same thickness [39, 44].

Gourdoupi et al. [48] observed that the doped main chain of PEM (PPyPO) based on the aromatic polyether containing polar pyridine groups possesses a higher molecular weight and is mechanically stable. Moreover, the PEM prepared has a high thermal and oxidative stability as well and can be used at a temperature higher than 120 °C. The mechanical and thermal characterisation of the obtained polymer showed the polymer glass transition temperature between 185 and 270 °C and the temperature for initial decomposition higher than 400 °C (Table 3). The PEM conductivity when doped with phosphoric acid at room temperature was 10^{-2} S/cm and on further increase in temperature, the conductivity reaches to 2.4×10^{-2} S/cm (Fig. 3). These values showed that the membrane chemical structure influences the properties of PEM as well as it fulfils the requirement used in high-temperature fuel cell.

An et al. [3] observed that sulfonated poly (ether ketone) (SPEEK) membrane treated with H_2SO_4 obtained a gradient cross-linking structure which restrain the

Table 3 T_g and T_d values for polymer and copolymer

Polymer/copolymer	T_g (°C)	T_d (5%)
PPy(90)coPO-iii	270	540
PQDPyEtO-iii	184	410
PQDPyEtO(50)coPSF-i	208	410
PPyPO-i	252	550

Reprinted with permission from [48]

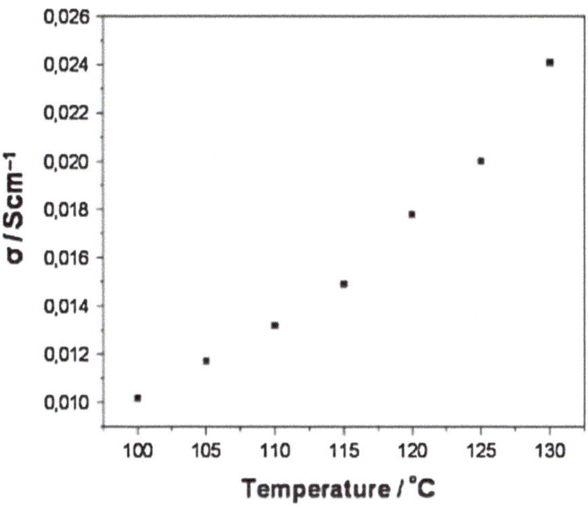

Fig. 3 Temperature dependence of ionic conductivity of the acid doped PPy(50) coPSF/PQDPyEtO(70)coPSF 50/50 blend with a doping level of 150 wt% H_3PO_4 (reprinted with permission from [48])

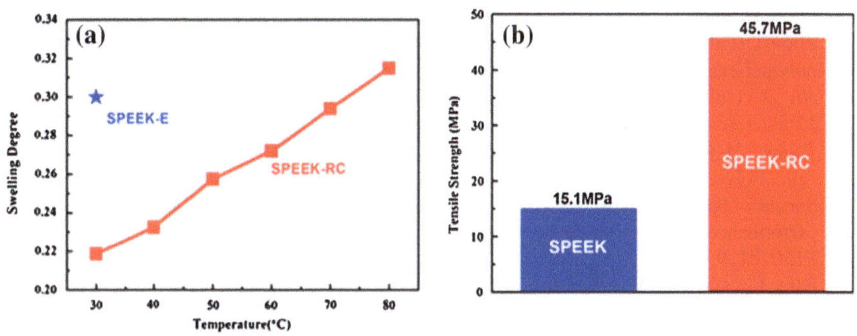

Fig. 4 A comparison of **a** swelling degree **b** tensile strength between gradient crosslinked SPEEK-RC and SPEEK/SPEEK-E membranes (modified and reprinted with permission from [3])

membrane from swelling and enhancing the membrane tensile strength. Figure 4a showed that SPEEK-RC (treated with H_2SO_4) showed a low swelling degree of 0.22 at 30 °C compared to SPEEK-E (pristine SPEEK membrane that has an equivalent degree of sulfonation with SPEEK-RC), swelling degree 0.30. While in Fig. 4b, pristine SPEEK showed a tensile strength of 15.1 MPa, three times lower than 45.7 MPa of SPEEK-RC. The low swelling degree and high mechanical strength of the membrane were due to covalent crosslinks which were added to the SPEEK molecular structure.

6 Conclusion

This chapter has reviewed the chemical and mechanical stabilities of polymer electrolyte membranes in terms of chemical and mechanical degradation of the membrane. Nafion® has been widely used in the fuel cell system because of its outstanding performance. However, it suffers from chemical and mechanical degradation that reduces the membrane stability and becomes a barrier for commercialisation in the PEMFC system. Therefore, it is crucial for researchers to find the solution and mitigation for the membrane degradation as well as developing a novel membrane that promises a better performance of fuel cell.

Acknowledgements The authors would like to acknowledge the financial support provided by University of Malaya under the Equitable Society Research Cluster (ESRC) Research Grant GC002A-15SBS and Postgraduate Research Grant PG057-2015B. The authors also thank the Ministry of Higher Education Malaysia for MyBrain 15 Scholarship.

References

1. Wang H, Li H, Yuan XZ (2011) PEM fuel cell failure mode analysis. CRC Press, Boca Raton
2. Zhou T et al (2015) A review of radiation-grafted polymer electrolyte membranes for alkaline polymer electrolyte membrane fuel cells. J Power Sources 293:946–975
3. An D et al (2016) Gradiently crosslinked polymer electrolyte membranes in fuel cells. J Power Sources 301:204–209
4. Tanaka M (2016) Development of ion conductive nanofibers for polymer electrolyte fuel cells. Polym J 48(1):51–58
5. Sanchez DG et al (2016) Analysis of the influence of temperature and gas humidity on the performance stability of polymer electrolyte membrane fuel cells. J Electrochem Soc 163(3): F150–F159
6. Fang J et al (2015) Electrochemical polymer electrolyte membranes. CRC Press, Boca Raton
7. Nasef MM et al (2016) Radiation-grafted materials for energy conversion and energy storage applications. Prog Polym Sci
8. Zakaria Z, Kamarudin SK, Timmiati S (2016) Membranes for direct ethanol fuel cells: an overview. Appl Energy 163:334–342
9. Smitha B, Sridhar S, Khan AA (2005) Solid polymer electrolyte membranes for fuel cell applications—a review. J Membr Sci 259(1–2):10–26
10. Peighambardoust SJ, Rowshanzamir S, Amjadi M (2010) Review of the proton exchange membranes for fuel cell applications. Int J Hydrogen Energy 35(17):9349–9384
11. Gubler L, Scherer GG (2010) Trends for fuel cell membrane development. Desalination 250 (3):1034–1037
12. Li Q et al (2015) High temperature polymer electrolyte membrane fuel cells: approaches, status, and perspectives. Springer International Publishing
13. Winter M, Brodd RJ (2004) What are batteries, fuel cells, and supercapacitors? Chem Rev 104(10):4245–4270
14. Letcher TM (2008) Future energy: improved, sustainable and clean options for our planet. Elsevier Science
15. Beguin F, Frackowiak E (2009) Carbons for electrochemical energy storage and conversion systems. CRC Press, Boca Raton

16. Pabby AK, Rizvi SSH, Requena AMS (2015) Handbook of membrane separations: chemical, pharmaceutical, food, and biotechnological applications, 2nd edn. CRC Press, Boca Raton
17. Zhang H, Shen PK (2012) Recent development of polymer electrolyte membranes for fuel cells. Chem Rev 112(5):2780–2832
18. Matos BR et al (2016) Nafion membranes annealed at high temperature and controlled humidity: structure, conductivity, and fuel cell performance. Electrochim Acta 196:110–117
19. Pan J et al (2015) Mechanically tough and chemically stable anion exchange membranes from rigid-flexible semi-interpenetrating networks. Chem Mater 27(19):6689–6698
20. Kim S-U et al (2015) Effect of sulfonated poly(arylene ether sulfone) binder on the performance of polymer electrolyte membrane fuel cells. J Ind Eng Chem 23:316–320
21. Yang Z et al (2016) Stability challenge in anion exchange membrane for fuel cells. Current Opin Chem Eng 12:22–30
22. Kim J et al (2016) Mesoporous ceria-silica/poly(arylene ether sulfone) composite membranes for durability of fuel cell electrolyte membrane. Microporous Mesoporous Mater
23. Cheng J et al (2016) Guanidimidazole-quanternized and cross-linked alkaline polymer electrolyte membrane for fuel cell application. J Membr Sci 501:100–108
24. Tanuma T, Itoh T (2016) Clarifying the chemical state of additives in membranes for polymer electrolyte fuel cells by X-ray absorption fine structure. J Power Sources 305:17–21
25. Zhang Y et al (2015) Recent developments on alternative proton exchange membranes: strategies for systematic performance improvement. Energy Technol 3(7):675–691
26. Kim DJ, Jo MJ, Nam SY (2015) A review of polymer-nanocomposite electrolyte membranes for fuel cell application. J Ind Eng Chem 21:36–52
27. Poornesh KK, Cho C (2015) Stability of polymer electrolyte membranes in fuel cells: initial attempts to bridge physical and chemical degradation modes. Fuel Cells 15(1):196–203
28. Marrony M et al (2008) Durability study and lifetime prediction of baseline proton exchange membrane fuel cell under severe operating conditions. J Power Sources 182(2):469–475
29. Büchi FN, Inaba M, Schmidt TJ (2009) Polymer electrolyte fuel cell durability. Springer, New York
30. Neburchilov V et al (2007) A review of polymer electrolyte membranes for direct methanol fuel cells. J Power Sources 169(2):221–238
31. Fiori C et al (2015) Critical review of fuel cell's membranes and identification of alternative types for automotive applications. Int J Hydrogen Energy 40(35):11949–11959
32. Zhang H, Shen PK (2012) Advances in the high performance polymer electrolyte membranes for fuel cells. Chem Soc Rev 41(6):2382–2394
33. Kraytsberg A, Ein-Eli Y (2014) Review of advanced materials for proton exchange membrane fuel cells. Energy Fuels 28(12):7303–7330
34. Zhang L et al (2012) Recent advances in proton exchange membranes for fuel cell applications. Chem Eng J 204–206:87–97
35. Haile SM (2003) Fuel cell materials and components? Acta Mater 51(19):5981–6000
36. Velan Venkatesan S et al (2016) Progression in the morphology of fuel cell membranes upon conjoint chemical and mechanical degradation. J Electrochem Soc 163(7): F637–F643
37. Shao Y et al (2007) Proton exchange membrane fuel cell from low temperature to high temperature: material challenges. J Power Sources 167(2):235–242
38. Tavassoli A et al (2016) Effect of catalyst layer defects on local membrane degradation in polymer electrolyte fuel cells. J Power Sources 322:17–25
39. Schmittinger W, Vahidi A (2008) A review of the main parameters influencing long-term performance and durability of PEM fuel cells. J Power Sources 180(1):1–14
40. Wu J et al (2008) A review of PEM fuel cell durability: degradation mechanisms and mitigation strategies. J Power Sources 184(1):104–119
41. Yuan XZ et al (2011) A review of polymer electrolyte membrane fuel cell durability test protocols. J Power Sources 196(22):9107–9116
42. Wang Y et al (2011) A review of polymer electrolyte membrane fuel cells: technology, applications, and needs on fundamental research. Appl Energy 88(4):981–1007

43. Hickner MA et al (2004) Alternative polymer systems for proton exchange membranes (PEMs). Chem Rev 104(10):4587–4612
44. Macauley N et al (2015) Favorable effect of in-situ generated platinum in the membrane on fuel cell membrane durability. J Power Sources 299:139–148
45. Bose S et al (2011) Polymer membranes for high temperature proton exchange membrane fuel cell: recent advances and challenges. Prog Polym Sci 36(6):813–843
46. Tripathi BP, Shahi VK (2011) Organic–inorganic nanocomposite polymer electrolyte membranes for fuel cell applications. Prog Polym Sci 36(7):945–979
47. Thiam HS et al (2011) Overview on nanostructured membrane in fuel cell applications. Int J Hydrogen Energy 36(4):3187–3205
48. Gourdoupi N et al (2008) New high temperature polymer electrolyte membranes. Influence of the chemical structure on their properties. Fuel Cells 8(3–4):200–208

Chapter 13
Electrospun Nanocomposite Materials for Polymer Electrolyte Membrane Fuel Cells

Shahram Mehdipour-Ataei and Maryam Oroujzadeh

Abstract Electrospinning is a low-cost versatile process with relatively high production rate for preparation of nanofibers that can be used in various fields. As the structure of electrospun mats is very similar to human tissues and organs, most of research activities in electrospinning are dedicated to biomedical applications. However, another interesting field for application of electrospun membranes is fuel cell application. The focus of this chapter is on the membrane of polymeric fuel cells and the methods of using electrospinning process for preparation of the membranes. Three important objectives of using electrospinning in membranes of polymeric fuel cells are reducing methanol crossover, improving proton conductivity, and suppressing water swelling, which will be explained in detail.

Keywords Proton exchange membrane fuel cells · Electrospinning · Nanofibers · Nafion

1 Introduction

Proton exchange membrane fuel cells (PEMFCs) are one of the most important types of fuel cells that convert chemical energy into the electrical energy directly and are used mainly in transportation and portable devices (Fig. 1). The proton exchange polymeric membrane is the heart of a PEMFC that transfers proton through the membrane and prevents crossover of electron and reactant gases [1, 2].

Perflourosulfonic acid (PFSA) based membrane commercially named Nafion is the most common polymeric membrane for PEMFCs. It shows excellent performance and properties at room temperature in the fuel cell environment but suffers from some disadvantages like dehydration and instability at temperatures higher

S. Mehdipour-Ataei (✉) · M. Oroujzadeh
Iran Polymer and Petrochemical Institute, P.O. Box 14965/115, Tehran, Iran
e-mail: s.mehdipour@ippi.ac.ir

M. Oroujzadeh
e-mail: m.orujzadeh@ippi.ac.ir

Inamuddin et al. (eds.), *Organic–Inorganic Composite Polymer Electrolyte Membranes*, DOI 10.1007/978-3-319-52739-0_13

Fig. 1 Schematic of a fuel cell

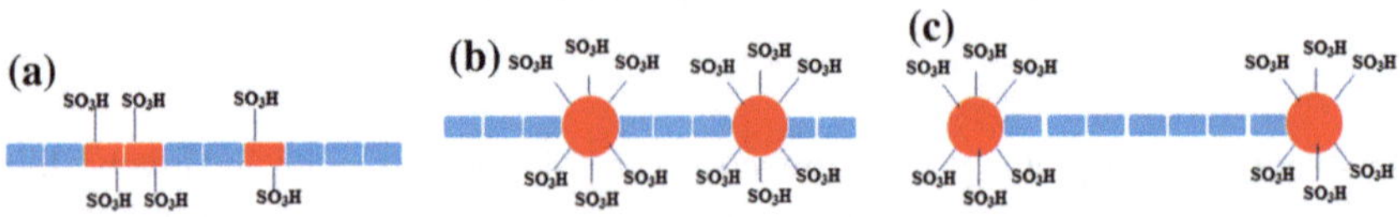

Fig. 2 Schematic structure of block copolymers **a** copolymers with accumulated sulfonic acid groups along the chain **b** and copolymers with accumulated sulfonic acid groups at two ends

than 80 °C, methanol permeability in direct methanol fuel cells (DMFCs) as well as high cost [3–5]. Alternative aromatic membranes such as sulfonated poly(imide)s [6, 7], sulfonated poly(ether ketone)s [8–10] sulfonated poly(ether sulfone)s [11–13], poly(benzimidazole)s [14–16], etc., that are thermally stable polymers with much lower methanol crossover for DMFCs have been introduced. However, almost all of the aromatic alternative membranes show lower proton conductivity compared to Nafion. Therefore, to overcome the disadvantages of Nafion and maintain its advantages simultaneously, a great deal of efforts has been made recently. Synthesis of block copolymers [17–20] and synthesis of polymeric structures with accumulation of sulfonic acid groups along the chain or just at the ends (Fig. 2) are some of attractive solutions [21–23]. The basic idea for most of these strategies is simulation of microstructure of Nafion in which distinct hydrophilic and hydrophobic separated domains exist [24]. The objective of all research activities in this field has been to have the membrane with high proton conductivity at high temperature and low relative humidity [25].

An interesting technique utilized recently is electrospinning process for preparation of efficient *proton exchange membrane fuel cells*. In this chapter, the electrospinning process and application of the process in proton exchange membrane fuel cells and also direct methanol fuel cells will be explained.

2 Electrospinning Process

Electrospinning is a low cost and versatile process with relatively high production rate of 70 m min^{-1} for preparation of nanofibers. The electrospinning process traces to Formhals patent in 1934 [26], but until 1990s when Reneker [27, 28] used it for preparation of nanofibers from some polymers did not receive much attention. Every soluble polymer with enough high molecular weight can be electrospun.

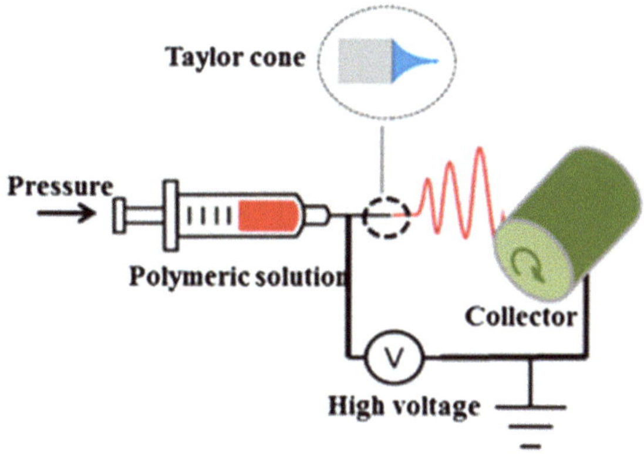

Fig. 3 Electrospinning apparatus

Fundamentally, in the electrospinning process, an electrical charge is used to draw submicron fibers from a polymeric solution with a specific concentration [29]. The typical electrospinning apparatus consists of a high voltage power supply, a nozzle, and a grounded collector (Fig. 3). As the high voltage applies, the components of polymer solution at the tip of the syringe charge and cause a repulsive force. During the process at the tip of the syringe a conical structure called Taylor cone forms. Consequently, at a critical voltage when the repulsive force overcomes surface tension of the polymeric solution, a jet erupts from the tip of the Taylor cone. If the voltage is high enough, the erupted jet which is called Rayleigh instability breaks up.

As the jet accelerates and passes the distance between syringe tip and the collector, thins, and split into the multiple filaments. The flying jet thins dramatically by almost three orders of magnitude while traveling the short distance between the nozzle and the collector. Along the way, the solvent also evaporates and fibers form on the surface of the collector [30–33].

Formation of fine fibers requires careful consideration of following parameters [34, 35]:

- Solution properties including conductivity, viscosity, and surface tension of the polymeric solution as well as dielectric constant and volatility of the solvent are very important in electrospinning process.
- Temperature and humidity of the electrospinning chamber are also affecting the quality of prepared fibers.
- Applied voltage, polymeric solution feeding rate, distance between nozzle and collector, and the speed of rotation of the collector also have significant effect on the performance of the nanofibers.

An important challenge with electrospinning process is to obtain uniform and defect-free fibers. As mentioned above, numerous parameters affect the electrospinning process and hence obtaining perfect fibers requires the adjustment among different conditions for electrospinning of each polymeric solution. Beads in fibers are the common problem. It has been found that higher concentration of polymeric

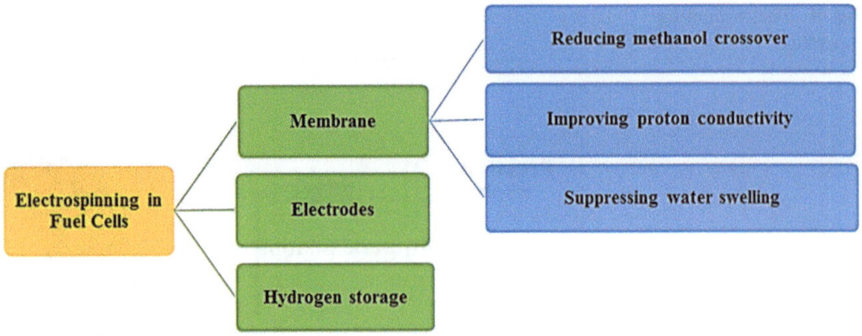

Fig. 4 Electrospinning in fuel cells

solutions results in fewer beads. Although the diameter of beads, if there are, increases at higher concentrations [36–38].

As the structure of electrospun mats is very similar to human tissues and organs, most of research activities in electrospinning of polymeric solutions are dedicated to biomedical applications. Nonetheless, second important application belongs to the filtration membranes and remaining publications about electrospinning involves other applications including fuel cell membranes [31]. There are publications about the use of electrospinning in fuel cells for production of electrodes, membranes, and also hydrogen storage (Fig. 4). Using the electrospinning process for fabrication of electrodes can reduce the nobel metal catalyst loading [39, 40]. In the case of hydrogen storage, hydrogen uptake in a reservoir is proportional to the surface area and pore volume. Small pores, 1 nm or below, are the most efficient for hydrogen sorption and smallest pores provide the highest storage capacity. Therefore, electrospinning process can be potentially used in this application [41].

In this chapter, after an introduction about electrospinning process and nanofibers, the application of electrospinning in preparation of proton exchange membranes for fuel cells will be considered in depth. The reasons of using this technique for membrane formation instead of routine membrane preparation techniques are: reducing methanol crossover in DMFCs, increasing proton conductivity, suppressing water swelling under high ion exchange capacities, and consequently reducing the cost. Although a great deal of publications has been with the aim of decreasing methanol crossover, every issue will be discussed separately in details in the next sections.

2.1 Electrospun Fibers

The most important aromatic polymers incorporated for preparation of electrospun mats in proton exchange membranes are reviewed in the following paragraphs.

2.1.1 Poly(vinylidene fluoride) (PVDF)

One of the most widely used polymers for preparation of tortuous path against methanol molecules is PVDF. It has been used as spun [42] or after some modifications, to improve bonding with Nafion matrix [43–45].

2.1.2 Poly(vinyl alcohol) (PVA)

PVA due to its higher affinity to water than methanol (55–10 wt%) is a promising modifier of Nafion for DMFCs [46]. Nanofibers of PVA are electrospun from aqueous solution of the polymer. In order to reducing surface tension of water and improving the spinnability of PVA, the solution can be mixed with a surfactant like cetyltrimethylammonium bromide (CTAB) [47]. Studies show that if PVA blends with Nafion and cast as routine, the optimum amount of adding PVA will be 5 wt% (causes slightly lower proton conductivity but three times less methanol permeability) and above this, the performance of DMFC dramatically decreases. On the other hand, electrospinning of PVA and impregnation of the mat with Nafion resin is more effective way for reducing methanol crossover. PVA fibers up to 10 wt% have no negative effect on performance of the DMFC. In this way and using PVA nanofibers, the thickness of the membrane can be reduced to about one-third of Nafion membrane with lower methanol crossover. Decreasing the membrane thickness reduces proton transfer resistance and improves the cell performance as well [46, 48].

2.1.3 Poly(phenylene oxide) (PPO)

Poly(phenylene oxide) fibers have also been used. Crosslinked bromomethylated PPO fibers have been used as a framework or support for sulfonated PPO matrix to suppress water swelling. In this manner, highly sulfonated PPO which completely swelled in water could be used as PEMFC membrane when reinforced with crosslinked bromomethylated PPO fibers [49].

2.1.4 Poly(arylene ether)s

Poly(ether ketone) and poly(ether sulfone) electrospun nanofibers were also used for reinforcing Nafion or other matrices. Lee and coworkers reported electrospinning of poly(ether ketone) solution containing SiO_2 nanoparticles and impregnation of that with Nafion resin. The proton conductivity of resulted membrane was 2.5 times higher than recast Nafion [50]. Sulfonated poly(ether sulfone) (sPES) electrospun nanofibers were also used for reinforcement of Nafion membrane for DMFC application. sPES nanofibers acted as methanol barrier and improved the performance of the cell [51]. There are some reports indicating the use of poly (arylene ether) electrospun fibers to reinforce other aromatic polymers. Zhang et al. reported impregnation of sulfonated poly(phthalazinone ether sulfone ketone) by sulfonated poly(ether ether ketone). The proton conductivity of resulted membrane was 1.3 times higher than casted membrane [52]. Choi and coworkers also electrospun sulfonated poly(arylene ether sulfone) and sulfonated POSS nanoparticles in which pores of the final mat had been filled with a crosslinkable polyurethane. The proton conductivity of the membrane was 2.4 times higher than that of Nafion although the water absorption of the membrane was much higher than Nafion as well [53]. Likewise, there are other reports highlighting the use of poly(arylene ether) fibers for fuel cell applications that will be discussed later [54–58].

2.1.5 Poly(imide)s

Kawakami and coworkers considered poly(imide) fibers for use in PEMFCs. They tried five- and six-membered ring poly(imide)s for this purpose. Because of the poor solubility of the six-membered ring poly(imide)s in organic solvents, five-membered ring poly(imide)s had been preferred. They reported that membranes composed of sulfonated poly(imide) fibers embedded in sulfonated poly (imide) matrix exhibited much higher proton conductivity in comparison to cast sulfonated poly(imide). They prepared uniaxially aligned sulfonated poly(imide) nanofibers which dramatically showed enhanced in-plane proton conductivity [59–61].

2.1.6 Poly(benzimidazole) (PBI)

Being thermally stable polymer has been very attractive for using as proton exchange membrane in fuel cell applications. PBI is a promising alternative of Nafion at high temperatures. Electrospun nanofibers of PBI have also been prepared. Li et al. electrospun polybenzoxazine modified PBI, then crosslinked nanofibers by the ring opening addition reaction between benzoxazine groups and imidazole rings. As addition of small molecules to the PBI solution reduces the viscosity and affects electrospinning condition, benzoxazine-containing polymer is used as crosslinking agent. The crosslinked mat impregnated with PBI and performed fuel cell membrane showed proton conductivity of 0.17 S cm^{-1} at 160 °C that was two times higher than solvent cast PBI membranes [62].

2.2 *Crosslinking of Electrospun Fibers*

When electrospun fibers are used as supporting porous structure and also even for impregnation with other polymeric solutions, they must have mechanically stable structure. Attaching nanofibers to each other by chemical bonding or crosslinking

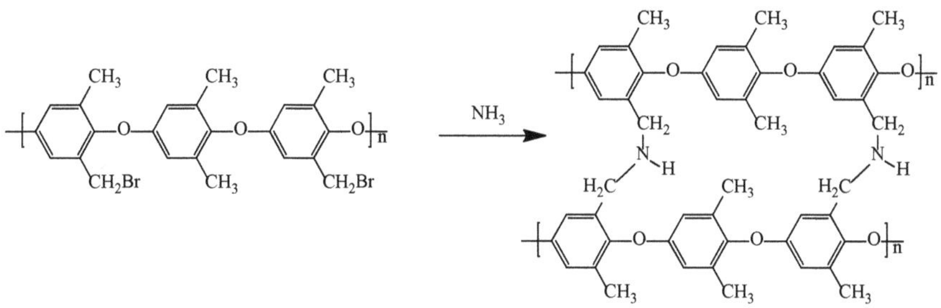

Fig. 5 Crosslinking of BPPO with ammonia

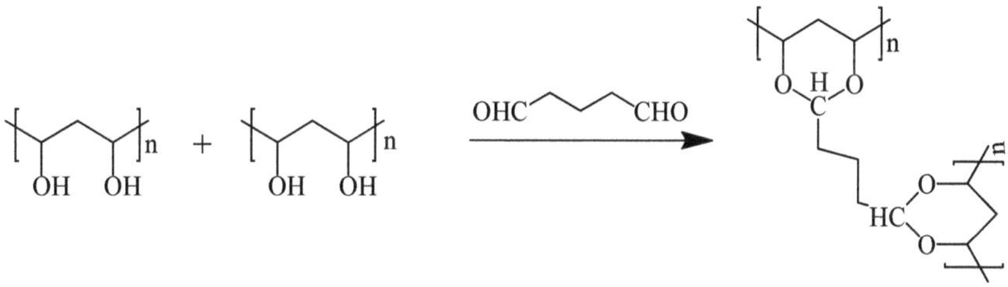

Fig. 6 Crosslinking of PVA with glutaraldehyde

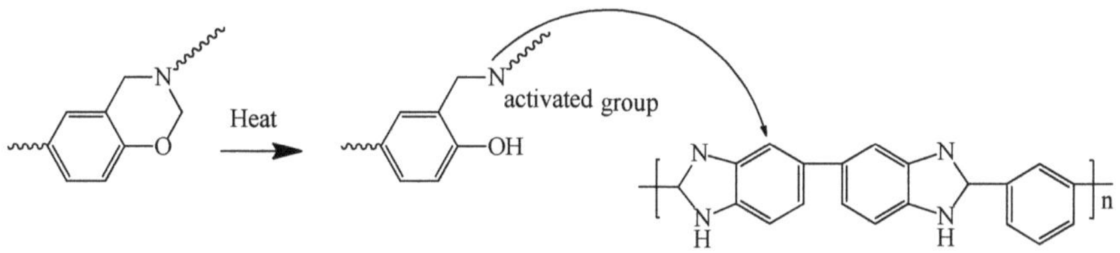

Fig. 7 Schematic of benzoxazine attack to the PBI

makes them more mechanically stable. For instance, for preparation of PPO electrospun nanofibers, bromomethylated PPO (BPPO) solution was electrospun and crosslinked after immersion of the mat into the ammonia solution. (Fig. 5) [49].

Crosslinking of PVA nanofibers by treating with glutaraldehyde vapor before impregnation with Nafion (Fig. 6) has been also reported [47, 48]. PBI fibers with the aim of improving mechanical strength and also increasing resistance of fibers have been crosslinked as well. For this purpose, the blend of polybenzoxazine and PBI is electrospun. The crosslinking reaction occurs between benzoxazine and imidazole rings (Fig. 7) [62].

2.3 Interface Bonding

In almost all publications with the aim of reducing methanol permeability of the membranes in DMFCs, Nafion has been used as a matrix and fibers of different polymers surround with Nafion resin; accordingly, the bonding strength between fibers and Nafion matrix is a challenging topic.

There are several methods to improve compatibility of Nafion with aromatic polymeric fibers for increasing the bonding strength. Therefore, development of chemical linkage between Nafion and polymeric fibers has been considered. Some examples are as follows.

Molla and Compan in their work sulfonated the external surface of electrospun poly(vinyl alcohol) in order to improve compatibility of Nafion and fibers as well as increasing the proton conductivity of the final membrane. 4-Formyl-1, 3-benzenedisulfonic acid disodium salt was used for sulfonation of the PVA fibers (Fig. 8) [47].

Introduction of poly(vinylidenefluoroide) (PVDF) fibers into the Nafion matrix has been studied. Surface-initiated atom transfer radical polymerization (ATRP) on PVDF electrospun fibers used for grafting poly(styrene sulfonic acid) chains on PVDF that improved compatibility of Nafion and fibers significantly. Proton conductivity of the prepared composites was 2.5 times more than Nafion membrane and their methanol permeability was lower than Nafion [44]. In another work, PVDF nanofibers functionalized with Nafion chains were found to improve interfacial compatibility between nanofibers and Nafion matrix. The reaction performed in three steps: The first reaction was surface-initiated radical polymerization of glycidylmethacrylate (GMA) on PVDF electrospun fibers, and the second was the addition reaction between maleimidobenzoic acid (MBA) and epoxide groups of grafted PGMA chains. The maleimide groups anchored on the surfaces of the modified PVDF nanofibers served as the active sites to react with the radicals-containing Nafion chains. Finally, reaction between Nafion and product of the second step resulted in the Nafion-functionalized PVDF nanofibers [45].

In another study conducted by Wang and Lin, blend of PVDF and PBI was electrospun and then impregnated with Nafion. Nevertheless, interaction between basic imidazole groups of PBI with sulfonic acid groups of Nafion improved

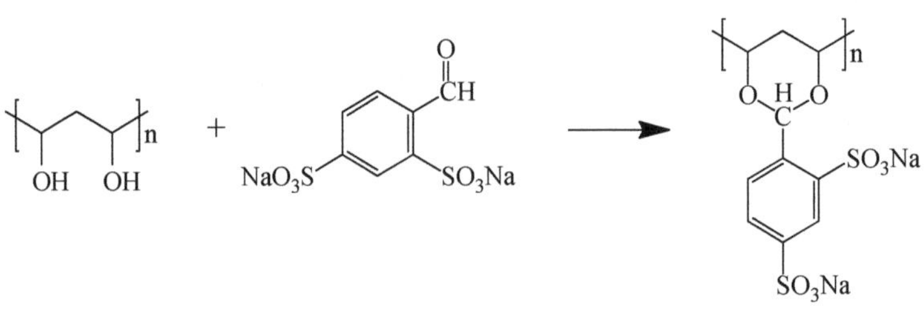

Fig. 8 Sulfonic acid functionalization of PVA fibers

interfacial bonding. Although the presence of PVDF–PBI nanofibers decreased the methanol permeability of the membrane, interaction of SO_3H groups of Nafion with imidazole rings decreased the proton conductivity. With the optimum amount of 5–10 wt% PBI, the performance of the DMFC did not undergo significant decrease [43].

3 Reducing Methanol Crossover

Direct methanol fuel cells are proton exchange type membrane fuel cells that feed with aqueous solution of methanol instead of hydrogen as a fuel. The most important problem with DMFCs is diffusion of methanol through the Nafion membrane. Methanol can crossover the membrane either by diffusion or electroosmotic drag via hydrophilic domains. Concentration gradient between anode and cathode and also interaction with ion exchange sites cause methanol crossover. Diffused methanol reacts with oxygen at the cathode and undergoes non-electrochemical oxidation.

Some simple but not completely effective suggestions have been tried previously for solving the problem: using low concentration methanol solution for feeding the DMFC to decrease the concentration gradient between cathode and anode. In this way, the methanol crossover and current density decrease simultaneously [39, 63]. Another common method is of using thicker Nafion membranes (commercial available Nafion 117) to prevent methanol crossover to some extent. However, thicker membrane increases the resistance against proton transport leading to higher ohmic resistance and lower performance of the DMFC. Thicker Nafion membranes cause higher cost of the final DMFC as well.

Nevertheless, there are many reports about modification of Nafion membrane by other polymers or particles. Blending of Nafion with low methanol compatible polymers such as poly(vinyl alcohol) [64], poly(ether ether ketone) [65], poly (benzimidazole) [66], polyaniline [67], poly(tetrafluoroethylene) [68] has been found useful. However, blending of low methanol permeable polymers with Nafion also reduces the proton conductivity of the final membrane hence there is an optimum concentration of modifier for blending to obtain highest possible performance [46, 48].

Preparing composites of Nafion and inorganic particles is another strategy in which inorganic particles act as physical barrier against methanol crossover. Hygroscopic metal oxide particles like SiO_2, TiO_2, and ZrO_2 enhance both water retention and thermal stability of the membrane [69–71]. Besides, there are reports of adding ionic functionalized inorganic fillers that show multifunction of enhancement of proton conductivity, water retention, and reduction of methanol permeability [47]. However, none of the mentioned solutions are so effective and therefore search for new methods to overcome the problem is continuing. Using electrospinning process for preparation of low methanol permeable membranes is one attractive way. Composite membranes prepared by impregnating the polymer nanofiber thin film, as a supporting material, with Nafion resin. Resulted thin membranes show very low methanol permeability, which reduces the cost and improves the performance significantly. A great deal of conducted research activities in application of electrospinning for fuel cells is with the aim of suppressing methanol permeability of Nafion membrane. In these membranes different sulfonated or non-sulfonated polymer nanofibers were used to reinforce Nafion. Almost all of them reveal successful decreasing methanol permeability without remarkable decrease in proton conductivity. In fact in most cases, due to decreasing the methanol crossover, in spite of reducing proton conductivity, the performance of the cell does not affect significantly [30, 42, 43, 46–48].

4 Improving Proton Conductivity

In spite of decreasing methanol crossover of the membrane, improving the proton conductivity using electrospun fibers has been another purpose of researchers. Accordingly, unlike previous section in which sulfonated or non-sulfonated fibers can be used for suppressing methanol crossover, in this part almost in all of reports just sulfonated polymers are electrospun and embed in sulfonated or non-sulfonated matrices.

It has been shown that proton conductivity of polyelectrolyte nanofibers is much higher than the bulk polymer. For example, the proton conductivity of Nafion nanofibers is 1.5 S cm^{-1} which is very higher than proton conductivity of the bulk Nafion membrane (0.1 S cm^{-1}) [45]. This behavior is due to the configuration of functional groups (SO_3H) in electrospun nanofibers. It has been proved that, in polyelectrolyte polymers separated hydrophilic and hydrophobic regions exist in different sizes, depending on the nature of the polymer [24]. Electrospinning of such polymers produces fibers in which hydrophilic and hydrophobic regions migrate to the inside and outside of the fiber (Fig. 9). Therefore, a proton conducting channel due to the orientation of sulfonic acid groups forms that leads to the more rapid proton transfer and consequently higher proton conductivity of the final membrane [72].

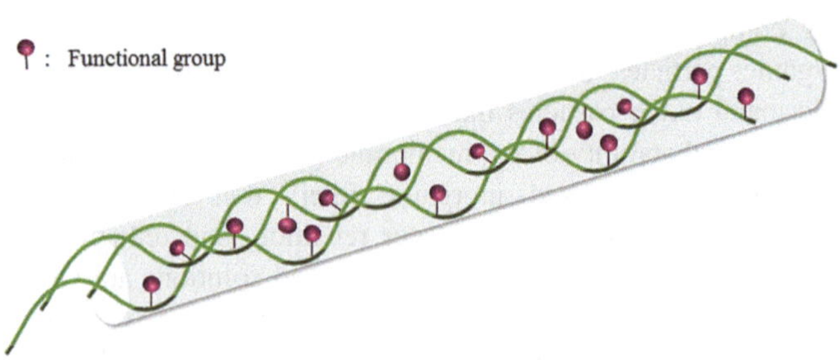

Fig. 9 Orientation of functional groups during electrospinning of polyelectrolytes

Due to the difficulty of fabrication of Nafion nanofibers, other sulfonated aromatic polymers were utilized for electrospinning process. Poly(ether sulfone)s [53, 54], poly(ether ketone)s [56–58], poly(imide)s [59–61], and poly(benzimidazole)s [62] were used to increase proton conductivity of non-perfluorosulfonic acid membranes.

To fill voids and spaces between electrospun nanofibers for obtaining gas impermeable membranes, fibers should be embedded into a matrix. Three ways have been reported for this purpose

- Immersion in a polymer solution, as mentioned in Sect. 3, the technique was used for impregnation of nanofibers with Nafion resin (Fig. 10). Other than Nafion, several functionalized or nonfunctionalized polymers were used as matrices like sulfonated polyimide [59–61], sulfonated poly(ether ether ketone) [52], and poly(benzimidazole) [62].
- Simultaneous electrospinning of two polymeric solutions and melting of one. In this method, dual electrospinning process is used. As represented in Fig. 11, the final electrospun mat containing fibers of both polymers heats under pressure, considering the melting point of two polymers. Accordingly, the fibers want to be the matrix, melt and flow between fibers of another polymer [54–58].
- Simultaneous electrospinning of two polymeric solutions and putting the mat in the presence of a solvent vapor that just solves one type of fibers (Fig. 12). Before placing the mat in suitable solvent vapor, the mat should be pressed to ensure physical contact between all fibers [54, 55].
- Embedding the sulfonated nanofibers in an inert matrix. Pintauro's group in their early works used this technique. They covered the fibers with an UV-crosslinkable polyurethane resin (Norland Optical Adhesive) and then crosslinked it to obtain dense and void free membrane [53, 73].

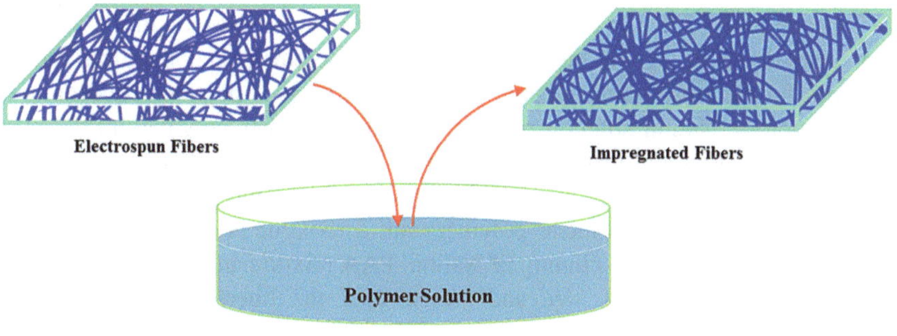

Fig. 10 Impregnation of electrospun fibers with polymer solution

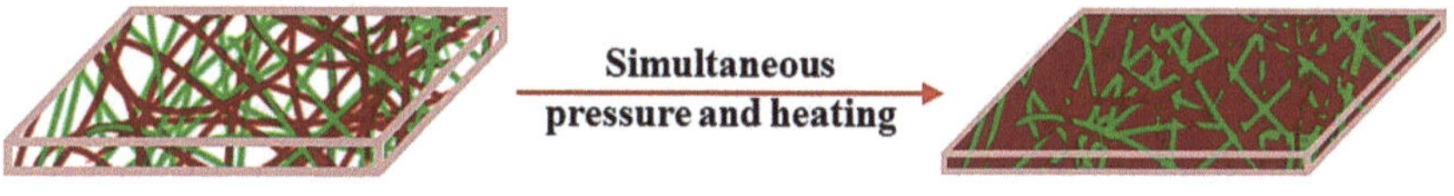

Fig. 11 Performing membrane from simultaneously electrospun fibers via hot-pressing

Fig. 12 Preparing membrane from simultaneously electrospun fibers with suitable solvent vapor

Fig. 13 Carrier polymers for electrospinning of Nafion

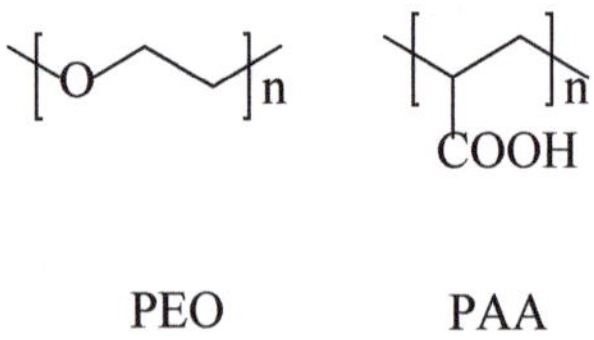

4.1 Electrospinning of Nafion

As previously mentioned, fabrication of nanofibers from Nafion via electrospinning is difficult. Nafion forms a micellar dispersion and not a real polymer solution in typical solvents. Thus, pure Nafion is not suitable for electrospinning due to the low viscosity and lack of sufficient chain entanglement. It has been reported that the viscosity of Nafion solution is similar to the viscosity of pure solvent and increasing the concentration of Nafion causes no significant change in viscosity of the solution. Low viscosity is a sign of negligible chain entanglement and therefore causes unsuccessful electrospinning of pure Nafion regardless of polymer concentration [74]. Therefore, Nafion solution must be mixed with other polymers to decrease aggregate formation and increase viscosity for proper electrospinning.

Applied carrier polymers are poly(acrylic acid) (PAA) and poly(ethylene oxide) (PEO) (Fig. 13). PAA is a linear polymer with carboxylic acid groups. Chen and coworkers reported electrospinning of Nafion–PAA mixture and solution properties including the morphology, size and structure of the fibers investigated. They claimed that continuous fibers of PAA were obtained from 5 wt% PAA in 3/1 isopropyl alcohol/water solvent. This PAA solution was blended with 5 wt% commercial Nafion solution and then electrospun. Smooth and defect-free fibers obtained were at PAA content of >12 wt%. Below this ratio, in spite of stable polymer jet, the nozzle blocked occasionally. The improved spinnability of Nafion in the presence of PAA was the result of increased viscosity and therefore increased chain entanglement [75]. In another study Pintauro group showed that by optimization of electrospinning condition, the PAA concentration can be dropped to

5 wt%. After impregnation of the Nafion–PAA nanofibers with the matrix polymer solution, PAA was removed by boiling the membrane in H_2SO_4 solution and deionized water subsequently [25].

Another carrier polymer in electrospinning of Nafion is poly(ethylene oxide) (PEO). Pintauro's group used PEO as a carrier for Nafion electrospinning at the concentrations of 1–0.3 wt% [55, 73]. Dong et al. showed that with high molecular weight PEO (molecular weight: 8000 kg/mol), Nafion could be electrospun by only 0.1 wt% of carrier polymer and high purity Nafion nanofibers could be obtained. With lower molecular weights of PEO, this result could not be attained [74]. Poly (vinyl alcohol) (PVA) also applies as a carrier for Nafion solution. It has been shown that 2 wt% PVA was needed to obtain defect and bead-free fibers [76].

4.2 *Aligned Nanofibers*

Investigations showed that uniaxially aligned nanofibers dramatically improve proton conductivity of membranes. Determination of proton conductivity parallel to the direction of fibers in aligned sulfonated polyimide fibers in sulfonated polyimide matrix shows that proton conductivity is 3–10 times higher than perpendicular direction [59]. Another work presented that the proton conductivity of such membranes in parallel direction to the aligned fibers increase with decreasing the nanofiber diameter and can be reached to 0.3 S/cm at 90 °C and 98% RH [61]. In spite of excellent proton conductivity of aligned nanofibers, there is no report on preparation of membranes with nanofibers array in through-plane direction. Actual proton conductivity in a fuel cell performs in through-plane direction although in experiments for simplicity, in-plane proton conductivity was reported.

5 Other Applications of Electrospinning in Fuel Cells

Although most of published work is on application of electrospinning in fuel cells is related to reducing methanol crossover and improving proton conductivity of the membrane, there are some other different reports as well.

It was shown that using electrospun fibers in a membrane causes suppressing water swelling under high ion exchange capacity (IEC). For this purpose, electrospun nanofiber mat acts as a framework for a matrix with high IEC value that is soluble or completely swollen in water [25, 49].

Another interesting application of electrospun fibers in fuel cells has been reported by Thiam et al. [77]. They tried not to prevent methanol crossover but also the oxidation of passed methanol. Pd–SiO_2 nanofibers prepared by electrospinning process, poly(vinyl pyrrolidone) (PVP) were used as carrier polymer and Pd–SiO_2 nanofibers were dispersed in Nafion polymer matrix during recasting of Nafion membranes. Inorganic fibers and Nafion matrix interacted through hydrogen bonding between Si–OH and SO_3H groups. Methanol molecules did oxidation reaction on Pd-nanoparticles attached to the silica fibers of this barrier layer and did not reach to the cathode.

6 Conclusion

Nanocomposite membranes consisting of electrospun nanofibers have been used as proton exchange membrane in PEMFCs/DMFCs. The main purpose was of reducing methanol crossover and also increasing proton conductivity of the membrane. There are also other reasons including decreasing water absorption and lowering the cost of the fuel cell. As mentioned, results show that using low methanol permeable polymers as nanofibers in structure of proton exchange membrane successfully decreases methanol permeability of the membrane and consequently increases the performance of the cell in DMFCs. In PEMCs, incorporation of sulfonated electrospun fibers improves the proton conductivity of the membrane significantly. However, the idea of using nanoscale fibers (functionalized or nonfunctionalized) in proton exchange membranes seems to be effective and attractive way that needs more attention from fuel cell researchers.

References

1. Hamrock SJ, Yandrasits MA (2006) Proton exchange membranes for fuel cell applications. J Macromol Sci, Part C: Polym Rev 46:219–244
2. Roziere J, Jones DJ (2003) Non-fluorinated polymer materials for proton exchange membrane fuel cells. Annu Rev Mater Res 33:503–555
3. Devanathan R (2008) Recent developments in proton exchange membranes for fuel cells. Energy Environ Sci 1:101–119
4. Hickner MA, Ghassemi H, Kim YS, Einsla BR, McGrath JE (2004) Alternative polymer systems for proton exchange membranes (PEMs). Chem Rev-Columbus 104:4587–4612
5. Zhang H, Shen PK (2012) Recent development of polymer electrolyte membranes for fuel cells. Chem Rev 112:2780–2832
6. Akbarian-Feizi L, Mehdipour-Ataei S, Yeganeh H (2010) Survey of sulfonated polyimide membrane as a good candidate for nafion substitution in fuel cell. Int J Hydrogen Energy 35:9385–9397
7. Kins CF, Sengupta E, Kaltbeitzel A, Wagner M, Lieberwirth I, Spiess HW et al (2014) Morphological anisotropy and proton conduction in multiblock copolyimide electrolyte membranes. Macromolecules 47:2645–2658
8. Chae K-J, Kim K-Y, Choi M-J, Yang E, Kim IS, Ren X et al (2014) Sulfonated polyether ether ketone (SPEEK)-based composite proton exchange membrane reinforced with nanofibers for microbial electrolysis cells. Chem Eng J 254:393–398
9. Gao Q, Wang Y, Xu L, Wei G, Wang Z (2009) Proton-exchange sulfonated poly(ether ether ketone) (SPEEK)/SiOx-S composite membranes in direct methanol fuel cells. Chin J Chem Eng 17:207–213

10. Oroujzadeh M, Mehdipour-Ataei S, Esfandeh M (2013) Preparation and properties of novel sulfonated poly(arylene ether ketone) random copolymers for polymer electrolyte membrane fuel cells. Eur Polym J. 49:1673–1681
11. Jung MS, Kim T-H, Yoon YJ, Kang CG, Yu DM, Lee JY et al (2014) Sulfonated poly (arylene sulfone) multiblock copolymers for proton exchange membrane fuel cells. J Membr Sci 459:72–85
12. Weiber EA, Takamuku S, Jannasch P (2013) Highly proton conducting electrolyte membranes based on poly(arylene sulfone)s with tetrasulfonated segments. Macromolecules 46:3476–3485
13. Oroujzadeh M, Mehdipour-Ataei S, Esfandeh M (2015) New proton exchange membranes based on sulfonated poly(arylene ether sulfone) copolymers: effect of chain structure on methanol crossover. Int J Polym Mater Polym Biomater 64:279–286
14. Aili D, Jankova K, Li Q, Bjerrum NJ, Jensen JO (2015) The stability of poly (2,2′-(m-phenylene)-5,5′-bibenzimidazole) membranes in aqueous potassium hydroxide. J Membr Sci 492:422–429
15. Aili D, Jensen JO, Li Q (2016) Polybenzimidazole membranes by post acid doping. In: High temperature polymer electrolyte membrane fuel cells. Springer, Berlin, pp 195–215
16. Cattaneo AS, Villa DC, Angioni S, Ferrara C, Quartarone E, Mustarelli P (2015) Structure and interactions in polybenzimidazole composite membranes for high-temperature polymer fuel cells: a full multinuclear solid-state NMR study. J Phys Chem C. 119:18935–18944
17. Kang K, Kim D (2015) Comparison of proton conducting polymer electrolyte membranes prepared from multi-block and random copolymers based on poly(arylene ether ketone). J Power Sour 281:146–157
18. Chen S, Chen K, Zhang X, Hara R, Endo N, Higa M et al (2013) Poly(sulfonated phenylene)-block-poly(arylene ether sulfone) copolymer for polymer electrolyte fuel cell application. Polymer 54:236–245
19. Shin DW, Lee SY, Lee CH, Lee K-S, Park CH, McGrath JE et al (2013) Sulfonated poly (arylene sulfide sulfone nitrile) multiblock copolymers with ordered morphology for proton exchange membranes. Macromolecules 46:7797–7804
20. K-I Liu, Lee H-C, Wang B-Y, Lue SJ, Lu C-Y, Tsai L-D et al (2015) Sulfonated poly (styrene-block-(ethylene-ran-butylene)-block-styrene (SSEBS)-zirconium phosphate (ZrP) composite membranes for direct methanol fuel cells. J Membr Sci 495:110–120
21. Matsumoto K, Higashihara T, Ueda M (2009) Locally and densely sulfonated poly(ether sulfone)s as proton exchange membrane. Macromolecules 42:1161–1166
22. Matsumoto K, Higashihara T, Ueda M (2009) Locally sulfonated poly(ether sulfone)s with highly sulfonated units as proton exchange membrane. J Polym Sci, Part A: Polym Chem 47:3444–3453
23. Higashihara T, Matsumoto K, Ueda M (2009) Sulfonated aromatic hydrocarbon polymers as proton exchange membranes for fuel cells. Polymer 50:5341–5357
24. Yang Y, Siu A, Peckham T, Holdcroft S (2008) Structural and morphological features of acid-bearing polymers for PEM fuel cells. Fuel Cells I 215:55–126
25. Choi J, Wycisk R, Zhang W, Pintauro PN, Lee KM, Mather PT (2010) High conductivity perfluorosulfonic acid nanofiber composite fuel-cell membranes. Chemsuschem 3:1245–1248
26. Formhals A (1934) US Patent 1975504.
27. Doshi J, Reneker DH (1993) Electrospinning process and applications of electrospun fibers. Industry Applications Society Annual Meeting, 1993, Conference Record of the 1993 IEEE: IEEE; 1993. pp 1698–703
28. Reneker DH, Chun I (1996) Nanometre diameter fibres of polymer, produced by electrospinning. Nanotechnology. 7:216
29. Ramakrishna S, Fujihara K, Teo W-E, Yong T, Ma Z, Ramaseshan R (2006) Electrospun nanofibers: solving global issues. Mater Today 9:40–50
30. Dong Z, Kennedy SJ, Wu Y (2011) Electrospinning materials for energy-related applications and devices. J Power Sources 196:4886–4904

31. Huang Z-M, Zhang Y-Z, Kotaki M, Ramakrishna S (2003) A review on polymer nanofibers by electrospinning and their applications in nanocomposites. Compos Sci Technol. 63:2223–2253
32. Hohman MM, Shin M, Rutledge G, Brenner MP (2001) Electrospinning and electrically forced jets. I. Stability theory. Phys Fluids (1994-present) 13:2201–2220
33. Fridrikh SV, Jian HY, Brenner MP, Rutledge GC (2003) Controlling the fiber diameter during electrospinning. Phys Rev Lett 90:144502
34. Subbiah T, Bhat G, Tock R, Parameswaran S, Ramkumar S (2005) Electrospinning of nanofibers. J Appl Polym Sci 96:557–569
35. Agarwal S, Jiang S (2015) Nanofibers and electrospinning. Encycl Polym Nanomater 1323–1337
36. Beachley V, Wen X (2009) Effect of electrospinning parameters on the nanofiber diameter and length. Mater Sci Eng, C 29:663–668
37. Ramakrishna S, Fujihara K, Teo W-E, Lim T-C, Ma Z (2005) Electrospinning process. An introduction to electrospinning and nanofibers. World Scientific Publishing, Singapore, pp 135–137
38. Yördem O, Papila M, Menceloğlu YZ (2008) Effects of electrospinning parameters on polyacrylonitrile nanofiber diameter: an investigation by response surface methodology. Mater Des 29:34–44
39. Chen P, Wu H, Yuan T, Zou Z, Zhang H, Zheng J et al (2014) Electronspun nanofiber network anode for a passive direct methanol fuel cell. J Power Sour 255:70–75
40. Bajon R, Balaji S, Guo S (2009) Electrospun Nafion nanofiber for proton exchange membrane fuel cell application. J Fuel Cell Sci Technol 6:031004
41. Thavasi V, Singh G, Ramakrishna S (2008) Electrospun nanofibers in energy and environmental applications. Energy Environ Sci 1:205–221
42. Choi S, Fu Y-Z, Ahn Y, Jo S, Manthiram A (2008) Nafion-impregnated electrospun polyvinylidene fluoride composite membranes for direct methanol fuel cells. J Power Sour 180:167–171
43. Wang S-H, Lin H-L (2014) Poly(vinylidene fluoride-co-hexafluoropropylene)/polybenzimidazole blend nanofiber supported nafion membranes for direct methanol fuel cells. J Power Sour 257:254–263
44. Li H-Y, Lee Y-Y, Lai J-Y, Liu Y-L (2014) Composite membranes of Nafion and poly(styrene sulfonic acid)-grafted poly(vinylidene fluoride) electrospun nanofiber mats for fuel cells. J Membr Sci 466:238–245
45. Li H-Y, Liu Y-L (2014) Nafion-functionalized electrospun poly(vinylidene fluoride) (PVDF) nanofibers for high performance proton exchange membranes in fuel cells. J Mater Chem A. 2:3783–3793
46. Lin H-L, Wang S-H (2014) Nafion/poly(vinyl alcohol) nano-fiber composite and nafion/poly (vinyl alcohol) blend membranes for direct methanol fuel cells. J Membr Sci 452:253–262
47. Mollá S, Compañ V (2011) Polyvinyl alcohol nanofiber reinforced nafion membranes for fuel cell applications. J Membr Sci 372:191–200
48. Lin H-L, Wang S-H, Chiu C-K, Yu TL, Chen L-C, Huang C-C et al (2010) Preparation of Nafion/poly(vinyl alcohol) electro-spun fiber composite membranes for direct methanol fuel cells. J Membr Sci 365:114–122
49. Yun S-H, Woo J-J, Seo S-J, Wu L, Wu D, Xu T et al (2011) Sulfonated poly(2, 6-dimethyl-1, 4-phenylene oxide) (SPPO) electrolyte membranes reinforced by electrospun nanofiber porous substrates for fuel cells. J Membr Sci 367:296–305
50. Lee C, Jo SM, Choi J, Baek K-Y, Truong YB, Kyratzis IL et al (2013) SiO_2/sulfonated poly ether ether ketone (SPEEK) composite nanofiber mat supported proton exchange membranes for fuel cells. J Mater Sci 48:3665–3671
51. Shabani I, Hasani-Sadrabadi MM, Haddadi-Asl V, Soleimani M (2011) Nanofiber-based polyelectrolytes as novel membranes for fuel cell applications. J Membr Sci 368:233–240

52. Zhang S, He G, Gong X, Zhu X, Wu X, Sun X et al (2015) Electrospun nanofiber enhanced sulfonated poly(phthalazinone ether sulfone ketone) composite proton exchange membranes. J Membr Sci 493:58–65
53. Choi J, Lee KM, Wycisk R, Pintauro PN, Mather PT (2010) Sulfonated polysulfone/POSS nanofiber composite membranes for PEM fuel cells. J Electrochem Soc 157:B914–B919
54. Ballengee JB, Pintauro PN (2013) Preparation of nanofiber composite proton-exchange membranes from dual fiber electrospun mats. J Membr Sci 442:187–195
55. Ballengee JB, Pintauro PN (2011) Composite fuel cell membranes from dual-nanofiber electrospun mats. Macromolecules 44:7307–7314
56. Oroujzadeh M, Mehdipour-Ataei S (2016) Anisotropic membranes from electrospun mats of sulfonated/nonsulfonated poly(ether ketone)s containing ion-rich paths as proton exchange membranes. Int J Polym Mate and Polym Biomate 65:330–336
57. Oroujzadeh M, Mehdipour-Ataei S, Esfandeh M (2015) Proton exchange membranes with microphase separated structure from dual electrospun poly(ether ketone) mats: producing ionic paths in a hydrophobic matrix. Chem Eng J 269:212–220
58. Oroujzadeh M, Mehdipour-Ataei S, Esfandeh M (2015) Microphase separated sepiolite-based nanocomposite blends of fully sulfonated poly(ether ketone)/non-sulfonated poly(ether sulfone) as proton exchange membranes from dual electrospun mats. RSC Adv 5:72075–72083
59. Tamura T, Kawakami H (2010) Aligned electrospun nanofiber composite membranes for fuel cell electrolytes. Nano Lett 10:1324–1328
60. Takemori R, Kawakami H (2010) Electrospun nanofibrous blend membranes for fuel cell electrolytes. J Power Sour 195:5957–5961
61. Tamura T, Takemori R, Kawakami H (2012) Proton conductive properties of composite membranes containing uniaxially aligned ultrafine electrospun polyimide nanofiber. J Power Sour 217:135–141
62. Li H-Y, Liu Y-L (2013) Polyelectrolyte composite membranes of polybenzimidazole and crosslinked polybenzimidazole-polybenzoxazine electrospun nanofibers for proton exchange membrane fuel cells. J Mater Chem A 1:1171–1178
63. Junoh H, Jaafar J, Norddin MNAM, Ismail AF, Othman MHD, Rahman MA et al (2015) A review on the fabrication of electrospun polymer electrolyte membrane for direct methanol fuel cell. J Nanomater 2015:4
64. Shao Z-G, Hsing I-M (2002) Nafion membrane coated with sulfonated poly(vinyl alcohol)-Nafion film for direct methanol fuel cells. Electrochem Solid-State Lett 5:A185–A187
65. Tsai J-C, Cheng H-P, Kuo J-F, Huang Y-H, Chen C-Y (2009) Blended Nafion®/SPEEK direct methanol fuel cell membranes for reduced methanol permeability. J Power Sour 189:958–965
66. Ainla A, Brandell D (2007) Nafion®-polybenzimidazole (PBI) composite membranes for DMFC applications. Solid State Ionics 178:581–585
67. Wang C-H, Chen C-C, Hsu H-C, Du H-Y, Chen C-P, Hwang J-Y et al (2009) Low methanol-permeable polyaniline/Nafion composite membrane for direct methanol fuel cells. J Power Sour 190:279–284
68. Chen L-C, Yu TL, Lin H-L, Yeh S-H (2008) Nafion/PTFE and zirconium phosphate modified Nafion/PTFE composite membranes for direct methanol fuel cells. J Membr Sci 307:10–20
69. Shao Z-G, Xu H, Li M, Hsing I-M (2006) Hybrid Nafion–inorganic oxides membrane doped with heteropolyacids for high temperature operation of proton exchange membrane fuel cell. Solid State Ionics 177:779–785
70. Jung UH, Park KT, Park EH, Kim SH (2006) Improvement of low-humidity performance of PEMFC by addition of hydrophilic SiO_2 particles to catalyst layer. J Power Sour 159:529–532
71. Chalkova E, Fedkin MV, Komarneni S, Lvov SN (2007) Nafion/zirconium phosphate composite membranes for PEMFC operating at up to 120 °C and down to 13% RH. J Electrochem Soc 154:B288–B295
72. Wu L, Zhang Z, Ran J, Zhou D, Li C, Xu T (2013) Advances in proton-exchange membranes for fuel cells: an overview on proton conductive channels (PCCs). PCCP. 15:4870–4887

73. Choi J, Lee KM, Wycisk R, Pintauro PN, Mather PT (2010) Nanofiber composite membranes with low equivalent weight perfluorosulfonic acid polymers. J Mater Chem 20:6282–6290
74. Dong B, Gwee L, Salas-de La Cruz D, Winey KI, Elabd YA (2010) Super proton conductive high-purity nafion nanofibers. Nano Lett 10:3785–3790
75. Chen H, Snyder JD, Elabd YA (2008) Electrospinning and solution properties of Nafion and poly(acrylic acid). Macromolecules 41:128–135
76. Laforgue A, Robitaille L, Mokrini A, Ajji A (2007) Fabrication and characterization of ionic conducting nanofibers. Macromol Mater Eng 292:1229–1236
77. Thiam H, Daud WRW, Kamarudin SK, Mohamad AB, Kadhum AAH, Loh KS et al (2013) Nafion/Pd–SiO_2 nanofiber composite membranes for direct methanol fuel cell applications. Int J Hydrogen Energy 38:9474–9483

Chapter 14
Fabrication Techniques for the Polymer Electrolyte Membranes for Fuel Cells

Aniruddha Chatterjee and D.P. Hansora

Abstract In this chapter, we summarize the various fabrication techniques for the development of polymer electrolyte membrane fuel cells (PEMFC). The purpose is to present a brief overview about different fabrication techniques such as plasma method, phase inversion method, sol–gel method, direct copolymerization, ultrasonic coating technique, ultraviolet polymerization, phase inversion method, in situ reduction, decal transfer method, and catalyst-coated membrane method. After reviewing the above all methods, cost-effective and simple fabrication technique would be useful in the synthesis of PEM depending on the applicability in FC.

Keywords Fabrication · Polymer electrolyte membrane · Fuel cells · Nanocomposites

1 Introduction

Polymer electrolyte membrane (PEM) fuel cells are energy devices, which can convert the chemical energy stored in hydrogen fuel, directly and efficiently, to the electrical energy with water as the only byproduct. These fuel cells are capable to reduce our energy consumptions, pollutant emissions, and dependence on fossil fuels. During the last couple of decades, great covenant of efforts has been made to advance the PEM fuel cell (PEMFC) technology and their fundamental research. Factors, such as durability and cost, are still the major barriers for commercialization of the FCs. Various fabrication methods for the development of PEMFC are

A. Chatterjee (✉)
Department of Plastic and Polymer Engineering, Marathwada Institute of Technology, Aurangabad 431010, Maharashtra, India
e-mail: aniruddha_chatterjee2006@yahoo.co.in; aniruddha.chatterjee@mit.asia

D.P. Hansora
University Institute of Chemical Technology, North Maharashtra University, Jalgaon 425001, Maharashtra, India
e-mail: 2568dharmesh@gmail.com

Inamuddin et al. (eds.), *Organic–Inorganic Composite Polymer Electrolyte Membranes*, DOI 10.1007/978-3-319-52739-0_14

summarized here for identifying a proper, cost-effective, and simple fabrication technique that would be useful while selecting the materials depending on the applicability of FCs. PEMFCs, made from sulfonated polystyrene-divinyl benzene copolymer, were first deployed in the Gemini space program during early 1960s. But, at that time, PEMFCs were extremely expensive and had had short lifetimes due to the oxidative degradation. DuPont commercialized Nafion by chloralkali processes and also demonstrated its potential interest in FC application. PEMFCs are generally found in three main applications: automotive, stationary, and portable power generation equipment. Later on, direct methanol fuel cells (DMFCs) came into the market because of their longer operating lifetimes and the ability of refueling versus the recharging the batteries [1–5].

Polymeric PEMs are the most widely used membranes for FC applications. Proton-conductive groups containing sulfonic acids are generally introduced into the main side chains of the polymer which can give thermal and chemical stability. Most newly developed polymeric PEMs are capable to work at an elevated temperature (>100 °C). Many polymeric PEMs can be operated at low relative humidity (<40%). But high cost associated with their synthetic processes is the biggest setback.

Different organic–inorganic hybrid nanocomposites have been prepared for PEMFC applications, which include organic materials with tin phosphate, silica, cerium phosphate, heteropoly acid, polyantimonic acid, titanium phosphate, titanium dioxide, zirconia and sulfated zirconia, zirconium phosphate, palladium, montmorillonite, zeolites, tungsten dioxide, phosphorous hydride-silicon dioxide, anhydrous phosphoric acid, inorganic/membrane impregnated with ionic liquids and heterocyclic solvents. On the other hand, polymer composites such as sulfonated poly(aryl ether ketone)s containing the hexafluoroisopropylidene diphenyl moiety, sulfonated poly(ether ether ketone)/polyaniline blends, cross-linked polyvinyl alcohol-silicon dioxide hybrids containing sulfonic acid groups, polyethelene oxide–potassium bicarbonate composites, nanosilica-polyethoxysiloxane within a sulfonated poly(ether ether ketone) matrix, titanium NP self-assembled aromatic polyamide thin films have also been used for PEM in FC application.

2 Recent Developments of PEM-Based on Organic–Inorganic Nanocomposites

Organic–inorganic nanocomposites based PEM have attracted a lot of attention during the last decade. Disparate characteristics of the two components can provide potential barriers for strategical preparation of convenient membrane, but simple processes were also reported for developing the highly efficient nanocomposite-based PEMs [1–4]. Wanga et al. [5] reviewed the technology, applications, and needs of fundamental research on PEMFCs [5]. PEMFCs based on ionomers were also developed [6]. Similarly, polyvinyl alcohol (PVA)/SiO_2 hybrid based PEMs, containing sulfonic acid groups, were also developed for direct methanol FC applications [7]. Organic–inorganic material based membranes were also prepared using polyether diamine and epoxy silane based composites [8]. Costamagn et al. [9] discussed about the fundamental and scientific aspects, including science and technology, related to PEMFC from the 1960s to the year 2000. Similarly, non-precious metal catalysts and self-supported electrocatalysts such as metal, metal alloys are another important supporting part of FC membranes [10–16]. PEMs are required for efficient FC performance due to their characteristics properties such as ionic conductivity, fuel permeation, mechanical stability, thermal stability, durability, and low fabrication cost [17]. Support materials for PEMFC and DMFC electrocatalysts [18], phosphotungstic acid doped catalysts [19], and microstructural pore analysis of the catalyst layer [20] have been also reviewed. Catalyst-coated membranes, prepared by screen printing method, were used in a PEMFC [21], while PEMFC catalysts can be electrodeposited by the use of a hydrogen depolarized anode [22]. Thin films of tungsten selenides were developed for electrocatalytic hydrogen evolution reaction using FC [23]. Various polymers and their nanocomposites are presently in use for industrial FC applications [17]. PEMFC composite comprising of triblock copolymer and heteropolyacid can also be used for FC applications [24]. Zaidi et al. [25] reported about research trends in PEMFCs. Various materials such as thin films of tungsten selenides [23] and phosphide-based materials [26] were used for electrocatalytic hydrogen evolution using FC. Organic–inorganic hybrid materials were also reported as PEMs for separation processes [27]. Carmo et al. [28] reviewed water electrolysis by PEMs. There are some alternatives toward proton-conductive anhydrous membranes for FC applications, such as heterocyclic protogenic solvents comprising polymer electrolytes. Anhydrous PEMs have almost no dependence on humidity, and they are able to maintain high proton conductivity at elevated temperature. Azole-based water-free polymer systems enable the formation of protonic defects and provide strongly labile proton donor–acceptor functions [29, 30]. Anion exchange membranes are also important for alkaline FCs. Merle et al. [31] reviewed and summarized various PEMs for alkaline FCs. These PEMs are made of heterogeneous and homogenous materials, IPN polymer based materials, and ion solvating polymer based materials. Degradation mechanisms and mitigation strategies of PEMFC durability were reviewed [32, 33], similarly, test protocols, [34] stack testing [35] and bipolar plates [36] related to PEMFC were also reviewed. PEMs for DMFCs have been also studied by Neburchilov et al. [37]. Polymer nanocomposites (e.g., PEO + $KHCO_3$) were used for PEMFC applications [38]. Approaches and recent developments of PEMFC operating above 100 °C were also studied [39, 40]. Silica nanocomposite based PEMFC systems, made of polyethoxysiloxane within a sulfonated PEEK matrices prepared by using the sol gel process, were also studied [41]. Various polymer systems were studied for the applications of proton exchange membranes [42, 43]. Various characteristics, of PEMFC such as electrical properties [38], surface morphology, proton mobility [41], effective transport properties with a focus on the gas diffusion layer [44], interfacial properties [45], and bactericidal anti-fouling [46], have been examined. Various models related to PEMFC were also studied [47, 48]. Sulfonated polymers such as polyaniline [3] poly(aryl ether ketone)s containing the hexafluoroisopropylidene diphenyl [49] moiety and factors affecting [50] the life of PEMFC were also investigated for PEMFC applications. Novel polyvinyl butyl based PEMs and their application in gel polymer electrolytes for lithium-ion batteries were reported [51]. Various polymeric nanocomposites based membranes were studied by several workers [52–58] for PEMFC applications. Most of proton exchange membranes [59, 60] were designed indicating their unit operation modeling [61–64], computational study [65], efficiency enhancement and cost reduction [66], self-passivating carbon film as bipolar plate based protective coating [67, 68], cooling flow designs [69], numerical study investigation of the effects of GDL compression and intrusion [70], PEM degradation modeling [71], improving gas diffusivity with bi-porous flow-field [72], and the properties effect on gas diffusivity [73] for PEMFC applications. Effects of idling temperature on high-temperature PEMFC degradation have been studied under simulated start/stop cycling conditions [74]. Hybrid inorganic–organic nanocomposites, made from polymeric nanocomposite based electrolytes, such as PVA/SiO_2 [7], sulfonated PEEK/silica nanocomposites [46], TiO_2 nanoparticle self-assembled aromatic polyamide [46], Nafion and fluorinated TiO_2 for PEMFCs [75], Nafion/inorganic nanocomposites [76], graphene supported platinum nanoparticles (NPs) [77], sulfonated PEEK/sulfonated NPs composite [78], UV polymerized 1-H-3-vinylimidazolium bis(trifluoromethanesulfonyl)imide [79], Nafion, [(ZrO_2)(HfO_2)0.25] and [(SiO_2)(HfO_2)0.28] NPs [80, 81], Nafion and fluorinated TiO_2 NPs [82], novel polymer-coated nanoparticles dispersed-carbon micro-nanofibers-based air-cathode [83], functionalized Al_2O_3 particles as additives in proton-conducting polymer [84], poly (ether-imide) [85], polytetrafluoroethylene (PTFE) [86] and high performance polymer based [87], metal-supported tubular solid oxide [88], SiO_2-ceramic nanoporous substrate-reinforced sulfonated poly (arylene ether sulfone) composite [89], poly(2,5-benzimidazole) (ABPBI)-based MEA assembly [90], poly(vinyl alcohol) (PVA) [91–93], partly fluorinated poly (arylene ether ketone sulfone) hydrophilicehydrophobic multiblock copolymers [94], multilayer-structured, SiO_2/sulfonated poly(phenylsulfone) composite [95], sulfonated bisphenol-A-polysulfone based composite PEMs containing tungstophosphoric acid [96], aligned polyaniline nanorods in situ grown on gas diffusion layer [97] were used for PEMFC applications. Feasibility of combining electrochemical impedance spectroscopy and synchrotron X-ray radiography was used to determine the influence of liquid water on PEMFC performance [98]. Detailed analysis of integrated steam ethanol reformer and different high temperature effects on PEMFC were studied by George et al. [99, 100]. Hydrogen generation from aluminum hydride for wearable PEMFC was studied [101]. Various selective membranes, made from polymer nanocomposites, were studied for hydrogen purification [102–105]. A galvanostatic analysis technique was developed by Lee et al. [103] as an in situ diagnostic tool for PEMFC single cells and stacks. Several polymer nanocomposites based PEMs, such as sulfonated poly(ether ether ketone) (PEEK)/polyaniline composites [3, 103–105] were also studied.

For efficient FC performance, PEMs are required having characteristics properties such as chemical and electrochemical stability, ionic conductivity, fuel permeation or low permeability, mechanical stability, thermal and hydrolytic stability, high durability, flexibility to operate with different fuels, fabrication and production cost. Criteria for selecting the inorganic NPs include hygroscopic characteristics, ceramic property, surface area, porosity, and pore connectivity [17, 24, 25, 29].

3 Fabrication Techniques for the Preparation of PEM

Organic–inorganic nanocomposite systems can be prepared by different synthesis routes, shimmering the various methods available to introduce each phase. During preparation of organic–inorganic nanocomposites, the organic component can be introduced as: (i) a precursor (monomer or oligomer), (ii) a preformed linear polymer (in molten, solution, or emulsion states), or (iii) a polymer network, physically (e.g., semicrystalline linear polymer) or chemically (e.g., thermosets, elastomers) cross-linked. The inorganic part can be introduced as: (i) a precursor (e.g., TEOS) or (ii) preformed NPs. Organic or inorganic polymerization generally becomes necessary if at least one of the starting moieties is a precursor [1–3]. Significant progress has been made in the current decade for the development of new organic–inorganic hybrid nanocomposite based membranes, depending on fundamental understanding of their preparation techniques, and applications in diversified areas.

Organic–inorganic hybrid nanocomposites can be prepared in different ways of blending the inorganic precursor (monomer and oligomer) and organic linear/network polymer matrix. With a monomer precursor, both polymerization and cross-linking of organic or inorganic segments are necessary. However, blending, sol–gel, and in situ polymerization techniques are important preparative methodologies (Fig. 1). The sol–gel process is used for the preparation of inorganic

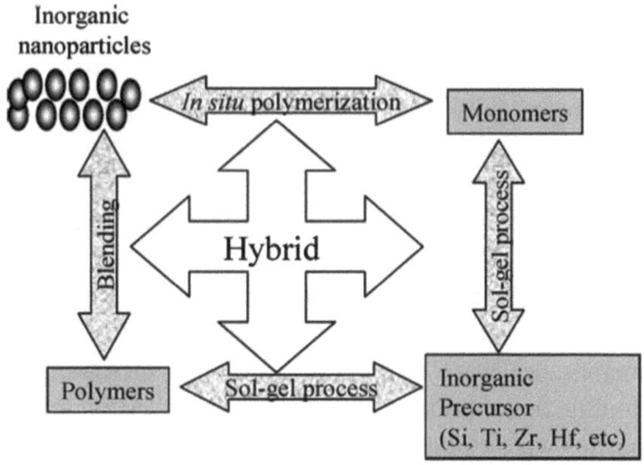

Fig. 1 Schematic representation of different routes for the preparation of organic–inorganic hybrid materials and PEMs (reprinted with permission of Ref. [4])

materials such as glasses and ceramics, and it is based on inorganic polymerization. The low temperature processing of a sol–gel technique provides a unique opportunity for tailoring well-controlled organic–inorganic hybrid materials [4–6]. Organic–inorganic polymer nanocomposites have been reported as an effective PEMs and also the significant improvements have been made on the PEM to empower the FC applications at high temperature and low humidity, for better water adsorption and retention, ionic conductivity, fuel crossover, thermo and mechanical properties, for better FC performance, durability, for ease of fabrication techniques [17, 30]. Lee et al. [56] reported various polymer membranes for high-temperature PEMFC applications. He also reported their synthesis and fabrication procedures for preparation of the membranes such as sulfonated aromatic hydrocarbon membranes, organic–inorganic composite membranes, polymer blend membranes, and PBI-based acid–base membranes [56]. In general the materials, used in synthesis of the PEMs, can be classified into three major groups: (a) perfluorinated ionomers (or partially perfluorinated), (b) nonfluorinated hydrocarbons (including aliphatic or aromatic structures), and (c) acid–base complexes. Mendes et al. [58] reviewed relevant literatures related to segmented PEMs published since 1998 focusing on the SFC design techniques: such as (i) printed circuit board, (ii) resistors network, and (iii) Hall effect sensors. Three techniques were described and their fundamental considerations for design, construction and electrochemical characterization were also provided. The effect of most important parameters on the current density distribution was also highlighted [58]. Rowshanzamir et al. [60] reviewed and summarized three types of membranes along with their preparation and modification methods. He also discussed about new materials, technologies, and research directions being pursued to meet the demanding performance and durability needs of the PEMFC industries. Several methods, which have been used to prepare PEM for FC applications, are described as follows.

3.1 Different Polymerization Routes

Polymerization routes such as in situ polymerization, UV polymerization, chemical polymerizations, and atom transfer radical polymerization (ATRP) are most widely used techniques for the preparation of PEM materials for FC applications. Sulfonated poly(ether ether ketone) (PEEK)/polyaniline (PAni) composite membranes, in order to prevent methanol crossover, were prepared by chemical polymerization of a thin layer of PAni in the presence of a high oxidant concentration on a single face modification. This cost-effective and simple fabrication technique can be used for the synthesis of such composite membranes [3, 4]. Infiltration techniques or doping routes have been used by introducing the inorganic nanoparticles (NPs) or precursors to modify PEM for better transport properties. Preformed membranes were swelled in solvent to increase the pore or void volume before doping or infiltration of inorganic NPs. After infiltration, the polymer nanocomposites were cured by heat, radiation or chemical grafting of inorganic precursor, to obtain covalent bonding inside the polymer matrix. Organic–inorganic nanocomposites can be obtained by infiltration of polymerizable organic monomers with oxide gels or by the mixing of polymers with a single or mixed metal alkoxides. Impregnation of porous oxide gels with organics followed by thermal or irradiation in situ polymerization has resulted in stable PEM for FC applications. Polymers can be trapped within the oxide gel network if the hydrolysis and condensation of metal alkoxide are carried out in the presence of preformed polymers [4, 6]. For the ideal ionic membrane, hierarchically ordered ionic channels/particles should orient vertical to the membrane surface. This can be achieved only via surface-initiated ATRP. Ordered two-dimensional macroporous silicon renders proton-conducting membrane by growing a thick uniform polyelectrolyte brush using surface-initiated atom transfer radical polymerization throughout the porous matrix [17, 24]. Hybrid organic/inorganic composite based PEM consisting of a triblock copolymer (tBC) and varying concentrations of heteropolyacid (HPA) were investigated for application in PEMFCs. An ABC triblock copolymer, that is, polystyrene-b-poly (hydroxyethyl acrylate)-b-poly (styrene sulfonic acid), PS-b-PHEA-b-PSSA, at 28:21:51 wt%, was synthesized via atom transfer radical polymerization (ATRP) and solution-blended with a commercial HPA [24]. Neburchilov et al. [37] reviewed various PEMs for DMFC applications, according to one report, chemical membranes were prepared by the copolymerization of tetrafluoroethylene with a vinylene monomer. Nafion-polyfurfural alcohol (PFA) nanocomposites membranes have been synthesized [37] by in situ polymerization with varying different wt% of PFA. Nafion-polypyrrole based membranes had been prepared by two methods: (i) the impregnation of polypyrrole by in situ polymerization and (ii) polymerization in hydrogen peroxide with Fe(III) as the oxidizing agent. A novel asymmetric membrane composed of a three-component acrylic polymer blend (TCPB) was reported to be prepared by polymerization of an acrylic polymer blend consisting of 4-vinylphenol-*co*-methyl methacrylate (P(4-VP-MMA)), poly(butyl methacrylate) (PBMA) and Paraloid B-82 acrylic copolymer resin [37]. Zhang et al. [43] also reviewed and reported about partially fluorinated acid ionomer membranes prepared via radiation-grafting polymerization, as shown in Fig. 2 [43].

And also this radiation-grafting polymerization method involves a cost-competitive option because of the use of inexpensive commercial materials.

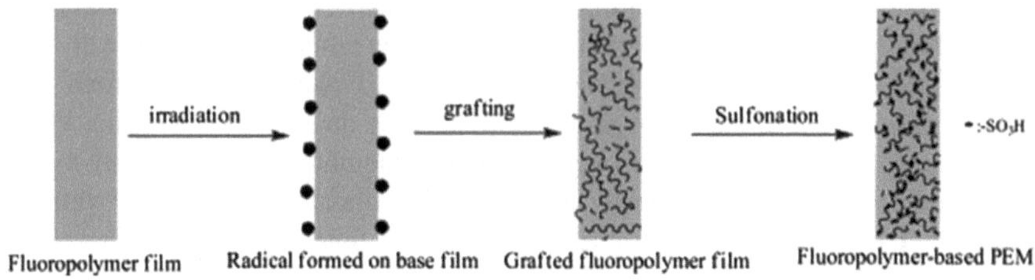

Fig. 2 Process for the preparation of PEMs by the radiation-grafting method (reprinted with permission of Ref. [43])

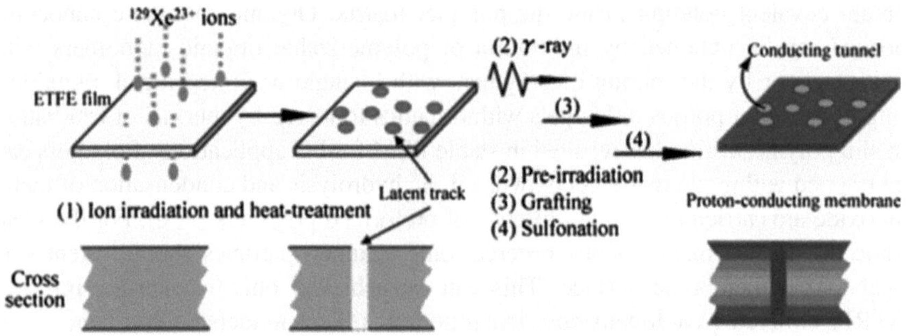

Fig. 3 Schematic diagram for the preparation process of Xe-irradiated ETFE-based proton-conducting membranes (reprinted with permission of Ref. [107])

There are three steps involved during the preparation procedure: (i) the pre-irradiation of the base matrix by ionizing sources, (ii) the grafting of a versatile monomer such as styrene onto the matrix, and (iii) the sulfonation of the grafted membrane (Fig. 2) [43]. The ETFE is widely used as base matrix for radiation-induced grafting PEM because of its high radiation stability and superior mechanical properties. Two modified radiation-induced grafting techniques were used to enhance the performance of ETFE-g-PSSA membranes [43, 106, 107].

In the first initial steps of these techniques, the ETFE film was pretreated by swift heavy ion-beams of Xe^{23+} to create latent tracks extending fully through the ETFE film (Fig. 3) followed by γ-ray pre-irradiation grafting of styrene. The ETFE-g-PSSA membrane exhibits highly anisotropic proton conductivity, with proton conductivity in the thickness direction 3 times greater than that in the surface direction because the styrene is readily grafted in the latent track region [43, 106, 107].

Kim and co-workers [108] explored the single-step synthesis of grafted fluoropolymer films via ATRP. However, the characteristics of their studies depend on the further modified technique to these grafted fluoropolymer films prepared through ATRP. P(VDF-co-CTFE)-g-PSSS films can be synthesized using ATRP, the terminal chlorine atoms on them are converted to end-functional azide groups and sequentially cross-linked by UV irradiation (Fig. 4) [43, 108].

The hybrid membranes composed of inorganic Si–O–Si linkages and organic polymer chains covalently immobilized with phosphonic acid (Fig. 5a) possess good thermal stability up to 200 °C [43, 109]. Saxena and co-workers [110] prepared highly conductive and stable organic–inorganic nanocomposites PEMs with controlled spacing between the inorganic segment and covalently bound sulfonic acid functional groups (Fig. 5b) through a more ingenious method. The resulting PEMs exhibited obvious improvement over the Nafion membrane as evidenced by the higher ratio of proton conductivity to methanol permeability (β-value), which is caused by the lower methanol permeability [43, 110]. Jiang et al. [111] reviewed the work appeared on the use of the sulfonic acid functionalized plasma polymerization techniques for the preparation of PEM for FC applications. These plasma polymerization techniques include glow discharge plasma polymerization, ion beam assisted

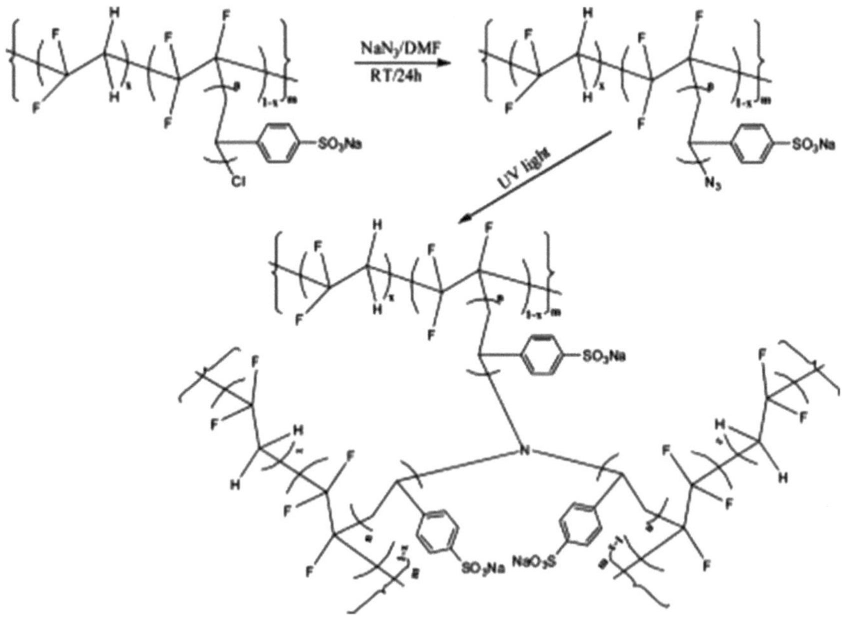

Fig. 4 Schematic procedure for the preparation of cross-linked P(VdF-co-CTFE)-g-PSSA membranes (reprinted with permission of Ref. [108])

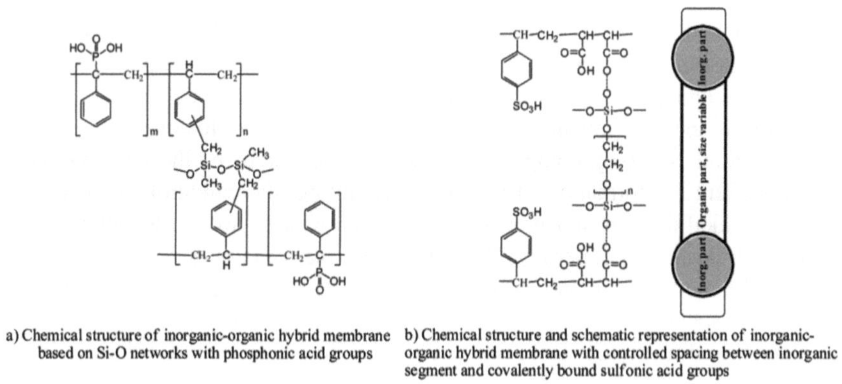

Fig. 5 Structures of two inorganic–organic hybrid membranes (**a**, **b**) (reprinted with permission from Refs. [109, 110])

plasma polymerization, low frequency after-glow capacitive coupled plasma discharge, pulsed plasma discharge polymerization, etc. [111]. The after-glow plasma discharge process used to prepare AEMFCs involved three steps, (i) plasma polymerization of 4-vinylbenzyl chloride (VBC) monomer into poly(vinylbenzyl chloride); (ii) quaternization of benzyl chloride groups into $–N^+(CH_3)_3Cl^-$ groups; and

(iii) alkalization of $-N^{+}(CH_3)_3Cl^{-}$ groups into $-N^{+}(CH_3)_3OH^{-}$ groups. The obtained plasma-polymerized AEMFCs exhibited a satisfactory thermal stability, chemical stability, and ionic exchange capacity [111]. Sulfonated poly(aryl ether ketone)s containing the hexafluoroisopropylidene diphenyl moiety were prepared by direct copolymerization using aromatic nucleophilic polycondensation of hexafluoroisopropylidene diphenol with 5,5′-carbonylbis(2-fluorobenzenesulfonate) and 4,4′-difluorobenzophenone (DFBP) at various molar ratios [49]. Pina et al. [79] represented the enhancement of durability of polymeric ionic liquid (PIL) membranes, prepared by ultraviolet (UV) radiation-induced polymerization, for high-temperature PEMFC applications. In particular, the copolymerization of 1-H-3-vinylimidazolium bis(trifluoromethanesulfonyl)imide [HVIm][TFSI] with divinylbenzene used as crosslinker (CL) or prepared by the "in situ" UV polymerization on a preexisting randomly porous polybenzimidazole (PBI) matrix as novel preparation methods, has been fully studied and compared as endurance strategies [79].

3.2 *Plasma Methods*

ETFE-SA based PEM can be prepared by plasma irradiation (with a total dose of 500 kGy and dose rate of 2.7 kGy s^{-1}) followed by sulfonation without grafts of other polymers. These PEMs were semicrystalline and the cross-linking effects of the crystallites result in the low swelling of the sulfonated PEMs [37]. Jiang et al. [111] reviewed in his paper about applications of the plasma technique for the preparation of PEM for FC applications. He reported that the PEM modified or directly synthesized by the plasma technique exhibited superior properties, such as higher ion conductivity, low fuel permeability, high thermal and chemical stability, which made them promising PEM for FC applications. However, the plasma polymerization is a very complicated process which involves the degradation of monomers and the formation of polymers. Therefore, the conditions used for the membrane modification and preparation must be well controlled to obtain membranes with desirable properties. The various plasma techniques, which have been used for the modification and for the preparation of PEM, are reviewed and along with their associated advantages and disadvantages. These include plasma reactors for surface modification, plasma depositing a polymer layer on the surface of PEM, modification of PEM by gas plasma, and plasma grafting techniques [111].

Won et al. [112] developed a selective thin layer on the top of polystyrene-block-poly(ethyleneran-butylene)-block-polystyrene (sSEBS) copolymer PEMs by simple plasma treatment in the presence of maleic anhydride (MA). The hydrophobic anhydride properties on the top of sSEBS membrane acted as a barrier for methanol permeability. Both the methanol permeability and the proton conductivity of the membrane, therefore, were observed to be decreasing gradually with increasing loading amounts of MA, due to the lack of proton exchange groups on the deposited layer. The recovery of the proton conductivity of the membrane was achieved

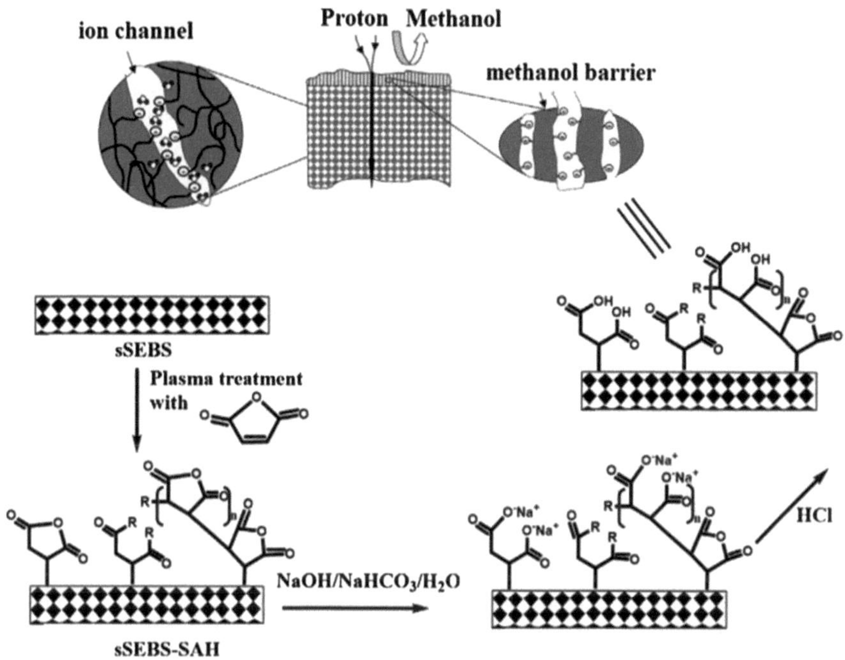

Fig. 6 Modification of the sSEBS membrane through the plasma treatment in the presence of MA and subsequent hydrolysis of anhydride groups to form the PEMs with high conductivity (reprinted with permission from Ref. [112])

through the hydrolysis of anhydride groups, which produced carboxylic acid groups, providing a facilitated transport site for the proton, as shown in Fig. 6.

3.3 Sol–Gel Method

The sol–gel method has been extensively used for the preparation of organic–inorganic nanocomposites as PEM in FC applications and also this method is environmentally friendly because of being low temperature process. Several comprehensive reviews are available on the sol–gel approach to prepare organic–inorganic hybrid materials. Sol–gel method includes mixing of organic polymer solution with organometallic precursors such as tetramethoxysilane (TMOS), tetraethyl orthosilicate (TEOS), Zr(IV)-propoxide, Ti(IV)- butoxide, etc., by hydrolysis and condensation in the presence of suitable catalysts, acid, or base [4–7, 17]. Synthetic techniques used in the sol–gel process to develop polymer/silica hybrid materials include in situ formation of an inorganic network in the presence of an organic polymer and the simultaneous formation of organic polymer and MO_2, forming polymer networks [4, 41]. The organic–inorganic nanocomposites made by sol–gel are known as "creamers/ormosils/ormocers" [4]. According to literature [4, 37], organic–inorganic nanocomposites prepared by sol–gel for PEM, include perfluorinated organic–inorganic nanocomposites (Nafion with modifiers such as Nafion-silica NPs, Nafion-ZrP, Nafion, Nafion-titanium dioxide, Nafion-thiophene, Nafion-metal oxides, Nafion-zeolite, Nafion-silica-PWA, Nafion-PAni nanocomposites), partially perfluorinated organic–inorganic nanocomposites (sulfonated styrene with PTFE backbone, functionalized PVDF-silica NPs), and styrene-based nanocomposites (PSMA-silica hybrid material prepared using styrene–maleic anhydride copolymer in the presence of TEOS as a coupling agent, TEOS grafted polystyrene-blocks in polystyrene-b-isobutylene-bstyrene block copolymers, TEOS grafted PSMA-PEG composites, sulfonated PTFE and grafted with TEOS). It has been reported [4] that hybrid materials can also be prepared by nonhydrolytic sol–gel method and because this method involves the reaction of a "metal" halide with an oxygen donor, such as an alkoxide, ether, alcohol, under nonaqueous conditions to form an inorganic oxide and alkyl halide byproducts. One important consideration in this process is the difference in reactivity of metals, such as silicon and transition metals. It was also reported [4] that conversions of polyethoxysilane into SPEEK-silica nanocomposite SPEEK-silica, SPEEK-BPO_4 and PEEK-TiO_2 nanocomposites based membranes, were prepared by sol–gel method. Similarly, polyamideimide/silica composite membrane compatibilized by sol–gel and thermal imidization was reported [4–6, 37, 41]. Organic–inorganic hybrids based on poly (vinyl alcohol) (PVA)/SiO_2 hybrid membranes containing sulfonic acid groups were prepared using the sol–gel process under acidic conditions. The PVA/sulfosuccinic acid (SSA)/silica hybrid membranes were fabricated from different SSA contents [7]. In this sol–gel method, aqueous 10 wt% PVA solutions were prepared by dissolving dry PVA in water and then allowed to reflux at 90 °C for 6 h followed by the addition of a TEOS mixture to the solution. The TEOS mixture was prepared by mixing H_2O, HCl, and TEOS in the molar ratio of 4:0.1:1, which was stirred at room temperature for 2 h. The 5 wt% of TEOS was added to the polymer weight. Then, the PVA/TEOS solutions were mixed together and then stirred at room temperature for 12 h. The solutions were poured onto a plexiglass plate, and casted by using the doctor blade process. The cast solutions were dried in the presence of air at room temperature. The fully dried membranes were peeled away from the glass plate, and then heated in a thermosetting oven at 120 °C for 1 h to induce the cross-linking reaction [7, 37].

Tang et al. [106] reported that a Nafion/SiO_2 composite membrane based on self-assembled Nafion–SiO_2 NPs (Fig. 7) can be formed through a reformative sol–gel process. The self-assembled Nafion–SiO_2 NPs are uniformly distributed in the Nafion/SiO_2 composite membrane and have an average particle size of 2.8 ± 0.5 nm. With the addition of 5 wt% SiO_2 NPs, the Nafion/SiO_2 composite membrane showed significantly improved performance in terms of stability and durability at cell/humidifying temperature of 60 °C [42, 43, 106].

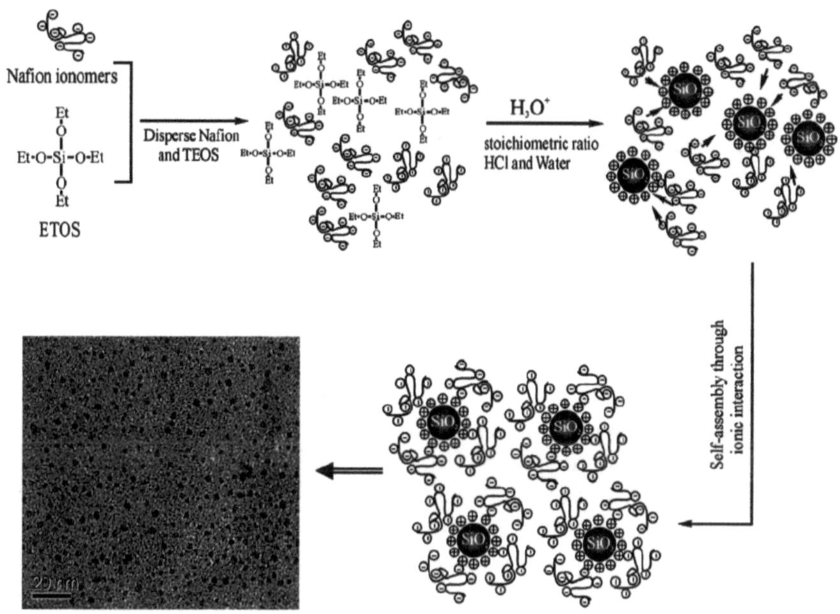

Fig. 7 Scheme for the formation of the Nafion–SiO_2 nanoparticles by self-assembly method. The inset shows the transmission electron microscopy (TEM) micrograph of the self-assembled Nafion–SiO_2 nanoparticles (reprinted with permission from Ref. [106])

3.4 *Ultrasonic Coating Technique*

Membrane electrode assemblies with Nafion/nanosize titanium silicon dioxide ($TiSiO_4$) composite membranes were developed by a novel ultrasonic spray technique and tested in PEMFC. Nafion/TiO_2 and Nafion/SiO_2 nanocomposite membranes were also fabricated by these technique and their characteristics and performances in PEMFC were compared with Nafion/$TiSiO_4$ mixed oxide membrane [76]. Nam et al. [78] also used ultrasound-assisted solution casting method for the preparation of sulfonated PEEK/sulfonated NPs composite for the FC application.

A modified catalyst-coated membrane (CCM), containing with phosphotungstic acid (PTA) in catalyst layer, can be fabricated using a heated ultrasonic spray method, reported by Wang et al. [19].

3.5 *Phase Inversion Method*

Partially perfluorinated organic–inorganic nanocomposites (PVDF-HFP-Nafion and PVDF-HFP-Al_2O_3) were prepared by a phase inversion method [4]. The polyvinyl acetal based polymer membranes were prepared by a phase inversion method using synthetic polyvinyl butyral (PVB) from polyvinyl alcohol (PVA) as the polymer matrix. The proportion of hydroxyl group in PVB was controlled by acid modification, and then appropriate polyurethane reaction occured between the residual hydroxyl groups of PVB and the 4,40-diphenylmethane diisocyanate (MDI) as cross-linking agent. As a result, three-dimensional network structures were obtained, which contributed to the chemical stability of PVB-based membrane in LiFP6-based liquid electrolyte [51].

3.6 In Situ Reduction

In situ impregnation process is a versatile technique for the preparation of low cost proton-conducting nanocomposite materials. The homogeneous membranes obtained at lower temperature following this technique influence their potential applications. The membranes were impregnated in the inorganic material's precursor solution and the process begins with its incorporation. The nucleophilic attack of the water, present in the membrane on the inorganic atoms, leads to the hydrolysis of the infiltrated inorganic precursor and as a result membranes have to be treated with necessary reactants for the condensation process

3.7 Catalyst-Coated Membrane by Screen Printing Method

Wang et al. reported [21] a new PEMFC catalyst paste based on direct screen printing (Fig. 8) method and developed a membrane electrode assembly (MEA) by catalyst-coated membrane (CCM) technique. A mixture of cyclohexanol and ethylene glycol was found to be the best solvent for the screen printed catalyst paste in terms of avoiding the swelling of K^+ form Nafion® membrane and mud-cracking in the catalyst layer during the CCM-MEA fabrication process and improving adhesion between the Nafion membrane and catalyst layers was also observed [21].

3.8 Solution Casting Method

Organic–inorganic polymer membranes were prepared by reacting 3-glycidoxypropyl trimethoxysilane with diamines containing polyether segments, followed by their hydrolysis and condensation with acid catalysis. Composite membranes were prepared by solution casting on Teflon petri dishes followed by evaporation of solvent and casting (dip-coating) the solution (just after 2 h heating at 80 °C) [8]. Solution casting is another simple method for the preparation of organic–inorganic hybrid nanocomposites. A few microliters of a dilute solution containing

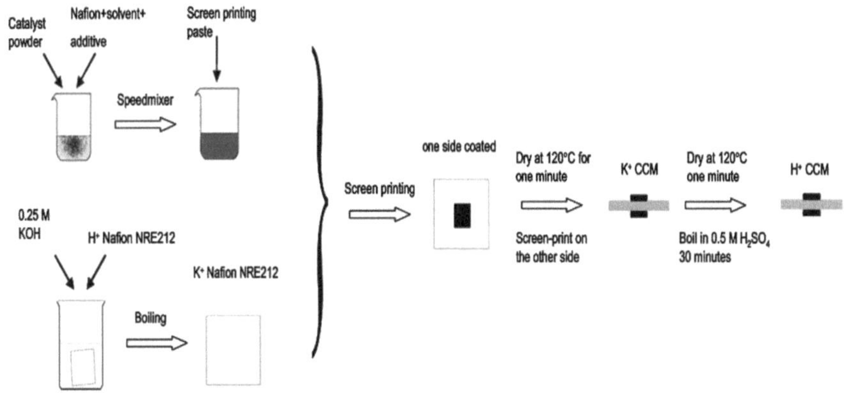

Fig. 8 Schematic of the CCM-MEA preparation procedure using direct screen printing method (reprinted with permission from Ref. [21])

the dissolved polymer and nanofillers can be film casted by mixing with the polymer solution under vigorous stirring followed by the evaporation of a solvent, because this enables direct incorporation of inorganic NPs into a polymer matrix. The thickness of the PEM could be easily controlled [17, 24]. The triblock copolymer composite membranes containing HPA in various weight ratios were prepared using a solution casting method. This method includes washing with water, vacuum drying, solution casting followed by drying [24]. Sulfonated polyphosphazene (sPPZ) membranes were prepared by dissolving benzophenone (a photo-initiator) in the membrane casting solution and then the resulting films, after solvent evaporation, were exposed to UV light. The films of ethane cross-linked polyphosphazene can be sulfonated from the surface inward. This prevents dissolution in water, thus allowing sulfonation to proceed through the material. The sPPZ membrane has the highest thermal and chemical stabilities and low methanol crossover in DMFC operation [37]. Noto et al. [75, 82] reported three hybrid inorganic–organic proton-conducting membranes based on a novel fluorinated titania labeled TiO_2F dispersed in Nafion prepared by solution casting method. Noto et al. [80] also reported the development of the hybrid inorganic–organic membranes, labeled as [Nafion/(ZrHf)x] and [Nafion/(SiHf)x], were prepared by a standard solvent-casting procedure [80].

3.9 *Other Methods*

Self-assembly structures of organic–inorganic nanocomposites are made of discrete organic and inorganic nanoscale components spontaneously organized with non-covalent interactions. These hybrid nanocomposites exhibit interesting characteristics because of constituent phases at nanoscale, high interfacial area, and synergetic properties. Tripathi et al. [4] reported that layer-by-layer self-assembly technique was remarkably adaptable approach to produce hybrid systems. Sulfonated PEEK-silicon dioxide-zirconium phosphate (ZrP) nanocomposites based PEMs can be prepared by electrochemical method in hydrophilic channels and voids of preformed and swelled membranes. In this technique, $ZrOCl_2$ was infiltered out followed by reaction with phosphoric acid or mesoporous ZrP/SiO_2 was blended after the incorporation of inorganic ZrP components. Polyimide (PI) and polybenzimidazole (PBI) based nanocomposites (PBI-phosphoric acid based composites) can be prepared by solution blending, while PBI-silicon dioxide nanocomposites can be prepared by solvent exchange method [4, 37, 42]. It was reported [4] that designing new PEMs can be done by modifications of aliphatic/aromatic polymers using suitable functionalization and introduction of inorganic NPs to obtain specific characteristics like: (i) mechanical strength and dimensional stability of PEM in hydrated state at intermediate temperatures; and (ii) water retention at higher temperature, to maintain the hydration and conductivity. Well-defined morphology and structure of PEMs with phase separation between hydrophobic as well as hydrophilic ionic domains along with stabilities for organic polymer chain containing sulfonic acid groups must be studied to assess their suitability [4]. Covalently cross-linked ionomer membranes via sulfinate alkylation, ionically cross-linked acid–base blend membranes materials as PEM were also reported [6]. An attempt has been made to develop a new polyethylene oxide (PEO) (PEO + $KHCO_3$) based solid PEM, by isothermal immersion technique. Using this conducting PEM, FC stack was fabricated with the configuration anode/polymer electrolyte/cathode [38]. Hybrid organic/inorganic reverse osmosis (RO) membranes, composed of aromatic polyamide thin films underneath titanium dioxide (TiO_2) nanosized particles, have been fabricated by a self-assembly process, aiming at the breakthrough of bio-fouling problems [46].

Consequently, the important problems in developing optimized organic–inorganic nanocomposite based PEMs are: (i) to identify the appropriate inorganic filler and dispersing conditions for selected polymer matrix; (ii) to optimize the concentration of filler best composite PEMs; (iii) to identify the synthesis and casting conditions; (iv) to acquire the knowledge of interactions between the NP, the polymer matrix and/or solvent.

4 Summary

Organic–inorganic hybrid nanocomposites based PEMs can be prepared in different ways by blending inorganic precursor (monomer and oligomer) and organic linear/network polymer matrix. With a monomer precursor, both polymerization and cross-linking of organic or inorganic segments are necessary. Various techniques for fabrication of PEMs are discussed, which include direct polymerization, plasma method, phase inversion method, sol–gel method, direct copolymerization, ultrasonic coating technique, ultraviolet polymerization, phase inversion method and in situ reduction, decal transfer method, and catalyst-coated membrane method. However, blending, sol–gel and in situ polymerization techniques are important preparative methodologies.

Acknowledgements Financial support from Science and Engineering Research Board (SERB), Department of Science and Technology (DST), New Delhi [Fast Track Project File No. SR/FTP/ETA-55/2011, dated November 11, 2012] is gratefully acknowledged for carry out this work.

References

1. Jiao K, Li X (2011) Water transport in polymer electrolyte membrane fuel cells. Prog Energy Combus Sci 37:221–291
2. Eikerling M, Kornyshev AA, Kucernak AR (2006) Water in polymer electrolyte fuel cells: friend or foe? Phys Today 59(10):38–44
3. Nagarale RK, Gohil GS, Shahi VK (2006) Sulfonated poly(ether ether ketone)/polyaniline composite proton-exchange membrane. J Membr Sci 280:389–396
4. Tripathi BP, Shahi VK (2011) Organic–inorganic nanocomposite polymer electrolyte membranes for fuel cell applications. Prog Polym Sci 36:945–979
5. Wanga Y, Chen KS, Mishler J, Cho SC, Adroher XC (2011) A review of polymer electrolyte membrane fuel cells: technology, applications, and needs on fundamental research. Appl Energy 88:981–1007
6. Kerres JA (2001) Development of ionomer membranes for fuel cells. J Membr Sci 185:2–3
7. Kima DS, Park HB, Rhim JW, Lee YM (2004) Preparation and characterization of crosslinked PVA/SiO_2 hybrid membranes containing sulfonic acid groups for direct methanol fuel cell applications. J Membr Sci 240:37–48
8. Sforc ML, Yoshida IVP, Nunes SP (1999) Organic-inorganic membranes prepared from polyether diamine and epoxy silane. J Membr Sci 159:197–207
9. Costamagn P, Srinivasan S (2001) Quantum jumps in the PEMFC science and technology from the 1960s to the year 2000 Part I. Fundamental scientific aspects. J Power Sour 102:242–252
10. Ma TY, Dai S, Qiao SZ (2016) Self-supported electrocatalysts for advanced energy conversion processes. Mater Today 19(5):27–265
11. Gasteiger HA, Kocha SS, Sompalli B, Wagner FT (2005) Activity benchmarks and requirements for Pt, Pt-alloy and non-Pt oxygen reduction catalysts for PEMFCs. App Cataly B: Envir 56:9–35
12. Othman R, Dicks AL, Zhu Z (2012) Non precious metal catalysts for the PEM fuel cell cathode. Int J Hydrogen Energy 37:357–372
13. Kim T, Popov BN (2016) Development of highly-active and stable Pt/C catalyst for polymer electrolyte membrane fuel cells under simulated start-up/shut-down cycling. Int J Hydrogen Energy 41:1828–1836
14. Cho YH, Lim JW, Park HY, Jung N, Ahn M, Choe H, Sung YE (2012) Performance of membrane electrode assemblies using PdPt alloy as anode catalysts in polymer electrolyte membrane fuel cell. Int J Hydrogen Energy 37:5884–5890
15. Lee KS, Kim DM (2012) Sputtering and heat treatment of pure Ni metal onto a carbon nanotube on carbon paper to fabricate electrocatalysts for the oxygen reduction reaction in PEMFC. Int J Hydrogen Energy 37:6272–6276
16. Oh SK, Kim MJ, Eom KS, Kyung JS, Kim DH, Cho EA, Kwon HS (2016) Design of MgeNi alloys for fast hydrogen generation from seawater and their application in polymer electrolyte membrane fuel cells. Int J Hydrogen Energy 41:5296–5303

17. Kumar GG, Nahm KS (2011) Chapter-27: polymer nanocomposites—fuel cell applications. In: Reddy B (ed) Advances in nanocomposites—synthesis, characterization and industrial applications, pp 639–660. InTech China, Available from: http://www.intechopen.com/books/advances-in-nanocompositessynthesis-characterization-and-industrial-applications/polymer-nanocomposites-fuel-cell-applications
18. Sharma S, Pollet BG (2012) Support materials for PEMFC and DMFC electrocatalysts—a review. J Power Sour 208:96–119
19. Chen GY, Wang C, Lei YJ, Zhang J, Mao ZQ, Guo JW, Wang JL (2016) Catalyst layer doped with phosphotungstic acid for degradation mitigation in polymer electrolyte membrane fuel cells. Int J Hydrogen Energy 1–6 (Accepted). http://dx.doi.org/10.1016/j.ijhydene.2016.04.222
20. Ghosh S, Ohashi H, Tabata H, Hashimasa Y, Yamaguchi T (2015) Microstructural pore analysis of the catalyst layer in a polymer electrolyte membrane fuel cell: a combination of resin pore-filling and FIB/SEM. Int J Hydrogen Energy 40:15663–15671
21. Wang W, Chen S, Li J, Wang W (2015) Fabrication of catalyst coated membrane with screen printing method in a proton exchange membrane fuel cell. Int J Hydrogen Energy 40:4649–4658
22. Mitzel J, Arena F, Natter H, Walter T, Batzer M, Stefener M, Hempelmann R (2012) Electrodeposition of PEM fuel cell catalysts by the use of a hydrogen depolarized anode. Int J Hydrogen Energy 37:6261–6267
23. Romanov R, Grigoriev S, Fominski V, Volosova M, Demin M (2015) Preparation and study of thin films of tungsten selenides for electrocatalytic hydrogen evolution. Phys Procedia 71:348–353
24. Choi JK, Lee DK, Kim YW, Min BR, Kim JH (2008) Composite polymer electrolyte membranes comprising triblock copolymer and heteropolyacid for fuel cell applications. J Polym Sci, Part B: Polym Phys 46:691–701
25. Zaidi SMJ (2009) Chapter-2: research trends in polymer electrolyte membranes for PEMFC. In: Zaidi SMJ, Matsuura T (eds) Polymer membranes for fuel cells, pp 7–25
26. Xiao P, Chen W, Wang X (2015) A review of phosphide based materials for electrocatalytic hydrogen evolution. Adv Energy Mater 5(24):1–13
27. Souza VC, Quadri MGN (2013) Organic-inorganic hybrid membranes in separation processes: a 10-year review. Braz J Chem Eng 30(04):683–700
28. Carmo M, Fritz DL, Mergel J, Stolten D (2013) A comprehensive review on PEM water electrolysis. Int J Hydrogen Energy 38:4901–4934
29. Ünügür S, Bozkurta A, Hosseini SS (2012) Alternatives toward proton conductive anhydrous membranes for fuel cells: heterocyclic protogenic solvents comprising polymer electrolytes. Prog Polym Sci 37:1265–1291
30. Wee JH (2007) Applications of proton exchange membrane fuel cell systems. Renew Sustain Energy Rev 11:1720–1738
31. Merle G, Wessling M, Nijmeijer K (2011) Anion exchange membranes for alkaline fuel cells: a review. J Membr Sci 377:1–35
32. Wu J, Yuan XZ, Martin JJ, Wang H, Zhang J, Shen J, Wu S, Merida W (2008) A review of PEM fuel cell durability: degradation mechanisms and mitigation strategies. J Power Sourc 184:104–119
33. Schmittinger W, Vahidi A (2008) A review of the main parameters influencing long-term performance and durability of PEM fuel cells. J Power Sourc 180:1–14
34. Yuan XZ, Li H, Zhang S, Martin J, Wang H (2011) A review of polymer electrolyte membrane fuel cell durability test protocols. J Power Sourc 196:9107–9116
35. Millera M, Bazylaka A (2011) A review of polymer electrolyte membrane fuel cell stack testing. J Power Sourc 196:601–613
36. Hermanna A, Chaudhuria T, Spagnol P (2005) Bipolar plates for PEM fuel cells: a review. Int J Hydrogen Energy 30:1297–1302
37. Neburchilov V, Martin J, Wang H, Zhang J (2007) A review of polymer electrolyte membranes for direct methanol fuel cells. J Power Sourc 169:221–238

38. Kumar KV, Sundari GS, Jyothi NK (2010) Fabrication and electrical characterization of PEM fuel cell based on (PEO + $KHCO_3$) polymer electrolyte membrane. Int J Inn Res Sci Eng Technol 2(10):5838–5847
39. Li Q, He R, Jensen JO, Bjerrum NJ (2003) Approaches and recent development of polymer electrolyte membranes for fuel cells operating above 100 °C. Chem Mater 15:4896–4915
40. Jiang Z, Jiang ZJ (2014) Plasma techniques for the fabrication of polymer electrolyte membranes for fuel cells. J Membr Sci 456:85–106
41. Colicchio I (2008) Silica-based nanocomposite membranes via the sol–gel process of polyethoxysiloxane within a sulfonated poly(ether-ether-ketone) matrix: morphology and proton mobility. Ph.D. theses, Universitäts professor Dr. rer. nat. Martin Möller
42. Hickner MA, Ghassemi H, Kim YS, Einsla BR, McGrath JE (2004) Alternative polymer systems for proton exchange membranes (PEMs). Chem Rev 104:4587–4612
43. Zhang H, Shen PK (2012) Recent development of polymer electrolyte membranes for fuel cells. Chem Rev 112:2780–2832
44. Zamel N, Li X (2013) Effective transport properties for polymer electrolyte membrane fuel cells with a focus on the gas diffusion layer. Prog Energy Comb Sci 39:111–146
45. Cho YH, Yoo SJ, Cho YH, Park HS, Park IS, Lee JK, Sung YE (2008) Enhanced performance and improved interfacial properties of polymer electrolyte membrane fuel cells fabricated using sputter-deposited Pt thin layers. Electrochim Acta 53:6111–6116
46. Kwak SY, Kim SH (2001) Hybrid organic/inorganic reverse osmosis (RO) membrane for bactericidal anti-fouling. 1. Preparation and characterization of TiO_2 nanoparticle self-assembled aromatic polyamide thin-film-composite (TFC) membrane. Environ Sci Technol 35:2388–2394
47. Gottesfeld S, Zawodzinski TA (1997) Polymer electrolyte fuel cells. Adv Electrochem Sci Eng 196–301 (edited by Alkire RC, Gerischer H, Kol DM, Tobias CW)
48. Springer TE, Zawodzinski TA, Gottesfel S (1991) Polymer electrolyte fuel cell model. J Electrochem Soc 138(8):2334–2342
49. Xing P, Robertson GP, Guiver MD, Mikhailenko SD, Kaliaguine S (2004) Sulfonated poly (aryl ether ketone)s containing the hexafluoroisopropylidene diphenyl moiety prepared by direct copolymerization as proton exchange membranes for fuel cell application. Macromol 37:7960–7967
50. Pei P, Chen H (2014) Main factors affecting the lifetime of proton exchange membrane fuel cells in vehicle applications: a review. Appl Energy 125:60–75
51. Lian F, Wen Y, Ren Y, Guan HY (2014) A novel PVB based polymer membrane and its application in gel polymer electrolytes for lithium-ion batteries. J Membr Sci 456:42–48
52. Mishra AK, Bose S, Kuila T, Kim NH, Lee JH (2012) Silicate-based polymer-nanocomposite membranes for polymer electrolyte membrane fuel cells. Prog Polym Sci 37:842–869
53. Litster S, McLean G (2004) PEM fuel cell electrodes. J Power Sourc 130:61–76
54. Prater KB (1994) Polymer electrolyte fuel cells: a review of recent developments. J Power Sourc 51:129–144
55. Couture G, Alaaeddine A, Boschet F, Ameduri B (2011) Polymeric materials as anion-exchange membranes for alkaline fuel cells. Prog Polym Sci 36:1521–1557
56. Bose S, Kuila T, Hien TX, Nguyen Kimc NH, Laua KT, Lee JH (2011) Polymer membranes for high temperature proton exchange membrane fuel cell: recent advances and challenges. Prog Polym Sci 36:813–843
57. Yuan W, Tang Y, Yang X, Wan Z (2012) Porous metal materials for polymer electrolyte membrane fuel cells—a review. Appl Energy 94:309–329
58. Perez LC, Brandaoa L, Sousa JM, Mendes A (2011) Segmented polymer electrolyte membrane fuel cells—a review. Renew Sustain Ene Rev 15:169–185
59. Zhang L, Chae SR, Hendren Z, Park JS, Wiesner MR (2012) Recent advances in proton exchange membranes for fuel cell applications. Chem Eng J 204–206:87–97
60. Peighambardoust SJ, Rowshanzamir S, Amjadi M (2010) Review of the proton exchange membranes for fuel cell applications. Int J Hydrogen Energy 35:9349–9384

61. Siegel C (2008) Review of computational heat and mass transfer modeling in polymer-electrolyte-membrane (PEM) fuel cells. Energy 33:1331–1352
62. Caglayan DG, Sezgin B, Devrim Y, Eroglu I (2016) Three-dimensional modeling of a high temperature polymer electrolyte membrane fuel cell at different operation temperatures. Int J Hydrogen Energy 41:10060–10070
63. Zhang Y, Verma A, Pitchumani R (2016) Optimum design of polymer electrolyte membrane fuel cell with graded porosity gas diffusion layer. Int J Hydrogen Energy 4:8412–8426
64. Fecarotti C, Andrews J, Chen R (2016) A petri net approach for performance modelling of polymer electrolyte membrane fuel cell systems. Int J Hydrogen Energy 1–19 (Accepted). http://dx.doi.org/10.1016/j.ijhydene.2016.05.138
65. Kim DK, Koh JS, Kim MS, Song HH (2015) Experimental and computational study on the dynamic interaction between load variation and back pressure control in a polymer electrolyte membrane fuel cell for automotive application. Int J Hydrogen Energy 40:12370–12381
66. Torkavannejad A, Sadeghifar H, Pourmahmoud N, Ramin F (2015) Novel architectures of polymer electrolyte membrane fuel cells: efficiency enhancement and cost reduction. Int J Hydrogen Energy 40:12466–12477
67. Wang Z, Feng K, Li Z, Lu F, Huang J, Wu Y, Chu PK (2016) Self-passivating carbon film as bipolar plate protective coating in polymer electrolyte membrane fuel cell. Int J Hydrogen Energy 41:5783–5792
68. Feng K, Guo X, Li Z, Yao C, Wu Y (2016) Investigation of multi-coating process treated magnesium alloy as bipolar plate in polymer electrolyte membrane fuel cell. Int J Hydrogen Energy 41:6020–6028
69. Alizadeh E, Rahgoshay SM, Esbo MR, Khorshidian M, Saadat SHM (2016) A novel cooling flow field design for polymer electrolyte membrane fuel cell stack. Int J Hydrogen Energy 41:8525–8532
70. Chippar P, Kyeongmin O, Kang K, Ju H (2012) A numerical investigation of the effects of GDL compression and intrusion in polymer electrolyte fuel cells (PEFCs). Int J Hydrogen Energy 37:6326–6338
71. Jereb LK, Sternig C, Fink C, Tatschl R (2016) Membrane degradation model for 3D-CFD analysis of fuel cell performance as a function of time. Int J Hydrogen Energy 1–13 (Accepted). http://dx.doi.org/10.1016/j.ijhydene.2016.05.229
72. Kozakai M, Date K, Tabe Y, Chikahisa T (2016) Improving gas diffusivity with bi-porous flow-field in polymer electrolyte membrane fuel cells. Int J Hydrogen Energy 1–10 (Accepted). http://dx.doi.org/10.1016/j.ijhydene.2016.05.131
73. Dhanushkodi SR, Capitanio F, Biggs T, Merida W (2015) Understanding flexural, mechanical and physicochemical properties of gas diffusion layers for polymer membrane fuel cell and electrolyzer systems. Int J Hydrogen Energy 40:16846–16859
74. Pinar FJ, Rastedt M, Pilinski N, Wagner P (2016) Effect of idling temperature on high temperature polymer electrolyte membrane fuel cell degradation under simulated start/stop cycling conditions. Int J Hydrogen Energy 1–12 (Accepted). http://dx.doi.org/10.1016/j.ijhydene.2016.05.091
75. Noto VD, Bettiol M, Bassetto F, Boaretto N, Negro E, Lavina S, Bertasi F (2012) Hybrid inorganic-organic nanocomposite polymer electrolytes based on nafion and fluorinated TiO_2 for PEMFCs. Int J Hydrogen Energy 37:6169–6181
76. Devrim Y, Erkan S, Bac N, Eroglu I (2012) Improvement of PEMFC performance with Nafion/inorganic nanocomposite membrane electrode assembly prepared by ultrasonic coating technique. Int J Hydrogen Energy 37:16748–16758
77. Sanlı LI, Bayram V, Yarar B, Ghobadi S, Gursel SA (2016) Development of graphene supported platinum nanoparticles for polymer electrolyte membrane fuel cells: effect of support type and impregnation-reduction methods. Int J Hydrogen Energy 41:3414–3427
78. Kim DJ, Choi DH, Park CH, Nam SY (2016) Characterization of the sulfonated PEEK/sulfonated nanoparticles composite membrane for the fuel cell application. Int J Hydrogen Energy 41:5793–5802

79. Lemus J, Eguizabal A, Pina MP (2016) Endurance strategies for the preparation of high temperature polymer electrolyte membranes by UV polymerization of 1-H-3-vinylimidazolium bis(trifluoromethanesulfonyl)imide for fuel cell applications. Int J Hydrogen Energy 41:3981–3993
80. Noto VD, Boaretto N, Negro E, Giffin GA, Lavina S, Polizzi S (2012) Inorganic-organic membranes based on Nafion, [(ZrO_2)–(HfO_2)0.25] and [(SiO_2)–(HfO_2)0.28]. Part I: synthesis, thermal stability and performance in a single PEMFC. Int J Hydrogen Energy 37:6199–6214
81. Noto VD, Boaretto N, Negro E, Stallworth PE, Lavina S, Giffin GA, Greenbaum SG (2012) Inorganic-organic membranes based on Nafion, [(ZrO_2)–$(HfO_2)_{0.25}$] and [(SiO_2)–$(HfO_2)_{0.28}$] nanoparticles. Part II: relaxations and conductivity mechanism. Int J Hydrogen Energy 37:6215–6227
82. Noto VD, Bettiol M, Bassetto F, Boaretto N, Negro E, Lavina S, Bertasi F (2012) Hybrid inorganic-organic nanocomposite polymer electrolytes based on Nafion and fluorinated TiO_2 for PEMFCs. Int J Hydrogen Energy 37:6169–6181
83. Singh S, Modi A, Verma N (2016) Enhanced power generation using a novel polymer-coated nanoparticles dispersed-carbon micro-nanofibers-based air-cathode in a membrane-less single chamber microbial fuel cell. Int J Hydrogen Energy 41:1237–1247
84. Branchi M, Sgambetterra M, Pettiti I, Panero S, Navarra MA (2015) Functionalized Al_2O_3 particles as additives in proton-conducting polymer electrolyte membranes for fuel cell applications. Int J Hydrogen Energy 40:14757–14767
85. Elangovan M, Dharmalingam S (2016) Preparation and performance evaluation of poly (ether-imide) based anion exchange polymer membrane electrolyte for microbial fuel cell. Int J Hydrogen Energy 41:8595–8605
86. Mack F, Morawietz T, Hiesgen R, Kramer D, Gogel V, Zeis R (2016) Influence of the polytetrafluoroethylene content on the performance of high-temperature polymer electrolyte membrane fuel cell electrodes. Int J Hydrogen Energy 41:7475–7483
87. Wei Z, Su K, Sui S, He A, Du S (2015) High performance polymer electrolyte membrane fuel cells (PEMFCs) with gradient Pt nanowire cathodes prepared by decal transfer method. Int J Hydrogen Energy 40:3068–3074
88. Han Z, Yang Z, Han M (2016) Fabrication of metal-supported tubular solid oxide fuel cell by phase-inversion method and in-situ reduction. Int J Hydrogen Energy 41:10935–10941
89. Seol JH, Wona JH, Yoon KS, Hong YT, Lee SY (2012) SiO_2 Ceramic nanoporous substrate-reinforced sulfonated poly(arylene ether sulfone) composite membranes for proton exchange membrane fuel cells. Int J Hydrogen Energy 37:6189–6198
90. Wanga S, Shang Y, Wanga Y, Wanga J (2013) Fabrication and electrochemical performance of poly(2,5-benzimidazole) (ABPBI)-based MEA by catalyst coated membrane (CCM) method for high-temperature polymer electrolyte fuel cells. Int J Hydrogen Energy 38:11060–11066
91. Zhong S, Cui X, Gao Y, Liu W, Dou S (2014) Fabrication and properties of poly(vinyl alcohol)-based polymer electrolyte membranes for direct methanol fuel cell applications. Int J Hydrogen Energy 39:17857–17864
92. Higa M, Hatemura K, Sugita M, Maesowa SI, Nishimura M, Endo N (2012) Performance of passive direct methanol fuel cell with poly(vinyl alcohol)-based polymer electrolyte membranes. Int J Hydrogen Energy 37:6292–6301
93. Gomes AS, Filho JCD (2012) Hybrid membranes of PVA for direct ethanol fuel cells (DEFCs) applications. Int J Hydrogen Energy 37:6246–6252
94. Chen Y, Guo R, Lee CH, Lee M, McGrath JE (2012) Partly fluorinated poly(arylene ether ketone sulfone) hydrophilicehydrophobic multiblock copolymers for fuel cell membranes. Int J Hydrogen Energy 37:6132–6139
95. Lee JR, Wona JH, Yoon KS, Hong YT, Lee SY (2012) Multilayer-structured, SiO_2/sulfonated poly(phenylsulfone) composite membranes for proton exchange membrane fuel cells. Int J Hydrogen Energy 37:6182–6188

96. Filho AAMF, Gomes AS (2012) Sulfonated bisphenol-A-polysulfone based composite PEMs containing tungstophosphoric acid and modified by electron beam irradiation. Int J Hydrogen Energy 37:6228–6235
97. Fu X, Wang S, Xia Z, Li Y, Jiang L, Sun G (2016) Aligned polyaniline nanorods in situ grown on gas diffusion layer and their application in polymer electrolyte membrane fuel cells. Int J Hydrogen Energy 41:3655–3663
98. Antonacci P, Chevalier S, Lee J, Yip R, Ge N, Bazylak A (2015) Feasibility of combining electrochemical impedance spectroscopy and synchrotron X-ray radiography for determining the influence of liquid water on polymer electrolyte membrane fuel cell performance. Int J Hydrogen Energy 40:16494–16502
99. George D, Suresh PV (2016) Detailed analysis of integrated steam ethanol reformer and high temperature polymer electrolyte membrane fuel cell. Int J Hydrogen Energy 41:1248–1258
100. Lee HJ, Kim BG, Lee DH, Park SJ, Kim Y, Lee JW, Henkensmeier D, Namb SW, Kim HJ, Kim H, Kim JY (2011) Demonstration of a 20 W class high-temperature polymer electrolyte fuel cell stack with novel fabrication of a membrane electrode assembly. Int J Hydrogen Energy 36:5521–5526
101. Thampan T, Atwater T, Cook C, Novoa J, Sutorik AC (2016) Hydrogen generation from aluminum hydride for wearable polymer electrolyte membrane fuel cells. Int J Hydrogen Energy 41:9402–9409
102. Chen HZ, Chung TS (2012) CO_2-selective membranes for hydrogen purification and the effect of carbon monoxide (CO) on its gas separation performance. Int J Hydrogen Energy 37:6001–6011
103. Lee KS, Lee BS, Yoo SJ, Kim SK, Hwang SJ, Kim HJ, Cho EA, Henkensmeier D, Yun JW, Nama SW, Lim TH, Jang JH (2012) Development of a galvanostatic analysis technique as an in-situ diagnostic tool for PEMFC single cells and stacks. Int J Hydrogen Energy 37:5891–5900
104. Seo DW, Lim YD, Lee SH, Jeong IS, Kim DI, Lee JH, Kim WG (2012) Preparation and characterization of sulfonated poly(tetra phenyl ether ketone sulfone)s for proton exchange membrane fuel cell. Int J Hydrogen Energy 37:6140–6147
105. Noto VD, Zawodzinski TA, Herring AM, Giffin GA (2012) Polymer electrolytes for a hydrogen economy. Int J Hydrogen Energy 37:6120–6131
106. Tang H, Wan Z, Pan M, Jiang SP (2007) Self-assembled Nafion–silica nanoparticles for elevated-high temperature polymer electrolyte membrane fuel cells. Electrochem Commun 9:2003
107. Kimura Y, Chen J, Asano M, Maekawa Y, Katakai R, Yoshida M (2007) Anisotropic proton-conducting membranes prepared from swift heavy ion-beam irradiated ETFE films. Nucl Instrum Methods Physc Res Sect B 263:463
108. Kim YW, Choi JK, Park JT, Kim JH (2008) Self-assembled Nafion–silica nanoparticles for elevated-high temperature polymer electrolyte membrane fuel cells. J Membr Sci 313:315
109. Kato M, Sakamoto W, Yogo TJ (2008) Proton-conductive sol–gel membranes from phenylvinylphosphonic acid and organoalkoxysilanes with different functionalities. J Membr Sci 311:182
110. Saxena A, Tripathi BP, Shahi VK (2007) Sulfonated poly(styrene-co-maleic anhydride)-poly (ethylene glycol)-silica nanocomposite polyelectrolyte membranes for fuel cell applications. J Phys Chem B 111:12454
111. Jiang Z, Jiang ZJ (2014) Plasma techniques for the fabrication of polymer electrolyte membranes for fuel cells. J Membr Sci 456:85–106
112. Won J, Choi SW, Kang YS, Ha HY, Oh IH, Kim HS, Kim KT, Jo WH (2003) Structural characterization and surface modification of sulfonated polystyrene–(ethylene–butylene)–styrene triblock proton exchange membranes. J Membr Sci 214:245–257

Chapter 15
Chitosan-Based Polymer Electrolyte Membranes for Fuel Cell Applications

Saiqa Ikram, Shakeel Ahmed, S. Wazed Ali and Himanshu Agarwal

Abstract Fuel cells have been receiving attention due to its potential applicability as a good alternative power source. Recently, cost-effective and eco-friendly biopolymer chitosan has been devised as membrane in electrolytes and electrodes in direct methanol fuel cell, alkaline fuel cell and biofuel cell applications. This paper reviews structure and property of chitosan with respect to its applications in fuel cells. In addition, different synthetic strategies to prepare chitosan-based polymer electrolyte membranes and their properties for the use in fuel cells are critically examined. These strategies include chemical modifications of chitosan, blending with other polymers and their composites for polymer electrolyte membranes among fuel cell applications. Recent achievements and prospect of its applications have also been included.

Keywords Chitosan · Polymer electrolyte · Membrane · Fuel cell · Applications

S. Ikram (✉) · S. Ahmed
Department of Chemistry, Faculty of Natural Science, Jamia Millia Islamia, New Delhi 110025, India
e-mail: sikram@jmi.ac.in

S. Ahmed
e-mail: shakeelchem11@gmail.com

S. Wazed Ali
Department of Textile Technology, Indian Institute of Technology Delhi, New Delhi
110016, India
e-mail: wazed@textile.iitd.ac.in

H. Agarwal
Department of Chemistry, LRPG Collge, Sahibabad, Ghaziabad 201015, India
e-mail: himan_007@rediffmail.com

Inamuddin et al. (eds.), *Organic–Inorganic Composite Polymer Electrolyte Membranes*, DOI 10.1007/978-3-319-52739-0_15

1 Introduction

Fuel cells have received considerable attention over the last two decades. Due to significant potential, different types of fuel cells have been developed [1]. Fuel cells are electrochemical devices, which convert chemical energy into electrical energy by the process which basically involves changing the electrolyte electrode system. Fuel cells are more efficient than conventional techniques of generating electrical power, which require many conversion steps before the production of actual electrical power. Fuel cells can achieve up to 60% efficiency by converting chemical energy directly to usable power and thermal energy [2]. Fuel cells generate electricity through direct electrochemical conversion of reactants, and as such show great promise for lowering the demand of hydrocarbon fuels thereby decreasing pollution and increase in energy conversion efficiency [3].

Operating in the low temperature range, polymer electrolyte fuel cells (PEFC) are particularly suited as a replacement for the combustion engine and batteries in automotive and portable power applications [4]. However, in order to achieve large-scale deployment, further development is required to bring the PEFC to a level where it becomes competitive with existing technologies in terms of cost, durability and performance. The focus of research thus far has been primarily on the reduction or replacement of expensive catalysts, improved durability of components, ameliorating water, heat management and increased overall electrical efficiency [5]. There are many parameters like operating temperature, pressure, adequate mechanical strength and stability, high proton conductivity to support high currents with minimal resistive losses chemical stability under operating conditions, which influence the efficiency of fuel cells [4]. The anion exchange membrane is one of the core components in solid alkaline fuel cell, which improves the performance of fuel cell. There are some requirements, which are needed for the smooth functioning of fuel cell membranes like durability, performance and cost. The desired properties for a successful commercial fuel cell development are directly linked to the performance of the membrane under operational condition of the AFC. The membrane in fuel cell acts as carrier for ion and barrier for gas and electrons. The membranes should have a low fuel crossover, e.g. transfer of hydrogen from the anode to the cathode in order to maximize the current efficiency and zero electronic conductivity. Ionic conductivity and mechanical stability are the important challenge for the preparation of an efficient anion exchange membrane [2].

In the past few years, CS has been extensively investigated as solid polymer electrolyte in low and intermediate temperature polymer electrolyte-based fuel cells, such as hydrogen–polymer electrolyte fuel cell (PEFC), direct methanol fuel cell (DMFC) and alkaline fuel cells. CS can serve as both polyelectrolyte membrane, and as matrix material for ion-solvating polymer composite membranes. In addition, CS has promising application in electrodes of polymer electrolyte-based fuel cells and also in biofuel cells. The advantages of chitosan for use as polymer electrolyte membrane include: its low-cost and eco-friendliness; the hydrophiliicity of chitosan which is a desirable property for use in high temperature and relatively low humidity environment; low methanol permeability; and available functionalities in chitosan backbone which allow chemical modification to tailor its properties [6].

2 Chitosan: An Overview

Among natural polymers, polysaccharides are among the best candidates due to their abundance in environment. Chitosan (CS), a polysaccharide bioresource, derived from chitin and, the second most abundant natural polymer after cellulose has been attracting considerable interest for a broad range of applications. Chitin occurs in cell walls of some fungi, but today it is commercially produced by partial chemical de-acetylation of chitin which is extracted from fungi by enzymatic extraction and from crustaceans (crabs, shrimp shell, insects or squid pen wastes) by deproteination, decalcification and decolorization [7, 8].

The structure of chitosan is very similar to that of cellulose (Fig. 1), only difference is the presence of amine group in absence of hydroxyl group [9, 10]. The structure of chitosan is primarily characterized by its molecular weight and degree of deacetylation and when the fraction of acetylated amine group is reduced to 50%,

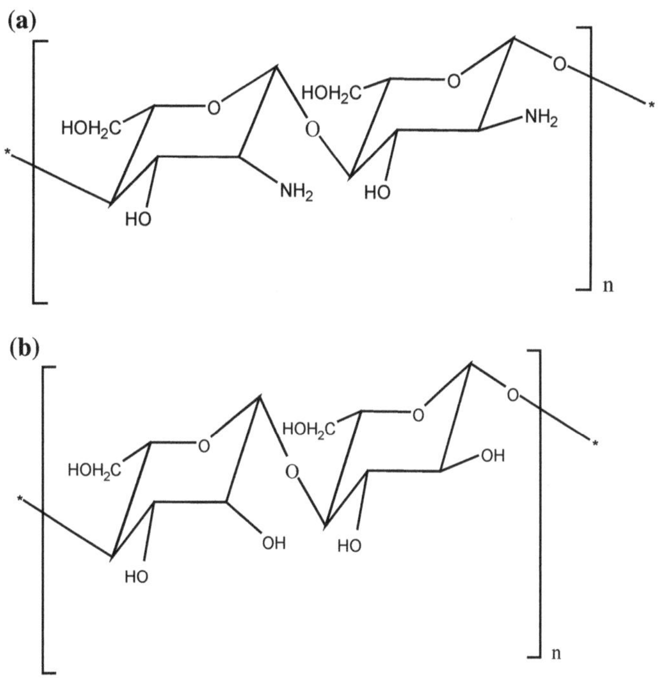

Fig. 1 Structures of **a** chitosan and **b** cellulose

and then resultant glucan is referred as chitosan. The solubility of chitosan depends on its origin, molecular weight and degree of acetylation [11]. Chitosan can be considered a strong base as it possesses primary amino groups with a pKa value of 6.3. Chitosan has a number of interesting physicochemical properties in aqueous solution due to its polycationic nature. The cationic nature is also thought to be slightly responsible for the broad range of biological functionalities such as antimicrobial activities and disease resistance activities in plants [12]. The antimicrobial activity of chitosan was reported to be further dependent on its molecular weight and degree of deacetylation.

3 Characterization of the Polymer Membrane and Their Desired Properties

In fuel cell, membrane is the core component. It must possess the desired properties like high proton conductivity to support high currents with minimal resistive losses and zero electronic conductivity; adequate mechanical strength and stability; moisture control in stack; low fuel or oxygen bypass to maximize coulombic efficiency; production costs compatible with intended applications and chemical as well as electrochemical stability under operating conditions. The performance of membranes in polymer electrolyte membranes also dependant on proton conductivity, which in turn depends on prevailing levels of hydration by reducing the membrane thickness, water drag or water crossover can be avoided thereby enabling an improvement in the fuel cell performance. Other advantages of reduced thickness include lower membrane resistance (enhancement in membrane conductivity), lower cost and rapid hydration [13].

Determination of water content, thickness, dimensional stability and tensile strength are the mechanical characterization of ion exchange membranes. The swelling capacity of a membrane not only affects its dimensional stability, but also its selectivity and electrical resistance. This important property of the membrane depends on several parameters such as the nature of the polymer and the ion exchange groups, counterions (swelling of the membrane increases after the alkalization step, i.e. the replacement of the haloid counter ions by the hydroxyl ions), charge density and crosslinking density of the membrane. The presence of water is necessary to obtain an efficient conductivity by enhancing the mobility of the ions [2].

4 Chitosan Based Membranes for Polymer Electrolyte

The commercial polymer electrolyte membranes (PEM) which are presently used in DMFC are Nafion 117 membrane [14]. It has the ability to provide high current densities at relatively low temperature and pressure but its methanol permeability is very high, hence it reduces the fuel cell performance. Among the many polymer electrolyte membranes studied, chitosan membranes have shown a better performance for their applications in low temperature fuel cells, hence it is being employed as one of the leading membrane forming material. It is extensively used in pervaporation for the dehydration of alcohols, because of its hydrophilic nature and very high affinity towards water. CS had also been devised as a favourable polymeric matrix for fuel cell applications considering desirable alcohol barrier property and proton conductivity as well as adequate thermal stability after cross-linking [15].

4.1 Chitosan Blend Polymer Electrolyte

Polymer blends are physical mixtures of two or more polymers with/without any chemical bonding between them. The objective of polymer blending is a practical one of achieving commercially viable products through either unique properties or lower cost as compared to other techniques might provide. Polymer blends generally combine desirable properties of each component within a near-homogeneous material, which are superior to those of component homopolymers. Blending technology also provides attractive opportunities for reuse and recycling of polymer wastes. CS is hydrophilic and thus has a high degree of swelling. An excessively high level of water uptake increases the fragility of the membrane, and makes it less durable in a fuel cell. To overcome the disadvantage of loss in mechanical strength in the wet state, CS is blended with tough polymer such as PVC, PVP, PE, etc. [6].

The blending of CS with sodium alginate and the viability of using polyion complex membranes for direct methanol fuel cell (DMFC) was investigated and the blend showed low methanol permeability (4.6×10^8 cm^2/s at 50 vol.% methanol concentration), excellent physicomechanical properties and relatively high proton conductivity (0.042 S cm^{-1}) [16]. *N*-Methylene phosphonic chitosan (NMPC)/PVA-based membranes exhibited mild cation selectivity and quaternized chitosan (QC)/PVA composite membranes had mild anion selectivity, while a blend of NMPC-QC/PVA membranes exhibited weak cation selectivity because of formation of zwitter ionic structure. The synthesized membranes, PC/QC-30 had the highest relative permeability, which may extend their suitability for electrolyte separations [17]. The CS–PVA–NH_4I electrolytes have been prepared by the solution casting method and their impedance of the electrolytes has been measured using Electrochemical Impedance Spectroscopy over the frequency range from 50 Hz to 1 MHz. The highest conducting sample 55 wt% (CS–PVA)—45 wt% NH_4I showing conductivity of 1.77×10^{-6} S cm^{-1} which is higher than the conductivity for the unblended electrolyte 55 wt% chitosan–45 wt% NH_4I found to be 3.73×10^{-7} S cm^{-1} and the activation energy for relaxation process is 0.87 eV. [18]. The blending of sulphonic groups grafted CS with pure chitosan opened up new arena as the blending leaded to cross-linking as an added advantage to increase the mechanical stabilities of the membranes. The dimensional swelling and the methanol crossover was suppressed successfully as well. The original chitosan membranes showed an area swelling value 39.3% decreased from 55.1% and methanol diffusion coefficient decreased from 1.0×10^{-6} cm^2 s^{-1} as compared to pure chitosan 4.7×10^{-7} cm^2 s^{-1} The so obtained blended cum cross-linked membranes showed reasonably good conductivity (0.03 S cm^{-1} at 80 °C) [19].

Cationic CS and anionic poly(acrylic acid) (PAAc) were blended and assessed for applicability in fuel cells as ionically cross-linked polyelectrolyte complex (PEC) membranes. These PEC membranes exhibited high ion exchange capacity, good proton conductivity, low methanol permeability and adequate thermal as well as mechanical stability. Among the blends used, the membrane blend with 50 wt% of CS and 50 wt% of PAAc, has been identified as ideal for DMFC applications as it exhibited low methanol permeability, excellent physicomechanical properties and comparatively high proton conductivity [20, 21].

4.2 Chitosan Cross-Linked Polymer Electrolyte

Polymer cross-linking leads to the formation of a permanent covalent network, which may allow the free diffusion of water/bioactive materials and also enhance the mechanical properties of the polymer. Chemical cross-links are formed by irreversible covalent links, as in covalently cross-linked chitosan. Moreover, they do not exhibit characteristics such as the modification of their properties in response to changes in their physiological environment (pH or temperature) that allow drug delivery to be efficiently controlled (http://www.sciencedirect.com/science/article/pii/S0008621505004969#bib5).

The reaction of cross-linking with primary amine groups in chitosan to produce covalent cross-linking is explored in many circumstances in order to detect the presence of free amine organic functions in simple or complex inorganic and organic compounds. A large and clear application of this procedure can be illustrated with many polymeric matrices [22]. The polymeric chains interconnect by cross-linkers, leading to the formation of a 3D network. They can be formed by complexation with another polymer, generally ionic, or by aggregation after chitosan grafting. Cross-linkers are molecules of MW much smaller than those of the chains between two consecutive cross-links [23]. Other components such as additional polymers to form a hybrid polymer networks (HPN) or semi- or full-interpenetrating polymer networks (IPN) can be added during the cross-linking reaction. However, the precise mechanism of reaction and the structure of the chemical compounds formed have not been studied in detail. The main disadvantage of cross-linking is the decrease in the overall flexibilities of the membranes under investigations which further reduces increase the risk with respect to their fragility; which needed to be addressed more seriously. The another disadvantage is the decrease in the open sites for the proton conductivities as it leads to the formation of more covalent bonds making the entire matrix as inert one. For the employment of the cross-linked membranes in fuel cell applications the ionic conductivities of the cross-linked polymer can be enhanced with the help of plasticization process. This process is applicable for the polymer electrolytes having low molecular weight substances. The fact is that plasticizers separate the polymer chains makes the assumptions true regarding their vital role in the increase of the ionic conductivity values. The polymer chain with mobile specific sites shows some ionic interactions but until now the mechanism is unknown.

Cross-linked-chitosan as proton exchange membranes are generally synthesized and characterized at room temperature. The cross-linking agents used were sulfuric acid; gluteraldehyde, boric acid, etc. The chitosan membranes were also employed as salt-complexed membranes such as with lithium nitrate. Since the cross-linking; reduces the flexibilities in the membranes therefore the cross-linking is generally followed by plasticization. Hence, salt-complexed membranes; plasticized (ethylene carbonate) and salt-complexed cross-linked and doped membranes with sulfuric acid have also been reported. Hybrid dual network membranes comprising CS/PVA networks cross-linked with sulfosuccinic acid (SSA) and glutaraldehyde (GA) and modified with stabilized silicotungstic acid (SWA) have been reported. Based on water sorption and proton conductivity measurements, the optimum content of 10 wt% SWA in the membrane has been established. The membrane–electrode assembly with 10 wt% stabilized SWA–CS–PVA hybrid membrane with SSA and GA as cross-linking agents delivers a peak power density of 156 mW cm^{-2} at a load current density of 400 mA cm^{-2} and 88 mW cm^{-2} at a load current density of 300 mA cm^{-2}, respectively, in DMFC at 70 °C [20].

The Nafion membranes are inferior to all of the membranes in ion exchange capacity and hydrogen gas permeability. An ideal amount of lithium nitrate salt can improve conductivity instead of ethylene carbonate being used to introduce conduction in Nafion membranes. For proton conduction in cross-linked membranes there should be the formation of a sulphate group, thus; for introducing the chitosan membranes containing 50% $LiNO_3$ followed by doping of sulphuric acid (4%) were fabricated and evaluated for their proton conductivities evaluated and found to be $(6.64 \pm 0.11) \times 10^{-2}$ and $(3.11 \pm 0.40) \times 10^{-2}$ S cm^{-1}, respectively before and after the cross-linking followed by the doping process [24].

In another study; cross-linked composite membrane having thin CS layer on microporous polysulfone substrate exhibiting high ion exchange capacity and proton conductivity in comparison to Nafion 117 had been reported at temperature above 100 °C [14].

Chemically cross-linked composite membranes of poly(vinyl alcohol-co-vinyl acetate-co-itaconic acid) (PVACO) and phosphomolybdic acid (PMA) have been fabricated and their proton conductivity as a function of PMA composition, cross-linking density and temperature has been investigated. The proton conductivity of the composite membranes was of the order of 10^{-3} S cm^{-1} and showing better resistance to methanol permeability than Nafion 117 under similar measurement conditions [25]. Blend membranes made of chitosan (CS) and poly(vinyl pyrrolidone) (PVP), were evaluated for their intermolecular interactions, methanol permeability and proton conductivity and compared to that of Nafion 117. In addition to being effective methanol barriers, the membranes have high ion exchange capacity and possess adequate thermal stability. The membranes consisting of quaternized poly(vinyl alcohol) (QAPVA) and quaternized chitosan

(2-hydroxypropyltrimethyl ammonium chloride CS) with glutaraldehyde as the cross-linking reagent have high conductivity (10^{-3} to 10^{-2} S cm^{-1}), and low methanol permeability at 30 °C [26]. Grafting of CS monomer with $–SO_3H$ leads to the formation of cross-linked CS sulfate membranes and then the polymers are cross-linked by the reaction between $–SO_3H$ groups in CS sulfate and the amido groups in pure CS [27]. Two series of cross-linked composite anion exchange membranes were prepared using different amounts of quaternized poly(vinyl alcohol) (QAPVA) and quaternized CS (2-hydroxypropyltrimethyl ammonium chloride chitosan) (HACC) using GA as the cross-linking reagent. The quaternary ammonium group grafted onto the matrix of QAPVA and HACC composite membranes have exchangeable anions [28]. The CS sulfate membranes have been developed by grafting the chitosan monomers with sulfonic groups, then cross-linking the polymers through the bond formation between the sulfonic groups in the CS sulfate and the amido groups in the pure CS monomers. The dimensional swelling as well as the methanol crossover of the chitosan membranes were suppressed successfully by the polymer blending, with area swelling value decreased from 55.1 to 39.3% and methanol diffusion coefficient decreased from 1.0×10^{-6} cm^2 s^{-1} of pure chitosan to 4.7×10^{-7} cm^2 s^{-1} of the membrane with chitosan sulfate content of $\sim$9.1 wt%. The thermal analysis indicated that the blending chitosan sulfate membranes were structure stable below 100 °C.

The cross-linked polymer blend of glutaraldehyde and sulfuric acid was effective in producing a reduction of methanol permeability from 9.2×10^{-8} cm^2 s^{-1} for CS/PVP blend to 7.3×10^{-8} cm^2 s^{-1} for cross-linked CS/PVP blend (GS-CS/PVP) and enhancing the conductivity from 0.019 S cm^{-1} for CS/PVP blend to 0.024 S cm^{-1} for GS-CS/PVP, thereby rendering it more suitable for a DMFC [29]. N-Methylene phosphonic chitosan (NMPC) was modified with poly(ethyleneglycol)-aldehyde (PEG-CHO) by reductive amination. The cross-linking with PEG-COH chains in NMPC increased the water swelling and hygroscopicity and the NMPC–PEG-CHO 8000 film leads to cell attachment and proliferation [30].

The high-energy conversion efficiency and environmental concerns of energy production make the polymer electrolyte membrane fuel cells (PEMFCs) ideal for future energy source. Polymer electrolyte membrane (PEM) is known as the vital element for PEMFC to be efficient. Only the significant reduction in methanol permeability has shown suitability of these polymers for DMFCs application. However, as far as proton conductivity is concerned, such membranes do not exhibit significant advantages over Nafion [19].

4.3 Chitosan Polymer Composite Based Polymer Electrode

Developing fully biodegradable polymer composites composed of biodegradable matrix and filler are attracting more attentions. Natural polymer composite materials are becoming increasingly important as membrane materials for fuel cell applications. Next generation composite materials should mimic the natural function as well as the mechanisms of polymer electrolyte membrane. These composites exhibit tailored physical, chemical, mechanical properties as well as predictable degradation behaviour. The appropriate selection of a particular composite for a given application requires a detailed understanding of relevant response. The main purpose of the preparation of composites with improved mechanical properties by adding suitable fillers for significant impact on the application of chitosan as polymer electrolyte membrane may be enhanced.

To investigate the application of membranes composed of PVA and its ionic blends with sodium alginate (SA) and chitosan in direct methanol fuel cells (DMFCs), were characterized for their ion exchange capacity (IEC) and swelling index values. Methanol permeability and proton conductivity were also estimated and the results were compared with those obtained for Nafion 117. The addition of small amounts of anionic polymer was particularly instrumental in the significant reduction of methanol permeability from 8.1×10^{-8} cm^2/s for PVA to 6.9×10^{-8} cm^2/s for the PVA–SA blend, which rendered the blend more suitable for a DMFC [31]. Sulfonated polysulphone (SPSF)/CS exhibited high IEC and proton conductivity higher than Nafion 117 at temperature above 100 °C. These membranes exhibiting satisfactory geometrical and thermal stabilities were considered as a potential alternative fuel cell membrane for high temperature operations. The modification of chitosan was carried out by introducing $–PO_3H_2$ group and composite membranes were prepared by blending the phosphonated CS and PVA of different compositions [15]. The proton transport number and conductivity of all these membranes found analogous to Nafion 117 membrane, but methanol permeability compared to Nafion 117 was quite low which renders their applications to direct methanol fuel cells (DMFCs). Furthermore, selectivity parameters for DMFC of all the synthesized membranes were found to be higher in comparison to Nafion 117 membrane [32]. Polymer electrolyte complexes of CS and poly vinyl phosphonic acid (PVPA) were prepared by in situ polymerization of vinyl phosphonic acid (VPA) in CS matrix [33]. Polybenzamidazole (PBI) and sulfuric acid (H_2SO_4) composite membranes were prepared but segregation in the phases was observed especially in PBI/H_3PO_4 composite membranes. Later; the membranes were cross-linked using HCHO as cross-linker by formal reaction which revealed affinity towards water and repelling behaviour for organic molecules such as methanol, which in turn to the hydration of polymer matrix with enhanced water retention [32] as an added advantage for their use in fuel cell applications.

Nanocomposite materials/membranes have been extensively studied for a long time; these are called nanocomposites if they employ nanosized inorganic phases. Organic–inorganic nanocomposites are generally organic polymer composites with inorganic nanoscale building blocks. Organic–inorganic nanocomposite polymer electrolyte membrane (PEM) constructs nanosized inorganic building blocks in organic polymer by molecular level of hybridization. This structural design has opened the possibility to combine in a single solid both the attractive properties of a mechanically and thermally stable inorganic backbone and the specific chemical reactivity, dielectric, ductility, flexibility, and processability of the biopolymer. A defining feature of polymer nanocomposites is that the small size of the fillers leads to a dramatic increase in interfacial area as compared with traditional composites. Moreover, they usually contain special properties of nanofillers leading to materials with improved properties. A biopolymer–clay nanocomposite has been reported which is obtained by assembling of chitosan onto sepiolite, a natural magnesium silicate. These materials were well-processed as self-supporting films and show recognizable potential interest as membranes for different processes related to separation of ions and gases, as well as components in electrochemical devices [34].

5 Chitosan for Fuel Cell

The world today is completely dependent on the use of energy which is produced from various technologies such as nuclear power and internal combustion engines. Although, such technologies are really expensive as well as not environmentally friendly since they emit greenhouse gases. The transportation industry is one of the major sources of air pollution; which leaded to a strong interest in the use of a polymer electrolyte membrane fuel cell (PEMFC) towards vehicle applications as for environmental concerns, fuel cell technology has been considered as an auspicious alternative for future energy requirements. Notable among the fuel cells is direct methanol fuel cell (DMFC), which has received much attention in recent years for its promising qualities.

Among the several kinds of fuel cells, the proton exchange membrane fuel cell (PEMFC) and the direct methanol fuel cell (DMFC) are known to utilize proton conducting polymer membranes [35]. The DMFC can be operated at low temperatures between 25 and 130 °C and its use of methanol limits pollution about 100 times below current vehicle emissions and within the national standards [36]. The major drawbacks for using a DMFC are its low power density and low fuel utilization which are related to poor reaction kinetics and methanol permeation though the polymer electrolyte membrane (PEM). Methanol permeation causes a mixed potential hence it contributes to a major fuel loss and as a result the DMFC performance decreases.

However, the progress in case of the PEM fuel cells seems to be slow due to the requirement in fabricating a suitable proton conducting polymer membrane that must be stable at fuel cell operating conditions. Owing to diverse energy demands, DMFC operating with liquid methanol as a fuel feed seems attractive as an alternative power source. In terms of applications, DMFCs are set to function as power sources for a range of stationary applications. At present, DMFCs employ Nafion, a proton conducting perfluorosulfonic acid polymer membrane, as electrolyte that also acts as a physical separator to prevent methanol crossover from the anode to the cathode. However, methanol crossover from anode to cathode across the Nafion membrane brings about a mixed potential at the cathode due to the loss of fuel and cell polarization.

CS is a potentially useful membrane material due to its non-permeability to alcohol, its films, in native state, exhibit high degree of swelling and very low conductivities. For accomplishment of better proton conductivities, high swelling levels are the prerequisite in the membranes, and the changes associated with these swellings affected the membrane's performance in terms of methanol permeability, dimensional stability and thermal stability. To solve the low conductivity and high swelling problems, CS has been either ionically cross-linked with sulfuric acid or incorporated into inorganic particles. In addition, different CS based PEMs examined so far have shown promising properties for application in the field of PEMFCs [37].

6 Chitosan for Biofuel Cell

A biofuel cell referred to as microbial fuel fell uses living micro-organisms to produce electricity. Broadly, it can be defined as those fuel cells, which use biocatalysts as well as include systems employing non-enzyme proteins. In addition, we would define biological fuel cells as devices capable of directly converting chemical energy to electrical energy via some electrochemical reactions involving biochemical pathways in micro-organism and solution/nanotubes concerned. Biofuel cell is constructed to fulfil the requirements of energy sources. The expansion of interest in fuel cells was triggered by the USA space program, in the late 1950s and early 1960s, which led to the development of microbial biofuel cells as possible technology for a waste disposal system for space flights that would also generate power [13].

Chitosan layer-based fuel cells have been reported and these fuel cells were made conductive by the addition of carbon nanotubes (CNTs). Chitosan serves as a material to immobilize the enzymes and pure chitosan without CNTs also works as an isolator between the conductive layers. This enables a metal free, thin, flat and scalable energy source [38]. Another study report the construction of a high performance biofuel cell based on the covalent immobilization of glucose oxidase (GOx) on chitosan-coated carbon cloth, which is used as an anodic catalyst. The GOx was coupled to CS using N-(3-dimethylaminopropyl)-N′-ethylcarbodiimide and N-hydroxysuccinimide as the condensing agents. The optimal immobilization condition at a reaction time of 50 min, pH of 5.9, and an enzyme/support ratio of 3 (w/w) was analysed by the Ridge Max Analysis. Under optimal conditions, the predicted and the experimental immobilized GOx activities were 34.42 ± 1.07 and 33.50 ± 0.92 U g^{-1} support, respectively. The power density was increased with GOx activity, and the maximum power of 1.87 mW cm^{-2} was obtained at a cell voltage of 0.44 V [38]. CaO immobilized onto chitosan beads was employed as a heterogeneous catalyst to produce biodiesel catalytically, through transesterification of soybean oil with methanol and were cross-linked with glutaraldehyde. This study developed an easy and energy-efficient process for preparing a chitosan-based immobilized CaO catalyst, which is promising for potential applications including biodiesel production [39]. Chitosan–carbon–nanotube–enzyme biocathode depicted a greatly enhanced and stable long-term current density of 0.19 mA mL^{-1}. By preventing the loss of the electrocatalytic activity of the enzyme, creating a protective microenvironment and facilitating good oxygen diffusion, it is possible to improve the performance of the biocathode, which in turn done by the fibrous microstructure of the electrode. The biocathode was able to deliver 0.17 mA mL^{-1} during 48 h. The result was reproducible as the three biocathodes retained 70–100% of the current output measured in phosphate buffer pH 7. The measured open circuit potential (OCV) in serum was between 0.48 and 0.5 V versus SCE for all electrodes. CS has been reported to stabilize laccase by the formation of hydrogen bridges between the CS and amino acids of laccase as well as electrostatic interactions [40].

6.1 *Microbial Biofuel Cell*

Production of energy from waste using microbial fuel cells for powering remote devices has gained considerable attention with interest in the last ten years. This is because the energy produced by this means is 'green' since they are derived from natural substrates such as carbohydrates. In a Microbial Fuel Cell (MFC), microorganisms act as catalyst for the generation of electricity via the anaerobic oxidation of organic and inorganic matters. Involvement of microorganisms in bioelectricity production, wastewater treatment and metal reduction makes MFC popular among biofuel cell industry. Electrons produced by the bacteria from these substrates are transferred to the anode (negative terminal) and flow to the cathode (positive terminal) linked by a conductive material containing a resistor, or operated under a load (i.e. producing electricity that runs a device). Simply, a classical MFC consists of anode and cathode chambers, with the microbes inoculated in the anaerobic anodic chamber while the oxidizing agents such as potassium permanganate are used in the cathodic chamber for oxygen reduction. By convention, current flows from the positive to the negative terminal, a direction opposite to that of electron flow. However, single chamber MFC has more advantages over dual chamber MFC [41, 42].

The chemical mediators such as neutral red or anthraquinone-2, 6-disulfonate (AQDS), can be added to the system to allow electricity production by bacteria unable to otherwise use the electrode. If no exogenous mediators are added to the system, the MFC is classified as a "mediator-less" MFC even though the mechanism of electron transfer may not be known [42]. In MFC, the organic substrate is degraded by bacteria, and the electrons released during the metabolic process are transferred to the extracellular electrode, converting the chemical energy to the electrical energy. However, the relatively low power density of MFC resulting from the sluggish electron transfer between bacteria cells and electrode limits its practical applications [43]. To improve MFC performance, extensive studies have been devoted to explore new anode materials for the transfer of electron between bacteria and anode directly [44–46]. It has been reported that graphene/polytetrafluoroethylene modified stainless steel mesh anode significantly improves the power density of *Escherichia coli*-catalyzed MFC, attributed to high specific surface area and good biocompatibility of the incorporated graphene for enhancing bacteria attachment on the anode [47]. In a report, the ice segregation induced self-assembly (ISISA) as a freeze casting technique was used to prepare a novel 3D chitosan/vacuum-stripped graphene scaffold with hierarchically porous structure composed of macropores formed by layered-branched architecture and meso/micropores from porous vacuum-stripped graphene embedded in the macropores. Porous vacuum-stripped graphene stimulates the *P. aeruginosa* to excrete more endogenous mediators and serves as primary unit in hierarchically porous architecture for larger specific surface area, which further increases the mediated electron transfer rate. All these factors, from the delicate anode structure design and anode material selection, contribute to the remarkable 78 times maximum powder density improvement [43]. The microbial cathode employing bacteria as biocatalysts to accept electrons from the cathode substratum offering a different path that avoids the use of noble or non-noble catalysts for oxygen reduction. In addition, it also enhanced the economic viability and environmentally sustainability of MFC systems. Moreover, biocathodes enable the use of alternate electron acceptors that can broaden the utility of MFCs and present potential opportunities for the microbially catalyzed conversion of electrical current into various value added products [31].

6.2 Enzymatic Biofuel Cell

Biofuel cells consist of either only enzyme or enzyme with an organism as catalyst and function under mild conditions such as 20–40 °C, near-neutral pH. Generation of high temperature or unwanted harsh reaction condition reaction make biofuel cells an attractive development prospect for their use in different applications. In fuel cell production, enzymatic biofuel cells have shown considerable potential advantages. The nanometre scale of enzymes greatly enhances the volumetric catalytic activity in comparison to microorganisms. Enzymatic biofuel cell development began in the 1960s. In 1964, Kimble and coworkers developed the first enzymatic biofuel cell. They constructed three different biofuel cells using glucose oxidase, amino acid oxidase or alcohol dehydrogenase at the anode and then compared the performance of each biofuel cell system. The oxidases proved viable and produced open circuit potentials of up to 350 mV, while alcohol dehydrogenase did not produce a positive open circuit potential. However, lack of efficient electron transfer mechanisms and minimal stability caused research to move toward metallic electrodes/electrocatalysts for biofuel oxidation rather than enzymes during the late 1960s and 1970s [48]. Kim et al. have reported enzyme stabilization and immobilization in nanoporous silica through the generation of single enzyme nanoparticles, chromotrypsin. This enzyme immobilization has been performed by the free radical initiated vinyl polymerization of a modified enzyme to methacryloxy propyl trimethoxysilane by exposure to 365 nm UV light. These single enzyme nanoparticles demonstrated remarkable enhancements in enzyme stability, improving the enzyme half-life from 12 h for chromotrypsin in solution to 143 days [49]. Furthermore, the immobilization of bilirubin oxidase for a cathodic system and glucose oxidase for an anodic system in addition to the incorporation of carbon nanotubes for the production of a biofuel cell have also been described. The decreases in electron transfer distance are mainly facilitated by DET incorporation of carbon nanotubes. In doing so, this membrane less biofuel cell is capable of generating approximately 120 mW cm^{-2} at 0.24 V at room temperature [50]. The incorporation of conductive nanomaterials greatly increase the surface area of the electrode allowing a higher concentration of enzyme to be located within electron tunnelling distances, and greatly increasing the potential for high-power densities [51]. For enhanced stabilization and charge transfer, Kim has employed covalent linking of oxidoreductase enzymes to carbon nanotubes in order to show respectable performance of fuel cell. His group produced a novel miniature fuel cell design that incorporates the immobilization of glucose oxidase clusters on carbon nanotubes, which are subsequently immobilized on a carbon felt electrode. This cell operated in a non-buffered environment at an open circuit potential of 0.33 V and at 0.55 V in a buffered system [52].

Therefore, to develop more efficient cell designs, one should have to use mixed fuel solutions and eliminate physical barriers between solutions. The traditional diffusion cell design has been modified by reducing the distance between the cathode and anode and the cathode is in direct contact with air, which leads to eliminate the limitations that arise from lack of oxygen. This design utilizes a gas diffusion electrode, which is hot pressed to a Nafion*s* 112 membrane [48]. Clauwaert et al. constructed a complete biological microbial fuel cell with bacteria at both the anode and the cathode and achieved a high cathodic potential, preventing the occurrence of cathodic hydrogen evolution and subsequent bacterial hydrogen consumption. The oxygen-reducing biocathode was further demonstrated with modifications to reactor architecture and electrode materials in order to reduce losses and increase power output [53]. In the case of biocathodes, nearly all of the current reports describe the role of microbial catalysis relative to abiotic cathodes. The apparent decrease in activation losses at oxygen-reducing biocathodes proved that bacteria acted as true catalysts [54].

7 Conclusions

In general, the significance and the relevance of chitin and chitosan are mainly sticking to zoologists, marine entomologists, and physiologists, but in the late 1970s, chemists all over the world devoted attention to chitin. In particular, it was apprehended that chitin, the second most abundant natural polymer after cellulose, is an ample source of chitosan. The unique cationic polysaccharide as such, was found superior to man-made cationic derivatives of cellulose and starch. Chitin and its primary derivative chitosan, biopolymers from renewable resources (shells of shellfish, the wastes of the seafood industry), offer a distinct set of advantageous biological and physicochemical characteristics that qualify them for a multiplicity of industrial, biotechnological, medical applications and in food industry. The absorption of metal ions as an adsorbent and the separations as well as relevance in energy devices such as fuel cell is a recent one, which are adding to cleaner industries establishment, hence a cleaner environment, water and food. The distinctive advantages of chitosan include availability, biocompatibility, biodegradability, non-toxicity, antimicrobial properties, heavy metal ions chelation, gel forming properties, ease of chemical modification, and high affinity to proteins. These characteristics of the chitosan make it widely exploited in the designing of environmentally benign membrane based applications for environment, biotechnology and energy devices such as fuel cell membranes. Polymer electrolyte membrane fuel cells have acquired due importance as they are best suited for applications where a quick start up is required such as in automobiles. Chitosan is receiving great interest as base materials for membrane electrolyte and as well as electrode in various fuel cells such as polymer electrolyte-based fuel cell including low to intermediate temperature polymer electrolyte fuel cells, direct methanol fuel cells, alkaline polymer electrolyte fuel cells and biofuel cells. The use of such eco-friendly, low cost naturally available polymer in fuel cell will take the technology to the new level and this Chapter looks at how chitin/chitosan materials can contribute to the development of membrane-based processes in setting up new dimension for energy devices.

Acknowledgements Shakeel Ahmed sincerely acknowledges University Grant Commission (UGC), New Delhi, India for its financial support in the form of JRF (NET) and SRF (NET).

References

1. Page KA, Rowe BW (2012) An overview of polymer electrolyte membranes for fuel cell applications. In: Polym Energy Storage Deliv Polyelectrolytes Batter Fuel Cells, pp 147–164 (American Chemical Society)
2. Merle G, Wessling M, Nijmeijer K (2011) Anion exchange membranes for alkaline fuel cells: a review. J Memb Sci 377:1–35. doi:10.1016/j.memsci.2011.04.043
3. Badwal SPS, Giddey SS, Munnings C et al (2014) Emerging electrochemical energy conversion and storage technologies. Front Chem 2:1–28. doi:10.3389/fchem.2014.00079
4. Smitha B, Sridhar S, Khan AA (2005) Solid polymer electrolyte membranes for fuel cell applications—a review. J Memb Sci 259:10–26. doi:10.1016/j.memsci.2005.01.035
5. Dodds PE, Staffell I, Hawkes AD et al (2015) Hydrogen and fuel cell technologies for heating: a review. Int J Hydrogen Energy 40:2065–2083. doi:10.1016/j.ijhydene.2014.11.059
6. Ma J, Sahai Y (2013) Chitosan biopolymer for fuel cell applications. Carbohydr Polym 92:955–975. doi:10.1016/j.carbpol.2012.10.015
7. Ahmad M, Ahmed S, Ikram S (2015) Adsorption of heavy metal ions: role of chitosan and cellulose for water treatment. Langmuir 2:280–289
8. Ahmed S, Ikram S (2015) Chitosan & its derivatives: a review in recent innovations. Int J Pharma Sci Res 6:14–30

9. Ahmed S, Ahmad M, Ikram S (2014) Chitosan: a natural antimicrobial agent—a review. J Appl Chem 3:493–503
10. Ahmed S, Ahmad M, Jayachandran M et al (2015) Chitosan based dressings for wound care. Immunopathol: Open Access 1:1–6
11. Ahmad M, Ahmed S, Swami BL, Ikram S (2015) Preparation and characterization of antibacterial thiosemicarbazide chitosan as efficient Cu(II) adsorbent. Carbohydr Polym 132:164–172. doi:10.1016/j.carbpol.2015.06.034
12. Qureshi MA, Khatoon F, Ahmed S (2015) An overview on wounds their issues and natural remedies for wound healing. Biochem Physiol: Open Access. doi:10.4172/2168-9652.1000165
13. Venkatesan PN, Dharmalingam S (2013) Characterization and performance study on chitosan-functionalized multi walled carbon nano tube as separator in microbial fuel cell. J Memb Sci 435:92–98. doi:10.1016/j.memsci.2013.01.064
14. Smitha B, Devi DA, Sridhar S (2008) Proton-conducting composite membranes of chitosan and sulfonated polysulfone for fuel cell application. Int J Hydrogen Energy 33:4138–4146. doi:10.1016/j.ijhydene.2008.05.055
15. Ramírez-Salgado J (2007) Study of basic biopolymer as proton membrane for fuel cell systems. Electrochim Acta 52:3766–3778. doi:10.1016/j.electacta.2006.10.051
16. Smitha B, Sridhar S, Khan AA (2005) Chitosan-sodium alginate polyion complexes as fuel cell membranes. Eur Polym J 41:1859–1866. doi:10.1016/j.eurpolymj.2005.02.018
17. Saxena A, Kumar A, Shahi VK (2006) Preparation and characterization of N-methylene phosphonic and quaternized chitosan composite membranes for electrolyte separations. J Colloid Interface Sci 303:484–493. doi:10.1016/j.jcis.2006.07.061
18. Buraidah MH, Arof AK (2011) Characterization of chitosan/PVA blended electrolyte doped with NH4I. J Non Cryst Solids 357:3261–3266. doi:10.1016/j.jnoncrysol.2011.05.021
19. Xiang Y, Yang M, Guo Z, Cui Z (2009) Alternatively chitosan sulfate blending membrane as methanol-blocking polymer electrolyte membrane for direct methanol fuel cell. J Memb Sci 337:318–323. doi:10.1016/j.memsci.2009.04.006
20. Smitha B, Sridhar S, Khan AA (2004) Polyelectrolyte complexes of chitosan and poly (acrylic acid) as proton exchange membranes for fuel cells. Macromolecules, 2233–2239
21. Liu YL, Hsu CY, Su YH, Lai JY (2005) Chitosan-silica complex membranes from sulfonic acid functionalized silica nanoparticles for pervaporation dehydration of ethanol-water solutions. Biomacromol 6:368–373. doi:10.1021/bm049531w
22. Monteiro OAC, Airoldi C (1999) Some studies of crosslinking chitosan-glutaraldehyde interaction in a homogeneous system. Int J Biol Macromol 26:119–128. doi:10.1016/S0141-8130(99)00068-9
23. Berger J, Reist M, Mayer JM et al (2004) Structure and interactions in covalently and ionically crosslinked chitosan hydrogels for biomedical applications. Eur J Pharm Biopharm 57:19–34. doi:10.1016/S0939-6411(03)00161-9
24. Soontarapa K, Intra U (2006) Chitosan-based fuel cell membranes. Chem Eng Commun 193:855–868. doi:10.1080/00986440500267337
25. Anis A, Banthia AK, Bandyopadhyay S (2008) Synthesis and characterization of polyvinyl alcohol copolymer/phosphomolybdic acid-based crosslinked composite polymer electrolyte membranes. J Power Sources 179:69–80. doi:10.1016/j.jpowsour.2007.12.041
26. Xiong Y, Liu QL, Zhang QG, Zhu AM (2008) Synthesis and characterization of cross-linked quaternized poly(vinyl alcohol)/chitosan composite anion exchange membranes for fuel cells. J Power Sources 183:447–453. doi:10.1016/j.jpowsour.2008.06.004
27. Stambouli AB, Traversa E (2002) Solid oxide fuel cells (SOFCs): a review of an environmentally clean and efficient source of energy. Renew Sustain Energy Rev 6:433–455. doi:10.1016/S1364-0321(02)00014-X
28. Wu H, Zheng B, Zheng X et al (2007) Surface-modified Y zeolite-filled chitosan membrane for direct methanol fuel cell. J Power Sources 173:842–852. doi:10.1016/j.jpowsour.2007.08.020

29. Smitha B, Sridhar S, Khan AA (2006) Chitosan–poly(vinylpyrrolidone) blends as membranes for direct methanol fuel cell applications. J Power Sources 159:846–854. doi:10.1016/j.jpowsour.2005.12.032
30. Ramos VM, Rodríguez NM, Henning I et al (2006) Poly(ethylene glycol)-crosslinked N-methylene phosphonic chitosan. Preparation and characterization. Carbohydr Polym 64:328–336. doi:10.1016/j.carbpol.2005.12.025
31. He Z, Angenent LT (2006) Application of bacterial biocathodes in microbial fuel cells. Electroanalysis 18:2009–2015. doi:10.1002/elan.200603628
32. Binsu VV, Nagarale RK, Shahi VK, Ghosh PK (2006) Studies on N-methylene phosphonic chitosan/poly(vinyl alcohol) composite proton-exchange membrane. React Funct Polym 66:1619–1629. doi:10.1016/j.reactfunctpolym.2006.06.003
33. Leela Mohana Reddy A, Tanur AE, Walker GC (2010) Synthesis and hydrogen storage properties of different types of boron nitride nanostructures. Int J Hydrogen Energy 35:4138–4143. doi:10.1016/j.ijhydene.2010.01.072
34. Darder M, López-Blanco M, Aranda P et al (2006) Microfibrous chitosan—sepiolite nanocomposites. Chem Mater 18:1602–1610. doi:10.1021/cm0523642
35. Basu S (2015) Proton exchange membrane fuel cell technology: india's perspective. Proc Indian Natl Sci Acad 81:865–890. doi:10.16943/ptinsa/2015/v81i4/48301
36. Singh D, Lu DM, Djilali N (1999) A two-dimensional analysis of mass transport in proton exchange membrane fuel cells. Int J Eng Sci 37:431–452. doi:10.1016/S0020-7225(98)00079-2
37. Ye Y-S, Rick J, Hwang B-J (2012) Water soluble polymers as proton exchange membranes for fuel cells. Polymers (Basel) 4:913–963. doi:10.3390/polym4020913
38. Bogner M, Schnaithmann M, Saegebarth J, Sandmaier H (2008) A simple biofuel cell based on chitosan films for lab on a chip application. Proc, PowerMEMS
39. Fu C-C, Hung T-C, Su C-H et al (2011) Immobilization of calcium oxide onto chitosan beads as a heterogeneous catalyst for biodiesel production. Polym Int 60:957–962. doi:10.1002/pi.3031
40. El Ichi S, Zebda A, Laaroussi A et al (2014) Chitosan improves stability of carbon nanotube biocathodes for glucose biofuel cells. Chem Commun 50:14535–14538. doi:10.1039/C4CC04862H
41. Huang L, Regan JM, Quan X (2011) Electron transfer mechanisms, new applications, and performance of biocathode microbial fuel cells. Bioresour Technol 102:316–323. doi:10.1016/j.biortech.2010.06.096
42. Logan BE, Hamelers B, Rozendal R et al (2006) Microbial fuel cells: methodology and technology. Environ Sci Technol 40:5181–5192. doi:10.1021/es0605016
43. He Z, Liu J, Qiao Y et al (2012) Architecture engineering of hierarchically porous chitosan/vacuum-stripped graphene scaffold as bioanode for high performance microbial fuel cell. Nano Lett 12:4738–4741. doi:10.1021/nl302175j
44. Qiao Y, Bao S, Li CM et al (2008) Nanostructured polyaniline/titanium dioxide composite anode for microbial fuel cells. ACS Nano 2:113–119. doi:10.1021/nn700102s
45. Rinaldi A, Mecheri B, Garavaglia V et al (2008) Engineering materials and biology to boost performance of microbial fuel cells: a critical review. Energy Environ Sci 1:417. doi:10.1039/b806498a
46. Logan B, Cheng S, Watson V, Estadt G (2007) Graphite fiber brush anodes for increased power production in air-cathode microbial fuel cells. Environ Sci Technol 41:3341–3346. doi:10.1021/es062644y
47. Zhang Y, Mo G, Li X et al (2011) A graphene modified anode to improve the performance of microbial fuel cells. J Power Sources 196:5402–5407. doi:10.1016/j.jpowsour.2011.02.067
48. Moehlenbrock MJ, Minteer SD (2008) Extended lifetime biofuel cells. Chem Soc Rev 37:1188. doi:10.1039/b708013c
49. Kim J, Grate JW (2003) Single-enzyme nanoparticles armored by a nanometer-scale organic/inorganic network. Nano Lett 3:1219–1222. doi:10.1021/nl034404b

50. Lim J, Malati P, Bonet F, Dunn B (2007) Nanostructured sol–gel electrodes for biofuel cells. J Electrochem Soc 154:A140. doi:10.1149/1.2404904
51. Kim J, Jia H, Wang P (2006) Challenges in biocatalysis for enzyme-based biofuel cells. Biotechnol Adv 24:296–308. doi:10.1016/j.biotechadv.2005.11.006
52. Fischback MB, Youn JK, Zhao X et al (2006) Miniature biofuel cells with improved stability under continuous operation. Electroanalysis 18:2016–2022. doi:10.1002/elan.200603626
53. Clauwaert P, Van der Ha D, Boon N et al (2007) Open air biocathode enables effective electricity generation with microbial fuel cells. Environ Sci Technol 41:7564–7569
54. Chen G-W, Choi S-J, Lee T-H et al (2008) Application of biocathode in microbial fuel cells: cell performance and microbial community. Appl Microbiol Biotechnol 79:88–379. doi:10.1007/s00253-008-1451-0

Chapter 16
Fuel Cells: Construction, Design, and Materials

D.P. Hansora and Aniruddha Chatterjee

Abstract In this chapter, the basic information about fuel cells (FC) has been encapsulated. The purpose is to summarize the available details about the types of FC, their design and constructions of FC, catalysts in FC, materials and the methods used for preparation of FC. Nonetheless, more emphasize is kept only on PEM for FC applications by considering the theme and contents of the book.

Keywords Fuel cells · Basics · Overview · History · Types · Preparation · Applications

1 Introduction

A fuel cell (FC) is an energy conversion device which converts the chemical energy (stored in fuels and oxidants) into electrical energy (or electricity) through electrochemical reactions [1]. Christian Friedrich Schönbein, (professor of physics and chemistry in Basel, Switzerland, observed the first FC effect indicating the generation of water and energy from a recombination of hydrogen and oxygen. William Robert Grove (professor of physics at the London Institution) published a brief study indicating the same effect and also designed the first FC generator of electrical current. Grove used a gas feed of hydrogen and oxygen to the corresponding electrodes and demonstrated the existence of a process, which was the reverse phenomena of electrolysis, on the other hand, Schönbein's research identified the

D.P. Hansora (✉)
Department of Plastic Technology, University Institute of Chemical Technology, North Maharashtra University, Jalgaon 425001, Maharashtra, India
e-mail: 2568dharmesh@gmail.com

A. Chatterjee
Department of Plastic and Polymer Engineering, Marathwada Institute of Technology, Aurangabad 431010, Maharashtra, India
e-mail: aniruddha_chatterjee2006@yahoo.co.in; aniruddha.chatterjee@mit.asia

Inamuddin et al. (eds.), *Organic–Inorganic Composite Polymer Electrolyte Membranes*, DOI 10.1007/978-3-319-52739-0_16

invention of guncotton and for the discovery of ozone [2]. In the twenty-first century, FC is in demand for generation of power with environmental protection, and it has also prompted an extensive research on energy conversion technologies. Efforts have been made toward commercialization of cost-effective hydrogen-based FCs to reduce dependence on oil and decrease pollution. The major application of polymer electrolyte membrane fuel cells (PEMFC) includes transportation (lightweight vehicles, buses) because of their potential impact on the environment, i.e., the control of emission of the greenhouse gases. Other applications include stationary and portable power generation. Most of major motor manufacturing companies generally use PEM-based FCs due to their high power density and excellent dynamic characteristics as compared with other types of FCs [1–9].

2 Different Types of Fuel Cells

A FC basically directly converts the chemical energy into electrical energy, i.e., heat and water via a redox reaction. The basic electrochemical FC comprises of an anode, a cathode, and an electrolyte. These FCs are generally connected in series as well as parallel by using bipolar plates to provide the desired voltage or current density. The fuel is oxidized at the anode, whereas the combustive (O_2) is reduced at the cathode (as an electrode). Hence, if the FC is provided with hydrogen, then FC is known as a clean energy producing device because no volatile organic compounds (VOCs) are emitted. However, the production, transportation, and distribution of hydrogen gas are not considered as clean processes. FCs are classified in different ways depending upon the nature of electrolyte to be used. The most common types of FCs are the polymer electrolyte membrane (also called the proton exchange membrane) fuel cell (PEMFC), direct methanol fuel cell (DMFC) (the same as PEMFC but instead of hydrogen, a methanol is used as the fuel), alkaline fuel cell (AFC) or alkaline anion-exchange membrane fuel cell (AEMFC), phosphoric acid fuel cell (PAFC), molten carbonate fuel cell (MCFC) and solid oxide fuel cell (SOFC), etc. These different types of FCs have various applications, i.e., PAFC, MCFC, and SOFC are used for stationary power generation, while PEMFC, DMFC, and AFC are mostly used for vehicular or portable applications. SOFCs are the other major type cells followed by PEMFC. These cells based on high-temperature O_2- ion conducting oxides are best suited for larger scale stationary applications. Hydrogen in PEMFCs and methanol in DMFCs are oxidized in the catalyst layer (CL) at the anode side of the FC. The resulting protons and electrons are forced onto separate paths to the cathode, and then they recombine with oxygen to produce water. PEMFCs have become promising alternative for future energy required for clean environment, power source for automotive, and power backup applications due to certain advantages. Because PEMFCs have been benefited from the advantages such as low operating temperature, high-power density, and zero/low emission and they do not require any fuel processing equipment. However, further improvements in terms of performance, durability,

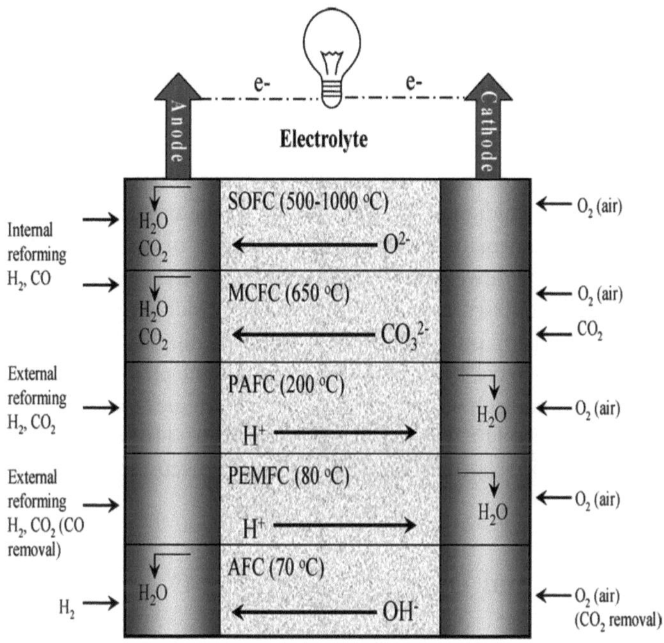

Fig. 1 Different types of FC and their operating conditions (reprinted with permission of Ref. [4])

Table 1 Operating and applicable properties of different FCs

Type of FC	Operating Temp. (°C)	Power density (mW/cm^2)	Fuel efficiency (Chem. to Elec.)	Lifetime (h)	Capital cost ($/kW)	Area of application
AFC	60–90	100–200	40–60	>10,000	>200	Space, mobile
PAFC	160–220	200	55	>40,000	3000	Distributed power
PEMFC	50–80	350	45–60	>40,000	>200	Portable, mobile, stationary
MCFC	600–700	100	60–65	>40,000	1000	Distributed power generation
SOFC	800–1000	240	55–65	>40,000	1500	Base load power generation

Reprinted with permission of Ref. [14]

and cost are necessary considerations for commercialization [1–5, 10–14]. Figure 1 illustrates different types of FCs along with their operating conditions.

Table 1 summarizes the operating and applicable properties of five main types of FC.

The lifetime of a PEMFC is a function of operating conditions, component materials, and degradation mechanisms. Wu and Yuan et al. [15, 16] have reviewed and summarized different degradation mechanisms and mitigation strategies, durability tests for better performance of PEMFC [17, 18]. Costamagn et al. [9] have nicely discussed the fundamental scientific aspects of PEMFC including their science and technology covering the period of 40 years starting from the 1960s to the year 2000 [9] and Zaidi et al. [11] have reported the newer research trends took place for the development of PEMFC [11]. The use of various PEMs in water electrolysis has been reviewed by Carmo et al. [19]. There have been some alternatives to proton conductive anhydrous membranes, including heterocyclic protogenic solvents comprising polymer electrolytes for FC applications [20–22]. Anion exchange membranes have been reported as important classification for alkaline FCs [23]. Degradation mechanisms and mitigation strategies of PEMFCs' durability have been also reviewed [15, 18], similarly test protocols [16] and stack testing [17] and bipolar plates [24] related to PEMFC were also discussed and reviewed.

3 Construction and Design of Different FC

3.1 PEMFC

Modern PEMFCs use proton-conducting PEM for power vehicular, portable, and small residential applications. The activation of kinetically hindered reactions such as oxygen reduction or methanol oxidation dissipates the energy at the electrodes. The migration of protons through the PEM and catalyst layer (CL) also dissipates the energy because of diffusion of reactants and products as well as permeation of liquid water.

Figure 2 depicts the reactions in a hydrogen-based PEMFC. The schematics of a single PEMFC and as a stack with three single cells are also presented in Fig. 2. A single PEMFC consists of an anode and a cathode (Fig. 2a) [1, 2, 4, 5, 25]. On the anode side, hydrogen flows into the flow channel through the gas diffusion layer (GDL) toward the CL, while in the anode CL, hydrogen splits into protons (hydrogen ions) and electrons. The protons pass through the membrane and flow toward the cathode CL. However, the electrons cannot pass through the membrane, but can easily pass to the cathode through an external circuit and hence it generates electricity. On the cathode side, oxygen from air flows into the flow channel through the GDL toward the CL, at the same time in cathode CL, oxygen reacts with the protons and electrons from the anode, produces water and heat. The protons transport across the membrane is basically due to the water concentration and pressure differences between the anode and cathode. This allows water to pass easily through the membrane in both directions. The reaction on the anode side

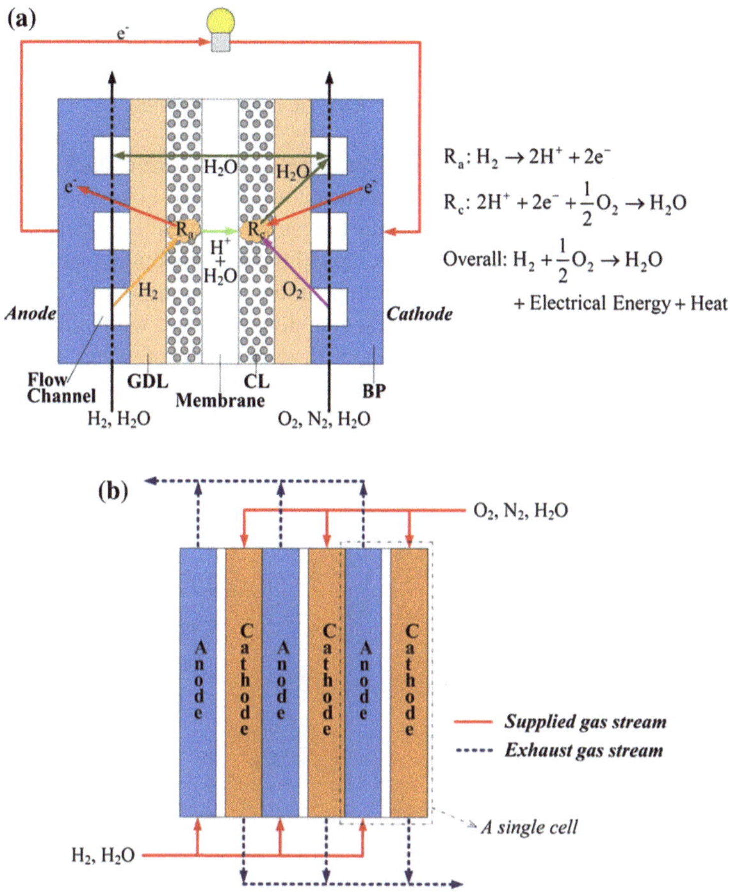

Fig. 2 Schematics circuits of **a** a single PEMFC and **b** a PEMFC stack with three single cells (reprinted with permission of Ref. [1])

indicates the splitting of hydrogen into protons and electrons, which is also known as a hydrogen oxidation reaction (HOR) [1, 2, 4, 5]:

$$H_2 \rightarrow 2H^+ + 2e^-$$

The reaction on the cathode side indicates that oxygen, protons, and electrons combinedly form water, which is also known as an oxygen reduction reaction (ORR):

$$0.5O_2 + 2H^+ + 2e^- \rightarrow H_2O$$

So, overall reaction suggests that simply hydrogen reacts with oxygen producing water, electrical energy, and heat:

$$H_2 + 0.5O_2 \rightarrow H_2O + \text{Electric Energy} + \text{Heat}$$

Generally, single cells are connected in series, which form a stack to produce higher voltages. Three single cells are connected in series, as shown in Fig. 2b. The supplied gases have to be distributed into the single cells through the inlet manifolds, and the exhaust gases have to be removed through the outlet manifolds. Due to the differences in materials and in local operating conditions among the different components of PEMFC, water can be present in different states with different phase. Many researchers have attempted to understand the state of water in PEMFC, because the state and phase change of water are observed different in membrane, i.e., particularly CL, GDL, and flow channel. Figure 3 shows the schematics of the

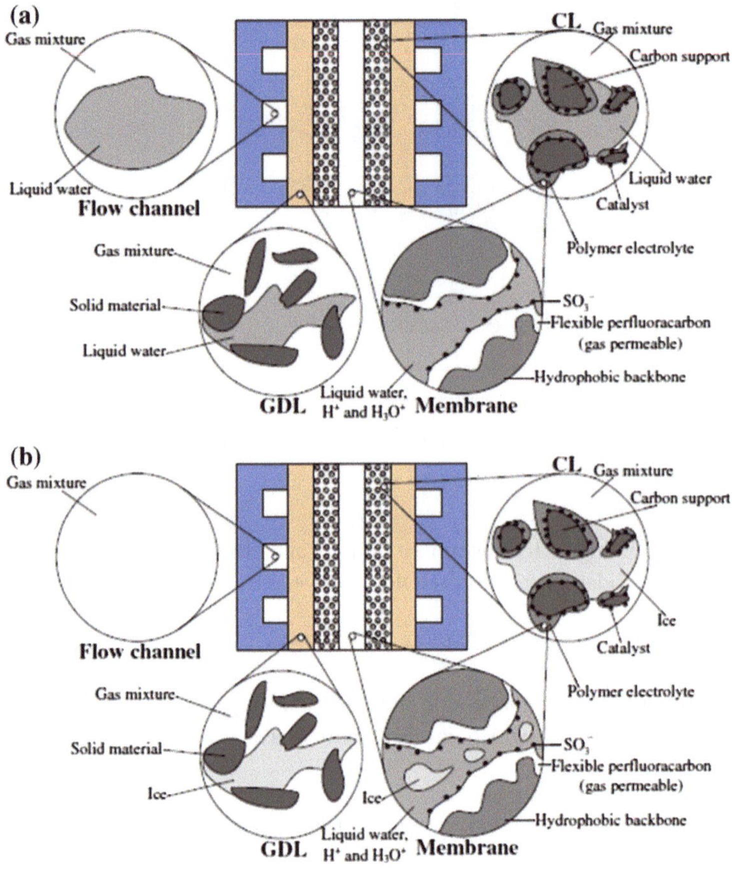

Fig. 3 Schematics of a single PEMFC with the structure of each cell component illustrated **a** normal operating condition; **b** cold start (reprinted with permission of Ref. [1])

structure of each cell component in a single PEMFC for both the normal operating conditions and cold start.

As shown in Fig. 3a, cell temperature generally ranges from 60 to 80 °C for normal operating conditions, and water exists in both vapor and liquid forms in the flow channel as well as pore regions of GDL and CL, the ionomer (polymer electrolyte) in membrane and CL absorbs much quantities of water in liquid state or bound to H^+ (e.g., H_3O^+). Due to this reason, PEMFC is oftenly used for automotive applications, but it should be unavoidable for vehicles driving below the freezing point of water in winter season, and also PEMFC must be able to successfully start up or "cold start" from subzero temperatures. During a "cold start" process for PEMFC, the initial cell temperature is usually equal to the surrounding temperature (usually under 0 °C in winter), and water mostly freezes [1, 6, 9]. As shown for PEMFC cold start (Fig. 3b), the formation of liquid water can be almost neglected since it freezes to ice. Therefore, water exists in vapor and ice forms but in the pore regions of GDL and CL (ice is formed if the local water vapor pressure is higher than the local water saturation pressure) for cold start of PEMFC. Since the ice can easily stick on the solid materials of CL and GDL and then difficult to move, the ice formation in flow channel might be neglected. Figure 3b also illustrates that water in the ionomer of membrane and CL may also freeze at subzero temperatures [1, 2, 4, 5, 9, 17].

As water is usually not desired in gas-supplying channels and pores then it should be quickly removed as possible because presence of water in the polymer electrolyte may serve as the proton shuttle. The water, as product, is efficiently coupled to the polymer electrolyte so it can improve the mobility of hydrogen ions that is required for the design of CL. In conventional PEM, polymeric side chains terminated by acidic hydrogen sulfonate groups are randomly tethered to hydrophobic polymeric backbones, as shown in Fig. 4 [2]. In the hydrated

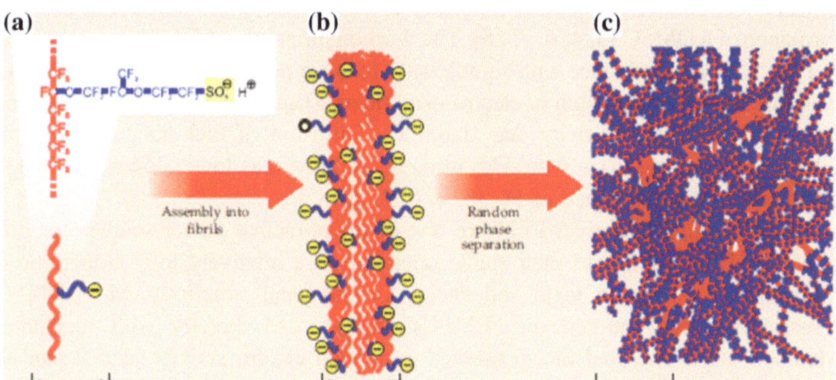

Fig. 4 A typical PEM. Nafion, consists of a hydrophobic fluoropolymer backbone (*red*) and a flexible side chain (*blue*) terminated by a hydrophilic sulfonate group (*yellow*). In the polymer electrolyte, the fibrils agglomerate and randomly self-assemble into regions that are hydrophobic (*mostly red*) and hydrophilic (*mostly blue*) (reprinted with permission of Ref. [2])

membrane state, mobile protons (H^+ ions) are formed from the dissociation of the HSO_3 groups which require a sufficient amount of water. Also water is not immobile, so proton migration is get coupled automatically to the transport of one or two water molecules in a phenomena called electro-osmotic drag.

This drag alongside the proton flux shuffles water molecules toward the cathode CL. A backward flux of water molecules toward the anode CL occurs from diffusion or pressure-driven flow that opposes the water shuffling toward the cathode. It was reported [2] that the water flux in the polymer membrane depends on its morphology and current density. Hydrated state, of the polymer electrolyte's parts even under drastic variations in operating conditions, is a key parameter of working of PEMFC. Designing PEM with reduced amount of electro-osmotic drag would be an advantage to the water management problem. This can be accomplished by partially immobilizing the water in the membrane without impairing its conductive abilities. High rates of water back flux can be achieved by using thin PEM with a higher concentration of hydrophilic regions but at the cost of less chemical stability, higher reactant permeability, and reduced mechanical robustness [2, 4, 5].

3.2 DMFC

DMFCs, prepared using PEMs, are potential alternative FCs for future cleaner energy requirements because these cells operated at relatively low temperatures and do not require any fuel processing equipment, as discussed earlier. However, limitations of DMFCs include: (i) slow kinetics of methanol oxidation at anode CL and (ii) diffusion of methanol from anode to the cathode side, across the PEM. The methanol does not only wastes fuel but also it can cause performance losses at the cathode due to the consumption of oxygen and catalyst poisoning. Thus, it is necessary to reduce the loss of methanol as fuel in PEM for improving the performance of a DMFC [3, 4, 6, 7, 26]. The working principle of DMFCs is presented in Fig. 5. New membranes with significantly reduced permeability of methanol and water transport (by diffusion or electro-osmotic drag) are required for DMFCs along with acceptable conductivity and stabilities. Diffusion of fuel (i.e., methanol in DMFC) from anode to cathode sides through the PEM may lower down the voltage efficiency and performance of FC [3, 4, 6, 7, 26].

DMFCs employing with PEM are one of the attractive power sources for different applications due to their stable operation at a relatively low temperatures, high energy generation yield and energy density, and simplicity. Most DMFC research has concentrated on PEMFCs that are fed directly with methanol. Methanol itself is a fuel that possesses significant electro activity, and it can be oxidized directly to generate carbon dioxide and water in a DMFC. The proton conduction through the PEM is associated with the water transport through the membrane. Methanol can also be transported to the cathode by an electro-osmotic drag (methanol crossover). This is due to the physical properties of methanol, such as its dipole moment, which consequently leads to a decrease in cell performance. It

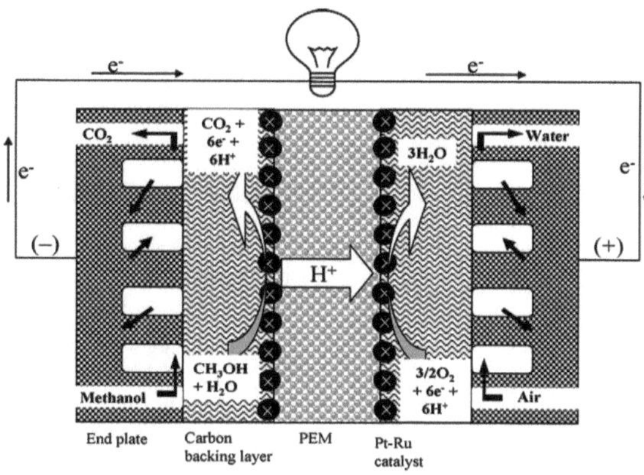

Fig. 5 Working principle of DMFC (reprinted with permission of Ref. [4])

is necessary to reduce the loss of fuel across the cell for improvement in the performance of a DMFC [4, 22, 26].

The work on various PEM materials for DMFC has been reviewed by Neburchilov et al. [26]. Unlike the membranes for hydrogen fuelled PEM fuel cells, perfluorosulfonic acid-based membranes showed complete domination, but the membranes for DMFC have numerous variations. The membranes have been discussed according to various properties, included methanol crossover, proton conductivity, thermal stability, and maximum power density. Hydrocarbon and composite fluorinated membranes showed the most potential for low cost membranes with low methanol permeability and high durability [18, 26, 27, 28].

3.3 AEMFC

To overcome draw backs of DMFCs, there has been a growing interest in developing direct methanol AEMFCs using alkaline anion-exchange membranes (AEMs) as PEM [29, 30]. The electrochemical reactions in AEMFCs are expressed as follows:

$$\text{At anode: } CH_3OH + 6OH^- \rightarrow CO_2 + 5H_2O + 6e^- \quad \text{(a)}$$
$$\text{At cathode: } 3/2O_2 + 3H_2O + 6e^- \rightarrow 6OH^- \quad \text{(b)}$$
$$\text{Overall: } CH_3OH + 3/2O_2^- \rightarrow 2H_2O + CO_2$$

In AEMFCs, OH^- ions, instead of protons, are transported across the membrane. In this case, the methanol crossover and the transportation of ions are in the opposite direction. This is expected to decrease the methanol crossover of the membranes and correspondingly to decrease the poisoning of the electrode catalysts by the permeated methanol, which maintains the catalytic activity of the electrode catalysts for longer time and achieving better FC performance [29, 30].

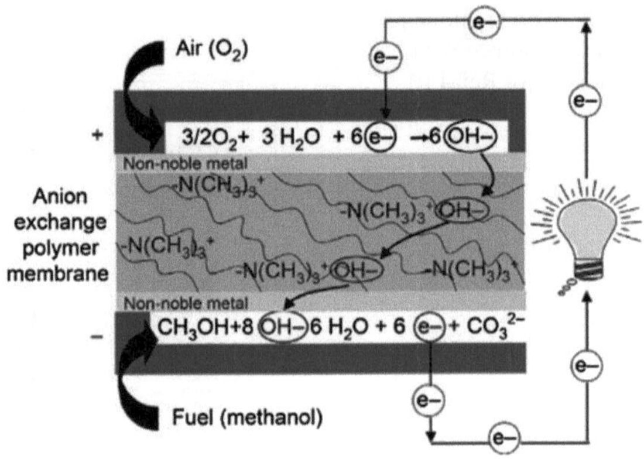

Fig. 6 Operating principle of AEMFCs (reprinted with permission of Ref. [30])

The working principle of AEMFCs is shown in Fig. 6. The expected advantages of AEMFCs, when compared to traditional anionic FCs (AFCs), are the possibility of easy handling and stocking liquid fuels as well as the better control on carbonatation due to the solid nature of the AEMFCs [12, 29, 30]).

3.4 *PAFC*

In the 1960s, PAFC was preferred for the alkaline battery, where the main problem was carbonation. The PAFC was reported as the first commercialized FC. ONSI Corporation, in the United States, and Gaz de France sold 200 kW FC for providing electricity to several houses. However, the high cost as well as the tendency of phosphoric acid to solidify are major drawbacks [30].

3.5 *SOFC*

Like PEMFCs, SOFCs have been widely investigated, especially in the United States, Japan and Europe. These are the most energetically efficient FCs. These are generally used for stationary devices: water, is extremely hot when leaving the FC, goes through turbines to drive generators to produce electricity. However, components have to exhibit a high thermal stability [30].

3.6 MCFC

MCFCs have been developed for stationary applications and considered for promising use of reformed methane or coal gas results in low NOx emissions. The MCFC differs from other FC as it does not require any noble metal as the catalyst, but uses mainly stainless steel. As no platinum is used, the poisoning of this metal by CO gas does not occur and many fuels can thus be used, such as gasified biomass or gasified wastes. Nevertheless, the impurities contained in fuels (H_2S, HCl, HF, NH_3) may enhance the corrosion of the electrodes, especially the cathode, because of the growth of oxides with a poor electrical conductivity. This issue has turned out to be a lock to the development of this MCFCs [30].

4 Catalysts for Different FCs

The CLs regulate the traffic and conversion of hydrogen gas in PEMFC or methanol in DMFC at the anode and oxygen at the cathode. The water's journey through a FC generally starts in the cathode CL where it is produced during the reduction reaction of the oxygen. Also oxygen reduction is a multistep process because it requires the migration of oxygen molecules, protons, and electrons through diffusion and conduction to the catalyst surface on which gas molecules adsorb and charge transfer occurs in the presence of water. Water as a product desorbs and diffuses away through the channels. During oxygen reduction reaction, hydrogen peroxide is an undesired intermediate which may be formed on either the anode or on the cathode; and being highly aggressive, hydrogen peroxide can degrade the PEM. The detailed stoichiometric balance of all those reaction processes can affect the overall rate of electrochemical conversion. Suitable catalysts must be stable in a hot, oxidizing, acidic, aqueous environment, while effectively driving the production of water and resisting deactivation caused by the poisoning effects of fuel contaminants and atmospheric pollutants [1, 2, 4, 5, 9, 14–16, 32–35].

Precious metals and alloys have shown the best balance of catalytic properties because catalysts should expose a large interfacial surface area so it can reduce the irreversible voltage losses during sluggish reactions such as oxygen reduction. Carbon has been preferred as supporting material because of its corrosion resistance, high conductivity and nanostructured morphology. CLs must be incorporated into a composite structure with a minimum of two distinct phases namely (i) the solid phase of carbon and platinum which can conduct electrons, (ii) Water phase filling the pore spaces in the composite which can transport protons, hydrogen and oxygen gas, and other water molecules. These kinds of two-phase thin (100–500 nm) composite layers work well for long because if they are too thick, the diffusion of dissolved reactant molecules and protons will be slow. Conversely, if they are too thin, they will be incapable to provide sufficient reaction surface. Thick layers (5–30 μm) of catalysts are typically used in PEMFCs. Solid carbon and platinum transport electrons to the catalyst, whereas a second, solid subphase, typically composed of polymer such as Nafion, provides sufficient proton conduction. To make the CL porous for creating the third percolation network, gases are allowed to diffuse readily through the catalysts. Generally, water competes with reactant gases in terms of space between the solid components and the degree to which liquid water penetrates into the CL. The ability of CL to direct water into the membrane material rather than the channel or to facilitate its evaporation affects the overall water and heat management of the cell. CL at cathode generates the maximum power density when roughly one-third of its volume is filled with proton conductor and the thickness is kept between 10 and 20 μm. A bimodal porous structures, as shown in Fig. 7, with intra- and inter-agglomerate pores, can provide the good results [2, 5, 15, 16, 33, 34]. Only 10–20% of the catalyst is utilized in three-phase composites to optimize the balance between large active areas and the transport of reactants. Thicker catalyst layers can provide larger areas to support reactions, on other side they generally impedes diffusion. Many researchers, reported proper design strives for new well-defined architectures based on nanometer scale template electrodes—using CNTs as building blocks, which can be used in the future by considering it as porous-electrode technology.

Nanostructured thin-based CL substrates were fabricated to incorporate organic whiskers with high surface-to-volume ratios so that 100% of the catalysts can be used [2, 5]. Wanga et al. [5] reported various catalysts [Pt and several of its alloys (Pt–Co, Pt–Ni, Pt–Fe, Pt–V, Pt–Mn and Pt–Cr)] used for PEM-based FC. Non-precious metal catalysts and self-supported electrocatalysts such as metal, metal alloys are another important part in PEM of FC [14, 15, 32–42]. Various materials such as thin films of tungsten selenides [42] and phosphide-based materials [43] were used for electrocatalytic hydrogen evolution using FCs [42, 43]. The lifetime of a PEMFC is a function of operating conditions, component materials, and degradation mechanisms of electrocatalysts or CL. Wu and Yuan et al. [15, 16] reviewed

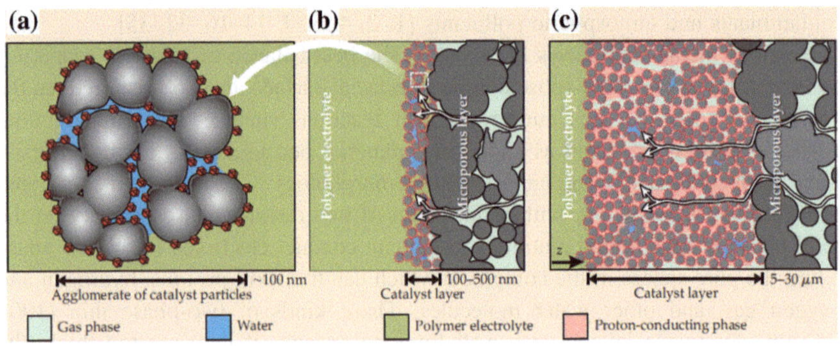

Fig. 7 Structure and composition of CL. **a** Carbon (*gray*) coated in platinum catalysts (*red*) based CL in a PEMFC. **b** If the pores in agglomerate fill with water (*blue*). **c** Secondary pores-based agglomerates in the course of the spontaneous aggregation of carbon particles (reprinted with permission of Ref. [2])

and summarized different degradation mechanisms, migration strategies, and FC durability test protocols for cell components [16–18]. Support materials for PEMFC and DMFC electrocatalysts [33], phosphotungstic acid doped catalysts [44], and microstructural pore analysis of the catalyst layer in a PEMFC [37] were also reviewed and summarized, which include different NPs such as CNTs, CNFs, graphene, nanodiamonds, TiO_2, TiN, TiB_2, SnO_2, WO_3 also conducting polymers.

Sanli and Gursel et al. reported the development of graphene supported platinum (Pt/graphene) as the electrocatalysts for PEMFC [45]. They demonstrated the comprehensive overview on synthesis, characterization, and FC performance of Pt/graphene electrocatalysts. These workers prepared graphene oxide (GO) supported Pt nanoparticles for PEMFCs by several impregnation reduction methods including ethylene glycol reflux, sodium borohydride reduction, and ascorbic acid reduction. The single FC tests showed that GO has the best performance compared to graphene nanoparticles and thermally reduced GO. Catalyst coated membrane prepared by screen printing method were also used in a PEMFC [46]. The PEMFC catalysts can be electrodeposited by the use of a hydrogen depolarized anode [47].

5 Materials and Methods for Preparation of PEM for Fuel Cells

Organic–inorganic nanocomposites-based PEMFCs contain inorganic building blocks at nanoscale in organic polymer by hybridization at molecular level. Intensive researches have been carried out to develop new PEM for DMFC applications. The PEMs in DMFC serve as barrier for methanol and conduct protons. Various PEMs have been used in DMFC, which include perfluorinated ionomer Nafion, polybenzimidazole/phosphoric acid blend membranes, Nafion membrane modified with inorganic phase such as silicate and zirconium phosphate. Due to major drawbacks (high cost, loss of conductivity at high temperature, and high methanol permeability) of perfluorinated polymers restrict their applicability. Therefore, non-perfluorinated PEM based on aromatic thermoplastics, such as polyimides (PI), poly(ether sulfone) (PES), polybenzimidazole (PBI), sulfonated PEEK, have been developed which have shown excellent chemical resistance, high thermo-oxidative stability, good mechanical properties and cost effectiveness [3, 14, 26, 48].

Tripathi et al. [4] have reviewed and summarized different types of organic–inorganic nanocomposites for better understanding of molecular-level chemistry, morphology, transport behavior, and polymer degradation in comparison with Nafion. They reported recent developments about new ionomers and hybrid membranes, containing inorganic NPs to control morphology and water retention by complexation of basic polymers with different oxoacids and inorganic fillers [49]. According to recent report [4], different organic–inorganic nanocomposites were used for PEM, which include perfluorinated organic–inorganic nanocomposites (Nafion with modifiers such as Nafion-silica NPs, Nafion-ZrP, Nafion-titanium dioxide, Nafion-thiophene, Nafion-metal oxides, Nafion-zeolite, Nafion-silica-PWA, Nafion-PAni nanocomposites), partially perfluorinated organic–inorganic nanocomposites (sulfonated styrene with PTFE backbone, functionalized PVDF-silica NPs, PVDF-HFP-Nafion, PVDF-HFP-Al_2O_3, PVDF-PSSA-Al_2O_3), non-perfluorinated organic–inorganic nanocomposites (sulfonated polymers such as PEEK, PI, PES, PBI, polyphynelenes and hydrophilic polymers such as PVA and chitosan), styrene-based nanocomposites (sulfonated co-polymers of styrene, sodium styrene sulfonated grafted polyacrylonitrile, TEOS-grafted polystyrene blocks in polystyrene-b-isobutylene-styrene block copolymers, TEOS grafted PSMA-PEG composites, sulfonated PTFE and grafted with TEOS PSMA-silica hybrid, PSSA-PWA by self-assembly technique), PEEK-based nanocomposites (sulfonated PEEK-silicon dioxide-zirconium phosphate (ZrP), SPEEK–silica, SPEEK-BPO_4 and PEEK-TiO_2 nanocomposites), polyether sulfone (PES) based nanocomposites, polybenzimidazole (PBI) based nanocomposites (PBI-phosphoric acid-based composites, PBI-silicon dioxide), P and polyetherimide-based nanocomposites (sulfonated PI, PI-phosphosilicate gel nanocomposites, PI-polypyrrole composites), chitosan-based nanocomposites (PVA-phosphonic acid modified chitosan nanocomposites, chitosan-silica nanocomposites), PVA-based nanocomposites (PVA-chitosan composites, PVA-H_2PO_4 composites, PVA-silicon dioxide nanocomposites) and layered NP or nanoclay-based nanocomposites (montmorillonite-nafion), finally CNT-based polymer nanocomposites (CNT or sulfonated CNT-based Nafion composites) [4–6, 10, 11, 14, 26, 27, 48, 50].

Several polymer nanocomposites-based PEM such as sulfonated poly(ether ether ketone) (PEEK)/polyaniline composites [3, 50] were also studied. Wanga et al. [5] reviewed the technology, applications, and the need of fundamental research on PEMFC [5]. PEMFCs based on ionomers have also been developed [6]. Similarly, polyvinyl alcohol (PVA)/SiO_2 hybrid based PEMs containing sulfonic acid groups were also developed for DMFC applications [7]. Organic–inorganic membranes were also prepared from polyether diamine and epoxy silane [8]. Development of PEMFC using organic–inorganic nanocomposites was reported in the literatures indicating several modifications such as: (i) improvement in the self-humidification of in membrane; (ii) reduction in the electro-osmotic drag and fuel crossover; (iii) enhancement in the mechanical and thermal strengths without deteriorating proton conductivity; (iv) increase in proton conductivity by introducing solid inorganic proton conductors; and (v) slow drying PEMs with high water retention capability [4]. Researchers have reported the different categories for the development of organic–inorganic nanocomposites based PEMFC which include (i) doping the inorganic proton conductors in PEMs; (ii) nanocomposites prepared by sol–gel method; (iii) covalently bonded inorganic segments with organic polymer chains; and (iv) acid–base PEMFC nanocomposites [4, 6].

Various polymers and their nanocomposites are in use, nowadays, for industrial FC applications [10]. PEMFC composite comprised of triblock copolymer and heteropolyacid have been also used for FC applications [51]. Organic–inorganic hybrid materials were also prepared for the development of PEMFC for separation processes [52]. PEO/$KHCO_3$ nanocomposites were used for PEMFC applications [53]. Efforts have been made to develop PEMFC operating above 100 °C [22, 29]. Silica nanocomposites-based PEMFC systems, prepared by the sol–gel process, of polyethoxysiloxane within a sulfonated PEEK matrix have been reported [54]. Various polymer systems have been studied for applications of proton exchange membranes [27, 48].

Sulfonated polymers [3], poly(aryl ether ketone)s containing the hexafluoroisopropylidene diphenyl moiety [55] and factors affecting the life of PEMFC [56] have been investigated for their PEMFC applications. An interesting study on novel polyvinyl butyl-based polymer membrane and its application in gel polymer electrolytes for lithium-ion batteries have been reported [57]. The suitability of various polymeric material-based nanocomposites membranes for PEMFC applications have examined by several workers [12, 30, 58–63]. Hybrid inorganic–organic nanocomposites made from polymer electrolytes such as PVA/SiO_2 [7], graphene supported platinum nanoparticles (NPs) [45], sulfonated PEEK/silica nanocomposites [64], TiO_2 nanoparticle self-assembled aromatic polyamide [64], nafion and fluorinated TiO_2 for PEMFCs [65], nafion/inorganic nanocomposite [66] sulfonated PEEK/sulfonated NPs composite [67], UV polymerized 1-H-3-vinylimidazolium bis(trifluoromethanesulfonyl)imide [68], Nafion, [(ZrO_2)/(HfO_2)0.25] and [(SiO_2)/(HfO_2)0.28] NPs [69, 70], Nafion and fluorinated TiO_2 NPs [71], novel polymer-coated nanoparticles dispersed-carbon micronanofibers-based air cathode [72], functionalized Al_2O_3 particles as additives in proton-conducting polymer [73], poly (ether-imide) [74], polytetrafluoroethylene (PTFE) [75] and high performance polymer based [76], metal-supported tubular solid oxide [77], SiO_2-ceramic nanoporous substrate-reinforced sulfonated poly(arylene ether sulfone) composite [78], Poly (2,5-benzimidazole) (ABPBI)-based MEA assembly [79], poly(vinyl alcohol) (PVA) [80–82], partly fluorinated poly(arylene ether ketone sulfone) hydrophilicehydrophobic multiblock copolymers [83], multilayer-structured, SiO_2/sulfonated poly(phenylsulfone) composite [84], sulfonated bisphenol-A-polysulfone-based composite PEMs containing tungstophosphoric acid [85], aligned polyaniline nanorods in situ grown on gas diffusion layer [86] were used for PEMFC applications. Various selective membranes made from polymer nanocomposites were studied for hydrogen purification [87]. A galvanostatic analysis technique was developed [88, 89] as an in situ diagnostic tool for PEMFC single cells and stacks.

6 Characterizations and Characteristic Properties of PEM for Different FC

The characteristic properties of polymer nanocomposites at the surface, interfaces, in bulk and under gas and supercritical fluid treatments have generated tremendous interest. The morphological and physicochemical properties of organic–inorganic hybrids depend on their composition, the size of the inorganic fillers and interfacial interactions. Potential characteristics such as mechanical, chemical, and thermal stability are vital for PEM. Stability of PEMs can be optimized by the degree of functionalization, interactions between organic–organic or organic–inorganic segments, controlled cross-linking, and chemical or surface modifications. Functionalization of the organic part or doping of polyelectrolytes in the matrix, are common methods, but suffer due to excessive swelling or leaching out proton carriers on prolonged use at elevated temperature [1–4, 49]. The PEEK/PAni based polymer composites were used for DMFC application considering optimum physicochemical and electrochemical properties, thermal stability as well as very low methanol permeability. Desired chemical compositions of PEM can be identified by NMR and FTIR spectroscopy. The chemical composition of PEM can also be obtained by EDX analysis. The inner structure of the PEM can be identified by TEM. Surface structure and 3D appearance of the PEM can be studied by SEM images. AFM is an effective tool for morphological characterization of PEM. Thickness of PEM can be measured by water uptake (swelling) test at particular temperature. Dimensional stability of PEM is also directly related to the water uptake, swelling properties and degree of functionalization. Organic–inorganic nanocomposites-based PEMs possess interesting route to reduce swelling by cross-linking and water storage via the inorganic segment. The surface functionalization enhances the compactness and thus reduces the swelling due to increased interactions. The hydrolytic and oxidative weight loss, obtained by refluxing the membranes under high temperature and pressure, serve as measures of chemical stability. Thus, knowledge of chemical degradation mechanisms is useful in the design of materials with improved stability and lifetime fuel cells environment. The degradation process and the thermal stability of the PEM were investigated using TGA and DSC. Data of coefficient of thermal expansion (CTE) and viscoelastic behavior of the PEM can be obtained by TMA and DMA or DMTA studies. The dynamic mechanical properties, such as storage modulus, loss modulus, and tan δ, give idea about molecular mobility transitions, such as the glass transition temperature. Ion-exchange capacity of the PEM was determined by salt splitting ion exchange method. Conductance of PEM can be measured by performing the resistance on potentiostatic two electrodes with the help of digital conductivity meter with AC mode. Proton transport number in the PEM phase can be estimated from the membrane potential measurements. Methanol permeability of the PEM can be determined using a diaphragm diffusion cell. The conductivity and methanol permeability of a PEM are crucial parameters, which give idea about their suitability for DMFC applications. To measure these properties, PEM are generally activated by chemical treatments (sulfuric acid and hydrogen peroxide) to remove impurities from the ionomer membrane and to activate sulfonic acid groups. The degree of crystallinity of nonlayered organic–ionorganic hybrid nanocomposites can be analyzed by Wide-angle X-ray scattering (*WAXS*) or wide-angle X-ray diffraction (*WAXD*)gives idea about the change in crystalline structure [3, 4, 6, 7, 58]. Various characteristics of PEMFC such as electrical properties [53], surface morphology, proton mobility [54], effective transport properties with a focus on the gas diffusion layer [34], interfacial properties [90] and bactericidal anti-fouling [64] have been extensively studied. Various models related to PEMFC were also been proposed [34, 91]. Most of proton exchange membranes [14, 92] were designed using their unit operation modeling [13, 93, 94], computational study [95], efficiency enhancement and cost reduction [96], self-passivating carbon film as bipolar plate based protective coating [97, 98], cooling flow designs [99], numerical study investigation of the effects of GDL compression and intrusion [100], PEM degradation modeling [101], improving gas diffusivity with biporous flow-field [102] and the properties effect on gas diffusivity [103] for PEMFC applications. Effects of idling temperature on high-temperature PEMFC degradation were investigated under simulated start/stop cycling conditions [104, 105].

7 Summary

The main aims and objectives behind studying about the various FCs are to give basic understanding of different types of FCs, design and constructions of FCs, materials and methods used for preparation of FCs, characterization and potential properties of FCs. Nonetheless, more emphasize is kept only on PEM for FC applications.

References

1. Jiao K, Li X (2011) Water transport in polymer electrolyte membrane fuel cells. Prog Energy Combus Sci 37:221–291
2. Eikerling M, Kornyshev AA, Kucernak AR (2006) Water in polymer electrolyte fuel cells: friend or foe? Phys Today 59(10):38–44
3. Nagarale RK, Gohil GS, Shahi VK (2006) Sulfonated poly(ether ether ketone)/polyaniline composite proton-exchange membrane. J Membr Sci 280:389–396
4. Tripathi BP, Shahi VK (2011) Organic–inorganic nanocomposite polymer electrolyte membranes for fuel cell applications. Prog Polym Sci 36:945–979
5. Wanga Y, Chen KS, Mishler J, Cho SC, Adroher XC (2011) A review of polymer electrolyte membrane fuel cells: technology, applications, and needs on fundamental research. Appl Ene 88:981–1007
6. Kerres JA (2001) Development of ionomer membranes for fuel cells. J Membr Sci 185:3–2
7. Kima DS, Park HB, Rhim JW, Lee YM (2004) Preparation and characterization of crosslinked PVA/SiO2 hybrid membranes containing sulfonic acid groups for direct methanol fuel cell applications. J Membr Sci 240:37–48
8. Sforc ML, Yoshida IVP, Nunes SP (1999) Organic–inorganic membranes prepared from polyether diamine and epoxy silane. J Membr Sci 159:197–207
9. Costamagn P, Srinivasan S (2001) Quantum jumps in the PEMFC science and technology from the 1960s to the year 2000 Part I. Fundamental scientific aspects. J Power Sources 102:242–252
10. Kumar GG, Nahm KS (2011) Chapter-27: polymer nanocomposites—fuel cell applications. In: Reddy B (ed) Advances in nanocomposites—synthesis, characterization and industrial

applications. InTech China, pp 639–660. Available from: http://www.intechopen.com/books/advances-in-nanocompositessynthesis-characterization-and-industrial-applications/polymer-nanocomposites-fuel-cell-applications
11. Zaidi SMJ (2009) Chapter-2: research trends in polymer electrolyte membranes for PEMFC. In: Zaidi SMJ, Matsuura T (eds) Polymer membranes for fuel cells. pp 7–25
12. Couture G, Alaaeddine A, Boschet F, Ameduri B (2011) Polymeric materials as anion-exchange membranes for alkaline fuel cells. Prog Polym Sci 36:1521–1557
13. Siegel C (2008) Review of computational heat and mass transfer modeling in polymer-electrolyte-membrane (PEM) fuel cells. Energy 33:1331–1352
14. Peighambardoust SJ, Rowshanzamir S, Amjadi M (2010) Review of the proton exchange membranes for fuel cell applications. Int J Hydrogen Energy 35:9349–9384
15. Wu J, Yuan XZ, Martin JJ, Wang H, Zhang J, Shen J, Wu S, Merida W (2008) A review of PEM fuel cell durability: degradation mechanisms and mitigation strategies. J Power Sources 184:104–119
16. Yuan XZ, Li H, Zhang S, Martin J, Wang H (2011) A review of polymer electrolyte membrane fuel cell durability test protocols. J Power Sources 196:9107–9116
17. Millera M, Bazylaka A (2011) A review of polymer electrolyte membrane fuel cell stack testing. J Power Sources 196:601–613
18. Schmittinger W, Vahidi A (2008) A review of the main parameters influencing long-term performance and durability of PEM fuel cells. J Power Sources 180:1–14
19. Carmo M, Fritz DL, Mergel J, Stolten D (2013) A comprehensive review on PEM water electrolysis. Int J Hydrogen Energy 38:4901–4934
20. Ünügür S, Bozkurta A, Hosseini SS (2012) Alternatives toward proton conductive anhydrous membranes for fuel cells: heterocyclic protogenic solvents comprising polymer electrolytes. Prog Polym Sci 37:1265–1291
21. Wee JH (2007) Applications of proton exchange membrane fuel cell systems. Renew Sustain Energy Rev 11:1720–1738
22. Li Q, He R, Jensen JO, Bjerrum NJ (2003) Approaches and recent development of polymer electrolyte membranes for fuel cells operating above 100 °C. Chem Mater 15:4896–4915
23. Merle G, Wessling M, Nijmeijer K (2011) Anion exchange membranes for alkaline fuel cells: a review. J Membr Sci 377:1–35
24. Hermanna A, Chaudhuria T, Spagnol P (2005) Bipolar plates for PEM fuel cells: a review. Int J Hydrogen Energy 30:1297–1302
25. Alizadeh E, Rahgoshay SM, Esbo MR, Khorshidian M, Saadat SHM (2016) A novel cooling flow field design for polymer electrolyte membrane fuel cell stack. Int J Hydrogen Energy 41:8525–8532
26. Neburchilov V, Martin J, Wang H, Zhang J (2007) A review of polymer electrolyte membranes for direct methanol fuel cells. J Power Sources 169:221–238
27. Zhang H, Shen PK (2012) Recent development of polymer electrolyte membranes for fuel cells. Chem Rev 112:2780–2832
28. Zamel N, Li X (2013) Effective transport properties for polymer electrolyte membrane fuel cells with a focus on the gas diffusion layer. Prog Energy Comb Sci 39:111–146
29. Jiang Z, Jiang ZJ (2014) Plasma techniques for the fabrication of polymer electrolyte membranes for fuel cells. J Membr Sci 456:85–106
30. Schieda M, Roualdès S, Durand J, Martinent A, Marsacq D (2006) Plasma polymerized thin films as new membranes for iniature solid alkaline fuel cells. Desalination 199:286–288
31. Noto VD, Zawodzinski TA, Herring AM, Giffin GA (2012) Polymer electrolytes for a hydrogen economy. Int J Hydrogen Energy 37:6120–6131
32. Ma TY, Dai S, Qiao SZ (2016) Self-supported electrocatalysts for advanced energy conversion processes. Mater Today 19(5):265–27
33. Sharma S, Pollet BG (2012) Support materials for PEMFC and DMFC electrocatalysts—a review. J Power Sources 208:96–119
34. Gottesfeld S, Zawodzinski TA (1997). In: Alkire RC, Gerischer H, Kol DM, Tobias CW (eds) Polymer electrolyte fuel cells. Adv Electrochem Sci Eng, 196–301

35. Othman R, Dicks AL, Zhu Z (2012) Non precious metal catalysts for the PEM fuel cell cathode. Int J Hydrogen Energy 37:357–372
36. Gasteiger HA, Kocha SS, Sompalli B, Wagner FT (2005) Activity benchmarks and requirements for Pt, Pt-alloy and non-Pt oxygen reduction catalysts for PEMFCs. App Cataly B: Environ 56:9–35
37. Ghosh S, Ohashi H, Tabata H, Hashimasa Y, Yamaguchi T (2015) Microstructural pore analysis of the catalyst layer in a polymer electrolyte membrane fuel cell: a combination of resin pore-filling and FIB/SEM. Int J Hydrogen Energy 40:15663–15671
38. Kim T, Popov BN (2016) Development of highly-active and stable Pt/C catalyst for polymer electrolyte membrane fuel cells under simulated start-up/shut-down cycling. Int J Hydrogen Energy 41:1828–1836
39. Cho YH, Lim JW, Park HY, Jung N, Ahn M, Choe H, Sung YE (2012) Performance of membrane electrode assemblies using PdPt alloy as anode catalysts in polymer electrolyte membrane fuel cell. Int J Hydrogen Energy 37:5884–5890
40. Lee KS, Kim DM (2012) Sputtering and heat treatment of pure Ni metal onto a carbon nanotube on carbon paper to fabricate electrocatalysts for the oxygen reduction reaction in PEMFC. Int J Hydrogen Energy 37:6272–6276
41. Oh SK, Kim MJ, Eom KS, Kyung JS, Kim DH, Cho EA, Kwon HS (2016) Design of MgeNi alloys for fast hydrogen generation from seawater and their application in polymer electrolyte membrane fuel cells. Int J Hydrogen Energy 41:5296–5303
42. Romanov R, Grigoriev S, Fominski V, Volosova M, Demin M (2015) Preparation and study of thin films of tungsten selenides for electrocatalytic hydrogen evolution. Phys Proced 71:348–353
43. Xiao P, Chen W, Wang X (2015) A review of phosphide based materials for electrocatalytic hydrogen evolution. Adv Energy Mater 5(24):1–13
44. Chen GY, Wang C, Lei YJ, Zhang J, Mao ZQ, Guo JW, Wang JL (2016) Catalyst layer doped with phosphotungstic acid for degradation mitigation in polymer electrolyte membrane fuel cells. Int J Hydrogen Energy 1–6 (Accepted) http://dx.doi.org/10.1016/j.ijhydene.2016.04.222
45. Sanlı LI, Bayram V, Yarar B, Ghobadi S, Gursel SA (2016) Development of graphene supported platinum nanoparticles for polymer electrolyte membrane fuel cells: effect of support type and impregnation-reduction methods. Int J Hydrogen Energy 41:3414–3427
46. Wang W, Chen S, Li J, Wang W (2015) Fabrication of catalyst coated membrane with screen printing method in a proton exchange membrane fuel cell. Int J Hydrogen Energy 40:4649–4658
47. Mitzel J, Arena F, Natter H, Walter T, Batzer M, Stefener M, Hempelmann R (2012) Electrodeposition of PEM fuel cell catalysts by the use of a hydrogen depolarized anode. Int J Hydrogen Energy 37:6261–6267
48. Hickner MA, Ghassemi H, Kim YS, Einsla BR, McGrath JE (2004) Alternative polymer systems for proton exchange membranes (PEMs). Chem Rev 104:4587–4612
49. Antonacci P, Chevalier S, Lee J, Yip R, Ge N, Bazylak A (2015) Feasibility of combining electrochemical impedance spectroscopy and synchrotron X-ray radiography for determining the influence of liquid water on polymer electrolyte membrane fuel cell performance. Int J Hydrogen Energy 40:16494–16502
50. Seo DW, Lim YD, Lee SH, Jeong IS, Kim DI, Lee JH, Kim WG (2012) Preparation and characterization of sulfonated poly (tetra phenyl ether ketone sulfone)s for proton exchange membrane fuel cell. Int J Hydrogen Energy 37:6140–6147
51. Choi JK, Lee DK, Kim YW, Min BR, Kim JH (2008) Composite polymer electrolyte membranes comprising triblock copolymer and heteropolyacid for fuel cell applications. J Polym Sci, Part B: Polym Phys 46:691–701
52. Souza VC, Quadri MGN (2013) Organic–inorganic hybrid membranes in separation processes: a 10-year review. Braz J Chem Eng 30(04):683–700

53. Kumar KV, Sundari GS, Jyothi NK (2010) Fabrication and electrical characterization of PEM fuel cell based on (PEO + $KHCO_3$) polymer electrolyte membrane. Int J Inn Res Sci, Eng Technol 2(10):5838–5847
54. Colicchio I (2008) Silica-based nanocomposite membranes via the sol gel process of polyethoxysiloxane within a sulfonated poly(ether-ether-ketone) matrix: morphology and proton mobility. PhD Theses, Universitäts professor Dr. rer. nat. Martin Möller
55. Xing P, Robertson GP, Guiver MD, Mikhailenko SD, Kaliaguine S (2004) Sulfonated poly (aryl ether ketone)s containing the hexafluoroisopropylidene diphenyl moiety prepared by direct copolymerization as proton exchange membranes for fuel cell application. Macromol 37:7960–7967
56. Pei P, Chen H (2014) Main factors affecting the lifetime of proton exchange membrane fuel cells in vehicle applications: a review. Appl Energy 125:60–75
57. Lian F, Wen Y, Ren Y, Guan HY (2014) A novel PVB based polymer membrane and its application in gel polymer electrolytes for lithium-ion batteries. J Membr Sci 456:42–48
58. Mishra AK, Bose S, Kuila T, Kim NH, Lee JH (2012) Silicate-based polymer-nanocomposite membranes for polymer electrolyte membrane fuel cells. Prog Polym Sci 37:842–869
59. Litster S, McLean G (2004) PEM fuel cell electrodes. J Power Sources 130:61–76
60. Prater KB (1994) Polymer electrolyte fuel cells: a review of recent developments. J Power Sources 51:129–144
61. Bose S, Kuila T, Hien TX, Nguyen Kimc NH, Laua KT, Lee JH (2011) Polymer membranes for high temperature proton exchange membrane fuel cell: recent advances and challenges. Prog Polym Sci 36:813–843
62. Yuan W, Tang Y, Yang X, Wan Z (2012) Porous metal materials for polymer electrolyte membrane fuel cells—a review. Appl Ene 94:309–329
63. Perez LC, Brandaoa L, Sousa JM, Mendes A (2011) Segmented polymer electrolyte membrane fuel cells—a review. Renew Sustain Energy Rev 15:169–185
64. Kwak SY, Kim SH (2001) Hybrid organic/inorganic reverse osmosis (RO) membrane for bactericidal anti-fouling. 1. Preparation and characterization of TiO_2 nanoparticle self-assembled aromatic polyamide thin-film-composite (TFC) membrane. Environ Sci Technol 35:2388–2394
65. Noto VD, Bettiol M, Bassetto F, Boaretto N, Negro E, Lavina S, Bertasi F (2012) Hybrid inorganic-organic nanocomposite polymer electrolytes based on nafion and fluorinated TiO_2 for PEMFCs. Int J Hydrogen Energy 37:6169–6181
66. Devrim Y, Erkan S, Bac N, Eroglu I (2012) Improvement of PEMFC performance with Nafion/inorganic nanocomposite membrane electrode assembly prepared by ultrasonic coating technique. Int J Hydrogen Energy 37:16748–16758
67. Kim DJ, Choi DH, Park CH, Nam SY (2016) Characterization of the sulfonated PEEK/sulfonated nanoparticles composite membrane for the fuel cell application. Int J Hydrogen Energy 41:5793–5802
68. Lemus J, Eguizabal A, Pina MP (2016) Endurance strategies for the preparation of high temperature polymer electrolyte membranes by UV polymerization of 1-H-3-vinylimidazolium bis(trifluoromethanesulfonyl)imide for fuel cell applications. Int J Hydrogen Energy 41:3981–3993
69. Noto VD, Boaretto N, Negro E, Giffin GA, Lavina S, Polizzi S (2012) Inorganic-organic membranes based on Nafion, [(ZrO2)$(HfO2)0.25] and [(SiO2)$(HfO2)0.28]. Part I: Synthesis, thermal stability and performance in a single PEMFC. Int J Hydrogen Energy 37:6199–6214
70. Noto VD, Boaretto N, Negro E, Stallworth PE, Lavina S, Giffin GA, Greenbaum SG (2012) Inorganic-organic membranes based on Nafion, $[(ZrO_2)\text{-}(HfO_2)_{0.25}]$ and $[(SiO_2)\text{-}(HfO_2)_{0.28}]$ nanoparticles. Part II: relaxations and conductivity mechanism. Int J Hydrogen Energy 37:6215–6227

71. Thampan T, Atwater T, Cook C, Novoa J, Sutorik AC (2016) Hydrogen generation from aluminum hydride for wearable polymer electrolyte membrane fuel cells. Int J Hydrogen Energy 41:9402–9409
72. Singh S, Modi A, Verma N (2016) Enhanced power generation using a novel polymer-coated nanoparticles dispersed-carbon micro-nanofibers-based air-cathode in a membrane-less single chamber microbial fuel cell. Int J Hydrogen Energy 41:1237–1247
73. Branchi M, Sgambetterra M, Pettiti I, Panero S, Navarra MA (2015) Functionalized Al_2O_3 particles as additives in proton-conducting polymer electrolyte membranes for fuel cell applications. Int J Hydrogen Energy 40:14757–14767
74. Elangovan M, Dharmalingam S (2016) Preparation and performance evaluation of poly (ether-imide) based anion exchange polymer membrane electrolyte for microbial fuel cell. Int J Hydrogen Energy 41:8595–8605
75. Mack F, Morawietz T, Hiesgen R, Kramer D, Gogel V, Zeis R (2016) Influence of the polytetrafluoroethylene content on the performance of high-temperature polymer electrolyte membrane fuel cell electrodes. Int J Hydrogen Energy 41:7475–7483
76. Wei Z, Su K, Sui S, He A, Du S (2015) High performance polymer electrolyte membrane fuel cells (PEMFCs) with gradient Pt nanowire cathodes prepared by decal transfer method. Int J Hydrogen Energy 40:3068–3074
77. Han Z, Yang Z, Han M (2016) Fabrication of metal-supported tubular solid oxide fuel cell by phase-inversion method and in situ reduction. Int J Hydrogen Energy 41:10935–10941
78. Seol JH, Wona JH, Yoon KS, Hong YT, Lee SY (2012) SiO_2 Ceramic nanoporous substrate-reinforced sulfonated poly(arylene ether sulfone) composite membranes for proton exchange membrane fuel cells. Int J Hydrogen Energy 37:6189–6198
79. Wanga S, Shang Y, Wanga Y, Wanga J (2013) Fabrication and electrochemical performance of Poly (2,5-benzimidazole) (ABPBI)-based MEA by catalyst coated membrane (CCM) method for high-temperature polymer electrolyte fuel cells. Int J Hydrogen Energy 38:11060–11066
80. Zhong S, Cui X, Gao Y, Liu W, Dou S (2014) Fabrication and properties of poly(vinyl alcohol)-based polymer electrolyte membranes for direct methanol fuel cell applications. Int J Hydrogen Energy 39:17857–17864
81. Higa M, Hatemura K, Sugita M, Maesowa SI, Nishimura M, Endo N (2012) Performance of passive direct methanol fuel cell with poly(vinyl alcohol)-based polymer electrolyte membranes. Int J Hydrogen Energy 37:6292–6301
82. Gomes AS, Filho JCD (2012) Hybrid membranes of PVA for direct ethanol fuel cells (DEFCs) applications. Int J Hydrogen Energy 37:6246–6252
83. Chen Y, Guo R, Lee CH, Lee M, McGrath JE (2012) Partly fluorinated poly(arylene ether ketone sulfone) hydrophilicehydrophobic multiblock copolymers for fuel cell membranes. Int J Hydrogen Energy 37:6132–6139
84. Lee JR, Wona JH, Yoon KS, Hong YT, Lee SY (2012) Multilayer-structured, SiO_2/sulfonated poly(phenylsulfone) composite membranes for proton exchange membrane fuel cells. Int J Hydrogen Energy 37:6182–6188
85. Filho AAMF, Gomes AS (2012) Sulfonated bisphenol-A-polysulfone based composite PEMs containing tungstophosphoric acid and modified by electron beam irradiation. Int J Hydrogen Energy 37:6228–6235
86. Fu X, Wang S, Xia Z, Li Y, Jiang L, Sun G (2016) Aligned polyaniline nanorods in situ grown on gas diffusion layer and their application in polymer electrolyte membrane fuel cells. Int J Hydrogen Energy 41:3655–3663
87. Chen HZ, Chung TS (2012) CO_2-selective membranes for hydrogen purification and the effect of carbon monoxide (CO) on its gas separation performance. Int J Hydrogen Energy 37:6001–6011
88. George D, Suresh PV (2016) Detailed analysis of integrated steam ethanol reformer and high temperature polymer electrolyte membrane fuel cell. Int J Hydrogen Energy 41:1248–1258

89. Lee KS, Lee BS, Yoo SJ, Kim SK, Hwang SJ, Kim HJ, Cho EA, Henkensmeier D, Yun JW, Nama SW, Lim TH, Jang JH (2012) Development of a galvanostatic analysis technique as an in-situ diagnostic tool for PEMFC single cells and stacks. Int J Hydrogen Energy 37:5891–5900
90. Cho YH, Yoo SJ, Cho YH, Park HS, Park IS, Lee JK, Sung YE (2008) Enhanced performance and improved interfacial properties of polymer electrolyte membrane fuel cells fabricated using sputter-deposited Pt thin layers. Electrochim Acta 53:6111–6116
91. Springer TE, Zawodzinski TA, Gottesfel S (1991) Polymer electrolyte fuel cell model. J Electrochem Soc 138(8):2334–2342
92. Zhang L, Chae SR, Hendren Z, Park JS, Wiesner MR (2012) Recent advances in proton exchange membranes for fuel cell applications. Chem Eng J 204–206:87–97
93. Caglayan DG, Sezgin B, Devrim Y, Eroglu I (2016) Three-dimensional modeling of a high temperature polymer electrolyte membrane fuel cell at different operation temperatures. Int J Hydrogen Energy 41:10060–10070
94. Zhang Y, Verma A, Pitchumani R (2016) Optimum design of polymer electrolyte membrane fuel cell with graded porosity gas diffusion layer. Int J Hydrogen Energy 4:8412–8426
95. Kim DK, Koh JS, Kim MS, Song HH (2015) Experimental and computational study on the dynamic interaction between load variation and back pressure control in a polymer electrolyte membrane fuel cell for automotive application. Int J Hydrogen Energy 40:12370–12381
96. Torkavannejad A, Sadeghifar H, Pourmahmoud N, Ramin F (2015) Novel architectures of polymer electrolyte membrane fuel cells: efficiency enhancement and cost reduction. Int J Hydrogen Energy 40:12466–12477
97. Wang Z, Feng K, Li Z, Lu F, Huang J, Wu Y, Chu PK (2016) Self-passivating carbon film as bipolar plate protective coating in polymer electrolyte membrane fuel cell. Int J Hydrogen Energy 41:5783–5792
98. Feng K, Guo X, Li Z, Yao C, Wu Y (2016) Investigation of multi-coating process treated magnesium alloy as bipolar plate in polymer electrolyte membrane fuel cell. Int J Hydrogen Energy 41:6020–6028
99. Fecarotti C, Andrews J, Chen R (2016) A petri net approach for performance modelling of polymer electrolyte membrane fuel cell systems. Int J Hydrogen Energy1–19 (Accepted). http://dx.doi.org/10.1016/j.ijhydene.2016.05.138
100. Chippar P, Kyeongmin O, Kang K, Ju H (2012) A numerical investigation of the effects of GDL compression and intrusion in polymer electrolyte fuel cells (PEFCs). Int J Hydrogen Energy 37:6326–6338
101. Jereb LK, Sternig C, Fink C, Tatschl R (2016) Membrane degradation model for 3D-CFD analysis of fuel cell performance as a function of time. Int J Hydrogen Energy 1–13 (Accepted). http://dx.doi.org/10.1016/j.ijhydene.2016.05.229
102. Kozakai M, Date K, Tabe Y, Chikahisa T (2016) Improving gas diffusivity with bi-porous flow-field in polymer electrolyte membrane fuel cells. Int J Hydrogen Energy 1–10 (Accepted). http://dx.doi.org/10.1016/j.ijhydene.2016.05.131
103. Dhanushkodi SR, Capitanio F, Biggs T, Merida W (2015) Understanding flexural, mechanical and physicochemical properties of gas diffusion layers for polymer membrane fuel cell and electrolyzer systems. Int J Hydrogen Energy 40:16846–16859
104. Pinar FJ, Rastedt M, Pilinski N, Wagner P (2016) Effect of idling temperature on high temperature polymer electrolyte membrane fuel cell degradation under simulated start/stop cycling conditions. Int J Hydrogen Energy 1–12 (Accepted). http://dx.doi.org/10.1016/j.ijhydene.2016.05.091
105. Lee HJ, Kim BG, Lee DH, Park SJ, Kim Y, Lee JW, Henkensmeier D, Namb SW, Kim HJ, Kim H, Kim JY (2011) Demonstration of a 20 W class high-temperature polymer electrolyte fuel cell stack with novel fabrication of a membrane electrode assembly. Int J Hydrogen Energy 36:5521–5526

Chapter 17
Proton Conducting Polymer Electrolytes for Fuel Cells via Electrospinning Technique

Mahadevappa Y. Kariduraganavar, Balappa B. Munavalli and Anand I. Torvi

Abstract Fuel cells are gaining a considerable attention as a clean and promising technology for energy conversion in the twenty-first century. One of the key benefits of fuel cells is the direct energy conversion that enables the achievement of high efficiency. Proton exchange membranes (PEMs) are the key components in fuel cell system and there is a considerable application-driven interest in lowering the membrane cost and extending the operating window of PEMs. Current proton exchange membrane fuel cells (PEMFC) technology is based on expensive perfluorinated PEMs that operate effectively only under fully hydrated conditions. To address this problem, electrospinning is a promising technique, which can produce nanoscale fibres. This chapter thus presents an overview of fuel cell technology and production of proton exchange membranes developed through electrospinning technique. An attempt was also made to discuss the recent progress made on the new materials, such as Nafion, poly(vinylidene fluoride) (PVDF), poly(ethylene oxide) (PEO), poly(vinyl alcohol) (PVA), etc. The unique three-dimensional network structures of the electrospun membranes offer adequate mechanical properties and proton conductivity. Among the nanofibres, sulfonated polyimide nanofibres showed improved membrane stability. Composite membranes composed of highly conductive and selective layer-by-layer (LbL) films and electrospun fibre mats are investigated for mechanical strength and electrochemical selectivity. At the end, we have discussed the present status and the future prospectus of electrospun nanofibres for fuel cell applications. To compile this chapter and to provide adequate information to the readers, we have explored all possible ways, such as research articles, reviews, books, book chapters and Google sites.

M.Y. Kariduraganavar (✉) · B.B. Munavalli · A.I. Torvi
Department of Studies in Chemistry, Karnatak University, Pavate Nagar, Dharwad 580003, Karnataka, India
e-mail: mahadevappak@yahoo.com

B.B. Munavalli
e-mail: bbmunavalli712@gmail.com

A.I. Torvi
e-mail: anandtorvi49@gmail.com

Inamuddin et al. (eds.), *Organic–Inorganic Composite Polymer Electrolyte Membranes*, DOI 10.1007/978-3-319-52739-0_17

Abbreviations

AFC	Alkaline fuel cells
BPPO	Bromomethylated poly(2,6-dimethyl-1,4-phenylene oxide)
CTE	Coefficient of thermal expansion
DC	Direct current
DMFC	Direct methanol fuel cell
DSC	Differential scanning calorimetry
EF	Electrospun fibre
EPM	Electrospun PVDF membrane
FCs	Fuel cells
FCV	Fuel cell vehicle
IEC	Ion exchange capacity
LBL	Layer-by-layer
LPG	Liquefied petroleum gas
MCFC	Molten carbonate fuel cells
MEA	Membrane electrode assemblies
NF	Nanofibre
NGO	Non-governmental organizations
NR	Nanorod
NW	Nanowire
OEM	Original equipment manufacturer
PA 6(3) T	Poly(trimethyl hexamethylene terephthalamide)
PAA	Poly(acrylic acid)
PAFC	Phosphoric acid fuel cells
PDAC	Poly(diallyl dimethyl ammonium chloride)
PEEK	Poly(ether ether ketone)
PEK	Poly(ether ketone)
PEM	Proton exchange membrane
PEMFC	Proton exchange membrane fuel cell
PEO	Poly(ethylene oxide)
PFSA	Perfluorosulfonic acid
PHR	Phenoxy resin
PLLA	Poly-L-lactide
POSS	Polyhedral oligosilsesquioxane
PPO	Poly(2,6-dimethyl-1,4-phenylene oxide)
PPO	Poly(2,6-dimethyl-1,4-phenyleneoxide)
PS	Poly(styrene)
PSSA-MA	Poly(styrene sulphonic acid-co-maleic acid)
PTFE	Polytetrafluoroethylene
PVA	Poly(vinyl alcohol)
PVC	Poly(vinyl chloride)
PVDF	Poly(vinylidene fluoride)
PVP	Polyvinyl pyrrolidone
RH	Relative humidity

SEM	Scanning electron microscopy
SOFC	Solid oxide fuel cells
T_c	Crystaline temperature
T_g	Glass transition
T_m	Melting temperature
WU	Water uptake

1 Introduction

Fuel cells (FCs) due to their unique properties are on the verge of creating a vast revolutionary change in the field of electricity. These are considered to be a promising technology for clean and efficient power generation in the twenty-first century. This technology was first invented by Sir Grove [1]. However, it has emerged as an alternative to traditional system for energy production only from 1960.

Fuel cell is a device that generates electricity through chemical reactions. Every fuel cell has two electrodes, one positive and the other one negative, called anode and cathode, respectively. The reactions that produce electricity take place only at the electrodes. Every fuel cell has also an electrolyte, which carries electrically charged particles from one electrode to other and a catalyst, which accelerates the reactions at the electrodes. Hydrogen is the basic fuel, but fuel cells also require oxygen. One interesting feature of fuel cells is that they generate electricity with very little pollution. In fuel cells, hydrogen and oxygen used for generating electricity ultimately combine to form a harmless by-product water. A single fuel cell generates a tiny amount of direct current (DC) electricity. In practice, many such fuel cells are usually assembled into a stack [2].

2 Classifications of Fuel Cells

Fuel cells are generally classified based on the employed criterion which typically depends on different parameters related to operating conditions and fuel cell structure. Fuel cell systems have different variables such as type of the electrolyte, type of the exchanged ion through the electrolyte, type of the reactants (e.g. primary fuels and oxidants), operating temperature and pressure, direct and indirect usage of primary fuels in the system and finally the primary and regenerative systems. Generally, fuel cells are classified according to the nature of electrolyte used in the fuel cell as [3, 4]:

(i) Alkaline fuel cells (AFCs) with alkaline solution as electrolyte;
(ii) Phosphoric acid fuel cells (PAFCs) with acidic solution as electrolyte;
(iii) Proton exchange membrane fuel cells (PEMFCs) with proton exchange membrane as electrolyte;
(iv) Molten carbonate fuel cells (MCFCs) with molten carbonate salt as electrolyte; and
(v) Solid oxide fuel cells (SOFCs) with ceramic ion as electrolyte.

The operating parameters and applicable properties of these fuel cells are summarized in Table 1. During the past few years, many advances have been achieved, but we still face many technical and economic obstacles in the commercialization of fuel cells. Recently, much of the research is focused on new materials for PEMFC membranes, which could be the next generation technology. Today's research is based on materials that can accelerate the commercialization of fuel cells and therefore serious efforts are being continued to develop proton exchange membrane with improved performance and durability [5–8].

Proton exchange membrane fuel cells work with a polymer electrolyte in the form of a thin permeable sheet. Efficiency is about 45–60% and operating temperature is about 50–80 °C. Cells output generally ranges from 50 to 250 kW. As far as possible, these cells operate at a low temperature to make them suitable for wide range of applications. To enhance the efficiency, platinum catalyst is used on both sides of the membrane. Hydrogen atoms are stripped of their electrons or ionized at the anode and the positively charged protons diffuse through one side of the porous membrane and migrate towards the cathode. The electrons migrate from anode to cathode through an external circuit and provide electric power along the way. At the cathode, the electrons, hydrogen protons and oxygen from the air combine to form water. The proton exchange membrane electrolyte must allow hydrogen protons to pass through while prohibiting the passage of electrons [9–11]. The schematic diagram of PEMFC is illustrated in Fig. 1 [12].

Anode electrode: $H_2 \rightarrow 2H^+ + 2e^-$
Cathode electrode: $\frac{1}{2}O_2 + 2H^+ + 2e^- \rightarrow H_2O$
The overall reaction of the cell [13]: $H_2 + \frac{1}{2}O_2 \rightarrow H_2O$

Proton conduction is the fundamental basis for proton exchange membrane fuel cells and is usually the first characteristic to be considered for evaluating the membranes efficiency for fuel cell applications. Resistive loss is proportional to the ionic resistance of the membrane and high conductivity is essential for the required performance especially at high current density [3, 14, 15].

2.1 Proton Exchange Membrane in Fuel Cell

Proton exchange membrane is the core component of the PEM fuel cell. It helps to carry the charge of protons to separate the reactant gases and electronic insulator

Table 1 Operating parameters and applicable properties of fuel cells

Type of fuel cell	Common electrolyte	Stack size	Operating temp. (°C)	Power density (mW/cm^2)	Fuel efficiency (Chem. to Elec.)	Lifetime (hr)	Capital cost ($/kW)	Applications
AFC	Potassium hydroxide	10–100 kW	60–90	100–200	40–60	>10,000	>200	Space, mobile
PAFC	Phosphoric acid	400 kW	160–220	200	55	>40,000	3000	Distributed power
PEMFC	Perfluorosulfonic acid	<1–100 kW	50–80	350	45–60	>40,000	>200	Portable, mobile, stationary
MCFC	Potassium carbonates	300 kW–3 MW	600–700	100	60–65	>40,000	1000	Distributed power generation
SOFC	Yttria stabilized zirconia	1 kW–2 MW	800–1000	240	55–65	>40,000	1500	Baseload power generation

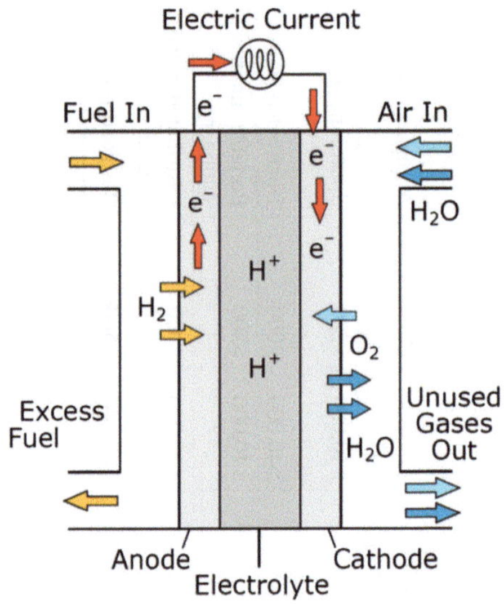

Fig. 1 Schematic representation of proton exchange membrane fuel cell

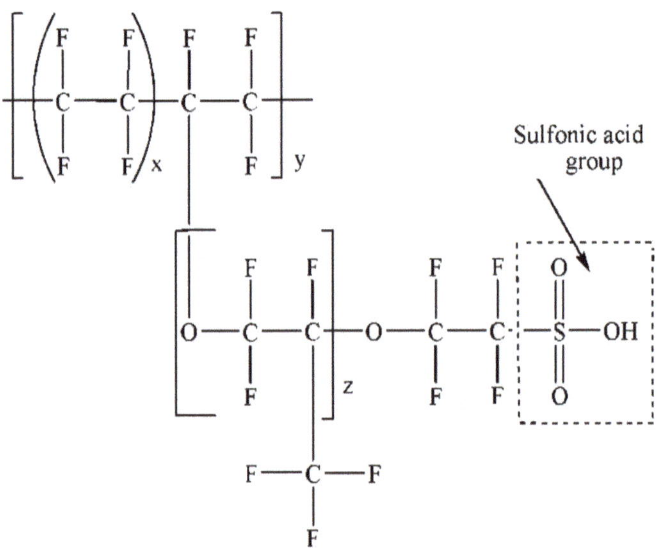

Fig. 2 Structure of Nafion with sulfonic acid group (x = 5–13.5, y = 1000, z > 1) (reproduced with kind permission from [18])

through the membrane [16, 17]. In 1970s, DuPont developed a perfluorosulfonic acid membrane called Nafion that has not only showed a twofold increase in the specific conductivity of the membrane, but also extended the life time by four orders of the magnitude. The structure of perfluorosulfonic acid is shown in Fig. 2.

Currently perfluorosulfonic acid (PFSA) membranes have been widely used as the proton exchange membranes due to their excellent chemical and thermal stabilities and high-proton conductivity in the hydrated state. However, there are some obstacles to use Nafion in fuel cell applications that include high cost, high methanol crossover, poor conductivity at low humidity and high operating temperature, complicated constructions of system and poisoning of anode catalysts above 100 °C by trace amounts of carbon monoxide [18–22].

In order to overcome these obstacles, alternative proton exchange membranes based on, modified perfluorinated materials, sulfonated polyaromatic hydrocarbons, acid–base blends and inorganic–organic composite materials have been developed. Acid-functionalized aryl-type hydrocarbon polymers like sulfonated polyimides, sulfonated poly(ether ketone)s, sulfonated poly(arylene ether sulfone)s, sulfonated poly(styrene) (sPS), sulfonated poly(benzimidazole)s, polyelectrolyte complexes and dendritic polymers [23–30] have been pursued as alternative membranes in recent years owing to their good mechanical property, excellent thermal stability, high-proton conductivity, acceptable cost, reasonable chemical durability and low methanol permeability [31, 32].

Recently, materials with nanometer-scale dimensions have attracted a great deal of attention. Nanomaterials and nanostructured materials, including nano-coatings, films and composites, are the prime choice as new and novel energy materials because of their reduced dimensions and the effect of surface properties on their behaviour while providing a high-volume fraction of interfaces and increased reaction rates. Therefore, exploring material chemistry down to the nanoscale to develop novel solutions for energy applications is nowadays possible by a systematic approach involving new methods for manipulation, assembly and characterization. Targeted development of new one-dimensional (1D) nanostructures, such as continuous nanofibres (NFs), large aspect ratio nanowires (NWs) and nanorods (NRs), is of much interest because of the dependence of their physical properties on directionality [33, 34]. The unique properties such as extremely large surface area, complex pore size and unique alignment on either woven or nonwoven fibre possessed by electrospun nanofibres make it practical in various applications especially in fuel cell system [35]. Figure 3 depicts the possible applications of electrospun nanofibres.

Other than electrospinning process, we can also find common methods such as drawing, template synthesis, phase separation and self-assembly. However, due to the versatility possessed by electrospinning process, it is more favourable to use in developing highly porous, patterned and nanosized fibres. Among the aforementioned preparation methods, electrospinning technique is the most versatile processing method, owing to the simple and easy production of polymer nanofibres. The electrospinning process can be advanced for mass production of long continuous nanofibres from various types of polymers. Furthermore, the electrospinning method can offer electrospun nanofibres with higher network ion exchange capability, which is crucial for proton transport process in fuel cell system. Thus, electrospinning is a top-down approach that is simple, versatile and cost effective [34, 36–38].

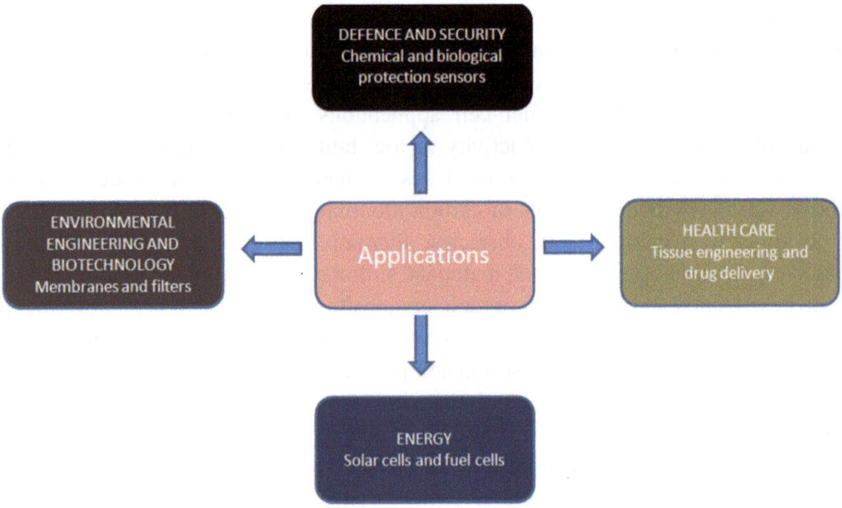

Fig. 3 Applications of electrospun fibres (reproduced with kind permission from [35])

3 Electrospinning Technique

The process of spinning fibres with the help of electrostatic forces is known as electrospinning. This is a unique approach using electrostatic forces to produce fine fibres. Fibre production method using electrostatic forces has invoked glaring attention due to its potential to form fine fibres. Electrospun fibres have small pore size and high surface area. There is also evidence of sizable static charges in electrospun fibres that could be effectively handled to produce three-dimensional structures. Compared to the conventional methods (wet spinning, dry spinning and melt spinning), the electrospinning method can produce fibres having much larger specific surface area and smaller pores size. Fibre diameter in the range of 10–1000 nm by adjusting the operating parameters can be achieved, whereas the conventional methods can only produce fibres with diameter in the range of 5×10^3 to 500×10^3 nm. Electrospinning can thus produce membranes with small diameter in the range of 10–1000 nm with high porosity due to random deposition on the collector [34, 39].

3.1 Mechanism of Electrospinning

Electrospinning is based on the application of an electric field to a drop of fluid polymer on the tip of a spinneret. In this technique, high-voltage electrostatic field is used to charge the surface of the polymer droplet and that will induce the ejection of a liquid jet through the needle tip. This is how, the electrostatic force can

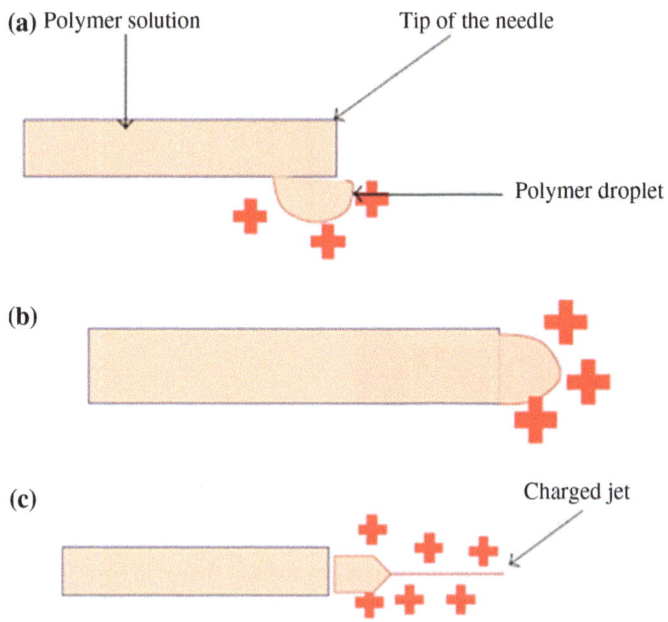

Fig. 4 Formation of Taylor cone with the equilibrium of electrostatic field (adopted with kind permission from [18])

overcome the surface tension of the droplet and form the Taylor cone at the tip of the needle, which eventually leads to charged jet (Fig. 4a, b). The formation of Taylor cone is proportional to the applied voltage, where the voltage keeps increasing until the equilibrium condition is achieved between the surface tension and the electrostatic force (Fig. 4c). The electric field controls the route of charged jet and the resulting solidified spun fibre will be collected on the rotating or stationary conductive collector [18].

3.2 Working Principle of Electrospinning

Electrospinning seems to be a very simple and sophisticated processing technique of producing nanofibres. The electrospinning process can be divided into several techniques, such as vibration electrospinning, magneto-electrospinning, siro-electrospinning and bubble electrospinning. The apparatus used for electrospinning is simple in construction, which consists of a high-voltage electric source with positive or negative polarity, a syringe pump with capillaries or tubes to carry the solution from the syringe or pipette to the spinnerette and a conducting collector like aluminium. The collector can be

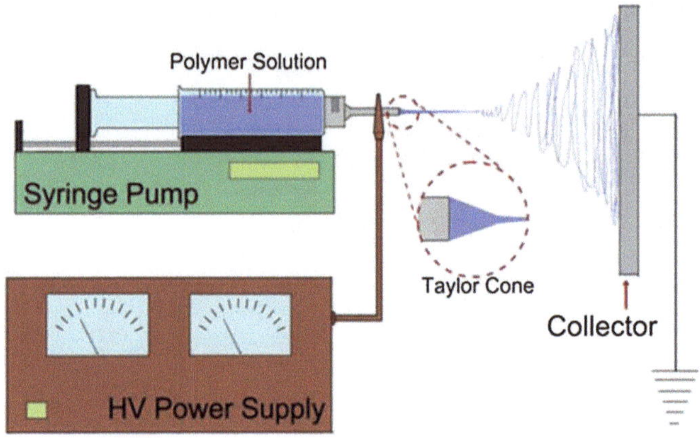

Fig. 5 Schematic diagram of electrospinning process (adopted with kind permission from [33])

made of any shape according to the requirements, like a flat plate, rotating drum, etc. A schematic diagram of the electrospinning process is shown in Fig. 5 [18, 34, 40].

Polymer solution is forced through a syringe pump to form a drop at the tip of the capillary. High voltage potential is applied to the polymer solution inside the syringe through an immersed electrode thereby inducing the free charges into the polymer solution. These charged ions move in response to the applied electric field towards the electrode of opposite polarity thereby transferring tensile forces to the polymer liquid. The pendant hemispherical polymer drop takes a cone-like projection at the tip of the capillary. However, when the applied potential reaches a critical value required to overcome the surface tension of the liquid, a jet of liquid is ejected from the tip of the cone. Most charge carriers in organic solvents and polymers have lower mobilities and hence the charge is expected to move through the liquid for larger distances. After the cone formation, the jet undergoes a chaotic motion or bending instability and is directed towards the oppositively charged collector. As the jet travels through the atmosphere, the solvent evaporates, leaving behind a dry fibre on the collecting device. For low viscosity solutions, the jet breaks up into droplets, while for high-viscosity solutions it travels to the collector as fibre [41–44].

3.3 Parameters of Electrospinning

Electrospinning is a simple and versatile method for generating an ultrafine fibre from several materials which include polymer, composite and ceramic. The electrospinning set-up consists of three major components which are of high-voltage power supply, syringe with metal needle and a conductive collector. The electrospinning process is affected by process, solution and ambient parameters. These operating parameters play decisive role in determining the desired quality of the electrospun fibre [45, 46].

3.3.1 Process Parameters

Applied Voltage Various instability modes that occur during the fibre forming process are expected to occur by the combined effect of both the electrostatic field and the properties of the polymer. It has been suggested that the onset of different modes of instabilities in the electrospinning process depends on the shape of the jet initiating surface and the degree of instability, which effectively produce changes in the fibre morphology. In electrospinning, the charge transport due to applied voltage is mainly because of flow of the polymer jet towards the collector and the increase or decrease in the current is attributed to the mass flow of the polymer from the nozzle tip. Deitzel et al. [41] have inferred that the change in the spinning current is related to the change in the instability mode. They experimentally demonstrated that an increase in applied voltage causes a change in the shape of the jet initiating point and hence the structure and morphology of fibres. Baumgarten [47] while carrying out experiments with acrylic fibres observed an increase in fibre length of approximately twice with small changes in fibre diameter with an increase in applied voltage. Megelski et al. [48] investigated the voltage dependence on the fibre diameter using PS. The PS fibre size decreased from about 20 to 10 μm with an increase in voltage from 5 to 12 kV, while there was no significant change observed in the pore size distribution. These results are in concurrence with the interpretation of Buchko et al. [49], who observed a decrease in the fibre diameter with an increase in the applied field while spinning silk-like polymer fibre with fibronectin functionality (SLPF). Generally, it has been accepted that an increase in the applied voltage increases the deposition rate due to higher mass flow from the tip of needle.

Nozzle Collector Distance The structure and morphology of electrospun fibres are usually affected by the distance employed between nozzle and collector. Buchko et al. [49] examined the morphological changes in SLPF and nylon electrospun fibres with variations in the distance between the nozzle and the collector. They showed that regardless of the concentration of the solution, lesser nozzle collector distance produces wet fibres and beaded structures. SLPF fibre morphology changed from round to flat shape with a decrease in the nozzle collector distance from 2 to 0.5 cm. This result shows the effect of the nozzle collector distance on fibre morphology. The work also showed that aqueous polymer solutions require more distance for dry fibre formation than systems that use highly volatile organic solvents.

Polymer Flow Rate The flow rate of polymer solution from the syringe is an important process parameter as it influences the jet velocity and the material transfer rate. In case of PS fibres, Megelski et al. [48] observed that the fibre and the pore diameters increased with an increase in the polymer flow rate. As the flow rate increased, fibres had pronounced beaded morphologies and the mean pore size increased from 90 to 150 nm.

Spinning Environment Environmental conditions around the spinneret, like the surrounding air, its relative humidity (RH), vacuum conditions, surrounding gas, etc., influence the fibre structure and morphology of electrospun fibres. Baumgarden [47] observed that acrylic fibres spun in an atmosphere of relative humidity more than 60% do not dry properly and get entangled on the surface of the collector. The breakdown voltage of the atmospheric gases is said to influence the charge retaining capacity of the fibres.

3.3.2 Solution Parameters

The morphology of fibre is greatly affected by concentration and conductivity of solution including volatility of solvent.

Solution Concentration Solution concentration determines the limiting boundaries for the formation of electrospun fibres due to variations in the viscosity and surface tension [41]. Low concentration solution forms droplets due to the influence of surface tension, while higher concentration prohibits fibre formation due to higher viscosity.

Solution Conductivity Polymer solutions are mostly conductive. With a few exceptions, the charged ions in the polymer solution are highly influential in jet formation. The ions increase the charge carrying capacity of the jet, thereby subjecting it to higher tension with the applied electric field. Baumgarden [47] showed that the jet radius varied inversely as the cube root of the electrical conductivity of the solution. Zong et al. [50] demonstrated the effect of ions by adding ionic salt on the morphology and diameter of electrospun fibres.

Volatility of Solvent As electrospinning involves rapid solvent evaporation and phase separations due to jet thinning, solvent vapour pressure critically determines the evaporation rate and the drying time. Thus, solvent volatility plays a major role in the formation of nanostructured fibres by influencing the phase separation process. Bognitzki et al. [51] found that the use of highly volatile solvents like dichloromethane yielded poly-L-lactide (PLLA) fibres with pore sizes of 100 nm in width and 250 nm in length along the fibre axis. Lee et al. [52] evaluated the effect of volume ratio of the solvent on the fibre diameter and morphology of electrospun poly(vinyl chloride) (PVC) fibres.

3.3.3 Ambient Parameters

Ambient parameters include humidity and temperature of surrounding environment of spinning area. Many studies have been conducted to examine the effects of ambient parameters. Generally, the average fibre diameter decreases with increasing temperature in the needle of electrospinning. With increasing humidity, some small circular pores may appear on the surface of the fibres. In the needleless electrospinning, the effect of relative humidity and temperature are different. Dao and Oldrich [53] investigated that fibre diameters are temperature dependent, but they are not proportional to the humidity in spinning process. Conversely, the non-fibrous property is humidity dependent, but it does not depend on the temperature in spinning process. The value of non-fibrous area increases strongly with the increase in humidity.

4 Electrospun-Based Proton Conducting Polymers

4.1 Nafion-Based Nanofibres

Nafion, a perfluorinated anionic (sulfonic acid) polymer, has been widely studied as a polymer electrolyte membrane in fuel cells. Nafion exhibits several key properties that result in optimal and long-term fuel cell operation such as thermal, mechanical and chemical stabilities and most importantly high water-saturated proton conductivities ($\sim$0.1 S/cm). Typically, Nafion is commercially available in several forms, including extruded and solution cast films of various thicknesses and equivalent weights (i.e. ion exchange capacities), dispersions in water/alcohol solutions and as pellets. Several recent results have indicated that scaling down Nafion to the nanometer size scale may impact its properties and subsequently its performance in devices. Nafion nanofibres have been investigated using scanning electron microscopy and were found to be natural form under certain fuel cell operating conditions. These results suggest that Nafion nanofibres improve the triple phase boundary and also enhance the proton transport from catalyst to the polymer electrolyte membrane by reducing the ohmic and mass transport resistance. Snyder and Elabd [54] have reported the improvement of fuel cell performance using Nafion nanofibres as fuel cell electrodes. Pan et al. [55] have demonstrated improved fuel cell performance with a Nafion nanofibre. Others have also reported the benefits of proton exchange nanofibres in fuel cells [56, 57].

Several researchers [58–60] have reported the electrospinning of Nafion blend nanofibres with fibre diameters less than 100 nm. However, all reports required the use of a carrier polymer (e.g. poly(ethylene oxide) (PEO), poly(vinyl alcohol) (PVA) or poly(acrylic acid) (PAA), where defect-free fibres could not be formed without at least 12 wt% of the carrier polymer. In order to get defect-free fibres, Dong et al. [61] successfully fabricated high-purity Nafion nanofibres via electrospinning only with 0.1 wt% of carrier polymer. For the first time, this offers the opportunity to investigate the properties of high-purity Nafion at the nanoscale. It is fact that proton conductivity is highly dependent on the percolated nanoscale ionic morphology of Nafion. Several investigators have also shown that conductivity can be affected by inducing some orientation to this morphology in bulk films through film extension methods. Proton conductivities as high as 1.5 S/cm were observed at

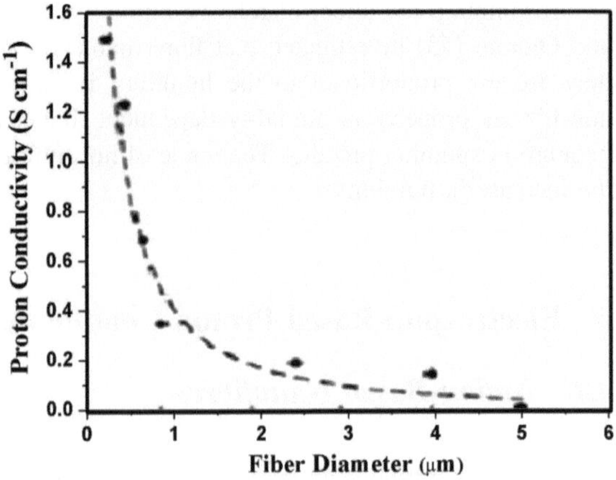

Fig. 6 Proton conductivity (at 30 °C, 90% RH) versus fibre diameter for high-purity Nafion nanofibres (adopted with kind permission from [61])

a fibre diameter of 400 nm at 30 °C with RH of 90% as shown in Fig. 6. The magnitude of this order is higher than that achieved from the bulk film properties ($\sim$0.1 S/cm).

This improvement is because of preferential alignment of interconnected ionic aggregates along the fibre axis direction. Fibre diameter increased from 125 (98 wt% Nafion) to 400 nm (99.9 wt% Nafion) by increasing the wt% of Nafion from 98 to 99.9 wt%. This study demonstrates that high-purity Nafion nanofibres can be produced using a high molecular weight carrier polymer. In electrospinning process, the concentration of PEO as a carrier polymer was found to be important. For example, successful electrospinning was achieved at a PEO solution concentration of 1.6 mg/ml at a fixed Nafion solution concentration of 110 mg/ml. However, lower PEO concentrations resulted in beaded fibres, while higher PEO concentrations resulted in poor electrospinning or an inability to collect a fibre mat even though some fibres could be collected. In this study, the shear force applied during the electrospinning process appears to elongate and orient ionic domains along the fibre axis direction. These aligned ionic structures result in higher conductivity and the extent of humidity at 30 °C for both a 600 nm wide single 99.9 wt% Nafion nanofibre (solid symbols) and a cast Nafion film (open symbols) as shown in Fig. 7. The expected result of film conductivity changing by approximately an order of magnitude for a change of 50–90% RH was observed. However, with the same humidity, twofold change in magnitude of proton conductivity was observed for the single nanofibre. This reveals that Nafion in nanofibre form is more sensitive to humidity changes. These results could have a practical impact on humidity sensors.

To study the effect of nature of polymer solution on formation of fibre and its structure, Chen et al. [59] investigated the electrospinning performance. It was found that solution properties of pure Nafion at various concentrations (5–35 wt%) and its blend with another polyelectrolyte, poly(acrylic acid) (PAA) produced beaded and smooth fibres for 8 and 12 wt%, respectively. This is because of the aggregate formation of Nafion in solution and lack of chain entanglement. In order

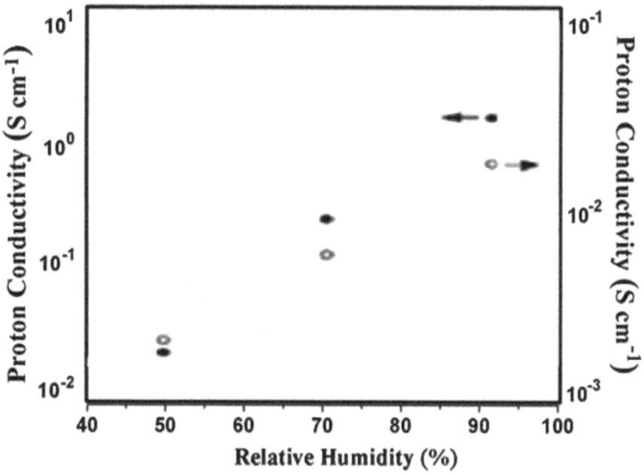

Fig. 7 Humidity dependent proton conductivity (30 °C) of a single high-purity Nafion nanofibre (*solid symbols*) and a cast Nafion film (*open symbols*) (adopted with kind permission from [61])

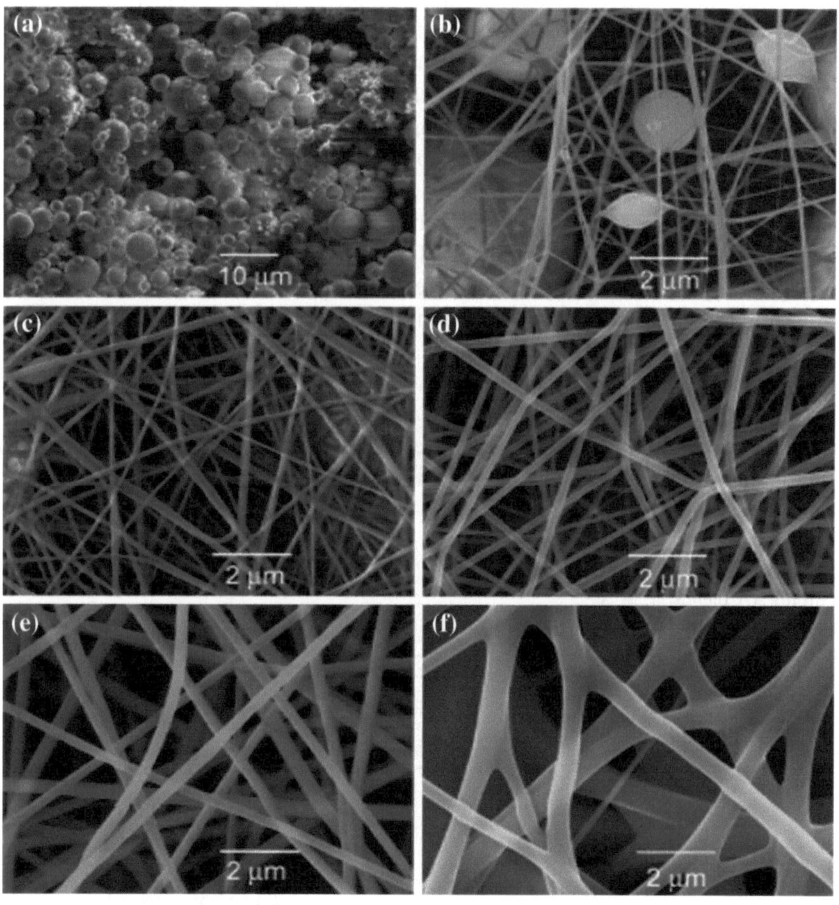

Fig. 8 SEM images of Nafion-PAA electrospun fibres with a PAA content of **a** 5, **b** 8, **c** 12, **d** 25, **e** 66 and **f** 100% (adopted with kind permission from [59])

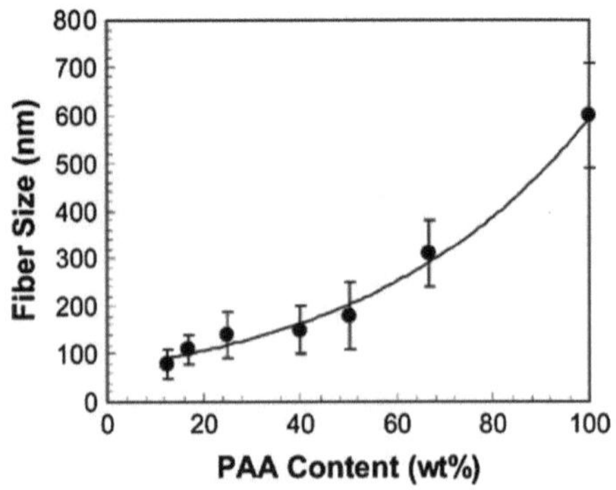

Fig. 9 Effect of PAA content on fibre size of electrospun Nafion-PAA (adopted with kind permission from [59])

to improve the spin ability, addition of linear polymer PAA is suggested for enhanced chain entanglement. This enhancement is sufficient only above 12 wt% of PAA that results in defect-free and smooth fibres as shown in Fig. 8. Fibre sizes of these blends increased from 90 to 600 nm with increasing PAA content as shown in Fig. 9. Electrospinning fibre formation from polymer solutions depends on the properties of solution, such as viscosity, surface tension and net charge density. For instance, electrospinning of cellulose acetate in a 2/1 acetone/dimethyl acetamide cosolvent produces fibres at surface tensions around 26 dyn/cm and viscosities between 1.2 and 10.2 P. High viscosities prohibit the electrospinning due to the lack of flow, which is caused by the high cohesiveness of solution. At low viscosities, only droplets are formed due to lack of chain entanglement. However, viscosity of 5 wt% solution of pure Nafion in a 3/1 isopropyl alcohol/water cosolvent is 6 cP. This value is similar to those of pure solvents and the viscosity increased by increasing the concentrations of Nafion in the same cosolvent. Similarly, 5 wt% of Nafion in methanol and DMF have viscosities of 8 and 10 cP, respectively. Viscosities of Nafion-PAA blend increased with increasing PAA content and showed decrease in conductivities as evident from Figs. 10 and 11. This is because of higher concentration of PAA content results in the formation of larger fibres and a broad distribution of sizes compared to its Nafion-PAA analogue. Surface tension

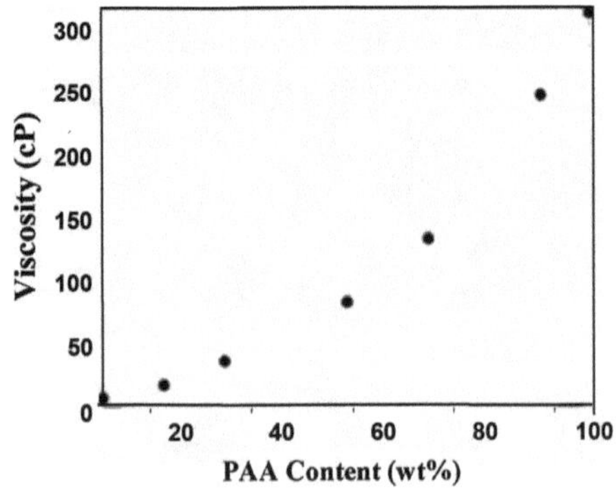

Fig. 10 Viscosity of Nafion-PAA blends at 24 °C (adopted with kind permission from [59])

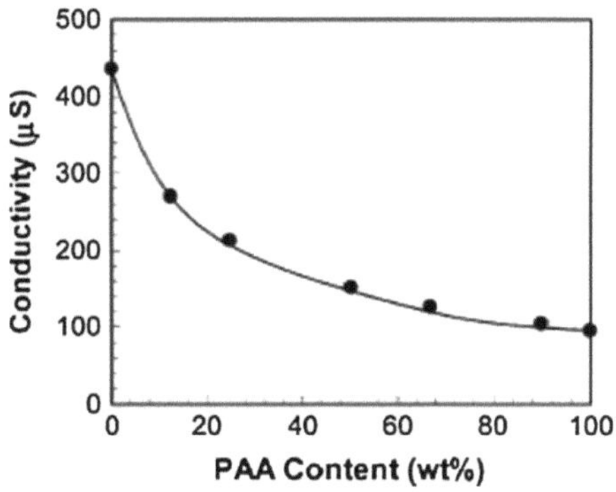

Fig. 11 Conductivity of Nafion-PAA blend solutions in 3/1 isopropyl alcohol/water at 24 °C (adopted with kind permission from [59])

of the polymer solution also directly affects the polymer jet formation, since the electrostatic forces in electric field must overcome the surface tension of the solution to form a polymer jet. The solvents that were used in electrospinning of pure Nafion have surface tensions (γ) between 22.9 and 43.8 mN/m. The observation of a smooth polymer jet in electrospinning suggests that these surface tensions are desirable for electrospinning.

Bajan et al. [58] have developed Nafion-polytetrafluoroethylene (PTFE) fibres from polymers such as PVA, PEO and polyvinyl pyrrolidone (PVP). Poly(ethylene oxide), when mixed with alcohol, solidifies immediately. However, when mixed with water, it formed a very steady sprayable solution as shown in Fig. 12a. The PEO/Nafion nanofibre mat was found to be difficult to remove from the collecting plate because of the low porosity. The mixture of PEO and standard Nafion produced steady and fine spray under the identical test conditions as shown in Fig. 12b. The 10% concentration PEO solution mixed with 40% (volume) Nafion-PTFE proved to be easily sprayable with multiple branching steady jets. However, because water was used as solvent in this case, the nanosized liquid jet did not dry properly before hitting the collecting plate. Although a fine mat was formed, the nanofibre was not well defined as shown in Fig. 12c, d.

Attempts of obtaining homogeneous mixture of PVA with alcohol proved unsuccessful as well. The PVA was insoluble even at 90 °C. A 5 wt% solution of PVA and water mixed easily, but the addition of Nafion made this solution too viscous to spray. A lower volume percent of Nafion in PVA is sprayable at very slow flow rate. The 5% PVA/water solution and 10% Nafion solution were mixed together at 60:40 volumetric ratio. Figure 12e, f show the electrospun Nafion/PVA nanofibre fabricated at a flow rate of 0.3 ml/h. The distance between the emitter tip and the collecting plate was 105 mm.

Similarly, a 5 wt% solution of PVP and water was found to be highly viscous after the addition of Nafion. A solution of PVP and alcohol formed a nice base for both Nafion and Nafion-PTFE compositions. Figure 12g, h show the PVP/Nafion composite fibre. A 10 wt% PVP in alcohol was mixed with Nafion solution in 3:1 volume ratio. Fibres with a diameter in the order of 10 μm were easily formed with smooth outer surfaces. During electrospinning, the standard Nafion solution formed

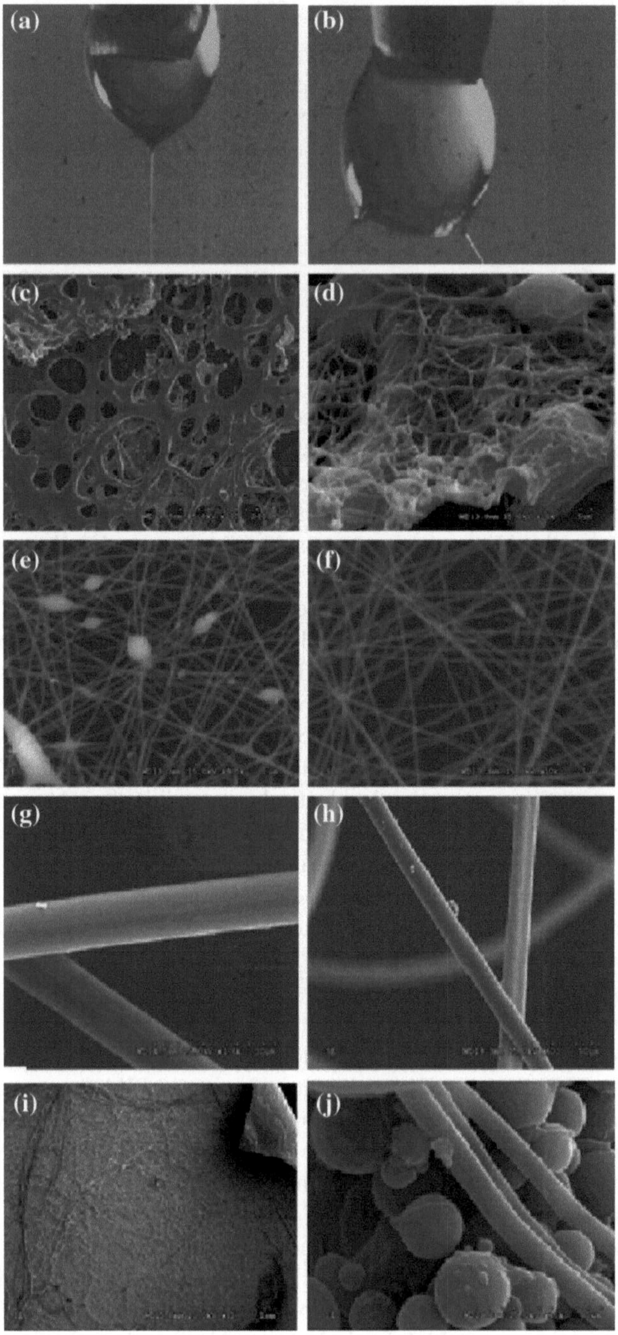

Fig. 12 **a** PEO forms a steady electrospin jet. **b** PEO/Nafion in steady electrospin. **c**, **d** PEO/Nafion-PTFE coating. **e**, **f** Electrospun Nafion/PVA nanofibres fabricated at 20 kV. **g**, **h** Electrospun PVP/Nafion composite fibres. **i**, **j** Electrospun PVP/Nafion-PTFE (adopted with consent from [58])

a tent-shaped, conical spider web with a large base on the ground collecting plate. These fibres formed as though they were being pulled from the spray needle onto one another. For this set of tests, the flow rates were between 0.5 and 2 ml/h, the voltage variation was between 10 and 15 kV and the distance between the emitter tip and the collecting plate was 100 mm.

For the Nafion-PTFE case, a 5 wt% PVP/alcohol-based polymer mixture produced an even mat, which could be peeled off from the electrode like a piece of paper. The mixture of PVP/alcohol to Nafion-PTFE has a volumetric ratio of 60:40. The mat was a fibrous-like cotton as shown in Fig. 12i. The mat was also slightly sticky. The mat formed from the combination of PVP and Nafion-PTFE was viewed under scanning electron microscopy (SEM) after being coated with platinum. The mat was dense and contained numerous layers of fibres. The fibres varied in size having diameter in the range of 500 nm 2 μm. The overall structure was good and contained very few beads within the fibres with the exception of small areas as shown in Fig. 12j. In order to consider the use of these nanofibres for a PEMFC application, their performance in water was tested. These results suggest that synthesized polymer based Nafion fibres or Nafion-PTFE fibres from polymers, such as PVA, PEO and PVP nanofibres can be used for new design of high power density PEMFCs.

To make potential applicability of Nafion nanofibres, Okafor et al. [62] developed proton exchange membrane for direct methanol fuel cell (DMFC). The nanofibre network composite membranes were prepared by interconnected network of Nafion (perfluorosulfonic acid) nanofibres that have been embedded in an uncharged and inert polymer matrix by electrospinning. The spinning solution was Nafion with a low concentration and high molecular weight PEO, as a carrier polymer. The total concentration of polymer spinning solution ranged from 5 to 25 wt%. The interconnected network of Nafion nanofibres with average fibre diameter in the range 160–700 nm was used to make the membranes, with the nanofibre occupying up to 85% of the membrane volume. The matrix polymer was crosslinked with Norland Optical Adhesive 63 under UV. The inter-fibre voids between fibres were filled with an inert/impermeable polymer to create a defect-free and fully dense nanofibre mat. Later is annealed, welded and compacted, and still contained inter-fibre voids, therefore those voids were filled with an inert and hydrophobic pre-polymer. As prepared nanofibre composite membrane exhibited credible proton conductivity and permeability within measured error analysis. The resulting membranes showed proton conductivity of 0.10 S/cm at 25 °C for 80% RH; and methanol permeability of 3.6×10^{-6} cm^2/s. The proton conductivities of the nanofibre composite film and Nafion 117 were in the same range. The results of methanol permeability indicated that the nanofibre composite film has lesser methanol permeability than Nafion 117 as shown in Fig. 13. Methanol permeability decreased without accompanying proton loss in Nafion nanofibre mat than Nafion 117, where nanofibre volume was higher.

DMFCs are attractive source for portable devices, but unfortunately DMFCs suffer from a high crossover of methanol fuel from anode to cathode through the Nafion membrane, consequent fuel waste and polarization loss at the cathode.

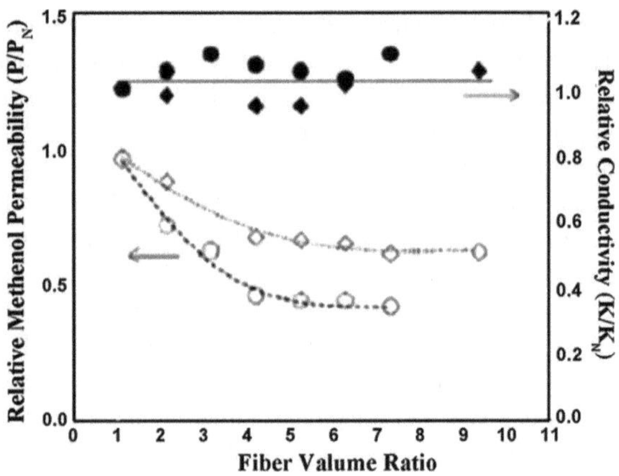

Fig. 13 Through-plane proton conductivity as a function of fibre volume ratio Nafion nanofibre composite and Nafion 117: ● proton conductivity of Nafion 117; ♦ Proton conductivity of Nafion nanofibre composite; ○ Methanol permeability of Nafion 117; ◇ Methanol permeability Nafion nanofibre composite (adopted with consent from [62])

To overcome this drawback of the Nafion membrane, many researchers have developed blend of Nafion and poly(vinylidene fluoride) (PVDF) and its copolymers. Although these membranes exhibit lower methanol crossover than Nafion, to overcome this Choi et al. [63] developed composite membranes consisting of poly (vinylidene fluoride) and Nafion by impregnating method and also studied the effect of impregnated Nafion content on the cell performance and methanol crossover. This is achieved by varying the amounts of Nafion (0.3–0.5 g) into the pores of electrospun PVDF (5 × 5 cm). The electrospun PVDF membrane is hereafter referred as EPM. The composite membranes with 0.3, 0.4 and 0.5 g Nafion are, hereafter referred as, EPM/Nafion-0.3 g, EPM/Nafion-0.4 g and EPM/Nafion-0.5 g. The developed composite membranes exhibited lower proton conductivity (Table 2) than the value of 3.20×10^{-2} S/cm found for Nafion 115 at 65 °C. Among the three composite membranes, the membrane with 0.4 g of Nafion (EPM/Nafion-0.4 g) showed the highest proton conductivity than that of membrane with 0.3 g. The lower conductivity of the latter could be due to the insufficient amount of Nafion present in the composite membrane, while the decrease in conductivity upon increasing the Nafion content from 0.4 to 0.5 g could be due to the segregation of a larger amount of Nafion into the surface of the latter (EPM/Nafion-0.5 g).

The thickness and morphology of the composite membranes differ slightly depending on the Nafion content. For example, the composite membrane with 0.3 g of Nafion (EPM/Nafion-0.3 g) consists of a single layer of Nafion-impregnated electrospun PVDF. This membrane has a rough surface due to a rapid evaporation of the solvent molecules during the drying step. In contrast, the composite membranes with 0.4 and 0.5 g of Nafion (EPM/Nafion-0.4 g and EPM/Nafion-0.5 g) showed additional Nafion layer on the surface of the Nafion-impregnated electrospun PVDF due to the excess amount of Nafion. The thickness of this additional Nafion layer was increased with increasing the amount of Nafion in the composite membrane. Interestingly, these two composite membranes with 0.4 and 0.5 g of Nafion have smooth surface morphologies compared to the membrane with 0.3 g of Nafion. It is anticipated that the composite membranes with smooth surface may

Table 2 Physical properties of the composite and Nafion 115 membranes

Sample name	Thickness (μm)	Wt. of Nafion in composite membrane (g)	Onset temperature (°C)	ΔH (J/g)	Crystallinity $(X_c)^a$	Proton conductivity (S/cm)
Nafion 115	125	–	–	–	–	3.20×10^{-2}
EPM/Nafion-0.3 g	100	0.3	164.3	28.6	5.71	1.55×10^{-3}
EPM/Nafion-0.4 g	105	0.4	165.5	26.2	5.24	2.25×10^{-3}
EPM/Nafion-0.5 g	110	0.5	165.0	24.0	4.81	2.25×10^{-3}

Reproduced with consent from Choi et al. [63]

[a] $X_c = \left(\Delta H_m^{\text{sample}} / \Delta H_m^{\circ}\right) \times 100$, $\Delta H_m^{\circ} = 104.7$ J/g

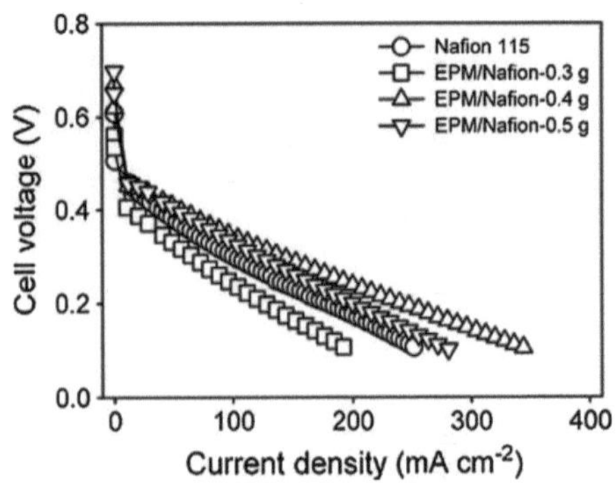

Fig. 14 Comparison of the polarization curves of the Nafion-impregnated electrospun PVDF composite membranes containing various amounts of Nafion with that of Nafion 115 membrane in DMFC at 65 °C in 2 M methanol solution. The data were collected with a methanol flow rate of 2.5 ml/min at the anode and an O_2 flow rate of 200 ml/min with a pressure of 20 psi at the cathode. The humidifier temperature for O_2 was same as that of the cell temperature. *Anode* 0.6 mg PtRu/cm^2; *cathode* 1.0 mg Pt/cm^2 (adopted with consent from Choi et al. [63])

provide better interfacial contact with the electrodes consisting of gas diffusion and catalyst layers, resulting in better cell performance. The electrochemical performances of the composite membranes with various amounts of Nafion at 65 °C in 2 M methanol solution with that of Nafion 115 membrane are shown in Fig. 14. The composite membranes 0.4 g Nafion/EPM and 0.5 g Nafion/EPM showed better cell performance than Nafion 115 membrane, which could be partly related to the additional Nafion layers on the surface of the composite membranes and resulting to better interfacial contact with the electrodes. The cell performance was increased significantly with increasing the content of Nafion from 0.3 to 0.4 g and then decreased slightly with further increasing the amount of Nafion (0.5 g). The drastic increase in performance from 0.3 to 0.4 g of Nafion could be due to the presence of an adequate amount of Nafion in the pores of PVDF and the significant increase in proton conductivity (Table 2). The decrease in performance from 0.4 to 0.5 g of Nafion could be due to a larger segregation of Nafion into the surface and a slightly increase membrane thickness. The cell performance decreased with increasing the membrane thickness due to the increase in ionic resistance.

4.2 Aromatic Hydrocarbon-Based Nanofibres

The perfluorosulfonated membranes, such as Nafion nanofibre, have been widely used because of their excellent oxidative and chemical stabilities as well as high-proton conductivity. The continued reliance on Nafion with its high cost, relatively high fuel crossover and low thermal stability due to their low T_g and high gas permeability, has proven to be a limiting factor for the development of direct methanol fuel cells. In recent years, alternative composite polyelectrolyte membranes have been investigated as substitutes for Nafion in PEMs, such as semi-interpenetrating polymer networks (IPNs) [64], sulfonated poly(ether ether ketone)/phenoxy resin (SPEEK/PHR) composites [65], poly(vinyl alcohol)/sulfonated polyhedral oligosilsesquioxane (PVA/sPOSS) [66] hybrid membranes and sulfonated poly(styrene)/poly(vinylidene fluoride) [67] blends compatibilized with block copolymers. More recently, researchers have attempted to fabricate a composite membrane based on sulfonated poly(2,6-dimethyl-1,4-phenylene oxide) (sPPO), reinforced by electrospun and crosslinked by bromomethylated poly (2,6-dimethyl-1,4-phenylene oxide) (cBPPO) [68] for a hydrogen fuel cell. While many of these alternative composite PEMs have shown promising results, there are still concerns regarding their mechanical durability, chemical stability and/or transport properties that prevent them from widespread use in DMFCs. In view of this, much of the efforts has gone into the development of novel polymer electrolyte membranes based on the sulfonated aromatic hydrocarbon nanofibre polymers for PEMFCs. The important technical focus is on developing polymer electrolyte membranes that are able to achieve high-proton conductivity, low gas permeability into the fuel and oxidant, sufficient thermal stability and long-term durability. If the polymer electrolyte membrane is composed of uniaxially aligned nanofibres having high-proton conductivity, the proton may be rapidly transported through the nanofibre. To achieve this, Tamura et al. [57] for the first time developed composite membranes composed of uniaxially aligned sulfonated polyimide nanofibres and sulfonated polyimide (solvent casting method) membranes for fuel cell. It was clear that the polyimides within nanofibre were significantly oriented or aggregated. As a result, the membrane stability (e.g. oxidative and hydrolytic stabilities) of the composite membrane was significantly improved with an increase in magnitude of nanofibre. Similarly, oxygen permeability of the composite membrane decreased compared to that of membrane without nanofibres. The proton conductivities and stabilities for the composite membranes of the sulfonated random co-polyimides and uniaxially aligned polyimide nanofibre were investigated. The proton conductivities of the composite membranes were measured in transverse direction, which is perpendicular to the membrane surface. The proton conductivity in the perpendicular direction of the composite membrane increased when compared to the dense membrane that was not containing nanofibres as shown in Fig. 15a, although the increased rate was lower than that of measured in the parallel direction. This may be because, the uniaxially alignment of nanofibres on the top surface was disordered due to a decrease of the electrostatic force formed between the charged nanofibre and the aluminium plate collector at the large amount of electrospun nanofibre and the proton conductivity in the perpendicular direction increased. It was found that the proton conductivity of the composite membrane strongly depends on the direction of an aligned nanofibre than that of the parallel direction. In addition, the proton conductivity measured in the parallel direction at low relative humidity of 30% for the composite membrane indicated 3 and 10 times larger value when compared to that in the perpendicular direction for the membrane and in the membrane without nanofibres (2.7×10^{-5} S/cm), respectively. This is due to

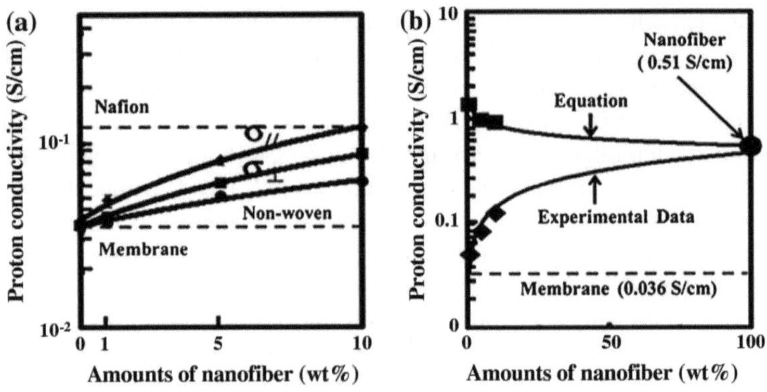

Fig. 15 **a** Proton conductivity of composite membrane containing aligned nanofibres for parallel (σII) and perpendicular ($\sigma^{\perp}$) directions at 80 °C and 98% RH; **b** apparent proton conductivity of nanofibre estimated by simulation (adopted with kind permission from [57])

Table 3 Coefficient of thermal expansion (CTE), oxidative and hydrolytic stabilities of composite membranes containing aligned nanofibres

Membranes	Wt. Ratio Membrane-Nanofibre	CTE[a] (ppm/K)	Oxidative Stability[b] (h)	Hydrolytic Stability[c] (h)
Membrane	100:0	29	10	750
Composite membrane	99:1	–	10	1000
Composite membrane	95:5	–	11	1400
Composite membrane	90:10	25	13	1600
Nanofibre	0:100	8.3	–	–

Reproduced with kind permission from [57]
[a]he measurement temperature was from 80 to 150 °C
[b]Fenton's stability was characterized by the time that the membranes was completely dissolved in 3% H_2O_2 containing 2 ppm of $FeSO_4$ at 80 °C
[c]Hydrolytic stability was characterized by the time that the membranes was completely dissolved in water at 80 °C

the fact that the composite membrane has a large amount of the bound water. The coefficient of thermal expansion (CTE), the oxidative and hydrolytic stabilities for the membrane and nanofibres prepared from polyimide were measured. As apparent from Table 3, the CTE value for the nanofibre was remarkably reduced when compared to that determined in the membrane. This suggests that the polyimides within nanofibre were significantly oriented or aggregated when electrospun. In addition, the composite membranes exhibited more stability for Fenton's reaction than that of membrane due to the orientation of the polyimides. The hydrolytic stability of the composite membrane was also improved remarkably owing to the presence of nanofibres.

Table 4 Ion exchange capacity (IEC), water uptake and number of water molecules per sulfonic group of composite membranes containing aligned nanofibres

Membranes	Wt. ratio Membrane-Nanofibre	IEC value[a] (Experiment/Theory)	Water uptake[b] (%)	λ^{c}	λ_f^{d}	λ_b^{e}
Membrane	100:0	1.41/1.50	22	8	2	6
Composite membrane	99:1	1.42/1.50	22	8	1	7
Composite membrane	95:5	1.42/1.50	27	10	1	9
Composite membrane	90:10	1.41/1.50	29	12	2	10

Reproduced with kind permission from [57]

[a]IEC values determined by titration method of membrane that immersed in 1 N NaOH

[b]Membrane was immersed in water at room temperature

[c]λ = Water sorption/(18 X IEC); $\lambda = \lambda_f + \lambda_b$...$\lambda$: the number of water molecules per sulfonic group

[d]λ_f: The number of free water molecules per sulfonic group

[e]λ_b: The number of bound water molecules per sulfonic group

There is a good correlation between the ion exchange capacity (IEC), water uptake (WU) and the proton conductivity. However, as apparent from Table 4, IEC and the water uptake values of the composite membranes are almost constant regardless of the amount of nanofibres, indicating that the polymer structure, such as phase separation formed in the nanofibre, influenced the proton conductivity.

The apparent proton conductivity of single nanofibre was calculated using the following equation [57]:

$$\delta = \delta_1\varphi_1 + \delta_2\varphi_2,$$

where δ is the proton conductivity measured using electrochemical impedance spectroscopy; δ_1 is the proton conductivity of the membrane without the nanofibre; δ_2 is the apparent proton conductivity of nanofibre; φ_1 is the volume fraction of the membrane and φ_2 is the volume fraction of nanofibre.

According to the above equation, the apparent proton conductivity of the composite nanofibres should have been increased with increasing the volume fraction of the nanofibre. However, this could not happen for the composite nanofibres as clearly seen in Table 5. This suggests that the above equation employed for the calculation of apparent proton conductivity of the nanofibre did not necessarily work well. This is because, the surface of the nanofibre partly undergoes dissolution in the polymer solution during the membrane preparation and thereby the apparent proton conductivity of the resulting composite nanofibre was decreased slightly with increasing the volume fraction of the nanofibre. Unfortunately, it is very difficult to measure the dissolved surface area precisely. Therefore, the proton conductivity of single nanofibre was simulated by considering the values of the apparent proton conductivity of the nanofibre calculated using

Table 5 Apparent proton conductivity of single nanofibre

Wt. Ratio (%) Membrane:Nanofibre	Apparent proton conductivity (δ) (S/cm)[a]
100:0	0.036[b]
99:1	1.3
95:5	0.93
90:10	0.88

Reproduced with kind permission from [57]
[a]At 80 °C and 98% RH
[b]Proton conductivity of membrane

above equation. The apparent proton conductivity of the nanofibre obtained by simulation in the parallel direction showed 15 times higher value than that of the dense membrane (without nanofibres) as shown in Fig. 15b. This is much superior to the conductivity obtained in the perpendicular direction. These findings suggest that the proton channel structure formed in the nanofibres rapidly and efficiently transports the protons.

Layer-by-layer (LbL) assembly is a versatile nanoscale fabrication technique that allows for the coating of any wet table substrate with a combination of two or more polymers possessing complementary interactions such as oppositely charged functional groups. LbL-based PEMs perform well in hydrogen and direct methanol fuel cells. The films can be tuned by adjusting the pH or salt content of the polymer solutions; the thickness per bilayer of films constructed by the LbL method ranges from as small as a few nanometers to over 100 nm. Matthew et al. [69] reported composite membranes of highly conductive LbL system composed of poly(diallyl dimethyl ammonium chloride) (PDAC) and sulfonated poly(2,6-dimethyl-1, 4-phenyleneoxide) (sPPO) which have shown particular promise for use in a DMFC. It has the highest ionic conductivity of any LbL assembled system to date at 70 mS/cm, comparable to that of Nafion, while possessing methanol permeability values less than 1/100 that of Nafion. A key weakness of this promising system is that it is mechanically deficient when hydrated. In fact, many electrostatically assembled polymer membrane systems present the same issues with regard to mechanical integrity. In view of this, the same group manipulates the structure of the underlying electrospun nanofibre scaffold to investigate the effects on the mechanical and functional performance of the composite proton exchange membrane. Figure 16 shows a diagram illustrating the main steps of composite PEM fabrication (electrospinning, spray-assisted LbL deposition, hot-pressing and LbL capping), as well as a representative cross-sectional SEM of a completed composite membrane. Thermal annealing of the prepared LbL electrospun fibre mats at temperatures near the glass transition was found to improve the mechanical response and dimensional stability of the coated PEMs. The transport properties of the composite systems are controllable by manipulation of the fibre formation and LbL deposition parameters. The thickness of each specimen was determined from the average of three measurements taken along the gauge length with a Mitutoyo

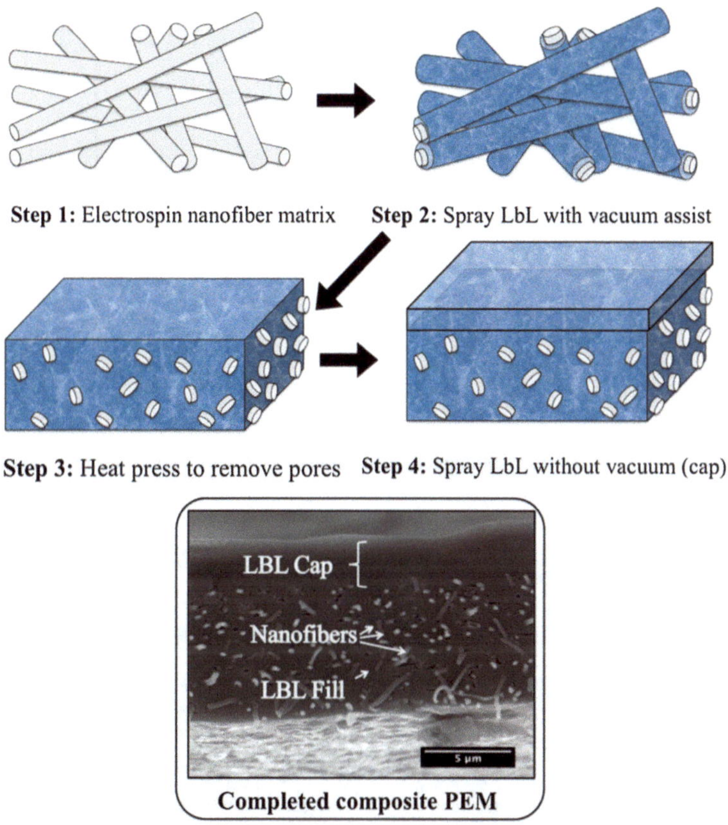

Fig. 16 Diagram showing the steps used in fabrication of the composite PEM: (1) Electrospinning of nanofibre matrix (with optional thermal annealing); (2) Spray-coating fibres with LbL polyelectrolytes with vacuum assist; (3) Heat pressing membrane to remove voids; (4) Coating LbL without vacuum to provide a capping layer to top of PEM. The SEM micrograph is a cross-section of a typical composite PEM. The scale bar of the SEM micrograph is 5 μm (adopted with kind permission from [69])

Digital Micrometer with a constant measuring force of 0.5 N. The annealing at temperatures close to the glass transition (T_g) of poly(trimethyl hexamethylene terephthalamide) (PA 6(3) T) was found to increase the mechanical strength while increasing the solidity of uncoated electrospun fibre (EF) by a factor of five to sixfold over the as-spun mats. These results showed that these improvements survive hydration and are reflected in the composite PEMs. The Young's module and yield stresses of the composite PEMs are greater than those of Nafion, all of the LbL systems exhibit brittle behaviour at 25 °C and 50% RH. Complete membrane electrode assemblies (MEAs) were constructed and used to evaluate composite PEM performance in an operational DMFC in comparison to Nafion.

For the application of PEM for fuel cell at higher operating temperature, Romklaw et al. [70] have developed electrospun fibres from poly(vinyl alcohol),

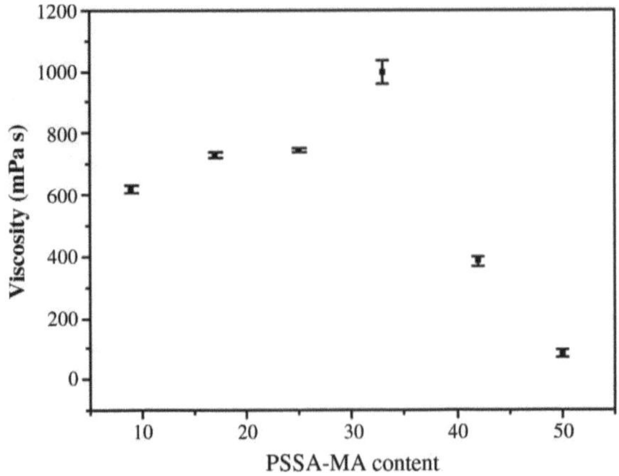

Fig. 17 Viscosities of polymer solutions as a function of PSSA-MA content (adopted with kind permission from [70])

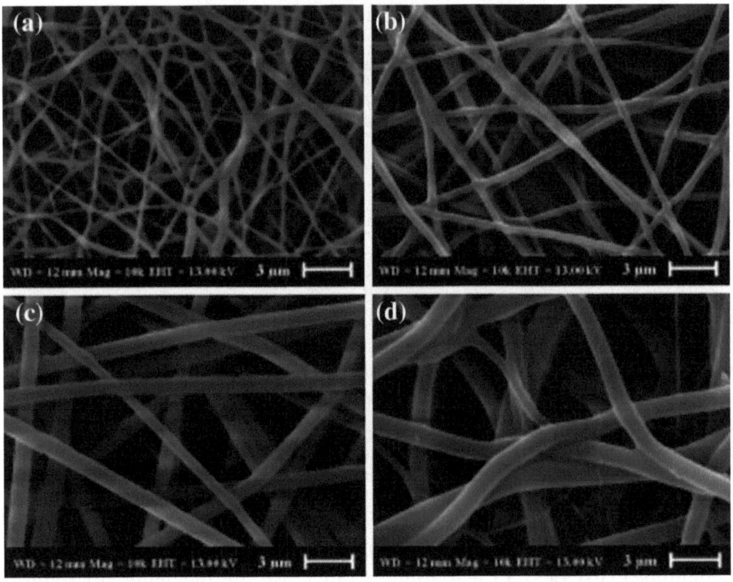

Fig. 18 SEM images of electrospun fibres with different mass% of PSSA-MA. **a** 9; **b** 17; **c** 25 and **d** 33% (adopted with kind permission from [70])

poly(styrene sulphonic acid-co-maleic acid) (PSSA-MA) and imidazole for potential use as proton conducting membranes at elevated temperatures. The effect of PSSA-MA content (9, 17, 25, 33, 42 and 50% by weight of PVA) on solution properties was investigated. Fibre mats were obtained from solutions with 9–33% PSSA-MA. Solution viscosity increased with increasing PSSA-MA until reaching to maximum value at 33% PSSA-MA. A drastic drop in the viscosity at higher

Table 6 DSC values of electrospun fibres and solvent-cast membrane with different mass% of PSSA-MA

Sample	PSSA-MA (%)	T_g (°C)	T_m (°C)	T_c (°C)
Electrospun fibres	9	82.2	207.8	176.1
	17	81.9	209.1	177.8
	25	81.4	207.8	171.3
	33	82.7	210.7	176.3
Solvent-cast membrane	25	80.4	215.1	185.0

Reproduced with kind permission from [70]

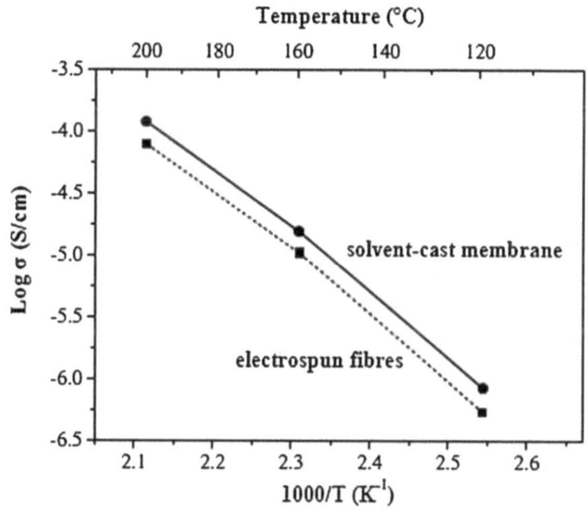

Fig. 19 Proton conductivities of electrospun fibres and solvent-cast membrane with 25% PSSA-MA (adopted with kind permission from [70])

PSSA-MA concentrations resulting from the phase separation is slightly responsible for the non-spin ability of polymer solution with 42 and 50% of PSSA-MA as shown in Fig. 17. The effect of PSSA-MA content on fibre size and morphology was studied using SEM as shown in Fig. 18. Fibre diameter was found to increase with increasing solution viscosity. A solution with 25% PSSA-MA provided uniform and bead-free fibres with an average diameter of 485 nm. Electrospun fibres were thermally stable up to 250 °C. All samples show similar T_g values of approximately 81 °C slightly lower T_m and T_c values were observed for fibres, indicating interruption of the crystallization of PVA as shown in Table 6. Proton conductivities of fibres and membrane with 25% PSSA-MA were determined within temperature range of 120–200 °C on cooling down cycle in non-humidified condition. Figure 19 showed the temperature dependence of proton conductivity as a function of reciprocal temperature. It was found that proton conductivity of solvent-cast membrane was slightly higher than that of fibres. In general, bulk conductivity is predominately governed by segmental mobility and charge carrier density. A small difference in conductivities of fibrous membrane and solvent-cast membrane with similar T_g and identical charge carrier content suggests that there is at least one more factor that influences proton conductivity. Since the measurement of proton conductivity was conducted in the direction of membrane thickness, a denser membrane (solvent-cast membrane) might provide a more continuous hydrogen pathway for proton transport. A maximum conductivity of 121 μS/cm was achieved from solvent-cast membrane at 200 °C.

To improve the proton conductivity of aromatic polymers, Maryam et al. [71] developed membranes containing separated hydrophobic/hydrophilic regions. Fully sulfonated and non-sulfonated poly(ether ketone)s (PEK)s were synthesized and then electrospun simultaneously. After that, non-sulfonated fibres were melted and the fully sulfonated fibres were surrounded by them as a matrix. Four membranes with different contents of fully sulfonated fibres were prepared in this way. For comparison, two membranes from random sulfonated PEKs with the same structures were prepared by usual solution casting method. To obtain separated microstructures, two polymeric solutions of synthesized PEK were used. One of them was fully sulfonated and each repeating unit of the polymer contained two sulfonic acid groups and the other was non-sulfonated PEK with the same structure. Simultaneous electrospinning of polymeric solutions on a rotating drum resulted a fibre mat consisting of fibres of fully sulfonated PEK and fibres of non-sulfonated PEK. Using the difference between melting points of the polymers, the non-porous membranes with different contents of sulfonated fibres were prepared. The schematic representation of the membrane preparation process is described in Fig. 20.

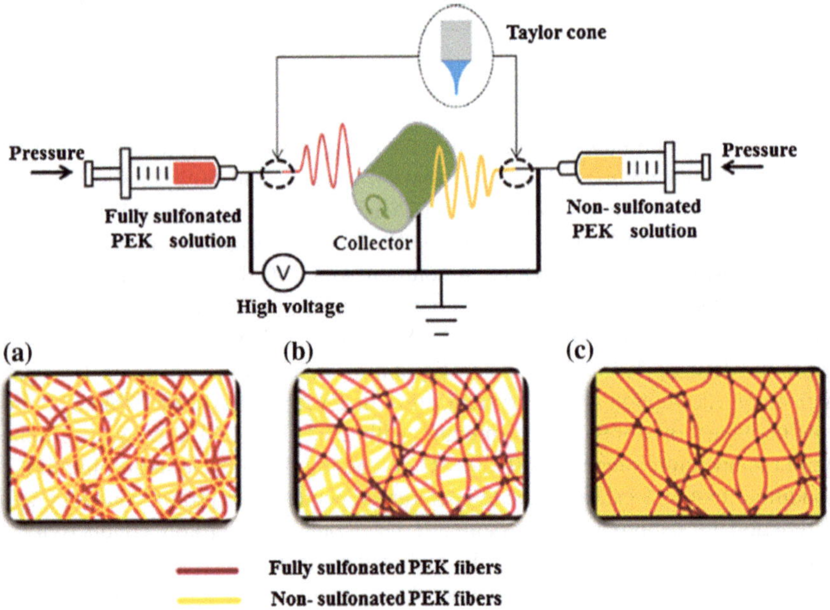

Fig. 20 Schematic representation of the membrane preparation process and dual electrospinning process **a** primary mat consists of fully sulfonated and non-sulfonated PEK; **b** the mat after pressing; **c** the membrane after hot-pressing (adopted with consent from [71])

Table 7 Properties of the membranes prepared by electrospinning (EL-samples) in comparison with membranes prepared from usual solution casting method (RAN-samples) and Nafion 115

Samples	IEC (meq/g) (measured)	IEC (meq/g) (calculated)	Proton conductivity (S/cm)				Swelling ratio (%)	Water absorption (%)
			(25 °C)	(80 °C)	(120 °C)	(140 °C)		
EL-20	0.11	0.097	0.0041	0.011	0.016	0.017	0.23	4.75
EL-25	0.15	0.17	0.014	0.030	0.039	0.038	0.20	5.96
EL-35	0.32	0.33	0.031	0.048	0.067	0.068	0.21	7.31
EL-45	0.71	0.72	0.053	0.112	0.168	0.170	0.49	13.2
RAN-1	0.76	0.86	0.025	0.028	0.042	0.038	0.6	11
RAN-2	1.39	1.44	0.080	0.097	0.120	0.108	1.5	21
Nafion	0.91	0.90	0.085	0.108	0.092	0.050	8.15	32.6

Reproduced with consent from [71]

The IEC values of electrospun samples fully sulfonated fibres (EL) and random sulfonated PEK samples produced by conventional casting method are listed in Table 7. From the table, it is observed that the IEC values increased from 0.11 to 0.71 meq/g with increasing sulfonation. The RAN-1 sample with the same structure and equivalent IEC and also RAN-2 sample with about twice the IEC value of RAN-1 followed the increasing trend.

Water absorption and proton conductivity in proton exchange membranes are closely linked together. Presence of water for transportation of proton is essential. However, excessive water absorption in proton exchange membranes may lead to decreasing the mechanical strength and dimensional stability of the membranes. As proton exchange membrane consists of combination of hydrophilic and hydrophobic regions and the hydrophilic regions are responsible for water absorption of membrane, the water uptake directly depends on the IEC of membrane as a measure of the number of sulfonic acid groups. As clear from the result present in Table 7.

The IEC of EL-samples increased with the increase in magnitude of their water absorptions and EL-45 showed the maximum amount of water absorption among EL-samples. Swelling ratio is another criterion for dimensional stability of a proton exchange membrane. Therefore, a membrane with high value of swelling ratio could not be used as a fuel cell membrane. As denoted in Table 7, the swelling ratios of all EL-samples are below one percent and are excellent for a proton exchange membrane. The swelling ratio for RAN-samples is higher than EL-samples which can be attributed to the dispersion of ionic groups in all parts of membrane whereas in EL-samples sulfonated fibres are covered with hydrophobic non-sulfonated matrix. As expected, the amount of swelling ratio for RAN-2 is more than RAN-1 of the higher number of ionic groups in copolymer chains. Moreover, all of membrane samples (composite membranes and RAN-samples) showed lower swelling ratio compared to Nafion 115. In-plane proton conductivity measurements for all EL-samples and RAN-samples were performed at 25 and 80 °C in water and the results are included in Table 7. Proton conductivity of EL-samples increased with increasing the content of fully sulfonated fibres. In the RAN-samples, this trend was also repeated. Meanwhile, the proton conductivity of all samples increased substantially with increasing the temperature from 25 to 80 °C, because at higher temperatures the energy of conducting ions increases and also the resistance against these ions decreases. Another conclusion was that at given temperature, the proton conductivity of EL-45 was much higher than RAN-1 sample with similar IEC and water absorption values. Surprisingly, in RAN-2 membranes with the IEC value of 1.39 (about twice the IEC of RAN-1 and EL-45 samples), the proton conductivity at 80 °C was less than that of EL-45 although at 25 °C, RAN-2 sample showed higher proton conductivity which may be attributed to its higher IEC value. Thus, it was found that formation of ion-rich channels in non-sulfonated PEK matrix accelerates proton transportation through generating paths for proton and no need to have higher IECs for obtaining high-proton conductivity. At temperatures higher than 80 °C, unlike Nafion 115, proton conductivity in all RAN and EL-membranes were increased. The proton conductivity of RAN-45 at 120 °C reached to 0.168 that was about twice of proton conductivity of Nafion 115 at that temperature. At 140 °C, the trend for EL- and RAN-samples was different. It is observed that the proton conductivity of EL-samples was found to be same though there was a much difference in measured temperature (120 and 140 °C). On the contrary, the proton conductivity of RAN-samples was decreased at 140 °C when compared to 120 °C. This is due to the fact that at high temperature, a less dehydration occurs in EL-samples as compared to RAN-samples, since the ion-rich fibres are embedded completely in hydrophobic matrix.

5 Present Status and Future Prospects

Hydrogen is the most abundant element in the universe and it could be converted into nonpolluting, zero-emission and renewable energy. Peter Hoffmann, in his pioneering book *Tomorrow's Energy* [72], covers the major aspects of hydrogen production, storage, transportation, fuel use and safety. He explained that hydrogen is not an energy source but a carrier-like electricity and introduces the concept of "*hydricity*", the essential interchangeability of electricity and hydrogen.

Electric energy is clean and convenient to use. Dry cells and lead batteries convert chemical energy into electric energy. A fuel cell, in principle, converts chemical energy into electric energy electrochemically. Since fuel cell generates electric energy as long as fuel is supplied continuously from outside, it could have many advantages over conventional generating systems. They are: low environmental pollution; highly efficient power generation; diversity of fuels (natural gas LPG, methanol and naphtha); reusability of exhaust heat; modularity; faster installation. Thus, fuel cells are expected to be used extensively in the future.

As we entered the twenty-first century, it is more urgent than ever to respond to energy and environmental issues, not only within particular countries, but also on a global scale. When we consider the problem of energy, it is important to not be distracted by nearsighted economic factors; we must also consider the medium and long-term environmental concerns. To address this challenge, fuel cells are actively being researched for use as distributed power units and vehicular power sources. The early adoption of fuel cells as new energy resource in practical applications is anticipated because of their high efficiency and friendliness to the environment.

At least eight of the world's largest automobile manufacturers have planned to bring fuel cell vehicles (FCVs) to market between 2013 and 2020. To assess the current status of automotive fuel cell technology and the plans for deploying of refuelling infrastructure. Interviews with the leading original equipment manufacturers (OEMs) and with governmental agencies and non-governmental organizations (NGOs) in the United States, Japan, South Korea and Germany were conducted in 2012. From the report, it is clearly indicated that fuel cell technology has progressed to the point where the leading manufacturers are convinced and they can bring FCVs to the market, which will satisfy their costumers' expectations for performance and durability. However, costs remain too high for success in the mass market. The progress in automotive fuel cell manufacturing costs to date with the future cost reductions believed necessary to compete effectively with internal combustion engine vehicles and estimates scale economies, rates of learning and rates of technological progress that would be sufficient to achieve future goals. Together with cost reduction, the development of an adequate hydrogen refuelling infrastructure is now OEMs' greatest concern. Refuelling infrastructure will be the key factor in OEMs' decisions about when and where to introduce mass-produced HFCVs.

Among the various fuel cells, polymer electrolyte fuel cells (PEMFCs) are being actively researched and developed, as they are suited for power generation systems and have a low waste heat temperature level. With the goal of building a high efficiency power generation system that runs on hydrogen fuel, Fuji Electric developed a stack module having an output capacity of several kilowatts. Thereafter, Fuji Electric worked to improve cell reliability and to miniaturize the size of the cell stack and performed run-time evaluation testing for more than 15,000 h with a 1 kW-class stack running on hydrogen fuel. Through the process of prototyping and evaluating 1 kW-class power generation systems, Fuji Electric has been able to identify potential problems, estimate cost reductions and investigate suitable applications for PEMFC power generation systems.

The development of PEMFC power systems has now been accelerated for vehicular and home-use applications. Due to the differences in the fields for application, it is believed that performance and cost specifications will also differ. The selection of an appropriate fuel is an issue for vehicular applications. At the moment, pure hydrogen fuel appears likely to be the main choice. On the other hand, home-use and stationary fuel cell systems use fuels such as town gas or liquefied petroleum gas (LPG), which have been used as the fuel in conventional cogeneration applications. Therefore, there is a need to develop reformers compatible with this fuel and fuel cells compatible with reformed gas. Although it is possible for vehicular and stationary fuel cell systems to use the same material for the separator in the main fuel cell unit, the cell unit design (especially the electrode) differs according to hydrogen or reformed gas and therefore these power generation systems require different types of development. Furthermore, it will be necessary to examine other suitable applications such as industrial applications that leverage the special features of stationary PEMFC power systems.

Development of new functional materials with targeted nano-material architectures is essential to meet the performance and durability targets of energy conversion and storage applications. To address this, nanofibres for fuel cells can potentially play a vital role in meeting the global energy demands in the coming century due to their unique structures and high surface area to volume ratios. Among the many fabrication methods, electrospinning is undergoing a renewed interest due to its low cost and high production rate for practical energy-related applications and provides great control over the fibre morphology, such as the homogeneity, porosity and a variety of structures including core–shell, hollow and aligned fibres.

There are still several challenges left to be overcome before the large-scale usage of electrospun materials in fuel cell applications. For nanofibrous applications in supporting of PEM materials, the developments are further needed to reduce the fibre diameters, improve the diameter uniformity and lower the cost. In the applications of composite PEM membranes, nanofibres have been widely investigated on mechanical stability, reinforcing component and proton conductor. With the further investigation of electrospinning technology, solutions with sufficient viscosity can be electrospun to achieve nanofibrous networks. Other fibrous structures composed of various materials, such as solutions of self-assembled small molecules, can also be electrospun to nanofibres and used as the filler for proton exchange membranes. Further work on electrospun nanofibres used in proton exchange membranes needs to focus on the progress through-plane proton conductivity and the improvement of fuel cell performance. Addressing these concerns requires dedicated, renewed, intense and extensive research on nanofibres from electrospinning, which will make their scale-up become practical for fuel cells from an industrial perspective in the near future.

6 Conclusions

The major attraction of a fuel cell power plant is its low environmental impact. Recent worldwide clean-air movements have greatly increased the importance of fuel cell development. Among the fuel cells for which early applications are promising, the development trends and future prospects have been described for PEMFC technology by positioning fuel cells as a key component in the fields of energy and the environment, Fuji Electric will continue to promote the commercialization and accelerate the development of PEMFCs. Cost reductions can be realized by expanding the market. Also, manufacturers must demonstrate the stable operability and social reliability of fuel cell plants. PEMFC development is showing remarkable progress towards practical applications in the near future.

Acknowledgements One of the authors (B.B. Munavalli) thanks UGC, New Delhi for awarding (RFSMS) fellowship to undertake research work. The authors also wish to thank UGC, New Delhi for providing the financial support under CPEPA Program [Contract No. 8-2/2008 (NS/PE)].

References

1. Grove WR (1839) On voltaic series and the combination of gases by platinum, philosophical magazine. J Sci 14:127–130
2. Sevim ÜC, Ayhan B, Seyed SH (2012) Alternatives toward proton conductive anhydrous membranes for fuel cells: heterocyclic protogenic solvents comprising polymer electrolytes. Prog Polym Sci 37:1265–1291
3. Kirubakaran A, Jain S, Nema RK (2009) A review on fuel cell technologies and power electronic interface. Renew Sustain Energy Rev 13:2430–2440
4. Yu X, Starke MR, Tolbert LM et al (2007) Fuel cell power conditioning for electric power applications: a summary. IET Electr Power Appl 1:643–656
5. Kim DS, Park HB, Rhim JW et al (2005) Proton conductivity and methanol transport behavior of cross-linked PVA/PAA/silica hybrid membranes. J Solid State Ionics 176: 117–126
6. Peighambardoust SJ, Rowshanzamir S, Amjadi M (2010) Review of the proton exchange membranes for fuel cell applications. Int J Hydrogen Energy 35:9349–9384
7. Rahman MK, Aiba G, Susan MA et al (2004) Proton exchange membranes based on sulfonimide for fuel cell applications. Electrochim Acta 50:633–638
8. Wang F, Hickner M, Kim YS, Zawodzinski TA et al (2002) Direct polymerization of sulfonated poly(arylene ether sulfone) random (statistical) copolymers: candidates for new proton exchange membranes. J Membr Sci 97:231–242
9. Chang JH, Hyeok JP, Park GG et al (2003) Proton conducting composite membranes derived from sulfonated hydrocarbon and inorganic materials. J Power Sourc 124:18–25
10. Elabd YA, Napadensky E, Sloan JM et al (2003) Triblock copolymer ionomer membranes: part I. Methanol and proton transport. J Membr Sci 217:227–242
11. Felix NCA (2015) Electric power management for a grid connected renewable energy sources. Int J Eng Sci 4:443–452
12. Dervisoglu R (2012): Solid oxide fuel cell protonic.svg. http://en.wikipedia.org/wiki/File: Solid_oxide_fuel_cell.svg. Accessed 3 May 2012
13. Bratsch SG (1989) Standard electrode potentials and temperature coefficients in water at 298.15 K. J Phys Chem 18:1–21
14. He R, Li Q, Bach A et al (2006) Physicochemical properties of phosphoric acid doped polybenzimidazole membranes for fuel cells. J Membr Sci 277:38–45
15. Qiao J, Saito M, Hayamizu K et al (2006) Degradation of perfluorinated ionomer membranes for PEM fuel cells during processing with H_2O_2. J Electrochem Soc 153:967–974
16. Costamagna P, Srinivasan S (2001) Quantum jumps in the PEMFC science and technology from the 1960 to the year 2000 part I. Fundamental scientific aspects. J Power Sources 102:242–252
17. La Conti AB, Hamdan M, Donald RC (2003) Mechanisms of membrane degradation. In: Vielstic W, Lamm L, Gasteiger HA (eds) Handbook of fuel cells fundamentals, technology, applications. Wiley, New York, pp 1741–1756
18. Hazlina J, Juhana J, Muhammad N et al (2015) A review on the fabrication of electrospun polymer electrolyte membrane for direct methanol fuel cell. J Nanomater 2015:1–16
19. Kariduraganavar MY, Kittur AA, Kulkarni SS (2012) Ion exchange membranes: preparation, properties and applications. In: Inamuddin, Luqman M (eds) Ion exchange technology I: theory and materials. Springer, New York, pp 233–276

20. Xuan C, Zheng S, Nancy G et al (2007) A review of PEM hydrogen fuel cell contamination: impacts, mechanisms and mitigation. J Power Sources 165:739–756
21. Yun SY, John R, Bing JH et al (2012) Water soluble polymers as proton exchange membranes for fuel cells. Polymers 4:913–963
22. Zhang N, Zhang G, Xu D et al (2011) Cross-linked membranes based on sulfonated poly (ether ether ketone) (SPEEK)/Nafion for direct methanol fuel cells (DMFCs). Int J Hydrogen Energy 36:11025–11033
23. Gode P, Hult A, Jannasch P, Johansson M et al (2006) A novel sulfonated dendritic polymer as the acidic component, in proton conducting membranes. Solid State Ionics 177:787–794
24. Hu H, Xiao M, Wang SJ et al (2010) Poly (fluorenyl ether ketone) ionomers containing separated hydrophilic multiblocks used in fuel cells as proton exchange membranes. Int J Hydrogen Energy 35:682–689
25. Kariduraganavar MY, Nagarale RK, Kittur AA et al (2006) A review on ion-exchange membranes: preparative methods for electrodialysis and fuel cell applications. Desalination 197:225–246
26. Mader JA, Benicewicz BC (2010) Synthesis and properties of random copolymers of functionalised polybenzimidazoles for high temperature fuel cells. Fuel Cells 11:212–221
27. Miyatake K, Zhou H, Matsuo T et al (2004) Proton conductive polyimide electrolytes containing trifluoromethyl groups, synthesis and DMFC performance. Macromolecules 37:4961–4966
28. Mulijani S, Dahlan K, Wulanawati A (2014) Sulfonated polystyrene copolymer: synthesis, characterization and its application of membrane for direct methanol fuel cell (DMFC). Int J Mater, Mech Manuf 2:36–40
29. Nakagawa T, Nakabayashi K, Higashihara T et al (2010) A high performance polymer electrolyte membrane based on sulfonated poly(ether sulfone) with binaphthyl units. J Mater Chem 20:6662–6667
30. Rachipudi PS, Kittur AA, Choudhari SK (2009) Development of polyelectrolyte complexes of chitosan and phosphotungstic acid as pervaporation membranes for dehydration of isopropanol. Eur Polym J 45:3116–3126
31. Seong YH, Won J, Kim SK et al (2011) Synthesis and characterization of proton exchange membranes based on sulfonated poly(fluorenyl ether nitrile oxynaphthalate) for direct methanol fuel cells. Int J Hydrogen Energy 36:8492–8498
32. Xu K, Oh H, Hickneret MA et al (2011) Highly conductive aromatic ionomers with perfluorosulfonic acid side chains for elevated temperature fuel cells. Macromolecules 44:4605–4609
33. Cavaliere S, Subianto S, Savych I et al (2011) Electrospinning: designed architectures for energy conversion and storage devices. Energy Environ Sci 4:4761–4785
34. Mitchell GR, Mohan SD, Davis FJ et al (2015) Structure development in electrospun fibres. In: Mitchell GR (ed) Electrospinning: principles, practice and possibilities. RSC, London, UK, pp 136–171
35. Seeram R, Kazutoshi F, Wee ET (2006) Electrospinning nanofibres: solving global issues. Mater Today 9:40–50
36. Bognitzki M, Hou H, Ishaque M et al (2000) Polymer, metal and hybrid nano- and mesotubes by coating degradable polymer template fibres (TUFT Process). Adv Mater 12:637–640
37. Li D, Xia Y (2004) Direct fabrication of composite and ceramic hollow nanofibres by electrospinning. Nano Lett 4:933–938
38. Reneker DH, Chun I (1996) Nanometre diameter fibres of polymer, produced by electrospinning. Nanotechnology 7:216–223
39. Audrey F, Ioannis SC (2003) Polymer nanofibres assembled by electrospinning. Curr Opin Colloid Interface Sci 8:64–75
40. Liu Y, He JH, Yu JY et al (2015) Controlling numbers and sizes of beads in electrospun nanofibres. Polym Int 57:632–636
41. Deitzel JM, leinmeser JK, Harris D et al (2001) The effect of processing variables on the morphology of electrospun nanofibres and textiles. Polymer 42:261–272

42. Taylor G (1969) Electrically driven jets. In: Proceedings of the Royal Society A. Mathematical, physical and engineering science. vol 313, pp 453–475
43. Yiu-in AE, Koombhongse S, Reneker DH et al (2001) Bending instability in electrospinning of nanofibres. J Appl Phys 89:3018–3026
44. Zhang YZ, Feng Y, Huang ZM et al (2006) Fabrication of porous electrospun nanofibres. Nanotechnology 17:901–908
45. Lu P, Ding B (2008) Applications of electrospun fibres. Recent Pat Nanotechnol 2:169–182
46. Maleki M, Latifi M, Amani-Tehran M (2010) Optimizing electrospinning parameters for finest diameter of nanofibres. Int J Chem Mol Nucl Mater Metall Eng 4:268–271
47. Baumgarten PK (1971) Electrostatic spinning of acrylic microfibers. J Colloid Interface Sci 36:71–79
48. Megelski S, Stephens JS, Chase DB (2002) Micro and nanostructured surface morphology on electrospun polymer fibres. Macromolecules 35:8456–8466
49. Buchko CJ, Chen LC, Yu S et al (1999) Processing and microstructural characterization of porous biocompatible protein polymer thin films. Polymer 40:7397–7407
50. Zong XH, Kim K, Fang DF et al (2002) Structure and process relationship of electrospun bioabsorbable nanofibre membranes. Polymer 43:4403–4412
51. Bognitzki M, Czado W, Frese T (2001) Nanostructured fibres via electrospinning. J Adv Mater 13:70–72
52. Lee KH, Kim HY, La YM et al (2002) Influence of a mixing solvent with tetrahydrofuran and N, N-dimethylformamide on electrospun poly(vinyl chloride) nonwoven mats. J Polym Sci (Part B) 40:2259–2268
53. Dao AT, Oldrich J (2010) Roller electrospinning in various ambient parameters. Olomouc, Czech Republic, EU 10:12–14
54. Snyder JD, Elabd YA (2009) Nafion nanofibres and their effect on polymer electrolyte membrane fuel cell Performance. J Power Sourc 186:385–392
55. Pan C, Wu H, Wang C et al (2008) Nanowire-based high-performance "micro fuel cells": one nanowire, one fuel cell. J Adv Mater 20:644–1648
56. Choi J, Lee KM, Wycisk R et al (2008) Nanofibre network ion-exchange membranes. Macromolecules 41:4569–4572
57. Tamura T, Kawakami H et al (2010) Aligned electrospun nanofibre composite membranes for fuel cell electrolytes. Nano Lett 10:1324–1328
58. Bajon R, Balaji S, Guo SM (2009) Electrospun nafion nanofibre for protonexchange membrane fuel cell application. J Fuel Cell Sci Tech 6:1–6
59. Chen H, Snyder JD, Elabd YA (2008) Electrospinning and solution properties of nafion and poly(acrylic acid). Macromolecules 41:128–135
60. Laforgue A, Robitaille L, Mokrini A et al (2007) Fabrication and characterization of ionic conducting nanofibres. Macromol Mater Eng 292:1229–1236
61. Dong B, Gwee L, David SC et al (2010) Super proton conductive high-purity nafion nanofibres. Nano Lett 10:3785–3790
62. Okafor C, Maaza M, Mokrani T (2014) Nafion nanofibre composite membrane fabrication for fuel cell applications. Int J Chem Nucl Mater Metall Eng 8:389–392
63. Choi SW, Fu YW, Ahn YR et al (2008) Nafion-impregnated electrospun polyvinylidene fluoride composite membranes for direct methanol fuel cells. J Power Sourc 180:167–171
64. Chikh L, Delhorbe V, Fichet O (2011) Semi-interpenetrating polymer networks as fuel cell membranes. J Membr Sci 368:1–17
65. Cai H, Shao K, Zhong S et al (2007) Properties of composite membranes based on sulfonated poly(ether ether ketone)s (SPEEK)/phenoxy resin (PHR) for direct methanol fuel cells usages. J Membr Sci 297:162–173
66. Choi J, Wycisk R, Zhang W (2010) High conductivity perfluorosulfonic acid nanofibre composite fuel-cell membranes. Chemsuschem 3:1245–1248
67. Piboonsatsanasakul P, Wootthikanokkhan J, Thanawan S et al (2008) Preparation and characterizations of direct methanol fuel cell membrane from sulfonated polystyrene/poly

(vinylidene fluoride) blend compatibilized with poly(styrene)-b-poly(methyl methacrytlate) block copolymer. J Appl Polym Sci 107:1325–1336
68. Yun SH, Woo JJ, Seo SJ et al (2011) Sulfonated poly (2, 6-dimethyl-1, 4-phenylene oxide) (SPPO) electrolyte membranes reinforced by electrospun nanofibre porous substrates for fuel cells. J Membr Sci 367:296–305
69. Matthew MM, David SL, Paula TH et al (2013) Mechanical and transport properties of layer-by-layer electrospun composite proton exchange membranes for fuel cell applications. J Appl Mater Interfaces 5:8155–8164
70. Romklaw B, Manus S, Pitak N et al (2014) Electrospun fibres from polyvinyl alcohol, poly (styrene sulphonic acid-co-maleic acid) and imidazole for proton exchange membranes. Sci Asia 40:232–237
71. Maryam O, Shahram MA, Masoud E et al (2015) Proton exchange membranes with microphase separated structure from dual electrospun poly(ether ketone) mats Producing ionic paths in a hydrophobic matrix. Chem Eng J 269:212–220
72. Hoffmann P (2012) Tomorrow's energy, revised and expanded edition, https://mitpress.mit.edu/books/tomorrows-energy-0

Index

Inamuddin et al. (eds.), *Organic–Inorganic Composite Polymer Electrolyte Membranes*, DOI 10.1007/978-3-319-52739-0

MIX
Papier aus verantwortungsvollen Quellen
Paper from responsible sources
FSC® C105338

If you have any concerns about our products,
you can contact us on
ProductSafety@springernature.com

In case Publisher is established outside the EU,
the EU authorized representative is:
Springer Nature Customer Service Center GmbH
Europaplatz 3, 69115 Heidelberg, Germany

Printed by Libri Plureos GmbH
in Hamburg, Germany